William and Lawrence Bragg, father and son

William and Lawrence Bragg, father and son

the most extraordinary collaboration in science

John Jenkin

Philosophy Program
La Trobe University
Victoria, Australia 3086
j jenkin@latrobe.edu.au

OXFORD
UNIVERSITY PRESS

OXFORD
UNIVERSITY PRESS

Great Clarendon Street, Oxford OX2 6DP

Oxford University Press is a department of the University of Oxford.
It furthers the University's objective of excellence in research, scholarship,
and education by publishing worldwide in

Oxford New York

Auckland Cape Town Dar es Salaam Hong Kong Karachi
Kuala Lumpur Madrid Melbourne Mexico City Nairobi
New Delhi Shanghai Taipei Toronto

With offices in

Argentina Austria Brazil Chile Czech Republic France Greece
Guatemala Hungary Italy Japan Poland Portugal Singapore
South Korea Switzerland Thailand Turkey Ukraine Vietnam

Oxford is a registered trade mark of Oxford University Press
in the UK and in certain other countries

Published in the United States
by Oxford University Press Inc., New York

First published 2008

British Library Cataloguing in Publication Data
Data available

Library of Congress Cataloging in Publication Data
Data available

Typeset by Newgen Imaging Systems (P) Ltd., Chennai, India
Printed in Great Britain
on acid-free paper by
Biddles Ltd., Kings Lynn, Norfolk

ISBN 978–0–19–923520–9

1 3 5 7 9 10 8 6 4 2

To
my parents, who made it possible
my wife, whose support and advice were essential
our children and their partners, who were always interested
and especially our grandchildren, who are the future

Preface

On 14 November 1915 two telegrams arrived at the Bragg family's new London address. One, for William Henry Bragg, had been redirected from their former home in Leeds, where he had been Professor of Physics. The second, for (William) Lawrence Bragg, his elder son, had been redirected twice: first from Leeds and then from Cambridge, where he had been studying. The latter read: 'Stockholm. Nobel Prize for Physics 1915 awarded to you and your father. Particulars by letter. Aurivillius, Secretary, Academy of Science'. Letters indeed followed, stating that the award ceremony had been adjourned, first to June 1916 and then indefinitely. It was wartime.

The circumstances were unusual. By 1914 the Prize had a high profile, but it had also acquired a nationalistic flavour as tensions grew. Initially the awards were deferred, but when hostilities continued it was decided to award prizes for 1914 and 1915: the German Laue for 1914 and the British Braggs for 1915. The Braggs are the only parent–child pair in history to share a Nobel Prize, and Lawrence remains the youngest ever to win the award.

William and Lawrence had halted their award-winning research on 'the analysis of crystal structures by means of X-rays'. William was engaged on a project to detect German submarines at sea by listening for the sound of their engines, Lawrence to locate German guns in Europe by recording the sound of their firings. Lawrence was stationed in the Belgian village of La Clytte: 'I remember', he wrote, 'that the friendly priest... brought up a bottle of Lachrymae Christi from his "cave" to celebrate'.[1] William's reaction is not recorded. His second son, Robert—Australian born and raised like the first—had been killed at Gallipoli.

In May 1920 William and Lawrence were invited to Stockholm to honour the prizewinners for 1915 through 1919. William refused and Lawrence also declined. William never delivered a Nobel Lecture; Lawrence gave his in 1922. For William the pain of the war had destroyed the joy of the award. But these men were already known and honoured for their personal modesty, integrity, and ability to enchant the general public with their science. Their lives had seemed blessed; now they were traumatized. And how was it possible for them to win a Nobel Prize just a few years after arriving in England from far-away Australia?

[1] W L Bragg, Autobiographical notes, p. 35.

Fig. 0.1 Swedish postage stamp commemorating the sixtieth anniversary of the Nobel Prize awarded to William and Lawrence Bragg, showing them as in 1915. (Courtesy: Sweden Post Stamps.)

Years later, in 1962, three men shared the Nobel Prize in Physiology and Medicine for 'discoveries concerning the molecular structure of deoxyribonucleic acid [DNA]', but there were two other people who were largely forgotten in the excitement: Rosalind Franklin, who had obtained the crucial data using the technique the Braggs had invented, and Lawrence Bragg, in whose laboratory Crick and Watson made the discovery. In the intervening years William and Lawrence Bragg had changed the face of many branches of twentieth-century science. The story of their lives is a fascinating and moving one. It deserves to be more widely known and appreciated.

Introduction

It is given to very few people to change forever the lifestyle and the welfare of the human condition. William Henry Bragg and his elder son, William Lawrence Bragg, were two such people. In 1912 they invented the scientific technique of X-ray crystallography and for the rest of their lives led the field as it multiplied and spread into most corners of science: physics, chemistry and biochemistry, materials science, metallurgy, mineralogy, molecular biology, medicine, and more.

Science affected the wider world rather little before 1900, but the twentieth century was different. The discovery of X-rays, radioactivity, the electron, relativity, and the structure of the atom in the years around 1900—and the advent of the strange, subatomic world of quantum mechanics in the decades thereafter—radically changed our view of nature and brought science to a prominence it had not anticipated and for which it was little prepared. The Second World War (1939–45) became the scientists' war, employing radar (radio detection and ranging), sonar (sound navigation and ranging), code breaking, penicillin, rocketry, the jet engine, plastics, and finally the atomic bomb. Of the last, the scientist who oversaw its development, Robert Oppenheimer, said, 'the physicists have known sin, and this is a knowledge which they cannot lose'.[2]

Much of this development was the work of physical scientists: physicists, mathematicians, chemists, specialist engineers. Soon after the war, however, things began to change markedly. In 1953 the structure of DNA was discovered and the biological sciences that had been quietly progressing in the background mushroomed into frantic activity and prominence. If the seventeenth century saw the first scientific revolution, then surely the twentieth witnessed the second, the first half dominated by the physicists and the second half by the biologists. The scientific, engineering, medical, and social implications have been enormous, and they will be even more profound in the future.

How did this dramatic shift occur mid-century? Primarily because, early in the century, William and Lawrence Bragg had invented a technique that employed X-rays to discover how atoms are arranged in the materials that make up our internal and external worlds. They invented X-ray crystallography, and it was this, for example, that both enabled the electronic revolution via solid-state components and yielded the structure of DNA. The confusions that have surrounded the identities and achievements of father and son remain, however, and already their fundamental contributions have been largely forgotten. A recent tome, *The Oxford Companion to the History of Modern Science*,[3] contains no specific entry for them, and recent books have either mentioned them only in passing or not at all.[4] In addition, much of the story leading to the discovery was played out in Australia, but very few Australians today recognize, or indeed care about, this family and its pivotal role in modern science. The present book has been written, in part, in the hope of rectifying some of these sad oversights. In addition, the story provides a rich example of science in a colonial setting, a topic of growing interest.

Biography is a study both popular with readers and of uncertain reputation in academe, and biography in science (science biography) is less well developed than other forms. Only a few scholars have written about science biography as a genre, notably in books edited by Shortland and Yeo and by Söderqvist,[5] and

[2] J R Oppenheimer, in lecture at MIT, 25 November 1947, quoted in M Ebison (ed.), *The Harvest of a Quiet Eye* (Bristol: Institute of Physics, 1977), p. 113.

[3] J L Heilbron (ed.), *The Oxford Companion to the History of Modern Science* (Oxford: OUP, 2002).

[4] See, for example, Sir Neville Mott (ed.), *The Beginnings of Solid State Physics* (London: Royal Society, 1980), a title that belies the content, and M Morange (transl. M Cobb), *A History of Molecular Biology* (Cambridge, Mass.: Harvard University Press, 1998), where William Bragg is mentioned in one sentence only (p. 112).

[5] M Shortland and R Yeo (eds), *Telling Lives in Science* (Cambridge: CUP, 1996), and T Söderqvist, *The History and Poetics of Scientific Biography* (Aldershot: Ashgate, 2007).

there is an unfortunate separation of the history of science from history more generally. Science biography is homeless, and there are also special problems associated with it. It can be so difficult to integrate the private life and the life in science that some authors have separated them into different chapters, or two authors have been employed on the one subject, one to write the science and the other to write the life.[6] Some scientists have disdained the merely personal or been quite unconscious of their inner self, and writing about them then poses special problems for the biographer. Following Frank Manuel's psychoanalytic biography, *A Portrait of Isaac Newton*,[7] science biographers have largely shunned the psycho-biographical approach.

A few of the essential and desirable characteristics of science biography have been agreed, however, some equally true of biography in general. In his classic 1979 article, 'In defence of biography: the use of biography in the history of science', Thomas Hankins wrote: 'A fully integrated biography of a scientist, which includes not only his personality but also his scientific work and the intellectual and social context of his times, is still the best way to get at many of the problems that beset the writing of history of science'.[8] Surely it is also the best way to undertake the biography itself. Other forms of biography have shown that psychoanalytic insights can be illuminating without necessarily overwhelming the text. Indeed, there are recent cases where an author has been able to weave together a narrative of life experiences, emotional life, and creative science.[9] I have no illusion about my ability to meet all these goals. I have tried and I have had great assistance, but the shortcomings are mine alone.

In considering retrospectively his substantial biography of Newton, *Never at Rest*,[10] Richard Westfall noted, 'Biography is indeed autobiography...It is impossible to portray another human being without displaying oneself'.[11] I am now intimately connected to the Bragg story. In the early 1970s I helped to establish a successful physics research group investigating the electronic properties of materials, using the photoelectric effect as a tool. It was all new and innovative according to international leaders in the field, but I also knew that the photoelectric effect was a pivotal piece of *fin de siècle* physics; after all, Einstein had won his Nobel Prize for understanding it. I wondered if the scientific leaders from that period had anticipated us, albeit with inferior apparatus, and indeed they had. In addition, my study revealed that Professor W H Bragg had done some important early experiments in Adelaide, and the city of my

[6] See, for example, M White and J Gribbin, *Einstein: A Life in Science* (New York: Dutton, 1993).

[7] F E Manuel, *A Portrait of Isaac Newton* (Cambridge, Mass.: Harvard University Press, 1968).

[8] T L Hankins, 'In defence of biography: the use of biography in the history of science', *History of Science*, 1979, 17:1–16, pp. 13–14.

[9] See, for example, T. Söderqvist, *Science as Autobiography* (New Haven: Yale University Press, 2003); R Keynes, *Darwin, His Daughter, and Human Evolution* (New York: Riverhead, 2001).

[10] R S Westfall, *Never at Rest* (Cambridge: CUP, 1980).

[11] R S Westfall, 'Newton and his biographer', in S H Baron and C Pletsch (eds), *Introspection in Biography* (Hillsdale, N J.: Analytic Press, 1985), p. 188.

birth, youth, and education called me back. I had done an honours science degree there earlier, with a major in physics, but no one had told me about Bragg! In addition, his son had been born and educated in my hometown, and together they had shown that it is possible to do outstanding research and win a Nobel Prize whether you are fifty-three or twenty-five years old. I wrote up aspects of the story and was seduced by the history of science.

As I explored, however, it became clear that their lives were not all sweetness and light. There were times of sadness and tragedy, of tension and insecurity, amongst the great achievements and wide acceptance. There are only two book-length studies of them: a charming biography of William by his daughter, Gwendolen Caroe, which avoids the difficult questions and from which Lawrence is largely absent, and a recent biography of Lawrence by Graeme Hunter, which says little about his father and which I have found flawed with respect to 'the life' if not 'the science'.[12] Neither author visited Australia or searched Australian archives. Both works now seem inadequate. I wrote articles and published a picture book to mark the centenary of William's arrival in Adelaide,[13] but I did not feel able to start a book. Because their lives were so intertwined it had become impossible for me to write a biography of either man alone, and yet a full, joint biography seemed neither realistic nor desirable. I also wanted to highlight the pivotal part of the story that had been largely ignored—the Australian years—and this provided a solution. Dick Freadman, Professor of English and authority on (auto)biography, assured me that a work covering both men, giving attention to the lives in their context as well as their science, including some psychoanalytic insight, and giving attention to the Australian period and the joint research that followed, was an acceptable project. Only in recent years have I understood this extraordinary family and how I might describe it sufficiently well to proceed.

Nomenclature

The long-standing confusion surrounding father and son, originating from the family tradition of naming the eldest son 'William', poses a problem for the modern writer. William Lawrence Bragg only used 'Lawrence' later, in his professional life, to distinguish himself clearly from his father,[14] and it was not something that his family or his friends used.[15] In later years, following the

[12] G M Caroe, *William Henry Bragg 1862–1942: Man and Scientist* (Cambridge: CUP, 1978); G K Hunter, *Light is a Messenger: The Life and Science of William Lawrence Bragg* (Oxford: OUP, 2004); J G Jenkin, [book review], *Historical Records of Australian Science*, 2005, 16(1):111–13. See also E N da C Andrade, 'William Henry Bragg, 1862–1942', *Obituary Notices of Fellows of the Royal Society*, 1942–44, 4:276–300; Sir David Phillips, 'William Lawrence Bragg, 1890–1971', *Biographical Memoirs of Fellows of the Royal Society*, 1979, 25:75–143.

[13] J G Jenkin, *The Bragg Family in Adelaide: A Pictorial Celebration* (Adelaide: Adelaide University Foundation, 1986).

[14] J G Jenkin, 'A unique partnership: William and Lawrence Bragg and the 1915 Nobel Prize in physics', *Minerva*, 2001, 39:373–92, 384.

[15] Caroe, n. 12, p. 35.

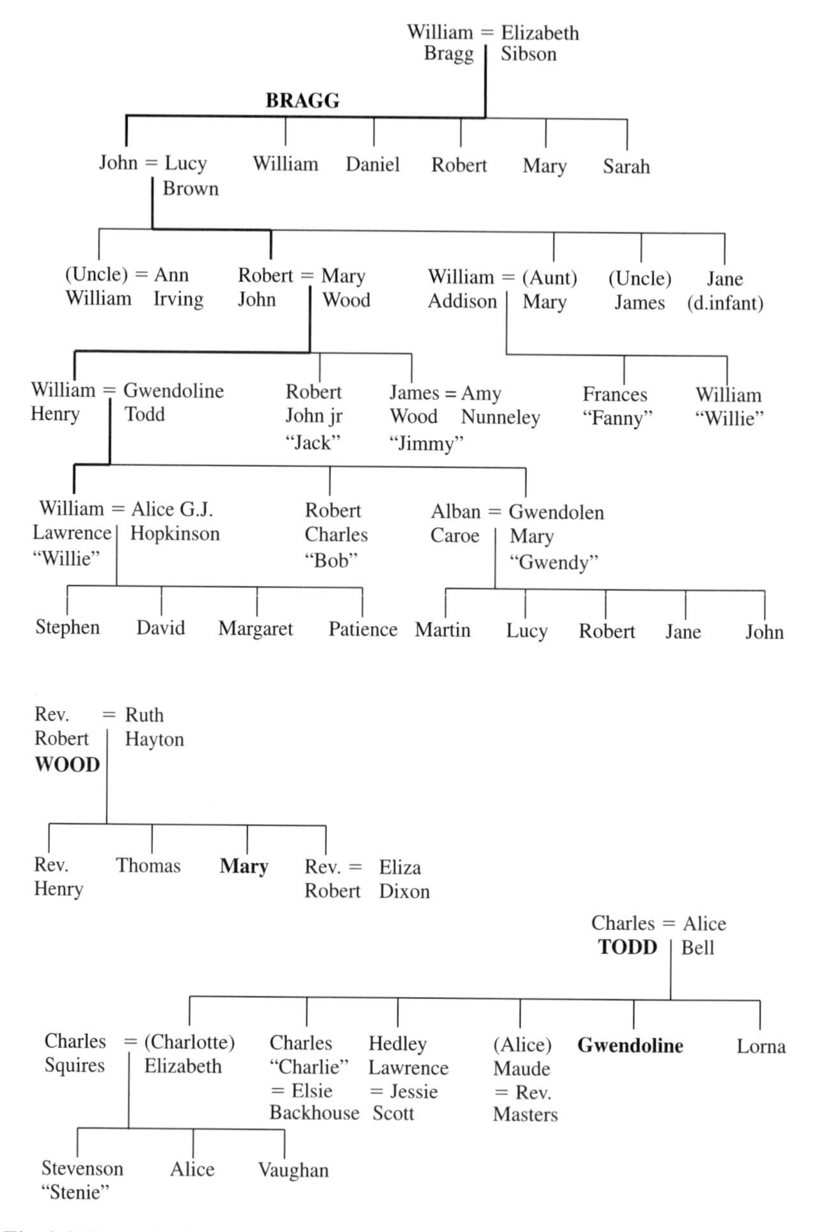

Fig. 0.2 Bragg family tree, showing people mentioned in the text (author).

use by their daughter/sister, Gwendolen Caroe (née 'Gwendy' Bragg), it has become common to use 'WHB' and 'WLB', but I find these labels impersonal. To those who knew them well, WHB was 'William' or 'Will', while WLB was 'Willie' or 'Bill'. To some 'Willie' sounds too familiar, but I find it congenial. I have therefore decided to use 'William' for the father, 'Willie' for the son as a child and youth, and 'Lawrence' when he himself began to use it professionally, since that is how he then became widely known, and to use Christian names for other family members, including their surname (Willie Addison) or family relationship (Uncle William) when confusion is possible. A 'family tree' is also provided to help avoid confusion.

Some explanation is needed regarding the notes at the end of each chapter. The papers of William and Lawrence Bragg in the Archives of The Royal Institution of Great Britain in London are abbreviated 'RI MS WHB box/file' and 'RI MS WLB box/file' respectively. In particular, the unpublished autobiographical notes of William, Lawrence, and Lawrence's wife, Alice, are RI MS WHB 14E/1 (c.1927, c.1937), 14F/1 (n.d.), RI MS WLB 87, and RI MS WLB 95 (*The Half Was Not Told*) respectively, and are denoted 'W. H. Bragg, Autobiographical notes', 'W L Bragg, Autobiographical notes', and 'A G J Bragg, Autobiographical notes' hereafter. Bragg papers in the care of Lady Adrian (née Lucy Caroe), Cambridge, UK, are abbreviated 'Bragg (Adrian) papers'; documents in the Archives of the University of Adelaide are written 'UAA, series number, item number'; documents in The National Archives (previously Public Record Office), London, as 'TNA (PRO), London'; and letters in the Rutherford correspondence in the Cambridge University Library, Add MS 7653, as 'CUL RC item number'. *The South Australian Register* newspaper is denoted *Register*; *The South Australian Advertiser* newspaper, *Advertiser*; and *The Adelaide Observer* newspaper, *Observer*.[16] Cambridge, Oxford, and Melbourne University Presses are represented by CUP, OUP, and MUP respectively.

Acknowledgements

Over more than twenty years, during which I have pursued the story on and off, I have incurred substantial institutional and personal debts but, unlike most authors, I ask to be excused their listing. I am afraid I shall forget some or give them less credit than they deserve; I fear that I shall hurt or annoy some who deserve better; I can thank none of them adequately. I hope they will understand. I have chosen a global approach as follows: libraries, archives, and their staff around the world have responded with unfailing courtesy and generosity to my many visits and inquiries; academic colleagues worldwide have encouraged and supported my research and changing career; university and government bureaucracies in Australia have provided support and financial assistance despite my criticisms of some of their philosophies and agendas; teachers throughout my education frequently encouraged and sometimes inspired me; my students have been interested in my teaching and research and have added to both; and editors have seen merit in my articles. I want to

[16] *The Observer* was the weekly edition of the *Register*, published for country and interstate readers who could not access the daily editions and who wanted a summary of the week's news and events.

mention specifically the staff of the University of Adelaide, the Royal Institution of Great Britain, and La Trobe University (especially those in physics, history, philosophy, and the Borchardt Library).

A few individuals deserve special mention, however. Members of the present Bragg family—especially Stephen, Maureen, and the late David Bragg, Margaret and the late Mark Heath, Patience and David Thomson, and Lucy and the late Richard Adrian—have been extraordinarily co-operative and helpful to the outsider who wanted to delve into their family past; Rod Home, the leader and mentor of the scholarly study of Australian science over many years, has quietly guided me throughout my journey in the history of science; the assistance of Ron and Margaret Gibbs has been essential regarding Australian history more broadly, and especially South Australian history; Harry Medlin and Susan Woodburn provided vital assistance in the earliest stages of the project and warm support thereafter; Dick Freadman advised me at a crucial time and subsequently; and my own family watched with surprise and concern, but always with affection, the late and unexpected journey I took from physics to the history of science.

A number of the topics addressed in this book have appeared earlier, in conference papers, journal articles, newspaper stories, and the like. I am grateful for permission from the publishers to use material from these publications. Likewise I am grateful for permission to quote from documents and reproduce illustrations from materials held by the people and the institutions listed throughout this volume. Last I owe a substantial debt to my publisher and its staff, particularly Sonke Adlung, Lynsey Livingston, Natasha Forrest, and Jonathan Rubery.

This book is dedicated, above all, to my four grandchildren, Katherine, William, Isobel, and Atticus. When I was educated in the post-WWII years, teaching and research blossomed and grew in the Australian universities in an atmosphere of scholarship, freedom, enthusiasm, and achievement. In my experience this continued into the 1980s, when in 1988 the University of Bologna celebrated its nine-hundredth anniversary and hosted a meeting of university leaders from around the world. A *Magna Carta* was signed, reaffirming the fundamental principals of universities in the European humanist tradition: that they must be autonomous institutions, that their teaching and research must be inseparable, and that they must have freedom in these two essential activities. In Australia in the same year, politics, economic rationalism, and managerialism began to destroy what had been built. I was part of that roller-coaster ride, and I deeply regret that we did not do more to protect what our parents had fought for and created through two World Wars and the Great Depression. My dedication carries the hope that a future generation will rediscover what we lost and will recapture the earlier spirit of university scholarship, for the benefit of my grandchildren, their peers, and all those who come thereafter.

Melbourne
April 2007

John Jenkin

Contents

1
Stoneraise Place

The day that William Henry Bragg was born in England—2 July 1862 at Stoneraise Place near Wigton in Cumberland—was a bright summer day. From time to time, however, dark clouds rolled in from the Irish Sea and there were sharp, heavy showers of rain.[1] It was a foreshadowing of the lives to come.

This far-northwest corner of England is isolated, bounded on the north by the Solway Firth, in the west by the Irish Sea and the Isle of Man, in the east by the Pennines, and on the south by the Cumbrian Mountains and the Lake District. In 1861 the population density of Cumberland was only one-third of the average for all England.[2] Nevertheless, the industrial revolution had arrived, principally in the extensive weaving of cotton and linen fabrics and the establishment of paper mills, breweries, tanneries, and soap and biscuit factories. Coal was plentiful, for local use and for export to Ireland from the seaports of Maryport, Workington, and Whitehaven. New iron-ore mines were opening regularly and smelting furnaces were being established, particularly a large plant for iron and Bessemer-steel production at Workington on the coast.

Squeezed between the Solway Firth and the Lake District was the Cumberland plain, with a mild, temperate climate and an annual rainfall in excess of thirty inches. The smaller agricultural holdings were generally in the hands of their owner-occupiers: 'estatesmen... long noted for their sturdy independence and attachment to routine husbandry'.[3] Mixed farming predominated, with horses, beef cattle, dairy breeds, pigs, sheep, pasture, turnips and potatoes, and some oats.

Bragg is a common name in Cumbria. Many of the tombstones now placed around the perimeter wall of the churchyard of St Michael's Parish Church in Workington, for example, bear the name. In particular, two large and impressive memorials have been set up outside the west door of the church, under the large, square bell-tower. One tombstone commemorates William's great-grandparents, William and Elizabeth Bragg, and four of their children. Great-grandfather William Bragg was lost at sea in 1805, and his sons, Robert and

[1] From 1974, Cumberland became part of an enlarged Cumbria.
[2] *Encyclopaedia Britannica* (Edinburgh: Adam and Black, 1875), ninth edition, vol. VI, pp. 696–9.
[3] Ibid., p. 698.

John (William's grandfather Bragg), suffered the same fate in 1820 and 1839 respectively. They were a seafaring family.[4]

In 1846, at age sixteen, William's father, Robert John Bragg, followed the tradition and left school to become an apprentice merchant seaman on the *Nereides* of Workington (530 tons). He made voyages from Liverpool to China, Bombay, and other ports in the Far East with a variety of cargoes. He was highly regarded by his masters by the time he received his mariner's ticket in 1851. On his next voyage as chief mate, Robert was one of only five men to survive the wreck of the *Nereides* in the mouth of the River Ganges near Calcutta. He continued to sail, more often in waters closer to home across the Irish Sea, but in the late-1850s he retired, perhaps not wishing to add further to the family's tragic seafaring record.[5]

Still only twenty-seven years old, Robert purchased Stoneraise Place freehold in 1858, a house and eighty acres of land in the parish of Westward near Wigton in Cumberland, to farm and settle down.[6] It is not clear how he was able to afford the £3,900 price; he had been careful with his pay and he may have received an inheritance. Stoneraise Place has an elegant and substantial Georgian farmhouse with a slate roof, extended in 1853, and a lush lawn and mature chestnut tree in front. Since 1990 it has carried a plaque in Westmorland green slate, engraved by Will Carter of Cambridge, commemorating the birth there of William Henry Bragg.

The rear of the house opens straight onto a yard surrounded by outbuildings, from which the land rises steeply through the home paddock to the top field. It was thick with recently cut stubble when I visited in 1987 and walked to the top of the hill, from where I could see northwest over the house to the Solway Firth and the Southern Uplands of Scotland. Turning south, the land drops away to a road, then a small brook, and finally rises again to the small clump of buildings making up the village of Westward. I was following the path that the young farmer took every Sunday to St Hilda's Parish Church, that his sons would later take to school there, and that later still William would describe: 'I suppose I soon managed to find my own way to school, for I don't think I was often conveyed. Across the home meadow and up over the top field, then on to the main road, across that and down a side road which crossed the Wiser [Wisa] beck at the bottom of the hill and then mounted again to the top of the rise where the school house stood, and the old church and grave yard'.[7]

On one of his first visits to St Hilda's, Robert Bragg noticed the vicar's only daughter, Mary, who played the barrel organ in the church gallery for services. She was a gracious figure, rather taller and more stately than her mother, and so kind, a niece recalled later.[8] Robert courted Mary and they were married by the Rev. Henry Wood, Mary's eldest brother, on 27 June 1861, in her father's

[4] Bragg (Adrian) papers; Stephen Bragg, Cambridge, personal communication.

[5] G M Caroe, *William Henry Bragg 1862–1942: Man and Scientist* (Cambridge: CUP, 1978), ch. 2.

[6] Documents in the care of Mr George Bainbridge and his wife, the last of three generations of owners of Stoneraise Place.

[7] W H Bragg, Autobiographical notes, p. 5.

[8] Caroe, n. 5, p.14.

church at Westward. Robert's best man was his younger brother, James, a kind and simple man, rather than his domineering elder brother, William. Mary's father, Rev. Robert Wood, had been appointed the perpetual curate at Westward in 1822 and would remain there, much loved, for more than sixty years until his death in 1883.[9] Mary had three brothers, two of whom followed their father into the Anglican ministry (Henry and Robert junior), so there was a high level of education and knowledge in the Wood family. Indeed, Robert Wood and William Dickinson of Workington 'were zealous co-workers in natural science for more than half a century, and left behind them valuable collections of dried plants as evidence of their industrious research'.[10]

Mary herself is remembered as having 'a natural bent for mathematics',[11] as evidenced, for example, by the large and colourful, intricate and precise quilt that she made and that remains in the family; and there is an unconfirmed suggestion that she taught mathematics at the small local school. The wedding had its excitements, as William later remembered: 'My father told me it took four rings to get my mother married to him. As he was walking up the road to the church he must have pulled the ring out of his pocket: anyway there was no ring at the critical moment. One had to be borrowed from Uncle Robert Wood; and this my father and mother would not return to him, so a ring had to be bought for my uncle to replace it; finally there had to be a real new wedding ring. Years afterwards my father was walking up the road to the church when he kicked up the original ring in the dust'.[12]

Two further sons were born at Stoneraise Place after William Henry: Robert John junior (called 'Jack') on 28 February 1864, and James Wood (called 'Jimmy') on 4 August 1866. The 1860s were a time of much distress for the local village of Wigton. There was a 'cotton famine' because of the American Civil War, and many small cotton-weaving families became destitute and left the town or stayed to rely on a soup kitchen.[13] Robert Bragg's farm prospered, however, and in 1867 he purchased another twelve acres nearby at Church Hill for £330. There were servants' rooms in the house and presumably farm labourers to assist with the many activities of a large, mixed farm. So successful was Robert, in fact, that he was able to embark on a major project, which illustrates the innovative mind and determination he must have possessed.

Threshing of grain had previously been accomplished on the farm with a horse-gin, where a horse drove the machinery by walking around a circular track. Robert Bragg determined to modernize the process by use of a large

[9] L M Laval, 'A short history of St Hilda's parish church Westward', locally printed, 1985; Rev. F B Swift, 'The parish church of St. Hilda, Westward', *Transactions of the Cumberland & Westmorland Antiquarian & Archaeological Society*, 1983, 83:151–6 (note 13 has dates 1824–84, but there is independent evidence that Wood died in March 1883; see letter R J Bragg to son William, March 1883, Bragg (Adrian) papers).

[10] H A Doubleday, *The Victoria History of the Counties of England: Cumberland, Vol. I* (London: Dawsons, 1968 reprint of 1901 original), p. 74.

[11] Sir David Phillips, 'William Lawrence Bragg 1890–1971', *Biographical Memoirs of Fellows of the Royal Society of London*, 1979, 25:75–143, 76.

[12] W H Bragg, Autobiographical notes, p. 3.

[13] T W Carrick, *History of Wigton (Cumberland): From Its Origins to the Close of the Nineteenth Century* (Carlisle: Thurnam & Sons, 1949), p. 125.

Fig. 1.1 (a) William Henry Bragg with his father, Robert Bragg, Cumberland, *circa* 1865. (b) William's mother, Mary Bragg (née Wood), Cumberland, *circa* 1865. (Courtesy: Dr S L Bragg.)

waterwheel. Open drainage ditches already running across the fields were tapped to obtain water, which flowed through a clay pipe towards the house and farm buildings. Here, on the steep land above, Robert excavated a large reservoir, approximately 120 by 60 feet by about 20 feet deep (36 × 18 × 6 m). A sluice valve and clay pipe fed water down through the old horse-gin building into a new adjoining pit, narrow but 20 feet long and 20 feet deep. This contained the 16-feet-diameter (5 m), wooden-spoked wheel, with buckets around its circumference, all below ground level. Water fed in at the top filled the buckets, turned the wheel, drove the threshing machine via a series of gears

Fig. 1.1 *Continued*

and flat belts, and left the bottom of the pit through the entrance to an underground conduit, which carried the water to a pond at the bottom of the farm. A trapdoor in the incoming water pipe gave the operator control of the wheel. It could also be used to drive a second, grain-crushing machine, which enabled the farm stock to digest the kernel of the grain that was subsequently fed to them. Robert Bragg had used the steep fall in the land to great effect, but it was nevertheless an extensive, labour-intensive, and expensive undertaking. It served the farm well for decades, until a paraffin engine replaced it.[14]

[14] Information on the water-wheel system and other aspects of the history of the farm was supplied to the author by Mr Bainbridge (n. 6), who himself used the system and later dismantled it in favour of a more modern arrangement.

Just a few scenes of these early days remained in William's memory, which he recalled lovingly in his autobiographical notes:[15]

> I think [my mother] must have been a sweet and kind woman. I remember how one day I was sitting on the kitchen table and she was rolling pastry and how I suddenly found I could whistle: and how we stared at one another for a quiet moment, amazed and proud of the new accomplishment. I remember something of a visit to the seaside at Allonby when Jimmy, four years younger than I, must have been just able to walk. For we were on the sands together, and being seized with the idea that bathing was the correct procedure at the seaside I succeeded in undressing him and putting him into a shallow pool. Fortunately we were seen from the hotel window. I remember going home again, and being at the railway station and seating myself on the edge of the platform with my feet dangling over the rails: my mother saw me just as the train was coming in and rushed to pull me back.
>
> I remember the day before I first went to school, I was playing on the floor of the little parlour at Stoneraise Place, and my father, coming home from market, threw down a little brown paper parcel on the floor beside me. My mother exclaimed at the carelessness: it was a slate and slate pencils, and I think there was some breakage. There was a little holder for slate pencils when they got short and I wondered what it was for. I must have been nearly five or just five. I could read moderately well, mother having taught me: I have a vague memory of that. Next day I was taken to the little school at Westward and as a test told to read the piece about 'George and his pony'. Having got through that I was told to sit still, and that no more would be wanted from me for that day....
>
> I think the schoolmaster, Hetherington, was a really good teacher: I have heard so since... I took kindly to his lessons, and before I left for Market Harborough in 1869, when I was just seven, I was fairly well on with the arithmetic. I was doing what was called 'practice' and the like. I suppose practice meant the compound addition, subtraction & multiplication of 'commercial practice'.

Next to the church the schoolhouse still stands, built by public subscription in 1828 and with a Mission Hall attached, built in 1924. The rectangular schoolroom, lit by large windows, heated by a stove, and surmounted by a small bell-tower, was used for all the children of the parish until 1969, when the complex became a community centre. Mr Hetherington was the schoolmaster for many years, and Rev. Robert Wood senior gave some lessons.[16]

The church stands on the site of one first erected there in the middle of the sixteenth century for the inhabitants of the forest or chase of Westward.[17] By 1770 it was in such a poor condition that funds were sought to demolish it and rebuild on the same site. This was done in 1785–6, and the exterior has

[15] W H Bragg, Autobiographical notes, pp. 4–5.
[16] Laval, n. 9.
[17] Swift, n. 9.

remained unaltered since that time: a plain rectangular building of sandstone, with a slate roof. The windows are all alike: three on the north and south sides and one at the east end, each a group of three slender lancets of equal height. Only two had stained glass when I visited in 1987. At the west end there is a small bell-tower and spire over the church door. In the 1860s William would have noticed a west gallery with the barrel organ and choir seats, a plaster ceiling, a three-decker pulpit and narrow box pews, each with two long seats facing one another.[18] In winter the church was cold, despite the stove being well stoked with Wigton charcoal. William's recollections continued:[19]

> My grandfather's vicarage was of course close by: I remember one day when I determined to go and have dinner with them instead of going home. It may have been, probably was, a hankering after my grandmother's famous potted meat...My grandfather was a fine old man, gentle and dignified...He was greatly beloved. The parishioners gave him a cabinet for his natural history specimens...He was a great collector, and quite a county authority on its plants and birds. Robert Wood, his son, inherited the passion: and I think Willie's and Gwendy's [William's surviving children] love of nature must have come from my grandfather...I remember my grandmother Wood as a dear little cherry-cheeked lady in a white cap. I have been told that as Ruth Hayton she had been one of the countryside beauties.

Carrick's *History of Wigton* is helpful in picturing William Bragg's childhood world beyond the farm.[20] Close to Stoneraise Place there was a Roman settlement on the road from the coast to Carlisle. Thereafter Celts, Angles, and Norsemen invaded the area, and it was heavily involved in the border wars between Scotland and England. Walter de Wigton succeeded to the local barony in 1258, his father having first used the name. In 1262 the privilege of holding a weekly market on Tuesdays and an annual fair on the vigil, day, and morrow of the nativity of the Blessed Virgin Mary was granted to Walter by King Henry III, traditions that continued through the nineteenth century. There was a fair held at the feast of St Thomas (21 December), when business in poultry and other Christmas specialities was brisk. The annual horse fair was held near Lady Day (25 March), and was known as Lady Fair. William and his family were surely part of many of these activities. He certainly remembered Wigton's covered market, where his mother sold her butter and eggs.[21]

In February 1868 'Master Bragg' (young William) received a valentine card from his cousin 'Fanny' (Miss Frances Addison) at Market Harborough, south of Leicester, in the centre of England.[22] Unbeknown to them all, William would soon join his distant and extended Bragg family there. He had two uncles (William and James) and one aunt (Mary) on the Bragg side, all born

[18] Ibid. The interior of the church was substantially altered in 1877.
[19] W H Bragg, Autobiographical notes, pp. 5–6.
[20] Carrick, n. 13.
[21] Caroe, n. 5, p. 5.
[22] Bragg (Adrian) papers.

at Workington in Cumberland. Uncle William had been trained as a chemist and since 1853 had had a shop in the centre of Market Harborough. He was a widower; his wife, Anne Irving, had died in 1860. Uncle James managed the adjoining grocer's shop and never married. Mary Bragg had married William Addison of Wigton, but that marriage had difficulties and Fanny Addison had been taken to live in Harborough with her uncles and her maternal grandmother, Lucy Bragg, who had come to Harborough following the death of Uncle William's wife. Assisted by his mother, Uncle William had willingly adopted the role of head of the family.[23] In April 1868 another card arrived at Stoneraise Place, a birthday card for William's mother from her youngest brother, Rev. Robert Wood. 'I wish you many happy returns of the day, and I hope that God will shower down his blessings upon you', he wrote. 'These birthdays remind us that as each one comes round we have one year less to spend on Earth'.[24] It was perhaps a typical reminder of the times, but it surely made Mary pause in the midst of her family duties.

Towards the end of the same year Mary again became pregnant. Like her own family, perhaps she was hoping for a daughter to join her three sons. Her earlier pregnancies had not been difficult and there was no cause for concern. But in the middle of February 1869 it became clear that all was not well, and towards the end of the month the family became deeply worried. William knew that something was badly wrong. Each day when he came home from school his mother was weaker and in greater distress, apparently in labour for days. His father was distracted and alarmed in a way he had not seen before, and the doctor who visited regularly seemed unable to help. William's mother deteriorated day by day; and then she died, at home, on the first of March.[25] It will be clear from his own later recollections quoted above that William was very close to his mother. The strong attachment of all young children to their mother had continued into his boyhood. Now his golden world of good health, loving parents, successful and enjoyable schooling, open-air freedom, and stability and security was torn asunder. The mother who had taught him to read and write, to whistle and sing, was gone forever.

William was a little less than seven years old, and we must ask what his mother's death might have meant to him. We have little evidence of what happened in the days following her death or how William reacted, but we do know something of his later personality, and there is modern understanding to guide us.[26] It is now widely acknowledged that even young children can be emotionally involved, and that the death of a parent has much more significance for young children than was formerly realized. Children can experience sorrow,

[23] Information from family sources listed above; also see family tree.

[24] Letter Robert Wood to Mary Bragg, 2 April 1868, Bragg (Adrian) papers.

[25] Death certificate from Wigton office of Registrar of Births, Marriages, and Deaths.

[26] See, for example, the following two works that bear on the question of children and the death of a parent: J Bowlby, *Attachment and Loss* (London: Hogarth, various dates), vol. I, *Attachment*, 1982 (2nd edn.), vol. II, *Separation*, 1973, vol. III, *Loss*, 1980; J W Worden, *Children and Grief: When a Parent Dies* (London: Guilford, 1996).

anger, loneliness, anxiety, and insecurity. In addition, a child's sense of time extends beyond its immediate experience and into the future. Children feel a close death not only through the immediate pain of separation but also in terms of the effect on their future.

We can be sure, therefore, that William asked questions, at least of himself: Why did it happen? Why couldn't the doctor help? Who would care for the family? Who would help with his school work? Who would continue his piano lessons in the drawing room?[27] The young Bragg family would have received some comfort from their Wood grandparents, with some answers in terms of their Anglican faith, and one of the Wood clergymen perhaps conducted the funeral service in the Westward church before officiating when his mother's body was interred in the church graveyard. If William attended, he surely watched in awe and disbelief, while the answers that came to his questions must have seemed unsatisfactory or beyond his understanding.

The certificate that formalized Robert's registration of his wife's death at Wigton twenty-two days later declared: 'Pleuritis 12 days, Premature Labour 8 days'. From this limited information it is difficult to determine precisely what went wrong.[28] The stated times indicate that the major problem was the pleuritis: inflammation of the pleura, the thin membrane covering the surface of the lungs, due to an underlying pneumonia, the cause of which could have been a pneumoniae organism or a tuberculosis. This then led to the premature labour that lasted eight days; although Mary probably died undelivered since such an infection-induced labour is often irregular and ineffectual. Furthermore, if the membrane had ruptured, an additional puerperal infection could have occurred, particularly if the attending doctor, as was common, did not observe the strictest cleanliness in attempting to assist his patient.[29] In summary, 'Mary Bragg probably died from a pleurisy due to an underlying pneumonia that could have been tuberculous'.[30]

Some time later Uncle William arrived at Stoneraise Place. As the head of the family he had a radical suggestion for his brother. 'You can't look after three young boys and run the farm by yourself', I hear him saying. 'I am unlikely to have children of my own and I have the assistance of our mother [Lucy Bragg] and a playmate [Fanny Addison] at Market Harborough. Young William is mature enough to come and live with me'. And so it was. When Aunt Mary's husband died unexpectedly, she and her son (Willie Addison) came to live at Stoneraise Place to care for her brother and his three boys. After a short time, however, William was packed off to Market Harborough in far-away Leicestershire to live with his extended Bragg family under Uncle William's dominance and stern Victorian discipline.[31]

[27] When the farm and its contents were sold later, there was a piano in the drawing room, n. 6.

[28] The following advice was provided to the author during February 1989 by the late Dr Harold Attwood, Medical History Unit, University of Melbourne.

[29] See, for example, I Loudon, *The Tragedy of Childbed Fever* (Oxford: OUP, 2000).

[30] Attwood, n. 28.

[31] W H Bragg, Autobiographical notes, pp. 8–9.

It would be important in any biographical study to examine the psychological effect on the subject of a mother's early death, but it is essential in the present case because the legacy that William carried impinged on his later life, his family, and his children. He was to gain scientific eminence in intimate association and collaboration with his elder son. Family relationships are therefore essential to any adequate understanding of the scientific advances that William and Lawrence Bragg made. The two works referred to earlier are helpful in this regard: the three-volume work by Bowlby has become a classic,[32] and the book by Worden provides a concise treatment.[33] Central to Bowlby's work is the principle: 'What is believed to be essential for mental health is that the infant and young child should experience a warm, intimate and continuous relationship with his mother'.[34] William had a strong attachment to his mother and, as the eldest son on a farm with a capable father, it can be assumed that his childhood relationship with his father was also close: 'such experience also promotes his sense of competence. Thenceforward, . . . personality becomes increasingly structured to operate in . . . resilient ways, and increasingly capable of doing so despite adverse circumstances'.[35]

In his second volume Bowlby deals with behaviours consequent on the loss of a mother-figure, and a major finding is that 'loss of mother-figure . . . is capable of generating responses and processes that [include] a blockage in the capacity to make deep relationships'.[36] Relevant conclusions are: first, that psychiatric disturbance is not the usual outcome,[37] and second, that 'among conditions known to mitigate the intensity of responses of young children separated from mother, the two most effective appear to be—a familiar companion [and] mothering care from a substitute mother'.[38] In this regard William was served well, for he did have a familiar companion in Fanny at Market Harborough, and there were substitute mothers, his aged grandmother and, four years later, his Aunt Mary.

Bowlby's final volume 'explores the implications . . . of the ways in which young children respond to a temporary or permanent loss of mother-figure',[39] a loss that is 'one of the most intensely painful experiences any human being can suffer'.[40] The responses of the child depend to a large extent on 'the experiences which a bereaved person has had with attachment figures during the course of his life, especially during his infancy [and] childhood'.[41] Since William had enjoyed a happy and fulfilling early childhood we may assume that his

[32] Bowlby, n. 26.
[33] Worden , n. 26.
[34] Bowlby, n. 26, vol. I, pp. xi–xii.
[35] Ibid., p. 378.
[36] Ibid., pp. xiii–xiv.
[37] Bowlby, n. 26, vol. II, p. 5. Worden notes that 'most children manage the tasks of mourning in a healthy fashion', n, 26, p. 16.
[38] Bowlby, n. 26, vol. II, p. 16.
[39] Bowlby, n. 26, vol. III, p. 1.
[40] Ibid., p. 7; also Worden, n. 26, p. 18.
[41] Bowlby, n. 26, vol. III, p. 76.

response to his mother's death was strong but not overwhelming. Furthermore, if we assume that William had seen the death of birds and animals on the farm and that he saw his mother buried in the Westward churchyard,[42] then he received the most crucial information he needed to mourn his mother with healing. What William soon lacked, however, was 'the comforting presence of his surviving parent... and an assurance that that relationship will continue'.[43] William's father had lost his own father when he was only nine years old; his experience and empathy could have been invaluable.[44]

Of the four common reactions of children to the loss of a parent listed by Worden—sadness, anxiety, guilt, and anger—we can assume that William experienced the first, and that, when Uncle William appeared to demand that he come to Market Harborough, William felt the second and third.[45] Young children, with a very limited amount of information with which to understand the world around them, introduce much illogical thinking into their attempts to understand cause-and-effect relationships: Why was he being sent away? Why was he being punished? Had he caused his mother's death? Such questions are common in young children in William's position.[46] In the middle of 1869, just when he needed acceptance and reassurance, William was disconnected and vulnerable.

In the long term, the legacy was a significant difficulty in making close personal relationships: 'He was always very reserved, but [in Adelaide] he had friends of the joking and teasing kind... [whereas] I do not think he made a single friend in England, though hundreds of acquaintances'.[47] Similarly, his children remembered, 'He was antipathetic to any kind of emotional strain. Personal problems arise from time to time in a family... and we had a tremendous struggle to get him to advise us... He would break off to talk of the weather, of his research, or garden, or the Royal Institution. If we stuck to it the advice would come, not orally, but generally in the form of a long wise letter'.[48] His daughter-in-law, Alice Bragg, wrote: 'father and son did not find it easy to communicate with each other, certainly not about their feelings'.[49]

[42] There is ample evidence that whole families traditionally attended funerals in the nineteenth centruy; also see Worden, n. 26, p. 23.

[43] Bowlby, n. 26, vol. III, pp. 270–1, 276.

[44] Worden, n. 26, pp. 16–17, 21, 36–7, has a discussion of mediators of the mourning process that supports the thrust of Bowlby's arguments.

[45] Worden, n. 26, ch. 4.

[46] Ibid., pp. 61–2.

[47] Letter W L Bragg to Sir Mark Oliphant, 13 October 1966, RI MS WLB 54A/27, p. 3.

[48] Sir Lawrence Bragg and Mrs G M Caroe, 'Sir William Bragg, F R S (1862–1942)' *Notes and Records of the Royal Society of London*, 1962, 17:169–82, 181.

[49] A G J Bragg, Autobiographical notes, p. 171.

2
Market Harborough

The year 1869 saw two other developments that would be relevant to William's future. His father rented the nearby Watch Hill property as a further extension of his farming land, which continued to prosper,[1] and in far-away Adelaide, Australia, Gwendoline Todd was born on 22 July that year.[2] William Bragg, however, was going to Market Harborough.

A delightful picture of the town at that time is given in *Yesterday's Town: Victorian Harborough*.[3] Roads lead out of the town west to Coventry, south to Northampton, and east to Kettering, but the road into town from the north dominates our story. Coming from Leicester, it broadens as it approaches the centre of Harborough into Upper High Street, where large cattle markets were held on the cobbled roadway every Tuesday. Proceeding slowly south, some buildings occupy the centre of the road until the Market Place or Church Square opens out, occupied by two buildings that William came to know intimately: the parish church of St Dionysius and the adjoining 'old grammar school'. The large church is dominated by its majestic tower and broach spire of grey ashlar, soaring high above the fourteenth-century ironstone church, the chancel in the Decorated style. Its Georgian trappings of three-decker pulpit, box-pews, and west gallery had been removed by the time William arrived.

The old grammar school is a timber-framed building with a large, first-floor schoolroom, covered by a steep gabled roof and supported on sturdy wooden posts, now a community facility.[4] It was founded by Robert Smyth, who had made his fortune in London after leaving his native Market Harborough, and it opened in 1614. Smyth had specified an open ground floor 'to keep the market people dry in times of foul weather', and it also served as a space for the weekly butter market.[5] Continuing south and after further buildings narrow the area, a large open space appears, the Sheep Market or The Square, where

[1] Bragg (Adrian) papers.

[2] Birth certificate from Adelaide office of Registrar of Births, Marriages, and Deaths.

[3] J C Davies and M C Brown, *Yesterday's Town: Victorian Harborough* (Buckingham: Barracuda, 1981).

[4] G K Brandwood and J C Davies, *Market Harborough Parish Church and the Old Grammar School* (Corby: Church Council, 1983); *Harborough District Council Official Guide* (Gloucester: British Publishing Co., n.d.), p. 15.

[5] J Anderson, *Bygone Market Harborough* (Leicester: Anderson, 1982), p. 4.

Fig. 2.1 Church and school at Market Harborough, Leicestershire, *circa* 1860. (Courtesy: Harborough Museum, Leicestershire Environment and Heritage Services.)

Uncle William's chemist shop was located at 2 The Square, on the corner of both Adam and Eve Street and St Mary's Road. Uncle James' grocery store was next door and their living quarters were upstairs on two floors over the shops. Young William's country world had contracted to the town.

The most significant events in early-Victorian Harborough had been the arrival of the railway, a branch line from the London–Birmingham track in 1850 to complement the branch canal of 1809, and the first issue of the *Market Harborough Advertiser* newspaper in 1854. With the railway came industrial expansion: in the corn market, in flour milling, in the clothing trades of glovers, dyers, and hat-makers, and in the businesses of the Symington brothers—from the manufacture of stays and corsets to coffee roasting and auctioneering. Ancient practices also continued: the regular cattle and sheep markets, the six fairs each year, and the work of the Town Estate of Market Harborough. The Town Estate was responsible for bridges and highways, for the supply and distribution of water, for some poor relief and schools for the poorer classes, and for apprentices and pupil teachers, all under the direction of a group of feoffees or trustees.[6]

Uncle William was a local identity, for not only did he own the chemist and grocer's shops in the centre of town, he had also acquired other property and was an oil-cake merchant and insurance agent. According to the 1861 Census he employed two men and a boy, and at both the 1861 and 1871 Census the family had a domestic servant. He was elected a Guardian of the Poor and later

[6] Davies and Brown, n. 3, *passim*.

a feoffee of the Town Estate, and he would be a churchwarden for many years. Young William later described his new family in these revealing paragraphs:[7]

> Uncle William was a fine character. He was rather domineering and was not always popular in his own family, still less in the town which he tried to push along in the ways which seemed to him right and generally were so. He used to lecture us [William, Fanny, and later Willie Addison] terribly, talking by the hour, and I suspect he was not to be shaken in his opinions by any one. But he had great ideals and he always wanted to make us share them. He was unsparing of himself and of us in trying to nerve us to do our best: there was to be no slackness. For all that he was very kind, and he had lots of humour. When in later years he had mellowed and I came down from Cambridge at intervals to spend vacations with him, the first evening was always uproarious because I had saved up all the jokes and stories for him and he sat drinking it all in.
>
> Uncle James was a dear kind man, very simple and earnest, repressed and overpowered by Uncle William. His education must have been cut short very early, and he had very little knowledge to build on. But all his life he tried to improve himself, and his chief reading was Cassell's Popular Educator, in which he struggled to learn French and other things all by himself. His chief pleasure was the long ride on one day of each week when he went on Black Bess to get orders at the country villages. The whole day was taken up with the journey and he had dinner at one of the furthest points on the round. I remember him coming home on cold winter days very ready for the hot supper that was ready for him. On the morning of his round our ante-breakfast ride was omitted.
>
> My grandmother [Bragg] was a real conservative, a Church person as against Chapel. She was terribly put out one day on finding that the maid-of-all-work had taken us to a service at the independent chapel to which she belonged. Grandmother thought our souls were in some danger, so in the evening when she was brushing Fanny's hair and I was reading alongside she went through the Apostles' Creed with us... item after item and we always agreed... Then she closed the book, reassured. We must have been eight years old at the time. Fanny was six months older than I was.

Robert Smyth, the founder of the grammar school, had served the Corporation of London for many years, rising to be Comptroller of the Chamber from 1597 until his death in 1623. He gave several sums of money to the City of London, as a result of which a number of annual payments were made to Market Harborough for charitable purposes, including the establishment of a school.[8] The City maintained responsibility for the school until about 1900, and the school badge is still that of the City. In the middle of the nineteenth century the building became dilapidated and the school inefficient, and it finally closed with the death of the headmaster in 1862. The inhabitants of Harborough repeatedly alerted the Charity Commissioners in London to the situation,

[7] W H Bragg, Autobiographical notes, pp. 10–12.
[8] Letter Deputy Keeper of the Records, Corporation of London, to author, 1 March 1983.

and a long report on the matter was presented to the Court of Aldermen in October 1867.[9] As a result the Charity Commissioners drew up a 'Scheme for the Government of the Grammar School, Market Harborough'. This appointed four trustees for the administration of The Grammar School Charity, including William Bragg Bragg (Uncle William), and gave considerable powers to the trustees for the regulation and management of the school, but it retained other powers, such as the appointment of the headmaster, with the Lord Mayor and Aldermen of the City of London.[10] The curriculum was specified in some detail at clause eighteen of the Scheme:

> 18. The secular instruction at the School shall comprise the Greek and Latin Languages, Mathematics, Algebra, Arithmetic, Land Mensuration and Surveying, and Elements of the Natural Sciences, English Grammar and Literature, General History, Geography and Writing, and such other subjects of useful knowledge as may from time to time be directed or authorized by the Trustees, after consulting with the Head Master. Religious instruction shall be given to the scholars according to the principles and doctrines of the Church of England; provided that no child shall be compelled to receive such instruction whose parents or next friends shall declare in writing that they entertain conscientious objections thereto.

A public subscription organized by the trustees raised enough money to renovate the building and add a rear brick portion to contain new amenities at a cost of about £400. The school reopened on Wednesday 4 August 1869 under its new headmaster, Rev. James Wood BA (London), previously Second and Mathematics Master of the Clergy Orphan School, Canterbury, who introduced the new curriculum and stayed until 1874.[11] This matched young William's arrival in Market Harborough just a few weeks earlier, an important and fortuitous coincidence. He later recalled:[12]

> Uncle William had in 1869 succeeded in re-establishing the old grammar school in Market Harborough. It is a quaint structure raised on wooden pillars.... The newly-appointed master, Wood by name, was an able man, I believe, and the school grew. I was one of the six boys with which it opened after a long interval. Perhaps it was because of my uncle's connection with the school that at the end of the first year [mid-1870] I was given a scholarship of £8 a year, exempting one from fees. At the prize-giving—there were many more than six boys at that time so that there was quite an assemblage—my name was called out and I went up to the desk to get the

[9] 'Market Harborough School (Smyth's Charity), Report to the Court of Mayor and Alderman from the General Purposes Committee, 22 October 1867', Corporation of London, Common Council Minutes and Reports, 1865, report no. 43.

[10] 'Scheme for the Government of the Grammar School, Market Harborough, 1868', sealed 18 February 1868, Corporation of London, Court of Aldermen papers, 16 June 1868, Board of Charity Commissioners for England and Wales.

[11] Market Harborough Grammar School, Minutes Book [of meetings of Trustees], vol. 1, April 1868–, held by the Parish Church, Market Harborough, Leicestershire.

[12] W H Bragg, Autobiographical notes, pp. 9–10.

> scholarship not knowing what it was: I was puzzled and disappointed to go back empty-handed. The school was quite good and I got on quickly enough: in 1873 I went up for the Oxford Junior Locals and was the youngest boy in England to get through. I got a 3rd class and was told that I would have done better but that the regulations forbade a higher class to any one who did not pass in Church History; in that I failed, as also in Greek.
>
> We lived a very quiet life in Harborough. Before breakfast Uncle James and I went out riding for an hour to an hour-and-a-half: we got to know all the villages round. Our longest rides would be... six miles away. I was not fond of riding for some reason, though I liked the morning air and I liked the pony. Ball games of all sorts have always interested me more than country sports: I enjoyed them very much, while hunting, fishing and the like did not attract me at all; moreover they never came my way. After the ride the day was filled with school and preparation for school: and an occasional walk. There were very few games in those days, as the school was a day-school, without grounds. At the end of my six years there we had a little football, which was a great delight.
>
> There were no parties for children: we never went to other people's houses, and no children came to ours. I think my uncle was too particular—he was indeed a refined and educated man—to let us fraternise with the children of the small shop-keepers and, as he was a shop-keeper himself, we were not asked to the houses of the lawyer, the parson and so on. I suspect too that he did not feel justified in entertaining much, as he had to work his way from times of poverty, and it was not until later that he was in quite comfortable circumstances.
>
> He was very good to me, taking infinite pains with my lessons, especially the Latin: we hammered out the Aeneid with great difficulty in the shop, while occasionally he broke off to serve a customer. Our scanning was quaint. There are a few unfinished lines in the Aeneid, and we tried to save a few syllables from the line before so that we could get two hexameters, not realising that the half line was incomplete. Mr Wood was greatly amused.

William did well in his studies. School accounts sent to his uncle survive, showing fees for the half year ending Christmas 1869, books totalling £3.11.3, and similar fees for the half year ending 23 June 1870, with examination results and a note on the back from headmaster Wood saying, 'I am very pleased to be able to inform you that Willie was one of two boys recommended by the Examiner for one of the Free Scholarships. Accordingly, the recommendation has been acted upon and Willie becomes one of the two first free scholars'.[13] The examiner was Rev. F M Beaumont, late Fellow of St John's College, Oxford, and Rector of East Farndon.[14]

There are a few letters of this period from William in Market Harborough to his father at Stoneraise Place, very formal and written in copperplate elegance,

[13] Bragg (Adrian) papers.
[14] Trustees' Minute Book, n. 11, meeting of 22 June 1870.

unlike his later economical style. We can imagine Uncle William insisting that his young nephew write to his father from time to time. 'Grandma is very kind and always pleased with me', William wrote. 'Uncle helps me with my exercises at night, so that I am in school before nine; the Master is very pleased with me, and I try to be a good boy and a good scholar. I like to learn Latin. Uncle says I count very well.... My dear Papa, I am your dear son, William Henry Bragg'.[15]

Both of William's grandmothers died during 1873, Ruth Wood in May and Lucy Bragg in November. He knew them both. There is a later photo of William and his two brothers standing together, reunited in Cumberland for a holiday with their father (see chapter 3). 'Each summer we went to Cumberland, Fanny and I and Willie Addison after he joined us. That was a delightful change; my chief memories are the wild cherries and helping with the harvest'.[16] William must have visited his Wood grandparents at Westward on such occasions. Grandmother Bragg, however, had been the lady-of-the-house in Harborough, and her loss was serious. Uncle William called his sister from Cumberland, and Mary and her son came to be reunited with their daughter and sister. William again:[17]

> Well, after Grandmother died, Aunt Mary came and brought Willie Addison, Fanny's brother, to live with us. Aunt Mary had been keeping house for my father at Stoneraise Place. My two brothers Jack and Jimmy had been going to school at Wigton, a second-rate grammar school I should think, not nearly as good as the one at Market Harborough. Willie Addison had been going with them. The change left my father alone at Stoneraise Place with the two boys: I suppose some sort of housekeeper was engaged.
>
> Aunt Mary was also a rather repressed personality: very afraid of Uncle William, shy and kindly and conventional... I don't quite know how my uncle managed to collar both Fanny and myself. I suppose our respective parents were talked to and forced to give us up. Fanny and I were very good friends: much in one another's society, of course, because... we were not on visiting terms with any other families.
>
> Willie Addison's arrival was not a happy event for me. In fact my life became, as I remember it, miserable to a degree. We did not fit and were never friends. He had a nasty temper, restless, unkind and sneering. He was constantly in trouble with the Uncle. I liked peace and was content to be alone with books or jobs of any sort, dreamy and lacking in enterprise outside the occupations I enjoyed.

In May 1873 young William (not yet eleven years old) wrote to his father regarding his forthcoming examinations for the Junior Certificate of the Oxford Local Examination: 'I do not think I shall pass but still I will try to do as well

[15] Letters W H Bragg to father, n.d., Bragg (Adrian) papers.
[16] W H Bragg, Autobiographical notes, p. 13.
[17] Ibid., pp. 11–12, 28.

as I can... I shall be examined in Leicester at the Vestry Hall... On Monday I shall be in examination from 2 till 8.30, Tuesday from 9 till 8.30, Wednesday the same; on Thursday and Friday from 9 to 12. I shall return, I think, on Friday afternoon... Your affectionate son, W H Bragg'.[18] He remembered later that 'Aunt Mary... took me to stay at Leicester during the week when I sat for the Oxford Local Exam. in the spring or summer of 1873. We felt we were on the loose and were companions: I had not known her as such before. We stayed at Cook's Temperance Hotel, the first hotel of the Great Cook. We nearly always got on well together'.[19] The certificate William later received contains the following information as to the subjects he passed and his overall result:[20]

Preliminary Subjects required of all Junior Candidates

I	1. Reading aloud	4. A short English Composition
	2. Writing from Dictation	5. Arithmetic
	3. Analysis and Parsing	6. Geography
		7. Outlines of English History

Subjects selected by the Candidate

II 1. The First Book of Samuel; the Gospel according to St. Luke

III 1. Latin

3. French

4. Mathematics

The Candidate was placed in the Third Division.

William remembered that he failed in church history and in Greek and that he was the youngest boy in England to obtain the certificate. Correspondence with the University of Oxford Delegacy of Local Examinations confirms that William was, indeed, the youngest boy to pass the examination in 1873, and that the items he failed were:[21] II Part 2. The Catechism, the Morning and Evening Services, and the Litany, and III 2. Greek. The regulations for the examination required candidates to satisfy the examiners in all the subjects of section I and in at least two, but not more than five, of the eight subjects available for selection by the candidate (the one in section II, 'The rudiments of faith and religion', and seven in section III, 'Optional subjects'). William attempted the maximum, passing all the preliminary subjects, half the religious subject, and three others. The examination schedule was clearly gruelling, and we might ask why William was pushed into the examination at the maximum level when he was exceptionally young for

[18] Letter W H Bragg to father, 25 May 1873, Bragg (Adrian) papers.
[19] W H Bragg, Autobiographical notes, p. 28.
[20] Bragg (Adrian) papers.
[21] Letters J P Holloway, Principal Administrative Assistant, University of Oxford Delegacy of Local Examinations, to author, 5 December 1984, 20 March 1985 and 16 May 1985, the second enclosing copy of 'Regulations for the year 1873—Examination of Junior Candidates'.

the ordeal. Perhaps we need look no further than Uncle William's earnest commitment to education and his desire to see his talented nephew excel.

The story of the Market Harborough Grammar School and the Oxford Local Examination is typical of many such schools in mid-nineteenth-century England.[22] The education of the middle classes had become weak and inefficient in the face of competition from the public schools at one end of the social spectrum and the national and British schools at the other end. What changed the situation was the introduction of public examinations, 'one of the great discoveries of nineteenth-century Englishmen. Almost unknown at the beginning of the century, they rapidly became a major tool of social policy'.[23]

The promotion of open competition as exemplified by examinations had three major objectives: first, to maintain and raise academic standards, second, to decide the fitness of candidates for public office or the professions, and third, social engineering—for example, to enable children from a lower social group to advance by attaining a higher level of education.[24] However, by mid-century there was no effective means of applying the external examination idea to the middle-class or secondary schools most in need of assistance, and it was this need that the creation of the Oxford and Cambridge Local Examinations in 1857–58 satisfied. They became of great value to education generally: the universities gave the guarantee of scholarship and impartiality that parents and teachers needed, causing teachers and communities to examine the quality of their local educational programmes, and they provided the universities with a priceless means of extending their pool of prospective students and of new and valuable friends.[25]

The movement led quickly to tangible outcomes, as in the case of the Market Harborough Grammar School. If the children of the local farmers, shopkeepers, merchants, and businessmen were to be educated and find rewarding places in the new competitive world, then the local grammar school needed to be repaired and reactivated. A headmaster of academic and personal integrity needed to be found, his standard of education needed to be measured regularly by the use of qualified external examiners, and his students needed to be prepared for the Oxford Local Examinations. This the leading citizens of Harborough accomplished admirably. As we have seen, young William was examined within the school before he faced the Oxford Locals, and this practice continued during his subsequent years in Market Harborough, when he was regularly mentioned in the printed reports of the annual external examiner.[26]

As the 1875 school year drew to a close and William completed the final year of the program offered by the school, his future education needed to be decided. Clearly he was intelligent and academically able and the only

[22] J. Roach, *Public Examinations in England 1850–1900* (Cambridge: CUP, 1971).
[23] Ibid., p. 3.
[24] Ibid., pp. 8–9.
[25] Ibid., p. 74.
[26] Bragg (Adrian) papers; William was specifically mentioned in the reports for 1871, 1872, and 1875.

question was where he should go. The family now had the funds needed and Uncle William had already entered him at Shrewsbury.[27] Founded around 1560, Shrewsbury had had an inspired teacher, innovator, but extraordinarily unpopular headmaster, Dr Samuel Butler, in the early nineteenth century. Butler is credited, for example, with discovering examinations as a method of competition that recognized and rewarded academic work and achievement.[28] By 1830 Shrewsbury was one of the nine 'great schools' of England, and its next headmaster further enhanced its reputation. Uncle William had apparently chosen well.

Unlike the previous occasion, however, this time William's father came to Market Harborough to confront his brother. William saw it thus: 'In 1875 my father came to Harborough and demanded me: he wanted to send me to school at King William's College, Isle of Man. I think he became alarmed lest he should lose me altogether... The College had a pretty good reputation: somehow the boys won a fair number of scholarships, one or two or three each year at Oxford or Cambridge. As a school it was poorly found: the fees were very low, board and tuition were about £60 a year'.[29] This time his father won, the main reason for the choice being clarified by William: 'The third son of my grandfather [Wood] was Robert Wood, [later] Vicar of Rosley. He did not go to Oxford, but was trained for the church at St Bee's. He was a very gentle, kindly man: of considerable ability but no force... He was a master at King William's College, Isle of Man, when Dixon was headmaster: and Aunt Eliza was the head's daughter. It was this connection with King William's that induced my father to send me there. Aunt Eliza was a woman of character, and in a large parish would have taken a lead to some purpose'.[30] Another reason for the choice was no doubt Robert Bragg's familiarity with the Isle of Man as a result of his earlier trips as a merchant seaman. William Henry Bragg's life was now to experience its second major upheaval; he was to go to an English public boarding school.

Uncle William's strong commitment to education in Harborough continued after William left. When a new headmaster arrived in 1887 and wished to develop new subjects, such as science, he found the old school building inadequate. However, there was opposition and indifference in the town to expenditure on a new school building, and an impasse developed that was only removed when Uncle William intervened in 1891. On his vacant paddock in Coventry Road he built a boarding house and schoolroom, and later a science laboratory, that he rented to the headmaster (see Figure 2.2). The school grew and prospered, and it now flourishes under the name of 'The Robert Smyth School'.[31]

[27] W H Bragg, Autobiographical notes, p. 13.

[28] J Gathorne-Hardy, *The Public School Phenomenon 597–1977* (Harmondsworth: Penguin, 1979), p. 41.

[29] W H Bragg, Autobiographical notes, p. 13.

[30] Ibid., p. 8.

[31] Davies and Brown, n. 3, pp. 58–9, 102–3.

Fig. 2.2 William Bragg Bragg ('Uncle William'), mellowed civic identity of later years, when his nephew was in Australia. (Courtesy: Mr M Brown and the Market Harborough Historical Society.)

William retained the affection he had developed in Harborough for his Aunt Mary and her daughter Fanny. Years later, after he had sailed to Australia as a professor and was courting a daughter of one of Adelaide's major public figures, he was given a picture frame by a member of her family. His girl-friend then wrote to her sister: 'My dear Maude Todd, The professor is highly delighted with the frame, and spent the whole of Sunday in trying to decide who to put in it and would take no suggestions of ours; at last he decided on his aunt and in it she reigns supreme. Gwennie Todd dear'.[32]

[32] Letter Gwendoline Todd to Maude Todd, n.d., Bragg (Adrian) papers.

His mother had given William a priceless legacy in the form of maturity, resilience and confidence in his own abilities, and these were critical to his life in Market Harborough and were enhanced by his experiences there. From his own description of Uncle William it seems highly unlikely that young William found him a warm and outwardly affectionate parent substitute, but he did provide a secure, predictable, and disciplined environment. Grandmother Bragg and Uncle James were constants, cousin Fanny was a companion, and Aunt Mary was affectionate. Life was centred on school and study, and for William this was reassuring and satisfying because he did well and he had an intellectual helpmate in Uncle William. One or two legacies remained, however: he 'liked peace and was content to be alone'.

3
King William's College

It was 1983 and it could have been 1875. Just outside Castletown, on the south coast of the Isle of Man, I stood on the sports fields of King William's College on a bleak September afternoon. Two hundred metres in front of me were the sloping sands and dark seas of Castletown Bay; a hundred metres behind the imposing, grey limestone walls of the main school building, its massive and forbidding clock-tower looming over the scene. I shuddered as I remembered what I had read about English public schools in the nineteenth century: of fagging and flogging, of the great Arnold of Rugby's reputation tarnished by hypocrisy, of Vaughan of Harrow's fame destroyed by pederasty and then restored by his refusal to accept high church office.[1] It was 1875 and I was momentarily alongside William Bragg at King William's College. The following day was different; it was sunny and peaceful in the crisp early morning. The lush turf was a brilliant, iridescent green, a colour I had not seen before. I was already feeling 'the character' of the modern College.[2]

The annual Founder's Day, the dining hall, the old boys' Society and the school magazine all commemorate the supposed origin of the school in the 1668 trust deed of endowment of Isaac Barrow, Doctor of Divinity, late Fellow of Peterhouse, Cambridge, uncle of the famous mathematician and divine of the same name, and successively Bishop of Sodor and Man and of St Asaph. As a royalist Barrow had been in retirement, but on 5 July 1663 he was consecrated Bishop in Westminster Abbey and thereafter Governor of the Isle of Man. He was 'one of the most respected of Manx bishops, and a great benefactor of the land'. Barrow was 'equally zealous for education, built and endowed schools, and required the clergy to teach'.[3]

Not until the King's birthday on 23 April 1830, however, were the foundation stones of the evangelical Anglican school and its adjoining chapel laid. The

[1] See, for example, J Chandos, *Boys Together: English Public Schools 1800–1864* (London: Hutchinson, 1984), and J Gathorne-Hardy, *The Public School Phenomenon* (London: Hodder & Stoughton, 1977).

[2] F Cain, 'The character of King William's', *Manx Life*, 1983, 12:33.

[3] T F Tout, 'Barrow, Isaac, D D (1614–1680)', in Stephen and Lee (eds), *Dictionary of National Biography* (Oxford: OUP, 1917–), vol. I, pp. 1218–19; a first full account of the history of the College has appeared recently: M Hoy, *A Blessing to this Island: The Story of King William's College and The Buchan School* (London: James & James, 2006).

building cost of £6,570 was met by accumulated funds from the Barrow Trust Fund, the mortgages of two Trust estates, and public subscriptions, and it took in its first students on 1 August 1833.[4] The College was named after King William IV, and in far-away South Australia there was another memorial to the King and his Queen. A new city, founded by British religious and political dissenters only in 1836, was named 'Adelaide' and its main thoroughfare 'King William Street'. William Bragg would come to know that city well only ten years after he had entered the only independent public school on the Isle of Man.

The early principals had great difficulties: Rev. Edward Wilson, Rev. Alfred Phillips, and Rev. Robert Dixon. In 1838 Wilson reported gloomily to the Bishop of Sodor and Man that, 'A large portion of ye interior walls remain to this day not even plaistered...The wet comes in thro' the whole S W front...we repeatedly catch 2 or 3 gallons of water of a rainy eveng in our drawing room...The chapel is neither completed outside nor decently fitted up within'.[5] He soon left to become a parish priest. Furthermore, tragedy struck on 14 January 1844 when fire destroyed the buildings, including the library, the chapel, the tower, and the Principal's house. Only the Vice-Principal's residence survived, but there was 'no loss of life or limb, nor any accident'. A circular reporting the fire, the survival of the walls without great damage, and requesting donations 'met with a handsome response', and the school was able to reopen in its refurbished buildings on 1 August 1845.[6]

James Wilson, a twin son of the first Principal and born at the school, a boarder there from 1848 to 1853, and later headmaster of Clifton College, remembered 'dirt and slovenliness...insufficient food...horrible bullying...indecencies indescribable...no bathroom...and teaching almost as bad as it could be...no one on the staff was a scholar and no one even a tolerable mathematician'.[7] After a master thrashed him 'till I roared', Wilson took his revenge: '[while] he was sound asleep...we put all his clothes into a footpan, and from every vessel in the room poured their contents into his boots and the footpan and retreated. We heard no more of it'.[8] The outdoor life was a salvation for Wilson and his friends, and use of their own study in the tower was 'a great joy'. He summarized his impressions of the school with the words: 'we were shamefully neglected...but in the two last years of the five we were left alone, and educated ourselves'.[9] Then things began to improve; during the long tenure of Rev. Robert Dixon the academic performance of the school rose, and in 1863 a new dining hall, hospital, dormitories, and kitchens helped to lift morale.

[4] F Cain, 'King William's College—150 years on', *Manx Life*, 1983, 12:23–25.

[5] P K Bregazzi, 'Introduction', in *King William's College 1833–1983* (Castletown: King William's College, 1983).

[6] 'Historical Retrospect' in H S Christopher (ed.), *King William's College Register 1833–1904* (Glasgow: Maclehose, 1905), pp. xi–xx.

[7] A T Wilson and J S Wilson (eds), *James M Wilson: An Autobiography 1836–1931* (London: Sidgwick & Jackson, 1932), pp. 8–9.

[8] Ibid., p. 11.

[9] Ibid., pp. 16, 18.

Fig. 3.1 King William's College, Isle of Man, *circa* 1905. (Courtesy: King William's College.)

A new Principal was appointed in 1866, and by August 1875, when William arrived, the school had improved further. Town gas and water were available throughout the College, a laboratory for practical chemistry had been fitted out, a new gymnasium had been completed, and the chapel had been transferred to the first floor, allowing four new classrooms to be constructed on the ground floor of the original building. Rev. Joshua Jones, MA, a first-class graduate and senior scholar in mathematics from Oxford, curate in Cheshire and Manchester, and headmaster of the Liverpool Institute before coming to King William's, was determined to continue the development begun by his predecessor.[10]

Only late-twentieth-century scholarship has revealed the true nature of the unreformed English public schools of the nineteenth century and the travesty of the pictures painted by Thomas Hughes in *Tom Brown's Schooldays* and by Frederic Farrar in *Eric, or Little by Little*.[11] Nevertheless and in spite of great resistance, the Palmerston government was persuaded by widespread criticism to institute a Public Schools Commission in 1861 to investigate the nine great

[10] Christopher, n. 6, p. xxvi.

[11] [T Hughes], *Tom Brown's Schooldays, by an old boy* (Cambridge: Macmillan, 1857); F W Farrar, *Eric, or Little by Little* (London: Black, 1858). Farrar had been at King William's College in the 1840s and probably drew on conditions there for his book.

public schools (later reduced to seven, King William's not among them). The commission became known by the name of its chairman, Lord Clarendon. Much has been written about its work and its findings and there is some disagreement as to its precise impact, but there is no doubt that, together with the influence of English society more widely, it changed forever the character and operation of these schools and those that tried to emulate them. The Public Schools Act of 1868 further regulated them.[12]

Over time the schools remodelled their governing bodies, changed their culture and curricula, and increased educational and physical facilities. Governing bodies were to have no financial stake in their schools, the freedom and independence of students to regulate and discipline themselves was curtailed by greater school control and a prefect system, and the study of sciences other than mathematics was introduced, along with additional new subjects outside the classical tradition. Food and living conditions improved, greater emphasis was placed on athleticism and games, more staff were employed to teach in separate classrooms, the house system became widespread, and formal examinations placed accountability on schools, staff, and students alike.[13] William was about to discover how much of this had reached King William's College:[14]

> I was at the School House under Scott. It was the plainest of the lot as far as living went: it was clean, however. Our meals consisted of: breakfast and tea, at which meals we each had one piece of butter, as much bread as we wanted, and tea. No jam, milk or cake or bacon or eggs unless we provided luxuries for ourselves, or our parents paid extra, and very few did that. Dinner consisted of meat and pudding, both very poor: supper, a piece of bread and butter. Baths once a week: and one thing I look back on with interest, namely that we were locked out of our dormitories all day, so that we had no chance of a change when we played football; we just took our coats off and put on jerseys as far as I remember. But the place was a very healthy one, and after the first year or two, when the bullying was rather unpleasant, I was happy enough.
>
> I stood high in the school and liked my work, especially the mathematics: and fortunately I was very fond of all the games and played them rather well. So, though I was a very quiet, almost unsociable boy, who did not mix well with the ordinary schoolboys, being indeed very young for the forms I was in, I got on well enough. As I grew older I became more at home, and at the end I was quite popular, I think, and as Head of the School I had some influence. John Kewley, now Archdeacon of Man, was my best friend.

[12] See, for example, Chandos, n. 1; E C Mack, *Public Schools and British Opinion Since 1860* (New York: Columbia University Press, 1941); T W Bamford, *Rise of the Public Schools* (London: Nelson, 1967); B Simon and I Bradley (eds), *The Victorian Public School* (Dublin: Gill & Macmillan, 1975); J A Mangan, *Athleticism in the Victorian and Edwardian Public School* (Cambridge: CUP, 1981).

[13] Ibid.

[14] W H Bragg, Autobiographical notes, pp. 13–18.

The masters were not very good: Hughes-Games the Headmaster [=Jones, see below] was fairly able, not very popular: Jenkins, our mathematics master in my later years was a good fellow, keen and a good teacher: he did well for us. There was a good classics man, Edwards, but I think he drank. Curiously enough the strongest character was the French master Pleignier, old 'Plan', of whom the boys were more afraid than of any one else. He had ideas too: once he tried to teach us senior boys to write good English essays, but I fear I disappointed him. He helped with the annual dramatic entertainment, which was great fun, the best event of the year. We made scenery, collected costumes, rehearsed at times when we might have been doing lessons, and generally broke away from the ordinary run. It was a great night when Castletown society came to see us act. There is a volume of The Barrovian among my books which contains some accounts of those days.

The incidents that come back to me in memory are not likely to interest you much. Perhaps the most remarkable was the occasion, about 1876, when the whole school was summoned for trespassing on the grounds of an old mill near Ballasalla. We all had to march down in crocodile form to the courthouse at Castletown and be tried by a Man jury, consisting, so we boys believed, of Manx farmers. Of course they found us guilty. They could not put us all in the box, but they picked out a few of us, of whom I was one. I owned up to having been there, and when asked what damage I had done I replied "nothing to speak of." This was warmly greeted by the prosecuting counsel, who now thought he had got hold of a principal villain. He was disappointed when he found that I had only got inside, with others, the big waterwheel and made it go round by walking up it in treadmill fashion. You see, I really did like the accurate, and my answer was strictly true. We were told we were all fined, and some of us, including me, more than the others, but I believe the fines were never collected.

During the first two years or so, we had long holidays twice a year only, the school periods being half-years. There was a break of a fortnight (I think) in the middle of the half-year when it was thought to be not worth while for us to go home, and a number of us stayed on. We enjoyed that: we idled and played games and went on picnics in which some of the masters' families would join us. Blackberrying was a common aim: we made blackberry squash by putting alternate layers of berries and sugar in jampots and squeezing them down: they were a substitute for jam.

In 1880 I went up to try for a scholarship at Trinity (Cambridge) and was awarded an Exhibition [Minor Scholarship, in fact]: this was in the spring and I was then 17. The authorities at Trinity thought, however, that I had better wait a year, so I went back to school. For that last year I was very much by myself in the work I did. I was Head of the School and one of the cricket eleven, so that, apart from the work, I was well in with the school doings. But I did not do well in the work in that year, and when I went up again to try for a scholarship at Trinity—hoping to get something better than that which I had won before—I did not do as well as in 1880 and was told that I would have won nothing had I not been successful the year before. I think it was bad for me to be so separate from the other boys in

> that last year: there was no competition and I was rather a difficulty to the masters because I needed special provision.

Some elements of these reflections invite examination and explanation. For many years the majority of boarders lived in the main school building, although the Principal and the various masters who lived in Castletown also took in a few boys. William was in the oldest and the plainest, the Principal's or School House. Clearly living conditions were basic, but perhaps tolerable and a substantial improvement on those of earlier years. The description of the meals resonates with the infinity of stories about college food that have travelled down the years. In winter the College environment, on the edge of the sea, can be bleak—windy, cold, and dark—but William makes no mention of it. The country boy was resilient and firm beneath his calm and gentle exterior.

A collection of his school reports has survived, and they are revealing despite their brevity. When he entered the school in August 1875 William was placed in the upper fourth form. Specific subjects studied were Mathematics, Arithmetic, and French, and the remainder were collected under the heading 'General Work', which included Religious Knowledge, Greek, Latin, English, and other subjects not specified. By Christmas William had topped his form of twenty-five boys in General Work, Mathematics, and Arithmetic, and was second in French. After only six months he was transferred to the lower fifth. At no stage did William study German, a deficiency that would hamper his research in later years.[15]

In the first half of 1876 he was mid-class in three of the four subjects but first or second in mathematics, with the result that at midsummer he was transferred to the sixth form mathematics class at the age of only fourteen. At Christmas the College Principal observed to his father that William was, 'The most promising pupil for his age in mathematics I think I have ever had. J J'[16] The J J here and in other contemporary documents represents the authorization of Joshua Jones as Principal and is curious, since all later college information refers to him as Joshua Hughes-Games. A letter in 1977 from the College to William's daughter and biographer, Gwendolen Caroe, reports that these names refer to the same person, who changed his name in the middle of his tenure at King William's, apparently because of 'some inter-family inheritance issue'.[17]

Topics covered in the sixth-form mathematics curriculum included arithmetic and algebra, geometry, trigonometry, mechanics and calculus, and 'problems', and there is clear evidence that the subject matter was advanced and well taught and that William had a special aptitude for it. These topics are precisely in line with the recommendations of the Clarendon Commission, and

[15] Bragg (Adrian) papers. The naming of the various school years and grades varied greatly over time and between schools, but the highest grade was usually called the 'sixth form', for boys in their last year of school and of about 18 years of age.

[16] Terminal Report, Christmas 1876, Bragg (Adrian) papers.

[17] Letter G R Rees-Jones to G M Caroe, 14 June 1977, Bragg (Adrian) papers; the signatures on William's school reports and other documents indicate that the change took place early in 1880.

the College also adopted the suggestion that 'mathematics classes should be settled by ability independent of normal class divisions'.[18] One set of William's sixth-form mathematics examination papers from the College, covering the five areas listed above, has survived in the Bragg (Adrian) papers. My academic colleagues who have personal backgrounds in English mathematics have assessed them.[19] All agree that the more difficult questions are of a very high standard, in excess of what many first-year undergraduate students in England would now be required to answer. This accords with the statement by Howson that the 'pressures... being exerted on the schools [after Clarendon] by the two ancient universities through their open scholarship examinations... were forcing schools to introduce the calculus—although at the university this was second-year work'.[20]

William's gift for and commitment to mathematics are confirmed by the fact that he was already near the top of the sixth-form mathematics class of fifteen boys from early 1877, and from Christmas 1878 he was top of this class for the next two-and-a-half years until he left the College. In addition, in October 1877 he was transferred to the sixth form for all his studies. At Christmas of that year, aged fifteen-and-a-half, the Principal reported of him: 'Gives the greatest promise of future success at the University in this subject. About on a level in attainment with boys three or four years older. J J'.[21] Similarly, at the conclusion of the 1879–80 academic year the Principal wrote to William's father: 'His attainments in Mathematics very high for his age. Possesses great mathematical ability. A most satisfactory student in every way. Examiner from Oxford spoke very highly of his work. Joshua Hughes-Games'. In all his school reports, William's general conduct was listed as 'Highly satisfactory'.[22]

Despite William's faint praise of him, the Principal had an excellent mathematics background, taught the subject in William's first years at the College, and knew his young student's potential first-hand. Howson's history, built on biographical studies of key players, has a chapter on James Wilson that also mentions Hughes-Games. Howson notes that, when some suggested that Euclid's classic geometry text was unsuitable as a school textbook and that geometry in English schools was backward compared with that on the Continent, Wilson was asked to 'go into the matter'.[23] Wilson produced his own textbook and precipitated a long battle with traditionalists, who asserted that there was little value in such a book if the examining bodies still insisted on the primacy of Euclid. An Association for the Improvement of Geometrical Teaching was established in 1871, and Wilson and Hughes-Games were elected joint Vice-Presidents. The Association produced two reports, but they were not successful and prompted little immediate change. For the present the new

[18] G Howson, *A History of Mathematics Education in England* (Cambridge: CUP, 1982), p. 130.

[19] King William's College mathematics examination papers for June 1878, Bragg (Adrian) papers. I owe special thanks to Rev. Dr John Scott and Dr Peter Stacey for advice on this matter.

[20] Howson, n. 18, p. 129.

[21] Terminal Report, Christmas 1877, Bragg (Adrian) papers.

[22] Bragg (Adrian) papers.

[23] Howson, n. 18, p. 131.

(1874) Oxford and Cambridge Schools Examination Board defined the content purely in terms of Euclid.[24] The desired changes came later, mainly during the period 1900 to 1914.

The appearance of Hughes-Games' name here is nonetheless revealing: he was apparently in the vanguard of change for school education in England after the Clarendon Commission Report. Furthermore, it was a master-stroke when, in 1876, he obtained the service of David Jenkins, M.A. and Wrangler in the 1870 Cambridge Mathematical Tripos,[25] to carry forward the mathematics teaching at King William's. Cambridge mathematics was extraordinarily influential at this period, both within the university and externally (see chapter 4), and Jenkins is an example of how the Cambridge system was reproduced at sites beyond its walls.[26] When he recorded his verdict of Jenkins in 1937, William had risen to the pinnacle of British science and did not hand out bouquets lightly. That he rated Jenkins 'a good fellow, keen and a good teacher' and that 'he did well for us' is high praise indeed. Jenkins himself formed a close bond both with William and with the school and, although he left in 1886 to take up a parish, having been ordained while at the College, he later wrote to William to congratulate him on his achievements. At his death he gave a legacy of £100 to the King William's College Society.[27] For his part, William kept Jenkins' letters and two photographs of him, one as a young man and one taken much later.[28] Perhaps the reasons for their mutual respect and affection are not hard to see. Teachers find satisfaction and reward in fostering all their students, but the thrill of watching William Bragg's success must have been tangible; and William clearly recognized a deep debt to the man who had guided his academic development more than any other.

The other masters who taught William at King William's are shadowy figures. R J Edwards had an M.A. from Lincoln College, Oxford, and taught classics from 1866 to 1886, while Victor Pleignier apparently had no tertiary qualifications according to the College *Register*, but he stayed at the school for many years from 1860 and taught French and English Literature. Both men played an active part in other school activities, encouraging the students in reading and literature, debating, and theatricals.[29]

One other subject that William undertook at King William's was Physical Geography and Geology. It was one of the subjects that William would pass at his 1881 Oxford and Cambridge Certificate Examination. He won the Byrom

[24] Ibid., pp. 130–136. The Association changed its name to the Mathematical Association in 1897 and is still active today.

[25] The Cambridge undergraduate degree in mathematics was called the Mathematical Tripos; those students performing well in the final honours examinations were termed Wranglers and were listed in order of merit. To be a high Wrangler, and especially the Senior Wrangler, was a notable and prestigious achievement (see chapter 4).

[26] See, for example, A Warwick, *Masters of Theory: Cambridge and the Rise of Mathematical Physics* (Chicago: Chicago University Press, 2003), particularly ch. 5.

[27] K S S Henderson (ed.), *King William's College Register 1833–1927* (Glasgow: Jackson Wylie, 1928), p. xxvi.

[28] Bragg (Adrian) papers.

[29] Henderson, n. 27, p. xxviii; *The Barrovian, passim.*

Geology Prize at the school in 1878,[30] and there is a letter from William to his father in which he says he is not sure of his geology examination results and the prize as yet, but that 'Mr Garside has examined a great number of my fossils and he says that they are all right so far'.[31] Finally, the first issue of the school magazine, *The Barrovian*, carried a letter to the editors from W H Bragg, in which he wrote rather boldly: 'Sirs—I would like to take advantage of the first publication of this Journal to ask a question about a subject to which, I think, attention should be paid. Why, if much trouble and care has been taken to form a collection of the fossils of this neighbourhood, is this collection allowed to be in a most neglected state? At present the fossils are of no use, valueless for exhibition on account of their dusty condition, useless as an object of instruction because they all require to be labelled and classified anew. The collection itself is almost unique and of great value. Surely, somebody could be appointed or obtained to clean and arrange it'.[32] We do not know the true state of the fossil collection, but in the next issue of *The Barrovian* William was obliged to write contritely: 'Sirs—I wrote last month to you about the fossils in the College library. I have found out, since then, that steps have already been taken to clean them, and as far as possible to re-label them. They are also, I believe, to be provided with new cases. I remain, yours truly, WHB'.[33]

Many years later, in 1935, William was both President of the Royal Society and President of The Science Masters' Association, and in his annual address to the latter body he noted: 'I am amazed at the growth in the substance and stability of science teaching, particularly if I think of my schooldays more than half a century ago, when, if the science masters and the boys of the "Modern Classes" did not develop an inferiority complex, it was not the fault of the more orthodox side of the school'.[34] Thanks to William and many others like him, science teaching had, indeed, made great strides during his lifetime.

King William's College had a number of scholarships tenable at the school and William won the Open Scholarship for 1876 as a result of outstanding results in his first two terms in the fourth form. In April of the next year his brother, Robert Bragg junior ('Jack'), entered the school and was soon awarded an Open Scholarship.[35] Tragically, Jack's academic progress was to be impeded and then finally cut short by illness. William recalled:[36]

> When Jack was very young—5 or 6?—he had a serious illness, something internal it must have been. He was fairly strong during his school life at Wigton. He came to KWC two years after me and got on splendidly. But he fell ill again and was in hospital (i.e. the sick room in the School

[30] Christopher, n. 6, p. 348.

[31] Letter W H Bragg to father, n.d., Bragg (Adrian) papers; Garside was an Assistant Master at the school from 1869 until 1895.

[32] *The Barrovian*, 1 March 1879, p. 10.

[33] *The Barrovian*, 1 April 1879, p. 12.

[34] Sir William Bragg, 'Presidential address; school science after school', in *The Science Masters' Association: Report for 1935*, pp. 8–16, 8, RI MS WHB 21/22.

[35] Christopher, n. 6, p. 336.

[36] W H Bragg, Autobiographical notes, pp. 24–5.

Fig. 3.2 Brothers together on holiday, (L to R) 'Jack', William, and 'Jimmy' Bragg, Carlisle, Cumberland, late 1870s. (Courtesy: Lady Adrian.)

House) for some time. I was allowed to sleep with him there, as there was no nurse. He was often in great pain. I remember going down to the kitchen (really forbidden ground) in the middle of the night to heat plates to put on his stomach. Once the cook heard the noise of some one moving and came down in deshabile and a state of great fright. The doctor thought he had an ulcer in his stomach: in these days he would no doubt have been operated on, but that was never done. He recovered and went on with his work. I asked two other boys to his room and we played whist. Unluckily

> the head-master came in and was horrified. I believe our house-master Trafford caught it badly for not looking after our morals more carefully! Cards were strictly forbidden.

The highlight of 1877 was the laying of the foundation stone for a new chapel, and in January 1879 it was consecrated. It included a stone pulpit, oak panelling and stalls, and tiling provided by donations and subscriptions. The space occupied by the old chapel in the main building was successively converted into two classrooms, a laboratory, three new dormitories, a library, and a museum.[37] In February 1878 the headmaster wrote to Robert Bragg to report: 'The boys are going on extremely well', but Jack's health remained a concern. 'If God spares them', he wrote, 'I expect to see them both very high Wranglers and Fellows of their College. I never saw more promising boys'.[38]

Another sign that the school was improving further was the institution of a school magazine that would prove to be a precious record of activities and achievements. It was named *The Barrovian* in honour of the founder, Bishop Barrow.[39] Its very first issue of 1 March 1879 contained a record of the first meeting of a school Chess Association, with leadership by students J M Walker, President, and W H Bragg, Secretary, and also news of a reorganization of the Literary and Debating Society under the guidance of masters Pleignier, Edwards, and Copas. A 'tourney' for the Chess Prizes was held over February–March. William and Jack both won in the first round, William was beaten by his younger brother in the second, and Jack had a win and a loss in round three, thereby finishing second overall to senior student James Walker. William successfully seconded the opposition to the debate, 'That Alexander the Great is worthy of the highest admiration both as regards his character and policy'.[40] A tennis club was also started at the beginning of second term 1879 and was 'very popular…greatly in vogue', with Rev. Trafford as President and W H Bragg as Secretary.[41] William's enjoyment of tennis would continue through his Cambridge studies and his early years in Adelaide, where he regularly filled in for the university team when students were unavailable during the long summer vacations.[42] He continued to be active in the Debating Society, and for the 1879–80 year was selected as a praepositor or prefect, a senior appointment that he would retain throughout his final two years at the College.[43]

William remembered the annual dramatic presentation as 'the best event of the year'. The second annual 'entertainment' organized by the Histrionics Committee was again held at the end of the Christmas term, this time fulsomely reported in *The Barrovian*.[44] It consisted of scenes from *The Merchant*

[37] Christopher, n. 6, pp. xx–xxi.
[38] Letter J Jones to R J Bragg, 5 February 1878, Bragg (Adrian) papers.
[39] A collection of issues of *The Barrovian* is held in the KWC archives.
[40] *The Barrovian*, 1 March 1879, *passim*, quotation pp. 6–7.
[41] *The Barrovian*, November 1879, p. 3.
[42] J G Jenkin. 'William Bragg in Adelaide: tennis too!,' *The Australian Physicist*, 1981, 18:69–70, 131.
[43] *The Barrovian*, November 1879, p. 8.
[44] *The Barrovian*, April 1880, pp. 10–14.

of Venice, a farce entitled *Turn Him Out*, and musical items. In the first, 'W H Bragg was very good as Bassanio', while the farce brought 'unceasing roars of laughter'. *The Barrovian* reporter listed the chief characters with approbation, including Nash as Julia, who 'would have looked highly lady-like if his trousers had not been so plainly visible beneath his dress', and then added, 'But the whole life of the piece was Bragg, as Susan, the maid of all work. From his first word "Lawks" to the end he kept the audience in continual fits of laughter. The farce decidedly appeared to be the most popular feature of the entertainment'.[45] The boy who thought himself 'very quiet, almost unsociable [and] who did not mix well with the ordinary schoolboys' hid behind a mask, shed his reserve, and allowed another side of his personality to emerge. It was 'great fun'.

James Bragg (Jimmy) entered the College in January 1880, so all three boys were now at the one institution for the first time. William remembered: 'Jimmy came to KWC about 4 years after me. He had been going to school at Wigton with Jack until then. He showed a lot of mathematical ability and ended by winning an exhibition at Emmanuel. Jim is more like our father than Jack or myself: much more sociable. I have found some of his letters from school to his father, delightful to read. I have never been much with Jim: I left Stoneraise when he was three: I saw him during summer holidays at Stoneraise and again when I used to go to the Isle of Man for holidays when I was at Cambridge. But we were apart most of the time. He is a most loveable person, as you all know'.[46] Three of Jimmy's letters to his father survive and are, indeed, delightful. Written in a strong, clear hand, the thirteen-year-old paints a vivid picture of his earliest days at King William's.[47] If he was indeed like his father, then these letters say much about Robert's warmth and sociability and support Gwendolen Caroe's assessment: 'I feel sure that Robert John Bragg, small and gentle, my grandfather, was "a dear man" '.[48]

> Dear Daddy 22 January 1880
>
> We arrived here safely on Wednesday night after a delightful passage, the sea being as calm as a mill pond, and I was not a trifle sick nor had even a feeling of it... When we got to Douglas we went down to the station and came to Castletown and then up to the College...
>
> The only thing I don't like here is the bed; although I had both my topcoat and my rug doubled I was not warm, but I suppose I will get used to it. Billy [William, his eldest brother] has been to town to buy some oatmeal to make some porridge tonight and is now busy making a lid for his wooden coalbox...
>
> How did the sale come off? Were the prices good? I hope they were. Poor old Toss. Has he gone to a good place?...

[45] Ibid., pp. 12–13.

[46] W H Bragg, Autobiographical notes, pp. 25–6.

[47] Letters J W Bragg to R J Bragg, 22 January, 8 February, 21 March 1880, in possession of family of Mr J A G Bragg, England.

[48] G M Caroe, *William Henry Bragg 1862–1942: Man and Scientist* (Cambridge: CUP, 1978), p. 28.

I think I have told you all the news of two days now, so with kind love and many kisses from your youngest son James W Bragg.

P S Please write soon.

[Again:]

Dear Father 8 February 1880

I think it is my turn to write to you this week, which has not been a very important one. I am top of the Fourth Form in Euclid, Arithmetic, Geography etc.... I went to Poolvash [Poyll Vaaish] last Wednesday (it being a given half) to get fossils but we only got a few. Could you please send us a Carlisle Journal with the list of the furniture in, please...

Do you know where about you are going to live at Harborough?

I have had a very bad cold lately but it is nearly better now. Jack has got a bad one now, and I think Billy is all right but swatting away for his exam in March... Billy makes use of his shaving dish to hold the subscriptions for the lawn tennis club; he has not got a razor yet...

I am, dear father, your longletterwriting, affectionate son J W Bragg

[Finally:]

Dear Father [at Market Harborough] 21 March 1880

I suppose I must satisfy that dear old uncle of mine about 24 out of 27. Well don't you see we take the number of boys in the class, that is 27; then the top boy in the form gets 27 marks, the second 26 and so on all the way down the class; then as I was fourth I got 24 marks. There now, aren't you happy, uncle?...

There was a fine kick-up this week. One of the boys boarding at Mr Edwards' came in a half-an-hour late one night, so Mr Edwards told him he would flog him next day; so in the morning when Mr Edwards was going to flog him the boy said he would defend himself and got out a tango (a piece of leather like a lace with a knot at the end and hurts awfully). So Mr Edwards told the Doctor [the Principal] about this, and the Doctor said, 'Bring him to my study and flog him', so the poor fellow was brought to the study. Then the Doctor, Mr Heaton & Mr Edwards all came to the study too, and were going to flog him when he brought out his tango again, kicked at Mr Edwards, hit Mr Heaton in the wind, knocked the breath out of him, and aimed the tango at the Doctor, but Mr Heaton managed to ward it off. Then they got him down and flogged him...

Do you mind sending me half-a-crown or so in stamps to get a plant press as there are no big books here that I can use like I could at Stoneraise?

I think that I have no more news now, so with kind love to all at Harborough

I remain your affectionate son J W Bragg.

Several points arise from these letters. First regarding conditions at the College, there was William's ability to leave the school to buy food items and to warm himself with coal, as well as the continuation of the flogging regime that was such a disgraceful characteristic of British schooling until relatively recent times. The sadism and barbarism of flogging, endemic in English public schools into the twentieth century and so horrendously described by Roald Dahl, is mind numbing.[49] Clearly it had the effect of frightening Jimmy Bragg. Then there is his delightful teasing of the formidable Uncle William; apparently 'that dear old uncle' could not intimidate this nephew as he had done to so many others in the family. Finally there are several references to a sale, the sale of Stoneraise Place, including all its animals and farm equipment and the substantial contents of the house.

With his three sons at King William's College, Robert Bragg, widower, was now alone on the large holdings he had farmed for nearly twenty years. As early as December 1876 William Banks, the owner of the major Wigton landed estate of Highmoor, had approached Robert with a view to buying Stoneraise Place and Beckbottom, which was owned by Robert Jefferson. In April the parcel of Stoneraise Place, the field at Church Hill, and the Beckbottom property were all sold to William Banks. Under the terms of the sale, however, Robert Bragg was to remain the 'occupier' of these properties, and he continued to farm them in the successful and profitable way that he had done in the past.[50] Subsequently, after being in the Banks family for more than thirty years, the estate was divided and sold in 1910, when Stoneraise Place returned to the care of an owner-occupier.[51]

When Robert finally decided to leave Cumberland in February 1880, he already had a healthy bank balance and needed merely to sell the majority of his household effects and move away. The sale took place on 4 February 1880 at the King's Arms Assembly Room, Wigton, and involved 'the whole of the valuable furniture and other effects, the property of Mr R J Bragg'. The contents were listed in *The Carlisle Journal*, and it was this issue of the newspaper that Jimmy asked his father to send to the College.[52] Robert went first to Market Harborough, but he did not plan to remain there. His love of the sea drew him back to a seaside location, and on the Isle of Man he would also be close to his sons. He chose Ramsey and settled into a comfortable and pleasant retirement there, buying several boats and associated equipment.

In March 1880 William went up to Trinity College, Cambridge, to try for a scholarship in anticipation of his graduation from King William's. Trinity already had a pre-eminent record in the undergraduate Mathematical Tripos examination, and it was destined to become similarly notable in physics in the decades ahead, providing most of the professors at the head of the Cavendish Laboratory. It had nurtured the greatest of all mathematicians and natural philosophers, Isaac Newton, whose large statue stood in the ante-chapel of the

[49] Examples abound; for example, by Winston Churchill (in *My Early Life*) and George Orwell ('Such, such were the joys' in *The Collected Essays, Journalism and Letters of George Orwell*), but the most distressing account I have read is Roald Dahl's *Boy: Tales of a Childhood* (London: Cape, 1984).

[50] Papers in the care of Mr George Bainbridge, England.

[51] For details of the Banks family and Highmoor estate and its house, see T W Carrick, *History of Wigton* (Carlisle: Thurnam, 1949), ch. XII.

[52] Bragg (Adrian) papers.

college chapel. The examination involved two mathematics papers, with some allowance being made for non-resident candidates, an essay on English literature, and Latin and Greek prose and poetry composition and translation. As a result of the examination, as William recalled, he was awarded one of the higher Minor Scholarships, worth £75 per annum for three years. This was commendable but a little disappointing to the school that had apparently expected a Foundation Scholarship.[53]

At the end of the Easter term of 1880 the Principal wrote to William's father saying, 'his success in winning an open Scholarship at Trinity College, Cambridge—the greatest distinction a school-boy can achieve—shows the extent & soundness of his attainments, as well as his mathematical ability, and bears out our former anticipations respecting him. It was well deserved. Joshua Hughes-Games D C L '.[54] This was followed in May by a letter reiterating the achievement and noting that 'he has won one at an unusually early age, and against unusually strong competition. As he is so young, we are inclined to think he had better defer going up to Cambridge for another year. Next year he may try for £100 [Foundation] Scholarship at Trinity: and failing that, he would keep his £75 one'.[55] Trinity and his father apparently agreed and William returned for a final year at King William's.

At the start of the new cricket season it was decided to do without the services of one or two masters and a professional in order to encourage more boys to gain experience in the First-XI. William was selected in the hope that, 'with more freedom and life in his play [he] will make a useful bat'.[56] In addition, blazers were introduced after much agitation and 'The members of the Eleven look very well in their new colours'.[57] But the cricket season proved disastrous for William. Opening the batting, he was unable to master the opposing opening bowlers. He was dismissed for a-pair-of-spectacles (zero runs) in the two innings of the opening game against West Derby Cricket Club, and he only accumulated nine runs in nine innings before being dropped from the team for the final match of the season. On the other hand he regularly secured a wicket or two as a change bowler. The College won two of the seven games it played in a competition that was clearly very uneven.[58]

The Barrovian of September 1880 recorded a theology essay prize for William and carried an excited report on the cricket match at Kennington Oval between Australia and England, the former without 'the demon Spofforth' due to injury, and the latter combining the best of the gentlemen and players of England. 'No cricket match ever excited so much interest or was so eagerly looked forward to', it was reported, and England won by five wickets after a

[53] 'Trinity College', *Cambridge University Calendar for the Year 1881*, pp. 520–1; personal communication from Trinity College Library.

[54] 'Remarks', *King William's College, Quarterly Report, Easter 1880, W H Bragg*, KWC archives.

[55] Letter Joshua Hughes-Games to R J Bragg, 14 March 1880, KWC archives.

[56] *The Barrovian*, April 1880, p. 23.

[57] *The Barrovian*, July 1880, p. 58.

[58] Ibid., pp. 83–91.

match that fully lived up to these expectations. It was, in fact, the first 'Test Match' on English soil.[59]

The annual prize day on 22 July incorporated verbal reports by the two external examiners from Hertford College, Oxford, who particularly praised the three upper boys of the sixth form and unusually suggested that two College Exhibitions to the University, each valued at £40 per annum for four years, be awarded to W H Bragg and J M Walker, whose marks made them inseparable. To this the school happily agreed. His Excellency the Lieutenant-Governor's Prize for Mathematics, consisting of the two volumes of James Clerk Maxwell's daunting text of 1873, *A Treatise on Electricity and Magnetism*, was awarded to W H Bragg. Jack won the Perspective Drawing Prize and Jimmy the Mathematics Prize for the fourth form.[60]

William was elected Head of the School for the 1880–81 academic year and was again prominent in several debates. Little can be known about the cricket season, since 'the scores... have been destroyed', but most of the team remained from the previous year, they had an excellent professional during the season, and they won six of their nine matches. 'W H Bragg was much improved in confidence and power... [and] played some very nice innings... [including] an eminently skilful innings of 17 not out'.[61]

The nature of William's final year at King William's College is best revealed by his own words quoted earlier. Academically it was not rewarding, for the masters had to find different work for him to do and he was thereby separated in class from his contemporaries. When he sat the Oxford and Cambridge Certificate Examination in July 1881 he passed in Greek, Mathematics, Physical Geography, and Geology, with distinction only in Additional Mathematics (algebra, trigonometry, statics, and dynamics).[62] While Hughes-Games predicted 'a most successful career at the University', William had to be content with his minor scholarship.[63] However, it was the religious revival in his last year that dominated William's recollections of his years at King William's College:[64]

> a much more effective cause for my stagnation was the wave of religious experience that swept over the upper classes of the school during that year. We were not singular, no doubt. Nerves are said to be liable to disturbance when boys are turning into men: and religious storms are common enough. Anyway we had it badly, in the sense that we were terribly frightened and absorbed: we could think of little else. We had prayer meetings and discussions. We were told that if we sought we should find: and we hoped that some how or other, at some time, we should suddenly be converted and know that we were saved, and avoid eternal damnation and hell fire. The issue was indeed quite simple. 'If we believed, we

[59] *The Barrovian*, September 1880, pp. 106–9.

[60] Ibid., pp. 132–9.

[61] *The Barrovian*, October 1881, pp. 122–6.

[62] 'Certificate for Oxford and Cambridge Examination, July 1881, Bragg, William H., aged 19', Bragg (Adrian) papers.

[63] 'Remarks', *King William College, Quarterly Report, Easter 1881, W H Bragg*, KWC archives.

[64] W H Bragg, Autobiographical notes, pp. 16–18; the star denotes a marginal addition.

> should be saved'. By 'saving' the words meant—so we thought—being delivered from an eternity in which we should be subject to pain worse than that of having teeth out without anaesthetic. Such an interpretation, if thoroughly absorbed, would, indeed must, make lunatics. That the Christian world has gone on for centuries accepting that interpretation and yet going about its ordinary business is one of the strangest facts of history. A few have felt its force and tried to save themselves by terrible acts of self-sacrifice of useless character, others by noble works in which self is forgotten: but the marvel is that there the words stand and are allowed to stand.
>
> The other word in the phrase is 'believed'. We took that to mean that we accepted the truth of all the statements made in the New Testament by Christ, by Paul and others, literally* (*On second thoughts I doubt if 'believing' meant to us anything so comprehensible: we were in fact not quite sure what it did mean: but it was something we had to do to be saved.) We were not sure we believed: sometimes we thought we did and thought we had attained to the 'peace that passeth all understanding', and any one of us who had got that far was the envy of the others. Then such a one would begin to have doubts about his believing, and would be in the soup again.
>
> It really was a terrible year. If a boy came to me now and told me that he was in trouble as we were in trouble, I should tell him that believing and saving could not mean literally what we thought they meant: our intellectual difficulty was no more than an intellectual difficulty, due to the fact that words mean different things to different people... Therefore I have always felt strongly that it is necessary to be continually emphasizing the evil effects of 'literal' interpretations...
>
> Jack—my brother—and I once summoned up courage to go to the headmaster and ask for counsel. He was very friendly and sympathetic and we all knelt down in his study while he prayed for us. But he did not resolve our difficulty.
>
> The storm passed in time, by sheer exhaustion and the fortunate distraction of other things, work and play. It lasted with me for some time after I left school, and then faded away. But for many years the Bible was a repelling book, which I shrank from reading.

The religious revival was one manifestation of the evangelical spirit that came to dominate nineteenth-century Anglicanism. In his review of the public school phenomenon, Gathorne-Hardy wrote in relation to public school reform and the Victorian moral climate: 'The essence of the religious revival started by Wesley and Whitefield... was feeling... [and] the feeling whipped up was guilt. Guilt over personal sin, sin which was to be conquered in desperate personal battles; sin which, since it was human, was also social... One of the most extraordinary and significant phenomena of this period is how this evangelical fervour and guilt swept first through the... non-conformist sects, then the Catholic sections of the English Church, the Church of England itself, until finally, by the middle of the [nineteenth] century, it had become a generalised social force, dominating every department of private and public life'.[65] The schools were part of this force, where the greatest sins were homosexuality and masturbation. One of the very few references to King William's College in Gathorne-Hardy's book is the following: 'Canon Farrar went to King William's College on the Isle of Man in the 1840s... a hundred years later, so strong, still, was the injunction against masturbation that all trouser pockets had to be sewn up as a bar to pleasure-seeking fingers. There were continuous spot inspections by the prefects. If one finger could be got in, one stroke of the cane; if two, two strokes; if the whole libidinous hand could be thrust in then you got a sound thrashing'.[66] Many books on this period refer to these evangelical revivals and their local or personal impact. Thus Chandos' chapter 14 is entitled 'A Demon Hovering'; Mack notes that, 'As was to be expected, since there are always more moralists in the public school world than believers in democracy and freedom,... the moral "volcano" on which "the masters of many schools are sitting"... received most attention'; and there are also whole books devoted to the topic of sex and the public schools.[67]

We can only guess what specific incidents prompted the Principal to launch the wave of religious experience that swept the upper classes of the school in 1880–81, but we can see that William was engulfed by it and that he focused his attention on what was meant by the words 'believed' and 'saved' in the phrase 'if we believed, we should be saved'. As the pre-eminent mathematician amongst the boys William had learnt that the solution of a problem depended first on the definitions of its individual elements. No progress towards a solution was possible until one understood the problem in all its facets. Applying the same method to his religious problem led to a dead end, because he could not understand the definitions of the individual words nor therefore the phrase overall; and nor could the headmaster solve the mystery. William and his peers were 'terribly frightened and absorbed... [to] avoid eternal damnation and hell fire', and they left the school with their fears unresolved.

William shunned organized religion for long periods during the rest of his life. A diligent search for evidence of church affiliation in William's early years in Adelaide, for example, has found nothing. The later years in Adelaide were notable for a strong and active church attachment, however, probably because a popular, fashionable Anglican church near the family's East Adelaide home had a broad, liberal philosophy and a thoughtful and much-loved priest, Canon Hopcraft.[68] Between 1899 and 1909, at one time or another, William was a seat-holder, sidesman, church warden, and lay reader at St John's Church, Halifax Street, Adelaide.[69] He participated in almost every church activity, from

[65] Gathorne-Hardy, n. 1, ch. 4, p. 79.

[66] Ibid., p. 100.

[67] Chandos, n. 1, ch. 4; Mack, n. 12, p. 157; for example, A. Hickson, *The Poisoned Bowl: Sex and the Public School* (London: Duckworth, 1995).

[68] D L Hilliard, 'The city of churches: some aspects of religion in Adelaide about 1900', *Journal of the Historical Society of South Australia*, 1980, 8:3–30; D. L. Hilliard, personal communications, 1982–4.

[69] *St John's Church Adelaide, Vestry Minute Book, 1840–1939*, State Records of South Australia, Adelaide, SRG 94/9/1, *passim*.

organizing 'motive power for the organ' to a memorial for Canon Hopcraft on his death and the selection of his successor.[70] When the Bragg family left Adelaide the annual vestry meeting of the church placed on record its 'high appreciation of the manifold services rendered to St John's Church by [Professor Bragg] and Mrs Bragg, and recently by their son, Mr Will. Bragg', and tendered them a farewell social.[71]

William's wife, Gwendoline, was President of the Sanctuary Guild, responsible for the decoration of the church, which she herself enhanced with a large, stencilled frieze—of a white, stylized grapevine against a green background—around the sanctuary.[72] This was in sympathy with a widespread Anglican movement in the late nineteenth century to refurnish and decorate old churches, regarded as a special work for the women of the congregation.[73] The frieze, long covered by layers of white paint, has recently been uncovered and restored.[74] The Braggs' elder son was remembered specifically because he had 'rendered excellent service as one of the teaching staff of the Sunday School'.[75] His young sister, Gwendolen Mary Bragg, was christened in the church on 3 June 1907.[76] Later, Willie was not religious in an orthodox sense but, prompted by the beauty of the world he discovered in nature, he did believe in a divine creator. His daughters told me, first: 'He once said that the thrill of scientific discovery was like that of a small boy admiring a steam train and suddenly being offered a ride in the cab by the engine driver—a powerful simile. My father was not indifferent to religion; he attended church services with my mother and was certainly not an atheist'. And second: 'He was certainly a deist and loved to talk about the "pattern and purpose" behind our lives that each must identify for himself... He read the King James' Bible aloud in the evenings, mostly the Old Testament and very much for the poetry. He knew the Bible extremely well'.[77]

During the remainder of William's life there is little to indicate a strong religious affiliation. Gwendoline recalled that, in Leeds, the family attended services at Leeds Parish Church, and that in London William would accompany his wife to Early Service; but that 'Never after his return from Adelaide did WHB take any active part in church affairs; increasingly he felt apart from organised religion'.[78] Caroe describes a number of occasions on which her father was drawn into public discussions of science and religion, but the impression left is that he was ill at ease in these situations and that he was often distressed by

[70] Ibid.

[71] 'Items of Interest', *St John's Parish Chronicle*, May 1908, p. 4 (copies held by the State Records of South Australia, Adelaide).

[72] R Biven, *Some Forgotten... Some Remembered: Women Artists of South Australia* (Adelaide: Sydenham Gallery, 1976), with notes on Gwendoline Bragg by her daughter Gwendolen Caroe.

[73] D L Hilliard, 'Religious practice and popular piety in South Australian Anglicanism, 1880–1960', copy of paper presented to History '84 Conference, Melbourne University, August 1984.

[74] The church and related buildings have been restored progressively by the Society of the Sacred Mission as St John's Priory.

[75] 'Valedictory', *St John's Parish Chronicle*, December 1908, p. 5.

[76] 'Baptisms', *St John's Parish Chronicle*, July 1907, p. 6.

[77] Margaret Heath (née Bragg) and Patience Thomson (née Bragg), personal communications.

[78] Caroe, n. 48, p. 165.

their outcomes. Most notably, during the early years of the Second World War, when he was President of the Royal Society, William was distressed to find himself embroiled in Moral Rearmament. He was one of a small group of leading British figures who, in a letter to *The Times* newspaper following a similar letter from a number of members of parliament, suggested that 'The real need of the day is therefore moral and spiritual rearmament... God's Living Spirit calls each nation, like each individual, to its highest destiny, and breaks down the barriers of fear and greed, suspicion and hatred'.[79] This was followed by a chapter in *Crisis Booklet No. 3* of the Student Christian Movement, in which William sought to say what the letter meant to him;[80] but after he and the other signatories were deluged by requests for further guidance and leadership, William felt compelled to withdraw:[81]

> Sir—The letter on moral rearmament which appeared on September 20 [sic] has been referred to in many subsequent letters addressed to you for publication and also in letters addressed to those who signed the original letter, of whom I am one. The references have all been extraordinarily generous and sympathetic. Some have asked for further guidance. I have not been in consultation with my colleagues as to whether a joint answer could or should be given. I write as an individual... It seems to me that the action in response to the appeal must be an individual action... Some will describe what they think and do in the form of words and some in another. But the purpose is the same, and no one need wait for a leader. I write this letter with great diffidence. I am, &c., W H Bragg.

Faced with the great anxiety of the war and the responsibility of leading the Royal Society and the Royal Institution, when age and poor health were reducing his energy and stamina, William's was a cry from the heart, but he lacked the religious commitment to carry it further. He tried to marry science and faith in an article in *The Hibbert Journal* and in a Riddell Memorial Lecture of 1941, but, while the purpose was honourable, his approach to these complex issues seems hopeful rather than realistic.[82] 'He presented Christianity as an "experimental religion" that was also willing to learn from experience, with dogma now being treated in the same way as a scientific hypothesis'.[83] In discussing religious faith, William's words lacked the clarity and conviction of his other writings, and they carried an uncertainty that remained from his last year at King William's College. He referred at length to his school experience in the lecture, and a reviewer highlighted it in the journal *Nature*.[84]

[79] Letter Baldwin of Bewdley et al. to *The Times*, 10 September 1938, p. 6.

[80] Sir William Bragg, 'The need of the day', in *Moral Rearmament* (London: SCM Press, n.d.), pp. 15–27.

[81] Letter W H Bragg to *The Times*, 27 September 1938, p. 8.

[82] Sir William Bragg, 'Science and the worshipper: in response to Sir Richard Tute', *The Hibbert Journal*, 1940, 38: 289–95; Sir William Bragg, *Science and Faith* (Oxford: OUP, 1941), being the text of the Riddell Memorial Lecture, University of Durham, 7 March 1941.

[83] P J Bowler, *Reconciling Science and Religion* (Chicago: University of Chicago Press, 2001), p. 52; however, I cannot agree with Bowler's view of Bragg as 'a devoted Anglican' (p. 36) or as an authoritative representative of Christian scientists (p. 52).

[84] 'Science and faith', *Nature*, 1941, 148:181–3.

How then are we to summarize William's time at the College? He had clearly done extraordinarily well in all facets of College life. His academic record was outstanding, he had taken a very full part in both the sporting and other extra-curricular activities, and he had earned the respect and affection of both the masters and his peers. Put in classes with older boys, he was 'very quiet' and 'did not mix well' in the early years, but he 'became more at home' later. When most of the school was prosecuted for trespassing in the old mill, William was picked out in court as a ringleader. Refusing to be intimidated, he answered honestly and confidently and secured the approbation of staff and students alike. Only the anguish of the religious revival permanently discoloured his memory of these school days. Otherwise he had learnt to focus on the things that mattered, had largely avoided those that did not, and had come through with flying colours.

In the future William would return to King William's College as its most notable 'old boy'. In 1933, the school's centenary year, he was a guest of honour at the annual prize-giving and other ceremonies.[85] He remembered 'the nice smell of white kid gloves, which all the boys had to wear in those days when they took prizes', as well as 'the coming of the first Rugby football', when previously 'they used to play with a round ball, and you had to bounce it as you ran'; and also 'how the soldiers came from Castletown to level the old cricket ground'.[86] In 1937 he presented the prizes, and 'Two old boys of King William's College, who were school chums there over 60 years ago—Sir William Bragg, OM, KBE, DSc, PRS, now the leading scientist of Great Britain, and the Ven. Archdeacon John Kewley, most revered of all living Manxmen—performed the opening ceremony of the Barrovian Hall and the new wing at the College'.[87] On one of these occasions William was accompanied by his daughter-in-law, to whom he confided that he had to be careful what he said so as not to resurrect memories of the dreadful conditions during his time there.[88] Finally, in 1983, the school's one-hundred-and-fiftieth year, the Isle of Man issued four stamps to honour the College and its alumni, one of which (for 28p) featured 'Sir William Bragg, O M , Nobel Prize Winner' and a drawing of his early X-ray crystallography apparatus.

The headmaster, Joshua Hughes-Games, was fulsome in his praise at the end of William's school career: 'His charge has given me the greatest possible satisfaction in every respect', he wrote. '[He] leaves school with the highest character, and with the esteem and affectionate regards of his masters. J. H.-G'.[89] It was a record of which to be proud, but a yet more demanding challenge lay ahead, the Mathematical Tripos at Cambridge.

[85] *The Barrovian*, November 1933, pp. 125–9.

[86] *The Barrovian: Centenary Report*, November 1933, pp. 125–6.

[87] Newspaper cutting entitled 'Founders' Day at King William's', *Isle of Man Examiner*, 30 July 1937, RI MS WHB 19A/5; Hoy, n. 3, chapter 6.

[88] Lady (Alice) Bragg, personal communication.

[89] 'Remarks', *King William's College, Terminal Report, Easter 1881, W H Bragg*, KWC archives.

4
Cambridge University

A book describing the major reorganization of British universities in the second half of the nineteenth century begins with the words: 'The [century] began with two decayed universities in England, highly restrictive in education and clientèle [Oxford and Cambridge]. It ended on the eve of the Great War with eight more universities and three university colleges having been founded in England... The pattern of British higher education... was never more radically reshaped than in the nineteenth century, and particularly in the period 1850 to 1914. University institutions were influenced by a greater range of social forces and in turn transmitted an impact into more new areas of national life than ever before'.[1]

William Bragg went up to Trinity College, Cambridge, in July 1881, close to his nineteenth birthday and during the university's long summer vacation, when the revolution within Cambridge was in full swing. He would carry away with him many of its features. As a college scholarship holder he was keen to settle in before the rush of new undergraduates in September, and to make an early start on the demanding mathematics that would dominate his life for the next four years. The college had been founded by King Henry VIII in 1546 by amalgamating two existing colleges, King's Hall and Michaelhouse, and it had been provided with substantial endowments from the dissolved monasteries. Architecturally it was dominated by the magnificent Great Court and the beautiful Library designed by Christopher Wren and decorated by Grinling Gibbons.

The mathematics program at Cambridge was unique and extraordinary, and William soon became aware of its form, history, and traditions. Basically its undergraduate teaching programme had been determined by the shadow cast by Isaac Newton's brilliance, two hundred years before. Newton's focus had been mathematics and natural philosophy, a term that then meant science but later became synonymous with physics. Having quickly absorbed the existing mathematical knowledge, Newton had pushed on to discover his own form of the calculus and other new mathematics. In natural philosophy he

[1]M Sanderson (ed.), *The Universities in the Nineteenth Century* (London: Routledge and Kegan Paul, 1975), pp. xi, 1.

Fig. 4.1 Undergraduate students in the Great Court, Trinity College, Cambridge, with its fountain, Great Gate, and Chapel in the mist; a familiar sight to William and his two sons, *circa* 1909. (Courtesy: Masters and Fellows of Trinity College, Cambridge.)

used his supreme talents as both mathematician and dextrous experimenter to make major discoveries in mechanics, astronomy, and optics. It was these topics, updated and in the case of optics much changed to a wave phenomenon, that came to dominate the Cambridge undergraduate curriculum during the eighteenth and nineteenth centuries for all undergraduates, not only those with a particular interest in mathematics.[2]

[2]The following gives a sample of the references available on the topic of Cambridge mathematics and physics: W W R. Ball, 'The Cambridge school of mathematics', *The Mathematical Gazette*, 1912, 6:311–23; id., *Cambridge Papers* (Cambridge: CUP, 1918); P M Harman (ed.), *Wranglers and Physicists* (Manchester: Manchester University Press, 1985); G Howson, *A History of Mathematics Education in England* (Cambridge: CUP, 1982); J Gascoigne, 'Mathematics and meritocracy', *Social Studies in Science*, 1984, 14:547–83; H W Becher, 'Radicals, Whigs and conservatives', *British Journal for the History of Science*, 1995, 28:405–26 and references therein to earlier articles by this author; D B Wilson, 'Experimentalists among the mathematicians', *Historical Studies in the Physical Sciences*, 1982, 12:325–71; R Sviedrys, 'The rise of physical science at Victorian Cambridge', op. cit., 1970, 2:127–51; id., 'The rise of physical laboratories in Britain', op. cit., 1976, 7:405–36; G Gooday, 'Precision measurement and the genesis of physics teaching laboratories in Victorian Britain', *British Journal for the History of Science*, 1990, 23:25–51. Two recent books provide good starting points: A Warwick, *Masters of Theory: Cambridge and the Rise of Mathematical Physics* (Chicago: Chicago University Press, 2003); D W Kim, *Leadership and Creativity: A History of the Cavendish Laboratory, 1871–1919* (Dordrecht: Kluwer, 2002).

Since the Elizabethan era, teaching in the colleges of the university had been directed towards the exercises that graduating students were required to complete. The most important of these were the 'acts', in which a student would propose a thesis, usually from philosophy, and would then defend it against objections in a Latin disputation or wrangle. By 1700, however, mathematics, astronomy, and optics had joined the classical subjects as elements in a possible curriculum leading to the Bachelor of Arts (BA) degree. Around 1730 the examination began to be held in the new Senate House in Cambridge and was not only conducted in English but also included the mathematics that would soon come to dominate the assessment.

By 1750 the examination had been extended and the final lists of successful candidates, in order of merit, had become authoritative and important. As a result, by 1770 written answers were required to a set of questions common to all honours candidates. By the end of the century the examination had been extended to three and then four days and alone determined the order of merit for graduating students. A high place in the list could lead to acclaim, a college fellowship, and an assured future. Teaching had always been primarily in the hands of the colleges rather than the university professors, but now a new element emerged. In order to secure as high a place as possible, serious students employed a private tutor to prepare them for the increasingly arduous final examination.

Continuing change occurred throughout the nineteenth century. Study became focused on those subjects that would be tested in the Senate House Examination, and there was increasing emphasis on the application of mathematics to problems of the external world. Around 1840, and to differentiate it from the Classical Tripos, the Senate House Examination was confined to mathematics, became known as the 'Mathematical Tripos', and increasingly included mathematical physics. The overall content was described as 'mixed mathematics'. There is uncertainty as to the origin of the term 'Tripos', but Rouse Ball contends that it originated in the earlier process, which included the participation of an 'ould bachilour', who sat on a three-legged stool or tripos and tested the candidate during the disputation.[3]

Around 1850 a Board of Mathematical Studies was established to supervise the study of mathematics in Cambridge. The extent of the final examination, that had grown to five days, then six, was now increased to eight and then nine days by 1873. As had been the practice for a hundred years, the successful candidates were graded and listed in order of merit in four categories: first 'wranglers', followed by 'senior optimes' and then 'junior optimes'– from the Latin compliment given to successful disputants, and finally 'poll-men'—from the description of this group as hoi-polloi. During the final days of the examination there was an extended range of subjects for those who wished to take honours. A further modification was introduced early in the 1880s, dividing the Tripos into three parts. The first two parts, of three days each, were taken

[3] Ball, *Cambridge Papers*, n. 2, p. 312.

at the end of a student's third year of study. For Part I the range of introductory subjects remained largely unchanged, after which the honours students were selected to proceed and the poll-men were awarded an ordinary BA degree. The second part of the examination was held soon thereafter, on advanced aspects of the earlier topics, when ranking of the candidates in order of merit in the three honours groups was determined. Part III was held just six months later, in the following January, to which only the wranglers were admitted. They were offered a range of advanced subjects from which to choose for examination, and they were then graded alphabetically in three classes: first, second, and third. The practical and developing subjects of heat, electricity, and magnetism, that had been removed from the examination in mid-century, were now restored as possible subjects for Part III.

Nearly every student now studied with a private tutor or 'coach', and two of them obtained a virtual monopoly on the studies of advanced honours students. The first was William Hopkins who, in the twenty-two years from 1828 to 1849, guided 175 wranglers, of whom seventeen were first or 'Senior'. The second was Edward Routh, himself the Senior Wrangler in 1854, who, in the thirty-one years between 1858 and 1888, taught between 600 and 700 pupils, many of whom became wranglers, twenty-seven being Senior. 'For all this time', his Royal Society obituary notice recorded, '[Routh] directed, almost without challenge, most of the intellectual activity of the *élite* of the undergraduate mathematical side of the University', while Warwick thought Routh 'probably the most influential mathematics teacher of all time'.[4] Rouse Ball described Routh's system of teaching as follows:[5]

> He gave catechetical lectures three times a week to classes of eight to ten men of approximately equal knowledge and ability. The work to be done between two lectures was heavy, and included the solution of some eight or nine fairly hard examples on the subject of the lectures. Examination papers were constantly set on Tripos lines (book-work and riders), while there was a weekly paper of problems set to all pupils alike. All papers sent up were marked in public...and, to save time, solutions of the questions were circulated in manuscript...The course for the first three years and the two earlier long vacations covered all the subjects of the Tripos—the last long vacation and the first term of the fourth year were devoted to a thorough revision. Of what is called cramming there was no trace; Hopkins and Routh might say that a particular demonstration was so long that it could not be required in the Tripos, but none the less they expected their pupils to master it. The system had faults, but it was under Hopkins and Routh that nearly all the best-known representatives of Cambridge mathematics in the nineteenth century were educated. The effectiveness of teaching of this kind was dependent on intimate constant personal intercourse, and the importance of this cannot be overrated.

[4] J L, 'Edward John Routh, 1831–1907', *Proceedings of the Royal Society of London*, 1910–11, 84:xii–xvi, xii; Warwick, n. 2, p. 231.

[5] Ball, 'The Cambridge school', n. 2, p. 321; also see Warwick, n. 2, ch. 5.

In Victorian Britain the Mathematical Tripos came to be the most prestigious of all degrees, with major cultural significance. It was thought to provide the ideal liberal education or general training of the mind, for future lawyers, physicians, natural scientists, and, indeed, for future leaders of Britain and her Empire in general. The top wranglers were accorded hero status and fêted around the country. It was also extraordinarily successful in producing the future leaders of British mathematics and physics: 'The reputations of many successful Cambridge mathematical physicists of the Victorian period were based on their remarkable powers of mathematical manipulation and problem solving; powers developed through years of progressive training under the highly skilled tutelage of a handful of brilliant mathematical coaches'.[6]

While the benefits of success were great, the pressure on honours students could be severe, particularly during the consecutive days of the gruelling final examination, upon which the entire success of their studies depended. Emotional as well as intellectual toughness was required, and there were casualties. James Wilson was Senior Wrangler in 1859. He had a nervous breakdown immediately after his examination and, during convalescence on the Isle of Wight, discovered that, 'everything he had learned at Cambridge had disappeared from his memory: "I could not differentiate or integrate; I had forgotten... all Lunar Theory and Dynamics; nearly the whole of Trigonometry... Happily Algebra and Euclid were safe"'.[7]

The Cavendish Laboratory, still young but maturing fast when William arrived in Cambridge, will also be important in our story. During the 1860s there was increasing agitation in Cambridge for the study of contemporary physics. Elsewhere, following the introduction of the first chemistry laboratories and engineering programmes, the first physics laboratories had been established during the 1860s after the example of William Thomson at the University of Glasgow. Cambridge had lagged behind all these innovations and now came under pressure to follow suit. A committee appointed in 1869 recommended the establishment of both a professorship of experimental physics and a physical laboratory to provide instruction in contemporary physics. Not welcomed by Cambridge traditionalists, it seemed that lack of funds would thwart the suggestion, until William Cavendish, the University Chancellor, offered to finance the recommendation himself. James Clerk Maxwell was already an examiner for the Mathematical Tripos and was appointed to the professorship in 1871. The new laboratory building in Free School Lane was opened in June 1874 and soon began receiving graduates from the Mathematical Tripos, who sought to add experimental experience to their mathematical qualifications.[8]

[6] A Warwick, 'A mathematical world on paper', *Studies in History and Philosophy of Modern Physics*, 1998, 29:295–319, 316.

[7] Howson, n. 2, p. 126.

[8] For the Cavendish Laboratory, see, for example, *A History of the Cavendish Laboratory 1871–1910* (London: Longmans Green, 1910); A. Wood, *The Cavendish Laboratory* (Cambridge: CUP, 1946); J G Crowther, *The Cavendish Laboratory 1874–1974* (New York: Science History Publications, 1974); Kim, n. 2.

Maxwell and his successor, Lord Rayleigh, stayed in the Cavendish for only eight and five years respectively, although they achieved much. Their successor in 1884, much to everyone's surprise, was Joseph John ('J J') Thomson. Thomson was educated first at Owens College, Manchester, before entering Trinity College in 1876 and graduating Second Wrangler and Smith's Prize winner in 1880. He was a Fellow of Trinity and a college and university lecturer in mathematics when appointed to the Cavendish chair. He was progressively turning his interests towards experimental physics.[9] The appropriateness and importance of the foundation of the Cavendish Laboratory was confirmed in the last two decades of the nineteenth century by the emergence of a remarkable group of theoretical and experimental physicists who would occupy many of the chairs of physics in British and Empire universities.

William was therefore embarking on a program with a long tradition but that was also undergoing significant change:[10]

> I went up to Cambridge in 1881, taking the rather unusual course of beginning work there in the Long [vacation]: I suppose I was in Cambridge six weeks or so, July and part of August. But I forget the exact date. I had rooms in Master's Court. I appreciated thoroughly the beauty of the whole place, and I liked going to Routh's classes. I was lonely, because I was doing the unusual thing: and I had no companions. But it was good all the same. As a scholar of the College I went up every Long afterwards: it was always a jolly time. Very few restrictions: just the regular classes three times a week with Routh, and the preparation for them. After that tennis a plenty: boating on the river above Cambridge and the summer weather, and Cambridge looking its best. I tried during that preliminary Long to get through an exam that would excuse me the Littlego [Previous Examination]: and I failed in Latin, which seems to me now to be very odd as I had studied Latin from the time I was seven, and given a lot of school time to it (and worked conscientiously too!). I had to take the Littlego in November after all.

The Master's Court, soon to be renamed Whewell's Court in honour of William Whewell, a former Master of Trinity and pivotal university scholar in mid-century, was directly across Trinity Street from the Great Gate, the main entrance to the ancient part of the college. Financed by Whewell, the court was 'gloomy' but nevertheless impressed the new student.[11] William was assigned room 3 on staircase O, with Henry Taylor as his tutor. Taylor had been Third Wrangler and Smith's Prize winner at the 1865 Tripos and was a Fellow and lecturer in mathematics of the college. His principal duties as a tutor, however, were to maintain discipline in the college and act as guardian

[9] W W R. Ball and J A Venn, *Admissions to Trinity College Cambridge, Vol. 5, 1851 to 1900* (London: Macmillan, 1913), p. 550; Lord Rayleigh, 'Joseph John Thomson 1856–1940', *Obituary Notices of Fellows of the Royal Society*, 1941, 3:587–609.

[10] W H Bragg, Autobiographical notes, pp. 18–19.

[11] G M Trevelyan, *Trinity College: An Historical Sketch* (Cambridge: Trinity College, 1972), p. 101.

and adviser to undergraduates in his care. In addition, William was accepted by Edward Routh, who taught during the summer vacation as well as throughout the academic year.[12]

William is listed in college records as a 'Pensioner'; that is, an ordinary fee-paying student. The fees were substantial. First, about £60 was required: for fees for admission, matriculation, examination, and graduation, for room furniture, outfitting, and crockery, and for a cap and gown (for all daily college and university activities) and a surplice (for chapel). In addition, expenditure for tuition, room-rent, meals, coal, room-attendant, laundress, and living expenses could easily exceed £100 per year. William's minor scholarship of £75 p.a. therefore provided welcome financial relief in addition to its kudos, entry to the College, and access to the best private tutor.[13] William's difficulties with the Previous Examination are reminiscent of his earlier failures in Greek while in Market Harborough, and in Latin at King William's College. Exemption from the Littlego was possible for students who had done well in the latter, but William had to sit the full Previous Examination in November of 1881 with 694 other students. He passed Part I in the Second Class (St Mark's gospel, Latin and Greek translation and grammar) and Part II in the First Class (Paley's *Evidences of Christianity*, mathematics).[14]

Outside Routh's classroom William was 'lonely'. Many of the rooms in the college were empty, but there were students in two other rooms on O staircase, there were students in Routh's class, and there was tennis and boating 'a plenty'. Shy in his new surroundings, his inability to form close relationships was now entrenched, but he did write regularly to Uncle William, visited Market Harborough during some vacations, and used to go to the Isle of Man for holidays when he was at Cambridge.[15] The new academic year and Michaelmas term began in October 1881. William recalled:[16]

> Cambridge gave me a good time of course: though I might have done much better if I had known more or been more easily sociable. I ought to have gone to lectures on other subjects than mathematics, and taken an interest in other things. It simply did not occur to me. I could not afford, or thought I could not afford, to join the Union or the Boating Club: which cut off a good many opportunities. I had none of those experiences of discussions of the world and its problems with other young men, which many men seem to look back upon with so much pleasure. I worked at the mathematics all the morning, from

[12] Information concerning Bragg's years at Trinity College has been obtained from: Ball and Venn, n. 9; *Trinity College Admission Book 1850*, Trinity College Library; *Trinity College: Room Rents 1871–1897*, op. cit.; 'Trinity College' in the annual *Cambridge University Calendar*; Bragg (Adrian) papers.

[13] Much useful information regarding student conditions can be found in *The Student's Guide to the University of Cambridge, Part I: General* (Cambridge: Deighton, 1882). Also see the annual *Cambridge University Calendar* for academic matters. William's KWC Exhibition of £40 p.a. assisted further.

[14] *Cambridge University Calendar for the year 1881*, pp. vi, 6–9; *Cambridge University Reporter*, 29 November 1881, p. 151, and 16 December 1881, pp. 206–12.

[15] Bragg (Adrian) papers; W H Bragg, Autobiographical notes, p. 26.

[16] W H Bragg, Autobiographical notes, p. 19.

> about 5 to 7 in the afternoon, and an hour or so every evening, and then to bed fairly early. Every afternoon I played a game, generally tennis, or went for a walk: my tennis was fairly good, so that I always found people ready to play.
>
> I changed my exhibition for a Major Scholarship in 1882, which gave me a standing in the College. I had the right then to join the Trinity Tennis Club without election, and wear the strawberry-and-cream blazer: which was a source of pride. I sat in the scholars' seat in chapel: and took my turn in reading the lessons.

William's exclusive focus on mathematics was encouraged by two common pieces of advice to matriculants, given in *The Students' Guide*. First, 'the candidates for honours... have only the examinations for that Tripos to pass, and they may devote the whole remaining time exclusively to the special subjects which they find themselves best able to master'; and second, 'It has therefore become... with Mathematical Honours men almost a universal practice to employ a private tutor... Accordingly, the greater part of a reading man's time may be occupied in preparation, not for lectures, but for his private tutor'.[17]

'I changed my exhibition for a Major Scholarship' is an extraordinary understatement. William showed such promise during the long vacation and the initial terms of his first year, that in April 1882 his Minor Scholarship was upgraded to a £100 p.a. Foundation Scholarship. In addition, there were college examinations every year for students not undergoing university assessment, and William duly took the Trinity College examinations in Easter week. He was one of five freshmen prizewinners in mathematics.[18] Three of the other prizewinners were William Sheppard, Walter Workman and William Cassie. Sheppard and Workman had won major scholarships before entering Trinity and all three were formidable competitors for William. Sheppard was an Australian, born in Sydney, although his secondary education had been at Charterhouse, in Surrey.[19]

An unexpectedly close relationship between mathematics and athleticism was a characteristic feature of life for most Cambridge undergraduates throughout the nineteenth century. A Cambridge wrangler of 1879, who went to Germany seeking intellectual enlightenment after graduation, returned disappointed, realizing that his hard study of mathematics, balanced by physical activities such as walking and hockey, had been especially beneficial and peculiar to Cambridge. Many other students reflected similarly on their Cambridge experience. In the face of the demanding Cambridge mathematics course, undergraduates used regular physical exercise to give their working day structure and to preserve robust health. The ideals of 'muscular Christianity'—clean living, discipline, and competition—had found a resonance in the lives of potential wranglers; they were 'mathematical athletes'.[20]

[17] *The Student's Guide*, n. 13, pp. 22, 75–6. See *also The Student's Guide to the University of Cambridge, Part II: Mathematical Tripos* (Cambridge: Deighton Bell, 1880).

[18] *Cambridge University Reporter*, 5 June 1882, p. 719.

[19] Ball and Venn, n. 9, pp. 647, 648, 664.

[20] A Warwick, 'Exercising the student body: mathematics and athleticism in Victorian Cambridge', in C Lawrence and S Shapin (eds), *Science Incarnate: Historical Embodiments of Natural Knowledge* (Chicago: University of Chicago Press, 1998), pp. 288–326, 310.

In 1882 William's brother Jack graduated from King William's College with a Trustees' University Exhibition. He participated in a range of school activities, and at the 1882 Prize Day he was singled out for special attention for having obtained full marks in all the sixth-form mathematics papers. He showed even more promise in mathematics than his elder brother and had won a scholarship to St John's College, Cambridge.[21] Like William he intended to undertake the Mathematical Tripos:[22]

> But he never took it up. He fell ill again, with the same complaint, and the Uncle took him in at Market Harborough. He was very fond of him, so was everyone of them in that house. He was anxiously looked after: I suppose that an operation was then looked on as a dreadful thing, and so no doctor advised it. I have an idea that there was a consultant from Leicester. It is difficult now to be sure of the conditions. Willie Addison[23] told me in 1898 that an operation ought certainly to have been carried out.[24] Jack had many interests during his invalid stay at Market Harborough. He had bees, there was a family bee company in fact: he collected stamps: he tried for prizes in the Truth competitions and won several. He took an interest in the news of the day, and must I think have introduced the idea of studying the daily paper. The weekly Market Harborough Advertiser had been the only source of news.

The next two years of William's life in Cambridge were more settled. He continued to study under Routh's expert direction and took daily exercise. In the Easter term of 1883 he again won a Trinity mathematics prize, along with Sheppard, Workman, and Cassie. In sport he added participation in hockey and lacrosse to his love of tennis. In the Bragg (Adrian) papers there are two cards, one giving the 'Rules of the Trinity Hockey Club' above the name 'W H Bragg, Secretary', the other the 'Rules of the Trinity College Lacrosse Club', as well as a booklet containing the 'Laws of Lacrosse'.[25] Caroe reports her father as saying that the students cut their hockey sticks from the hedges, and that he carried a scar on his head inflicted by the Duke of Clarence at hockey.[26] Years later, in 1908, a photograph of 'The Cambridge University [Lacrosse] Team of 1884' was published in the racy Adelaide weekly, *The Critic*, because it contained a number of Adelaide identities. The caption included: 'Professor Bragg and Mr. P A Robin, also of Adelaide, were absent when the photograph was taken'.[27] Also in the Bragg (Adrian) papers are cards indicating that William took some

[21] *The Barrovian*, October 1882, p. 291, and July 1882, p. 282.

[22] W H Bragg, Autobiographical notes, p. 25.

[23] Of Willie Addison, William says: 'He went to Uppingham School and afterwards to Caius [Cambridge]. He was cox of a Caius boat, and was, I believe, happy in his College life. He was to train for medicine, but my Uncle made him take the Mathematical Tripos first... Willie got through his medical course all right and went to a practice in Tenterden [Kent]: afterwards to the Scilly Islands', W H Bragg, Autobiographical notes, p. 28.

[24] When William was back in England from Australia on leave (see chapter 9).

[25] Bragg (Adrian) papers.

[26] G M Caroe, *William Henry Bragg 1862–1942: Man and Scientist* (Cambridge: CUP, 1978), p. 23.

[27] *The Critic*, Adelaide, 5 August 1908, p. 22.

interest in the activities of The Footlights Dramatic Club. In particular, there are two cards showing that he participated in play-readings by 'The Gypsies': for example, as Polonius, the first clown, and Fortinbras in a reading of *Hamlet*. He now felt at home. He was enjoying much of what Cambridge had to offer.

The year 1884 saw the culmination of William's childhood, youth, and education. As the academic year came towards its mid-year close, he faced the daunting Mathematical Tripos, several days of uninterrupted examination that would require all his intellectual, physical, and psychological powers. Now twenty-two years old, he would soon have to make crucial career decisions, based on the results he achieved. The examination programme for Part I began on Monday 26 May 1884 and continued for three days, with examinations in the mornings (9 a.m. to 12 noon) and the afternoon (1:30 to 4:30 p.m.). Part II, including use of the calculus and the methods of analytical geometry, began on Thursday 5 June and continued for another three days under the same conditions.

Rev. Dr John Scott has commented on the Tripos examination papers for Parts I and II of 1884 as follows: 'What strikes me about the papers is the general style of the questions. They are basically straightforward applications of reasonably elementary mathematics... Anyone who obtained a good mark would have demonstrated great manipulative skills and logical thought, but would not really have progressed that much into the great world of higher... mathematics. The other feature that is striking is the missing parts of what is now essential in... mathematics. There is no Analysis, the part of mathematics that puts concepts like limits, differentiation, etc. on a logical basis. This is now a fundamental part of first-year university mathematics. There is no mathematical logic, fluid mechanics, statistics, projective geometry among many other gaps. The papers display... a desire to test manipulative skills and not dig deep into underlying theory'.[28]

After Part I, the *Cambridge University Reporter* recorded that, 'The following candidates have acquitted themselves so as to deserve Mathematical Honours', and there followed an alphabetical list including 'Bragg Trinity'.[29] The honours grades were released after Part II: 'Mathematical Tripos, Parts I and II, 1884. Wranglers: Ds Sheppard Trinity [the Australian], 2 Workman Trinity, 3 Bragg Trinity, 4 Young Peterhouse, 5 Cassie Trinity'.[30] William remembered:[31]

> At the end of the three years, I took the Tripos: I was rather run-down and a little frightened, especially when I could not sleep the night before: a novel experience which shows that I was not really in a bad way. My Uncle got alarmed at my letters, and came to Cambridge to reassure me. I

[28] 'Mathematical Tripos Examination Papers 1884', Cambridge University Library; Howson, n. 2, pp. 220–1 gives a useful summary of the topics examined in 1884; Rev. Dr J. F. Scott, Cambridge Mathematical Tripos graduate, Emeritus Professor of Mathematics of the University of Sussex, UK, and past Vice-Chancellor of La Trobe University, Australia, personal communication.

[29] *Cambridge University Reporter*, 10 June 1884, p. 848.

[30] Ibid., 14 June 1884, p. 862.

[31] W H Bragg, Autobiographical notes, pp. 19–20.

> was afraid I had not done well in the exams: I remember the anxious mind as I walked up Senate House passage to hear the results. When I heard my name called out as Third Wrangler I was really amazed. I had never expected anything so high, not even when I was in my most optimistic mood. I was fairly lifted up into a new world. I had a new confidence: I was extraordinarily happy. I can still feel the joy of it! Friends congratulated me: Whitehead (of Harvard now) came and shook me by the hand saying, 'May a fourth Wrangler congratulate a third?' He had been fourth the year before.[32] As for the Uncles!

Like so many of his fellow students William clearly felt the strain of this pivotal examination, but his growing maturity and tenacious approach kept his nervousness within bounds. He communicated his misgivings to Uncle William. As we have seen, William's autobiographical notes are characterized by modesty and self-deprecation, sometimes to the point of obscurity, but his expression of unrestrained joy and elation at his third position is convincing and genuine. It was a crucial turning point in his life. From now on he would make decisions and plot his career with a new level of assurance. The young natural philosopher had left behind his Cumberland childhood, his Leicestershire school-days and his Manx youth, even if one or two important legacies remained, hidden deep within (see Figure 4.2).

Later in 1884 William's youngest brother, Jimmy, graduated from King William's College and came up to Cambridge. He had participated in a wide range of school activities and was a prefect in his final year. He, too, had won a mathematics scholarship, to Emmanuel College, and he graduated BA in 1887. Less academic than his brothers, he later farmed in New Zealand and then built a successful import–export business between Australia and England.[33]

For William the future beckoned, but in which direction? As Third Wrangler there was no doubt that he would go on to take Part III of the Tripos, but this involved a choice between four possible areas of study: Group 1, higher mathematics alone; Group 2, some mathematics and more complex parts of Newton's Principia; Group 3, some mathematics together with thermodynamics, electricity, and magnetism; and Group 4, hydrodynamics (including waves and tides) and wave motion (sound, physical optics, and the vibrations of elastic solids such as strings and bars). Contrary to earlier opinion, based upon William's own self-deprecating remark that he 'had never studied Physics',[34] I believe it was at this time that William carefully considered his future. He pondered where a career might lie and accordingly made a deliberate decision to study experimental physics.

This is supported by the fact that William did not prepare an essay in an attempt to win one of the two Smith's Prizes that were available after Part III of the Tripos and that ambitious graduates were keen to win. These prizes had been established

[32] This refers to Alfred North Whitehead, later to become very well known as a mathematician, joint author with Bertrand Russell of *Principia Mathematica*, philosopher of science, etc.

[33] King William's College, *School Lists 1887, passim*; *The Barrovian, passim*; G M Caroe, n. 26, p. 22.

[34] W H Bragg, Autobiographical notes, p. 1.

Fig. 4.2 William Bragg, Third Wrangler, Cambridge Mathematical Tripos, 1884. (Courtesy: King William's College.)

at Cambridge in 1768 by the will of Robert Smith, sometime Master of Trinity College. They were designed to foster interest in applied mathematics.[35] Previously judged by examination, from 1885 the prizes were determined by the quality of an essay on a subject of the candidate's own choice. This requirement tested different skills and encouraged research in applied mathematics. In 1886 the second (Workman) and seventeenth wranglers won the two Smith's Prizes for the student cohort to which William belonged; but by then he was on his way to Australia.

William's decision to enlarge his career horizon by including experimental physics was prompted by two major considerations. First, as he said himself, 'I might have wanted to be amongst books and people in Cambridge: I might have wanted to work for a [Trinity College] fellowship, though, as a matter of fact, my chances did not look well, because in 1883 the 2nd, 3rd, 4th &

[35] J Barrow-Green, '"A corrective to the spirit of too exclusively pure mathematics": Robert Smith and his prizes at Cambridge University', *Annals of Science*, 1999, 56:271–316.

5th Wranglers were all Trinity men, and in my year the 1st (Sheppard), 2nd (Workman), 3rd (myself) and 5th (Cassie) were all Trinity men'.[36] Then, as David Wilson has observed, 'The average wrangler interested in a career in science and certified as lacking the powers of a Stokes, Kelvin or Maxwell might decide to cultivate the experimental side of his abilities. If he hoped to teach... formal work in experimental physics would enhance his candidacy'.[37]

The other ingredients in William's choice were Richard Glazebrook and J J Thomson. Glazebrook—Fifth Wrangler in 1876 and now Fellow of Trinity College, university lecturer in mathematics, and demonstrator in experimental physics—had himself gone straight into the Cavendish Laboratory after graduation, had trained as an experimental physicist, and was now a successful researcher and outstanding teacher of laboratory-based physics. He was about to publish, with his colleague Napier Shaw, their ground-breaking textbook *Practical Physics*.[38] Furthermore, it is clear from a reference that Glazebrook later wrote for him that William had attended some of Glazebrook's lectures in preparing for Part II of the Tripos, when he had otherwise decided to avoid such classes: 'Mr W H Bragg of Trinity College attended several courses of my lectures while preparing for the Mathematical Tripos'.[39]

William also knew J J Thomson well from their common residency in Trinity College, and he had noted Thomson's rapid elevation to the Cavendish professorship of experimental physics. We know J J played the card game whist and took regular exercise through walking and organized sports.[40] William had played whist with his brother Jack in the King William's College sickroom, and he was a keen sportsman. 'I knew him [J J] pretty well at that time [1885]: he and Carey Wilberforce and I used to play tennis together', William later recalled.[41] It had been 'Maxwell's view of the function of the laboratory that it should be a place to which men who had taken the Mathematical Tripos could come, and, after a short training in making accurate measurements, begin a piece of original research'.[42] This scheme continued during the early years of Thomson's tenure, for it was Thomson's strongly held view that 'most of the students... who are studying applied mathematics would be much better equipped for research in [physics] if they came into touch with the actual phenomena in the Laboratory'.[43]

[36] W H Bragg, Autobiographical notes, p. 22.

[37] Wilson, n. 2, p. 365.

[38] Lord Rayleigh and F J Selby, 'Richard Tetley Glazebrook 1854–1935', *Obituary Notices of Fellows of the Royal Society*, 1936, 2:29–56; R T Glazebrook and W N Shaw, *Practical Physics* (London: Longmans Green, 1885).

[39] Letter Agent-General to Registrar, 18 December 1885, enclosing copy of Board of Selection's decision, Bragg's letter of application and his three testimonials, UAA, S200, docket 5/1886.

[40] Lord Rayleigh, *The Life of Sir J J Thomson, O.M.* (Cambridge: CUP, 1942), p. 10.

[41] W H Bragg, Autobiographical notes, p. 21; 'Carey' Wilberforce would appear to be L R Wilberforce, Trinity College and Cavendish student during Bragg's years there and later Professor of Physics at Liverpool: see J A Venn, *Alumni Cantabrigienses* (Cambridge: CUP, 1940), vol. VI, pt II.

[42] J. J. Thomson, *Recollections and Reflections* (London: Bell, 1936), p. 95.

[43] Quoted in Wilson, n. 2, p. 352.

Glazebrook was responsible for Group 4 of the possible subjects for Part III of the Tripos, and it was this that William chose. He thereby avoided mathematics alone, as well as the classical physics of Newton and the confused and rapidly evolving area of electricity and magnetism. Glazebrook's hydrodynamics and wave motion were more settled, amenable to elegant theorizing and enchanting experimentation, and very practical. These were fields to which William would return time and again during his career, sometimes as its central focus and sometimes nearer the edge of his attention. The farmer's son, who had enjoyed geology, sport, and the out-of-doors as well as mathematics, was finding a niche for himself in a combination of mathematics and experimental physics:[44]

> During the autumn of 1884 I worked for Part III of the Tripos as it then was. I believe none of us did too well: but we nearly all got Firsts because the Senior Wrangler did not do any better than we did, and they could not give him a Second. I was terribly proud because a publisher came and asked me my terms for solving the problems in Smith's 'Conics', to go into a book of answers. I had other things to do and had to say no, but I remember that in my mind I declined an offer which I thought might bring me in £5! Why, £150 would have been nearer the mark! It just shows how little I knew of matters outside my own line of work. I was, in fact, very much shut in on myself, unventuresome, shy and ignorant. And yet I enjoyed my life at Cambridge tremendously: I missed much no doubt, being the sort of young man that I was, but I gathered in a lot. University life is spacious and beautiful. Cambridge is a lovely place, and Trinity is something to be very proud to belong to. I loved it all, the work and the games, the place itself and the country round and all the incidents. In my last year or two I had a delightful set of rooms over the old Combination room [fellows' common room]; the staircase was just opposite the entrance doors of [the dining] Hall.

Success in the Tripos brought two immediate rewards: first, early in 1885 William moved into a spacious set of rooms in the Great Court of Trinity College (number 1 on staircase S).[45] Second, his desire to continue his training in experimental physics in the Cavendish Laboratory was welcomed by Glazebrook and Thomson. Glazebrook's later reference for William stated: 'Mr W H Bragg of Trinity College attended several courses of my lectures while preparing for the Mathematical Tripos, and since that time he has worked under my suggestions at the Cavendish Laboratory while studying practical physics. In his preparation for the third part of the Mathematical Tripos I supervised his reading as University Lecturer in the branch he was taking up'.[46]

William worked in the Cavendish Laboratory for the remainder of 1885, almost a whole year. We have no precise information on what he did during this time. Glazebrook published a number of articles during 1884 and 1885 on

[44] W H Bragg, Autobiographical notes, pp. 20–1.
[45] *Trinity College: Room Rents 1871–1897*, Trinity College Library.
[46] See n. 39.

both theoretical and experimental topics, but there is no indication that William assisted with them.[47] It seems more likely that he undertook a number of the experiments that were set up in the teaching laboratory and that are described in Glazebrook's laboratory textbook with Shaw. For example, the Tripos group for which Glazebrook was responsible had corresponding chapters in the textbook, entitled: 'Mechanics of liquids and gases', 'Acoustics', 'Reflexion and refraction—mirrors and lenses', and 'Mechanics of solids'.[48]

On 16 May 1885 William's father, Robert John Bragg, died at Ramsey. After he had sold Stoneraise Place but was still occupying it as a tenant farmer, Robert had made his will with a Wigton solicitor. He bequeathed all his estate and effects to his two brothers, William and James Bragg in Market Harborough, as joint trustees and executors: 'to pay and apply such interest dividends rents and annual proceeds in equal portions for the benefit maintenance and education of my three sons until they respectively attain the age of twenty one years. And when and so soon as they respectively attain that age to convey transfer and pay over to them their respective shares of the said Estate and accumulations'.[49] The total assets amounted to £1,727, arising from the sale of a number of boats and their equipment, household furniture, and shares in the Manx Northern and Foxdale railways, and from several cash deposits. After funeral expenses, £1,520 remained for the later benefit of the three boys.[50] Robert Bragg was buried with his wife, Mary, in the Westward churchyard. He had made the best provision he could for his sons, and his affections called him back to Cumberland as his final resting place. He had been small and gentle, a largely uneducated seaman-farmer, and a distant spectator only of William's growth and development. William did not record his father's death in his autobiographical notes. Perhaps it was too painful, or perhaps the memory had faded by the time he wrote. Father and son had drifted apart.

An unexpected opportunity

Late that year an apparently chance event occurred that was to change profoundly the future course of William's life:[51]

> At the end of 1885 I was going, one morning, along the King's Parade to attend a lecture by J J Thomson at the Cavendish, and was joined on the way by the lecturer himself. I knew him pretty well at the time: he and Carey Wilberforce and I used to play tennis together. He asked me if Sheppard was going in for the Adelaide post. This was the professorship

[47] 'A list of memoirs containing an account of work done in the Cavendish Laboratory', in *A History of the Cavendish Laboratory*, n. 8, pp. 288–91.

[48] Glazebrook and Shaw, n. 38.

[49] Copies of will and associated documents of Robert John Bragg, Bragg (Adrian) papers and papers in the possession of Stephen Bragg, Cambridge.

[50] Ibid.

[51] W H Bragg, Autobiographical notes, pp. 21–4.

in mathematics and physics which Horace Lamb was just resigning. He had been in Adelaide since the foundation of the young university in 1877(?) [in fact, founded 1874, first students 1876] and wanted to get back to England. I had seen the advertisement, and the magnificent offer of a salary of £800 a year. I said to J J that I had heard nothing of any such intention on Sheppard's part... I was astonished at the question: it had not occurred to me that any one so young might be eligible. Also the salary seemed too big for such untried people—I had a vague idea that £300 a year was more our style. Then I asked J J whether I might have any chance... and he said that he thought I might! So when the lecture was over I went and telephoned an application—it was the last day of entry.

A few days later I was summoned to an interview in London, to the office of the Agent-General for South Australia. I found that I was one of three who had been sent for to be interviewed. One of the three—the name was Adair, I think, I had not heard of him—could not come, he was ill. The other, beside myself, was my late examiner in the Tripos, Graham! That was a queer situation, he and I sitting together in the waiting room.

The interviewers were Lamb, J J and the Agent-General, Sir Arthur Blyth. They knew all about me, and the interview was short. I remember that they asked me if I regretted having applied, and I said with some astonishment 'Certainly not'. I think that if I had been more sophisticated I might perhaps have been less positive. I might have wanted to be amongst books and people in Cambridge: I might have wanted to work for a fellowship, though as a matter of fact my chances did not look well... So I am glad that no sophistication prevented me.

I went back to Market Harborough, and that evening as Fanny and I were playing about on the piano, a telegram was brought to me. 'As new professor of mathematics and physics in Adelaide University, would I give some particulars of my career'. Well! You can imagine my delight!... An assured position, a salary beyond all expectation, a new country with all the adventure of going abroad to it, a break away from being a subject, to be now my own master. I took the telegram across to my Uncle [William] at the shop: he read it, finished without a word the posting that he was doing, took me home across the square in the dark, and on the way he broke down. It had not occurred to me that the glorious success would mean to him a parting that he would feel so badly. But I hope that his own pride in the result of what he had always worked for through me carried him through. People used to stop and ask him if it was really true about his nephew, and he could answer and speak about his 'nephew the professor'! Perhaps, too, his excitement and pride were rather a strain on his feelings...

By the way I forgot to say before this that the electors could have sent out a Senior Wrangler of great ability, but he was not safe with the bottle. They thought, however, that they had better consult an Adelaide man who happened to be in London, and he was in favour of the young man who so far had kept off the drink. The Adelaide man was my future father-in-law [Sir Charles Todd].

The next three weeks was a grand time! Preparations for the passage, new clothes, new outfit altogether: and there was a grant of £150 from the

> Agent-General to cover it all. Visits of my Aunt and myself to the outfitters in Cornhill (Silver & Co.), visit to the shipping office with Sir Arthur Blyth, interviews, cleaning up at Cambridge, farewells to friends and so on. I got a book or two on South Australia and read with eagerness about the place and its history. Then finally the Aunt and Uncle William came up to London the day before I sailed. I had been staying there for a short time. Next day they saw me off at Tilbury and there I was away on the great adventure, thrilled by it. The Aunt and Uncle William, when they came to London, brought the news that my brother Jack had just died. After seeing me off they must have gone straight back to his funeral. I had not known he was so ill when I said goodbye to him.

However, neither William's invitation to apply, nor his selection for the Adelaide post, was as haphazard or as fortuitous as these notes suggest. The unusual foundations of the colony of South Australia, its capital city Adelaide, and its early university, will be discussed in the next chapter. Since the colony was young and the number of prospective students small, the university had begun teaching in 1876 with only four foundation professors and two princely private benefactions. The four chairs embraced classics and comparative philology and literature, English language and literature and mental and moral philosophy, pure and applied mathematics, and natural science (including geology, mineralogy, and chemistry). No junior academic staff members were appointed for a number of years.[52]

The foundation Elder Professor of Pure and Applied Mathematics at the University of Adelaide was Horace Lamb, Second Wrangler in the Mathematical Tripos of 1872 and Fellow and lecturer in mathematics of Trinity College, Cambridge. Lamb had married in 1875 and had thus been required to resign his college position. Although his formal responsibilities in Adelaide were confined to mathematics, Lamb voluntarily instituted and gave courses in natural philosophy (physics) at all three levels of the arts and science degrees. Furthermore, in so far as space and apparatus would allow, he also held regular laboratory classes for his natural philosophy students. He became a beloved teacher, popular public lecturer, and respected member of Adelaide society. He carried a large teaching and examining load, saw six of his children born there, and wrote and published the first edition of his famous classic text on hydrodynamics.[53]

In December 1883 Lamb had written to the University Registrar requesting a year's leave-of-absence in order to travel to Britain to recharge and enhance his academic knowledge and skills. A long and distressing debate followed, between Lamb and the University Council, at the end of which Lamb was

[52] W G K Duncan and R A Leonard, *The University of Adelaide 1874–1974* (Adelaide: Rigby, 1973); *Calendar of The University of Adelaide*, Adelaide,1877–84 (annually).

[53] J G Jenkin, 'The appointment of W H Bragg, FRS, to The University of Adelaide', *Notes and Records of the Royal Society of London*, 1985, 40:75–99; J G Jenkin and R W Home, 'Horace Lamb and early physics teaching in Australia', *Historical Records of Australian Science*, 1995, 10:349–80; R B Potts, 'Lamb, Sir Horace (1849–1934), mathematician', in B Nairn, G Serle, and R Ward (eds), *Australian Dictionary of Biography* (Melbourne: MUP, 1974), vol. 5, pp. 54–5; H Lamb, *A Treatise on the Mathematical Theory of the Motion of Fluids* (Cambridge: CUP, 1879), subsequent editions entitled simply *Hydrodynamics* and still in print from the nineteenth century.

granted leave after the Council had instituted additions to the university statutes regarding leave-of-absence. It was the first such provision in Australia. By the time of Lamb's departure in mid-1885, however, there was considerable doubt that he would return. With the aid of his old Trinity College colleague (and William's Trinity tutor), Henry Taylor, Lamb had applied for the chair of pure mathematics at Owens College, Manchester, to which he was formally appointed after interview in England. Despite the difficulties, Lamb and the University of Adelaide parted amicably, and Lamb subsequently served the university in various honorary capacities for many years. For its part, the university determined that, if a replacement would soon be required, it would seek a professor who could cover experimental physics as well as pure and applied mathematics.[54]

When the university received a telegram confirming Lamb's Manchester appointment and his Adelaide resignation, arrangements for the selection of a successor were implemented immediately. A plan had been drafted between Lamb, the University of Adelaide, and the Agent-General for South Australia in London, Sir Arthur Blyth. Accordingly, on 5 October Blyth wrote to J J Thomson, asking him 'to aid the university in the selection of a successor to Professor Lamb', and to 'name the newspapers in which you think the advertisement should appear'.[55] Thomson agreed, suggested six publications, and with Lamb and Blyth formed the Board of Selection, with full authority to make the appointment without further reference to Adelaide.[56] Such an untrammelled procedure is a vivid illustration of the reliance Australian universities then placed on Oxbridge. There was one notable Australian applicant for the position, William Sutherland, MA (Melbourne), BSc (London), who later became an outstanding theoretical chemical physicist.[57] He had to send his application to London. The conditions as set out in the advertisement were as follows:[58]

> The University of Adelaide
>
> Elder Professor of Mathematics and Experimental Physics
>
> The Council invite applications for the above Professorship. Salary £800 per annum. The appointment will be for a term of five years, subject to renewal

[54] Jenkin, 'The appointment', n. 53; much of what follows is taken from this source.

[55] Copy of letter from Agent-General to J J Thomson, 5 October 1885, in 'Letter book of Agent-General for South Australia regarding University of Adelaide (1878–1904)', State Records of South Australia, Adelaide, GRG 55/7/1; a few copies of letters received by the Agent-General are also included.

[56] Copy of Thomson's reply, suggesting *The Times*, *Nature*, *Cambridge University Reporter*, *Oxford University Gazette*, *The Athenaeum* and *The Academy*, ibid.

[57] W A Osborne, *William Sutherland: A Biography* (Melbourne: Lothian, 1920); T J Trenn, 'Sutherland, William', in C C Gillispie (ed.), *Dictionary of Scientific Biography* (New York: Scribner's Sons, 1976), vol. XIII, pp. 155–6; R W Home, 'Sutherland, William (1859–1911)', in J Ritchie (ed.), *Australian Dictionary of Biography* (Melbourne: Melbourne University Press, 1990), vol. 12, pp. 141–2.

[58] Copy of letter Agent-General to advertising agents Messrs G Street & Co., London, 7 October 1885, SRSA, n. 55.

> at the discretion of the Council. Salary will date from the 1 March, 1886, and the Professor will be expected to enter on his duties on that date. An allowance will be made for travelling expenses. Applications, with testimonials, should reach Sir Arthur Blyth... not later than 1 December 1885.

The circumstances of William's application have been vividly portrayed by his own words quoted above. His letter of application is dated 1 December 1885, the date on which Thomson spoke to him, the closing date.[59] It was all a great piece of luck; or was it? Thomson may have known, or could have guessed, that the majority of applicants lacked strong qualifications in both mathematics and physics. It was a position for which the students that were his special focus—those high wranglers he was encouraging to enter the Cavendish Laboratory—would be ideally suited. William Henry Bragg was precisely one such young graduate. Thomson asked Bragg about possible applicants; Bragg asked Thomson if he might have a chance; Thomson told Bragg he thought he might![60]

'The total number of candidates is twenty three', the Agent-General reported to the Adelaide Registrar, 'but one of these has sent in an informal application which cannot be entertained'.[61] Even without him the field was impressive: fourteen Cambridge graduates, of whom thirteen were wranglers and two Smith's Prize winners, two Oxford graduates, two London, one Trinity College Dublin, and three whose background I have been unable to trace. Thomson and Lamb met Blyth in London and drew up a short-list for interview: John Adair, Christopher Graham and William Bragg. We may wonder why the only Senior Wrangler and First Smith's Prize winner on the list, Thomas Harding (BA 1873), was not invited to attend. William gave one possible answer, and Harding had already abandoned school teaching for a legal career. Adair had been schooled at Trinity College Dublin (BA 1873), was Seventh Wrangler in 1878, had taught briefly, and was planning to study in the Cavendish Laboratory. Graham had also come from Trinity College Dublin (BA 1873), was Third Wrangler and winner of the Second Smith's Prize in the same year as Adair, had senior school-teaching experience, and had been Senior Moderator for William's Mathematical Tripos examinations in 1884 and 1885. Both Adair and Graham were Irish and about thirty-four years old.[62] From the information we have, those chosen for interview seem to have been three of the strongest candidates, although Thomas Lyle might have been included, given his exceptional undergraduate record at Trinity College Dublin in both mathematics and experimental science (BA 1883 with

[59] See n. 39.

[60] In addition, there is a story that Thomson had picked out Bragg earlier. Lamb had come to see Thomson about the Adelaide vacancy and, looking out of his Trinity College room into the quadrangle, Thomson had remarked, 'That's the young man for you: Third Wrangler last year, he has taken a first in Part III of the Tripos this year'. The story was told by Victor Edgeloe, long-serving Registrar of the University of Adelaide, in a radio talk in 1980: UAA, V A Edgeloe, 'Seven talks for 5UV', Adelaide, June 1980, p. 14.

[61] Letter Agent-General to Registrar, 4 December 1885, UAA, S200, docket 3/1886.

[62] Venn, n. 41; J Foster, *Alumni Oxonienses* (London: Foster, 1891); R W Home, 'Lyle, Sir Thomas Rankin (1860–1944)', in B Nain and G Serle (eds), *Australian Dictionary of Biography* (Melbourne: Melbourne University Press, 1986), vol. 10, pp. 172–4.

two gold medals). Adair was ill and could not attend in London, although he would have been known to the committee and was a strong candidate.

William supplied three testimonials. That by Glazebrook has been quoted above in part and continues, 'I have also examined him in various College Examinations. I have thus had ample opportunity of becoming acquainted with Mr Bragg's powers and I have no hesitation in recommending him most strongly to the Electors for the Professorship of Mathematics and Experimental Physics at Adelaide as being extremely well qualified to discharge the duties of the post and likely in every respect to give satisfaction'. Taylor, his college tutor, summarized William's Cambridge record and added, 'he was a most diligent and exemplary student...his work in my experience was always characterized by neatness and accuracy, points of great importance in a teacher'. Finally, Routh certified that William had 'great mathematical talent. He read with care and attention and thus made rapid progress...I am therefore glad to recommend him to the electors & believe that he will prove an efficient Professor'.[63]

A telegram broke the exciting news at Market Harborough that evening, and Uncle William unexpectedly broke down. It is hard to imagine a loving bond between the austere Victorian disciplinarian and his young nephew in William's early years in Market Harborough, but now a deep affection had grown up between them. Uncle William had mellowed, his businesses had prospered and he was now more secure, and the considerable effort he had invested in William's growth and development had paid off more handsomely than he could have imagined. For his part William had come to realize that his uncle had had the best of intentions, that he had made considerable sacrifices for his nephew, and that they had travelled a rocky road together and emerged triumphant. The journey had sometimes been painful but the culmination was a 'glorious success'. Uncle William had become the father William had lost.

The next day Lamb hastened to give the Adelaide University Chancellor 'some account of the manner in which we have discharged our stewardship'. He reported:[64]

> At our first meeting we had little difficulty in reducing the list to three, and then adjourned in order to give these three the opportunity of waiting personally on us. Yesterday the interviews were held and—after some slight hesitation between two of the candidates—we unanimously recommended that the Prof'p be awarded to Mr Bragg of Trinity College, Cambridge...It is evident that his math'l abilities are of the highest, and he has also worked at Physics in the Cavendish Laboratory under my coadjutor in the appointment (Prof. J J Thomson), who says that his work is very good. I was up at Cambridge a week before our last meeting and...Mr Bragg bears a high reputation in every way—and I may add for the satisfaction of the Univ'y on a really important point—that

[63] See n. 39.

[64] Letter Lamb to Chancellor, 18 December 1885, UAA, S280, in 'Envelope 162: Correspondence re Professors'.

> his bearing at our interview yesterday was unexceptional—and that he contrasted favourably in this respect with another candidate whose academ'l qualifications were of about equal value. As far as I can judge, the only possible source of misgiving as to the propriety of our choice is Mr Bragg's youth; he is only 23. Personally, I do not think much of this. I cannot but remember that I was myself not much older when I went to Adelaide—and that I did not (to my knowledge) find it any drawback....
>
> I can testify also that Prof. J J Thomson took great care and trouble in this matter, and shewed the greatest anxiety to come to a fair decision.
>
> With kind regards, I am, my dear Chief Justice
>
> Yours very sincerely Horace Lamb.
>
> P S The most curious incident in the award was a letter from Lord Carnarvon (Viceroy of Ireland) arguing that there might be a danger that 'justice to Ireland' would not be done unless some Irish Math'n of repute was put on the Board to look after the interests of Irish candidates. Sir A Blyth sent a very dignified reply.

Two years later Thomson confided to his friend Richard Threlfall, by then Professor of Physics at the University of Sydney, regarding another application from Adair, this time for a demonstrator position in Sydney: 'I do not think he has a very extensive knowledge of the book-work of Physics, but he is a good Mathematician (in fact, he nearly got Bragg's appointment)...he is a gentleman, but an Irish one, and this is my chief doubt, as Sir Arthur Blyth told me Irishmen were very unpopular in Australia'.[65] Graham was also an Irishman, as was Lyle. Lyle had the added disadvantage that he had completed his studies in Dublin, although he was soon to follow William to Australia as Professor of Natural Philosophy at the University of Melbourne. It is not surprising, therefore, that the Earl of Carnarvon and Lord Lieutenant of Ireland had written to the Agent-General for South Australia supporting claims by Trinity College Dublin that 'Irish Candidates for Educational posts have been frequently overlooked by the Colonial authorities...in mathematics especially...as these appointments are practically in the hands of Cambridge men'.[66] Blyth replied that his instructions from Adelaide did not permit him to accede to the request, but he promised to forward the correspondence to Adelaide. Blyth was sensitive to local prejudices. South Australians were predominantly English and Welsh and very strongly non-conformist. There existed a lower proportion of Irish immigrants in Adelaide than in other Australian cities and those Irish men and women who had emigrated were predominantly working class, unskilled, and generally disliked.[67]

[65] Letter J J Thomson to R Threlfall, 7 August 1887, Cambridge University Library, Thomson correspondence, Add MS 7654, T19.

[66] Letter Agent-General to Registrar, with enclosures, 2 December 1885, UAA, S200, docket 2/1886.

[67] See, for example, C Nance, 'The Irish in South Australia during the colony's first four decades', *Journal of the Historical Society of South Australia*, 1978, 5:66–73; D L Hilliard, 'The city of churches: some aspects of religion in Adelaide about 1900', ibid., 1981, 8:3–30.

Lamb had been only twenty-six years old when appointed to Adelaide and Thomson himself was only twenty-eight when awarded the Cavendish chair. Thomson had been chosen from a field of outstanding applicants, several of who had seniority and achievement over him. Glazebrook had buried his disappointment and remained in the Cavendish Laboratory until the new professor established himself. Thomson was in his debt, and Glazebrook clearly had a high opinion of William. Lamb knew the Adelaide situation intimately and knew what was required. Thomson and Bragg were near-contemporaries in Trinity College. In retrospect the selection is not surprising. As J J said to William many years later, 'I remember advising you to go in for the Professorship. I have always congratulated myself on having done such a good piece of work'.[68]

William's reference to Charles Todd in his reminiscences heralds the arrival of a man who was to play a very important role in William's development and maturation: as a family man, teacher, and prominent public figure in Adelaide and beyond. Todd, born (1826), raised, and educated in England (privately and at the Greenwich and Cambridge Observatories), had gone to South Australia in 1855 as Superintendent of Telegraphs and Government Astronomer. In 1870 the onerous duties of Postmaster-General were added. Happily combining the three roles, Todd became one of Australia's greatest public servants, best remembered as the architect and builder of the transcontinental Overland Telegraph Line. Constructed during 1870–72, over 1,980 hostile miles (3,180 km) traversed only once before by Europeans, it was an outstanding piece of engineering and became a communication lifeline between Britain and Australia. Many other successes followed, not least Todd's leading role in the development of meteorology in Australia.[69]

Todd toured England and the Continent with his eldest daughter, Elizabeth ('Lizzie'), during 1885–86 and kept a diary of their travels.[70] From this it is clear that, although the trip was prompted by medical advice to take a long rest, Todd was intent on learning as much as he could about developments in the fields for which he was responsible, attending a Berlin conference, renewing old acquaintances, and making new ones. Based in Stockport, he visited London, Manchester, Liverpool, and Birmingham during October–December 1885. He saw Horace Lamb, Sir Arthur Blyth, and Sir William Preece, Engineer-in-Chief of the British Post Office, on several occasions. He and

[68] Letter J J Thomson to W H Bragg, 27 December 1936 (in reply to Bragg's letter of congratulation on Thomson's 80th birthday), Bragg (Adrian) papers,.

[69] See, for example, G W Symes, 'Todd, Sir Charles (1826–1910)', in G Serle and R Ward (eds), *Australian Dictionary of Biography* (Melbourne: Melbourne University Press, 1976), vol. 6, pp. 280–2; W H Bragg, 'Sir Charles Todd, KCMG, 1826–1910', *Proceedings of the Royal Society of London*, 1911, 85:xiii–xvii; A Thomson, *The Singing Line* (London: Chatto and Windus, 1999); R W Home and K T Livingston 'Science and technology in the story of Australian federation: the case of meteorology, 1876–1908', *Historical Records of Australian Science*, 1994, 10:109–27.

[70] Diary of C Todd of a tour on the Continent and in England... 1885–86, State Library of South Australia, Adelaide, PRG 630/6.

Lizzie also saw family in Cambridge, where Lizzie met the solicitor Charles Squires, whom she subsequently returned to marry.[71] In 1886 Cambridge University conferred an Honorary MA on Charles Todd in person. William's recollection that the selection committee had consulted Todd regarding one of the applicants is therefore confirmed. Reminded of this by William on the occasion of his eightieth birthday, Thomson added 'I had forgotten the interview with Sir Charles Todd, but if that had anything to do with your marriage, I feel I had builded still better than I thought'.[72]

The selection committee may have wondered if he would accept the appointment, but William had no doubt. Its attractions were clear to him. Australia was well known, and service in the colonies was a well-trodden path for capable Englishmen. The Cambridge colleges had a surprisingly large number of Australian undergraduates at this time, and many of the prominent Adelaide families were represented.[73] The University of Adelaide had already made a particular impression on William, as he later recalled: 'From the very beginning the University of Adelaide has been known in the scientific world. Of all the textbooks on hydrodynamics, the best is that of Professor Lamb, the first preface of which was dated from Adelaide. The subject deals with the motions of masses of water, with waves and tides, the movements of ships in the sea, and numbers of other important problems. When I was a Cambridge student, and glanced often at the title page of this book, I used to be quite fascinated by the "Adelaide, South Australia" which was printed thereon. I wondered what sort of a place it was, what sort of conditions they were, under which the book was written, and whether there would ever be any chance of my obtaining a position like that of Professor Lamb'.[74] It was common for the young Australian universities to ask their professors to cover more than one discipline and joint lectureships in mathematics and physics were widespread until the 1920s. The salary of £800 p.a. was princely, second only to the Professor of Natural Science (£1,000) and above that of the other Adelaide professors (£600).[75]

Congratulations flowed in to Market Harborough for the new professor. His King William's College mathematics master, Jenkins, wished him well; his Trinity tutor, Taylor, hoped that he would return to England, 'unless you tie yourself more permanently to Adelaide'; and his fellow student, Workman, suggested that 'it is a fine thing indeed to have been chosen above the heads of [the other applicants]'. Lamb wrote generously, commenting on the 'the glorious blue sky and brilliant sunshine of Australia', offering advice about the university facilities he would find and the need to enhance them, and about recently published textbooks, and finally offering assistance 'in any way...I

[71] Thomson, n. 69, p. 266.

[72] See n. 68.

[73] D van Dissel, 'The Adelaide Gentry 1880–1915', unpublished MA thesis, University of Melbourne, 1973.

[74] *Register*, 1 February 1908, p. 9.

[75] Salary Sheets, 1888–1920, UAA, S114.

still feel a warm interest in the old place, and in all connected with it'.[76] Aunt Mary took William shopping for 'new clothes, new outfit altogether'. He had to clear out his room at Trinity College and, in fact, pay rent for the Lent term of 1886 (January to Easter), since he had not been able to forewarn the college of his imminent departure. And there were friends and family to farewell. His last winter Christmas for some time was celebrated at Market Harborough, but it was spoilt by 'a severe cold'.[77] The stress and excitement had taken their toll.

Uncle William and Aunt Mary came to London to see William off, but it was not a happy departure. The RMS *Rome*, of 5,013 tons and built on the River Clyde, entered the Australian service in 1881 and was then the largest and best equipped liner in the P&O fleet of streamers. It had compound engines and one of the first refrigerated compartments; it would be a comfortable floating home for six weeks.[78] The sadness that hung over the trio at Tilbury on that sailing day, 14 January 1886, was caused by the condition of William's brother Jack. He was seriously ill, and now his condition had deteriorated even further. He had not died, as William later thought, but he would do so two days later, on 16 January, at just twenty-one years of age. Uncle William and Aunt Mary did, indeed, go 'straight back to his funeral'. William added, 'They must have missed Jack terribly when he died, he was such an effective member of the household. Of course I had long been but a visitor at vacation times: but it must have made a gap when we were both gone'.[79]

Given what is known of William's mathematics education, especially its extensive content of mathematical physics and of his deliberate decision to add experimental physics to his skills, how are we to understand his constant insistence that: 'As I had never studied Physics (or Chemistry), I tried to learn some on the way out',[80] and 'Although I had never done any of the latter [physics], nor worked at the Cavendish Laboratory except for a couple of terms after I had taken my degree, it was supposed by the electors that I would probably pick up enough as I went along to perform my duties at the Adelaide University'.[81] Even after his arrival in Adelaide William referred to himself as 'Professor of Mathematics' alone until about 1899, despite his increasing dedication to physics.[82] In fact, this is just one example of similar statements that have misled historians and other writers over the years, including his own son.[83]

[76] Bragg (Adrian) papers.

[77] Letter Agent-General to Registrar, 24 December 1885, enclosing copy of letter W H Bragg to Agent-General, 23 December 1885, explaining that his medical certificate will be delayed by a few days due to 'a severe cold', UAA, S200, docket 6/1886.

[78] See, for example, *Lloyd's Register of British and Foreign Shipping: from 1 July 1885 to 30 June 1886* (London: Lloyd's, 1886); M R Gordon, *From Chusan to Sea Princess: The Australian Services of the P&O and Orient Lines* (Sydney: Allen & Unwin, 1985).

[79] W H Bragg, Autobiographical notes, p. 25.

[80] Ibid. (RI MS WHB 14F/1), p. 1.

[81] W H Bragg, Autobiographical notes, p.30.

[82] In his Adelaide correspondence and in the annual *Calendar of The University of Adelaide* until 1899.

[83] Sir Lawrence Bragg, 'William Henry Bragg', *New Scientist*, 1960, 7:718–20 ('he had no training in physics', p. 718).

The relatively new Natural Sciences Tripos at Cambridge, with its emphasis on chemistry, geology, and aspects of biology, did not fully embrace physics until 1873.[84] Most of its graduates were therefore not yet strong in either mathematics or physics. William was a product of the Mathematical Tripos, still the pre-eminent qualification in Britain and with a good coverage of theoretical physics. He felt especially deficient in his knowledge of electricity and magnetism, which was emerging as an area of both theoretical and practical importance and that he suspected might be significant in a new and developing city. He had won a copy of Maxwell's classic textbook at King William's College but had hardly opened it in the intervening years. Part of its difficulty was the way it was written. One contemporary remark said: 'it would have been an immense improvement to Maxwell's 'Electricity' to have been written by Routh'![85] William would study these topics on the long voyage to Australia.

However, the overwhelming ingredient in William's insistence that he knew no physics was his excessive humility. He had pre-empted his mathematical contemporaries in terms of the qualifications that would increasingly be useful for employment, but he certainly did not want to make them envious or jealous. He wanted to believe that he was just lucky to get the job, and he was unsure of his ability to handle its substantial demands. He did not want to be recognized as an expert in both mathematics *and* physics. He wanted no attention; he found notoriety embarrassing. Years later his Trinity College and Adelaide colleague, Sydney Talbot Smith, said: 'Well, we all know how clever men can delight to exaggerate their own shortcomings. As William always humorously told the story, he just bought some books on physics, studied them on the voyage, and... was only about two jumps ahead of his students'.[86] William's life was to be characterized by 'humility and disinterest in himself; concern for others and, of course, for science; a wish not to impose his personal views (about people) on others; and a withdrawal from too close personal contacts'.[87] This loving yet penetrating appraisal is by Lady Adrian, the elder daughter of Gwendolen Caroe, William's only daughter and biographer, and the grandchild who was closest to him in his late years.

[84] Wilson, n. 2.

[85] Quoted in Warwick, n. 2, p. 306.

[86] S Talbot Smith, 'Memories of Sir Wm. Bragg', *The Mail* (newspaper), Adelaide, 4 April 1942, p. 7.

[87] Lucy Adrian, personal communication.

5
Adelaide: early years

A leading historian of South Australia has written: 'The story of the development of South Australia is often forgotten or misunderstood. As a colony and state of Australia, it has been neither the biggest nor the most prosperous, and it was settled after most of the others. Yet it has an interesting and at times exciting history, differing from that of its eastern neighbours with their better known stories of convicts, gold rushes and wool booms'.[1] Australia is usually considered to be a young country. Its white settlers arrived only two hundred years ago, yet dark-skinned people, the first Australians, had reached its shores very much earlier. Even after Europeans discovered it, Australia remained of little importance to the great explorer nations. Only an occasional European ship reached its shores, and left unimpressed by the poor country and its shy inhabitants. Captain James Cook's discovery of the east coast in 1770 made his voyage the most important of the early expeditions, since it prompted the British government to become seriously interested in developing the strange land. White people first arrived as settlers in 1788, beginning the British colony of New South Wales, a gaol for convicts under the charge of Governor Arthur Phillip. It was 1815 before the colony showed significant signs of success, as Sydney town prospered and as squatters, now including free settlers, pushed westwards and found new grazing land. Further settlements followed.

Early Dutch and French navigators who saw a little of the south coast of Australia were also unimpressed, and most of it was still unknown in 1800. The British sent out Matthew Flinders to explore and map 'the unknown coast', where he met his French counterpart Nicolas Baudin in Encounter Bay, before successfully circumnavigating the Australian mainland. South Australia's other great discoverer was Charles Sturt, who left Sydney in 1829 hoping to solve the puzzle of Australia's inland rivers, and subsequently travelled along the Murrumbidgee River into an even bigger waterway, the River Murray. This led him into South Australia, where he met many Aborigines, saw the Mount Lofty Range and productive land that abounded with kangaroos, and followed the Murray to its mouth at Lake Alexandrina and the Southern Ocean. Even by 1836, however, the year in which the mainland of South Australia was first settled, the more remote regions of Australia were still unexplored and uncolonized.

[1] R M Gibbs, *A History of South Australia from Colonial Days to the Present* (Adelaide: Southern Heritage, 1984), p. 1.

There were few British colonies that had as extensive a set of plans and theories for its establishment as South Australia, and the colonizing plan of the difficult but creative Edward Gibbon Wakefield was central to the scheme. Land was crucial, and Wakefield had noted that the lack of a fair and open distribution of land had been a major cause of serious problems in earlier colonies. In addition, most settlers wanted to avoid convicts and be spared the established religion, class structure, and industrial poverty of their homeland. They wanted social and religious freedom and the opportunity to progress through their own hard work. Wakefield believed that planned emigration could be achieved by selling the land at a reasonable, fixed price. The money raised could then be used to bring out new emigrants, who would work on the land before accumulating enough funds to buy their own acres. Wakefield and his followers, the Colonial Reformers, called it 'systematic colonisation'.

The British House of Commons passed an act in 1834 to establish South Australia, including the appointment of Governors; but control of the system of land sales and emigration was vested in a Board of Commissioners, headed by Colonel Robert Torrens. Initially the plan was followed, but friction between the Governor and the Resident Commissioner, as well as land speculation and the diversion of funds from emigration to essential services, undermined it. The establishment of a new settlement in a harsh and unknown environment was a question of compromise and survival rather than of religious adherence to a plan developed half-a-world away.

Yet, after some very difficult early years, the experiment succeeded, thanks to the commitment and courage of the colonists and the vision of a few of their leaders. The plan for the site and disposition of the city of Adelaide, developed by Colonel William Light, the colony's first Surveyor-General, and George Kingston, left a legacy that modern citizens still celebrate. They placed the city astride the only fresh water in the vicinity, the River Torrens, although it could degenerate into a series of stagnant pools in a hot summer. The two parts of the city had wide streets, laid out on a north–south, east–west rectangular grid, with open squares to beautify and open them to the sky. But their greatest gift was the figure-eight of wide and generous 'Park Lands' that still surround and embrace North Adelaide and (South) Adelaide. The city site was six or seven miles east of the Gulf St Vincent (Golfe Joséphine the French called it). Further east the Mount Lofty Range ('the Adelaide hills') limited the flat Adelaide plain.

By the time William arrived, Adelaide had 130,000 inhabitants, gas lighting, two reservoirs providing piped water, the first drainage and sewerage system in Australia, and good postal, telegraph, and telephone communications. The widest and most important thoroughfare in Adelaide was King William's Street, with Victoria Square and the Supreme Court at its centre and with elegant government buildings and banks lining its sides. The two most prominent buildings, each with an impressive tower announcing its importance, were the General Post Office and the Adelaide Town Hall. The city had commercial and entertainment areas to its north and just a little slum-like housing in the West End; otherwise there were many solid houses and empty blocks of land. North Terrace was

Fig. 5.1 King William Street, Adelaide, with the towers of the Town Hall (L) and General Post Office (R) but without the horse-tram tracks and telegraph and telephone wires that soon intruded, *circa* 1880. (From an original print in the possession of the author.)

notable for its public and cultural buildings. Spacious suburbs grew beyond the Park Lands, assisted by a developing system of trains and horse-trams.

By the 1850s South Australia had achieved formal separation of church and state, responsible self-government with triennial parliaments, manhood suffrage and the secret ballot, and had introduced the Torrens title system to simplify land transactions. The Education Act of 1875 required compulsory attendance at government primary schools for most children, and education would become free, compulsory, and non-religious by the end of the century. The Advanced School for Girls and the Agricultural School were the only public secondary schools; otherwise secondary education was in the hands of private, denominational colleges. A university had been established in 1874, taking its first students in 1876.

In the country, wheat and grazing prospered and provided agricultural self-sufficiency, although inadequate rainfall and degradation of the soil soon became serious problems. Mining was promising, particularly of the rich copper deposits discovered north of Adelaide and extracted by immigrant Cornishmen, and the earliest vineyards were planted by English and German settlers. Wages were high because of a shortage of labour, and food was relatively cheap.[2] The Englishman Anthony Trollope visited Australia in 1871–72 and was pleased by what he saw. In his *Australia* he devoted seven chapters to South Australia, noting that 'South Australia has a peculiar history of its

[2] Ibid., these paragraphs are based on the early chapters of Gibbs; see also D L Johnson, 'The Kingston/Light plan of Adelaide and founding the city', *Journal of the Historical Society of South Australia*, 2004, 32:5–18.

own, differing very much from those of the other Australian colonies', and that 'Adelaide is a pleasant, prosperous town, standing on a fertile plain'.[3]

There was much for Professor William Bragg to discover when, just fifty years after the first settlers had come ashore at the same spot, he arrived at Glenelg, the seaside village named after Lord Glenelg of the Colonial Office, London. William wrote two accounts of his voyage; one as a series of letters to Uncle William, who then 'made a journal of it all and sent it to the Barrovian, the KWC Magazine' (italics below),[4] and the other as part of his autobiographical notes (plain text):[5]

> Now to go back to my start for Australia in January 1886. The boat was the Rome, then the largest boat in the P&O fleet. *She is commanded by Capt. Cates, the commodore of the fleet. One thing in her arrangements which struck me as strange was that the best cabin, and the accommodation for the first-class passengers, is forward of mid-ships, indeed quite into the bow. Here, by Sir Arthur Blyth's advice, a good cabin, 10 feet by 8, beautifully furnished, and with every accommodation commensurate with space, was allotted to me, all to myself too. It was explained to me that the ship's rapid progress caused a draught from bow to stern, so that in warm climates I would have first use of the air; that the smoke from the funnels and the vibration of the screw were less offensive forward.*
>
> It was a great adventure to me of course, it was a new life, and I was my own master, all by myself, and I enjoyed it to the full. There were nice people on board. One of them... was particularly kind to me. I had a single berth cabin away up in the bows, and the Bay of Biscay lifted me up and down a degree or twenty feet each pitch of the ship: so it felt anyway. I was ill and Rendall fetched me out and took me to a spare berth in his own cabin amidships. *I had a very fair night. In the morning the ship was rolling tremendously. Of course I could not get up for breakfast. I lay all that Saturday morning just comfortably miserable. In the afternoon I suddenly resolved to get up. To my surprise I walked the deck for two or three hours feeling quite well... I retired for the night, which was passed after a fairly comfortable miserable fashion: but I rose on the morning breaking, went on deck, amused myself somehow, and was never sick again.*
>
> Then on Monday morning (we had left on Friday) we were all a cheery crowd, sitting on deck in the sunshine, watching the coast of Spain go by. We saw Gibraltar from the sea. We landed at Malta and toured the island and dined on shore. *The island is simply a rock. Here and there a scanty covering of soil, on which a carpet of light green verdure shows off by contrast the natural bareness in alternating large grey patches.*

[3] P D Edwards and R B Joyce (eds), *Anthony Trollope: Australia* (Brisbane: University of Queensland Press, 1967), pp. 621, 636.

[4] W H Bragg, Autobiographical notes, p. 30; *The Barrovian*, April 1886, pp. 37–40; July 1886, pp. 82–8; December 1886, pp. 123–8; April 1887, pp. 28–33.

[5] W H Bragg, Autobiographical notes, pp. 29–31. The 'Rendall' mentioned was Charles Rendall, who became a prominent Melbourne schoolmaster, teaching at Scotch College and then opening Haileybury College, named after his own English school, which thrived and is now well known.

People are out of doors everywhere... with soft handsome Italian faces, women with great black cloaks over head and shoulders, and priests, lots of them, with black garments and shovel-hats.

We saw Pt Said and all the clouds of pink-lined flamingoes in Lake Menzalah [Manzala]. *It was a queer sight this coaling at Port Said... One plank led from the barge through a hole in our ship's side in connection with a second return plank from the ship. A string of the Arabs with baskets full of coal on their heads was for ever toiling up one plank; another string with empty baskets under their arm running down the other... The Canal is a wide ditch cut in the sand... All this is very monotonous in description; it is monotonous in experience. Not a tree! Sand, sand everywhere...*

I propose to describe a day of my life on board ship. Well, I begin by being roused by the barber in a very sleepy and lethargic state about 6.30 or 7 a.m. Shaving over, and the man of beard gone, I begin to wonder when I shall start for the baths... I get into the bath; very good ones they are; sea water without limit, of course, and a little fresh to pour over your head... Coffee, tea, porridge, and six or seven hot dishes await you... My seat at the table is near that of the Captain, my immediate society is that of two elderly gentlemen, and we four frequently sit out all the others... After breakfast we go on deck and read or promenade. I have managed to do some work, but we are all very lazy.

I tried to learn some physics on the way out: I was professor of mathematics *and* physics. Although I had never done any of the latter, nor worked at the Cavendish except for a couple of terms after I had taken my degree, it was supposed by the electors that I would probably pick up enough as I went along to perform my duties at the Adelaide University. So I read some 'Deschanel's Electricity and Magnetism'.[6]

Lunch, or tiffin, comes off at 1, and we get all kinds of cold meat etc. The afternoon is a repetition of the morning. At 4 tea parties are formed in the saloon... At 5.30 it is then custom to retire to dress for dinner... Six o'clock dinner consists of about seven courses: vegetable always forms one; another is curry; dessert and coffee are always on the table. We have, after dinner, a concert or dancing every other evening... Sunday morning church parade is a great institution. All the officers, men and Lascars appear in their best.[7] White uniforms are everywhere. A Lascar's love of colour comes out in his turban and his sash: red, green, yellow etc. The Seediboys (stokers and firemen from Africa) come out in gorgeous waistcoats over their white robes, and wear tremendous turbans. So much did we get accustomed to tropical heat that, after leaving Colombo, we felt the wind cold at 72°F... The wind was right ahead and blowing very hard as we rounded Cape Leeuwin. Land was sighted at 11 a.m. on February 24th, and it was with interest I took my first look at Australia, as seen in cliffs, gum trees, and dried-up grass, showing in black and white patches on the slopes.

[6] While he may have read the original French edition, it seems much more likely that Bragg had a copy of J D Everett's translation, *Elementary Treatise on Natural Philosophy by A Privat Deschanel* (London: Blackie, 1873–), Part III: Electricity and Magnetism.

[7] Lascars were Asian, African, and other foreign seamen serving on British ships.

> The six weeks voyage came to an end and I was landed by tender at Glenelg [on Saturday 27 February 1886]. *I was standing on the gangway, when I was asked to point out—myself… the author introduced himself as a newspaper correspondent… I then, for the first time, underwent the process of being interviewed, and I trust my replies were such as to give satisfaction to the public, who are supposed to be interested in what I have to say.*
>
> It was late at night, and the Registrar took me to stay at the hotel because the last train for Adelaide had gone… He thought we might go and see 'Old Jack Morgan'—I think the name is a sufficient description. We went up to his bedroom, and my first welcome to Australia was to help myself out of the whiskey bottle which I should find in a corner.
>
> Next day was a Sunday: I had got a room at the York Hotel, an old fashioned sort of place, comfortable enough: it was very hot and the mosquito curtains were necessary. Curiously enough I never used them but there. A friendly doctor, Lendon, called for me and took me on his round in his Victoria. We called at Dr Way's and I was refreshed with green figs, lovely I thought. We went to supper at the Observatory, and I met the Todd family for the first time. Such a jolly lot they were! Mrs Todd made the household, of course. I was much impressed by the calm statement that she did not think she could go to the Government House party because she had not a dress fit for it. Such open and unconventional a confession was a surprise to me. I was marvellously fortunate in being thrown into a society of the Todds and people like them, so open and kind and good-natured. The whole thing, the going to Australia to a new work and an assured position, the people I met there, the sunshine and fruit and flowers, was a marvellous change for me. I know that I had been lucky enough in England, but I am not ungrateful when I say that going to Australia was like sunshine and fresh invigorating air.

Some captains, arriving near Adelaide and wishing for a fast turn-around of their ship, ignored the better facilities at Port Adelaide and simply anchored offshore. Passengers, mail, and cargo were loaded into small boats and taken to the pier that jutted out into the sea from the sandy beach, and then transported to Adelaide by steam train. William's first night was spent at the Pier Hotel on the beachfront. There has been a hotel on this corner ever since 1856, and the present tram track still follows the old railway line along Jetty Road, through the Glenelg shops, and on to King William Street and Victoria Square in the centre of Adelaide.

William's long journey had provided a useful transition from midwinter Britain to the torrid southern summer in the heat and openness of the Adelaide plain. Indeed, during his first week in Adelaide and in a typical burst of hot weather between two 'cool' changes, maximum daily temperatures ranged from 74° up to 99° and back down to 76°F (23, 37, 24°C), while the corresponding overnight minima were 56°, 79° and 63°F (13, 26, 17°C).[8] William stayed briefly at the York Hotel, just five-minutes' walk from the university.[9]

[8] *Register*, 6 March 1886, p. 4.
[9] M Burden, *Lost Adelaide: A Photographic Record* (Melbourne: OUP, 1983), p. 215.

That first Sunday, Alfred Lendon took William on his medical round in his horse-drawn carriage. It was one of the most important days in the twenty-three years William spent in Australia. Lendon was born in England and educated at King's and University Colleges, London (MD, MRCS), and at twenty-eight was just a little older than William and had arrived just a little earlier (1883). He was also a bachelor and would soon be appointed to the staff of the university as lecturer in forensic medicine. Lendon was remembered later for his long service to the Adelaide Children's Hospital and nursing, for his many executive and honorary medical positions, and for his writings on medical and other topics.[10] He occupied a two-storey terrace house on North Terrace, directly opposite the university, where he lived and conducted a joint practice with Dr Davies Thomas. William gladly accepted Lendon's offer to board there, moved in a few days later, and stayed until his marriage in 1889. Lendon would be best man at the wedding, and William would later be godfather to Lendon's elder son. William had found a kindred spirit on his first full day in Australia.

After a few medical calls the couple arrived at Dr Way's and were 'refreshed with green figs, lovely I thought'. Edward Way was also a physician with accommodation and rooms on North Terrace, and in 1887 he would join the university as lecturer in obstetrics and diseases of women.[11] He was part of the very prominent Way family. Rev. James Way had been sent to South Australia by the Bible Christian wing of the Methodist Church and in 1850 had become the first Superintendent of its Adelaide District.[12] There were two boys and three girls in the family that Way and his wife successively brought to South Australia: Edward was the younger son, Samuel the elder. Substantially self-taught, Rt. Hon. Sir Samuel Way would be, for many years, Chief Justice of South Australia, Vice-Chancellor and then Chancellor of the University of Adelaide, and Lieutenant-Governor of the State.[13]

Finally on this momentous day, Lendon and Bragg trotted across the central city to a neat clump of buildings in the West Park Lands, one of the few establishments to be allowed on this precious ground. Here the ample two-storey home of Charles Todd and his family looked out over West Terrace, with observatory and telegraph buildings dotted about, and here the two young bachelors had been invited for supper. Todd was genial and friendly, and he possessed a bountiful sense of humour that revelled in puns, spoonerisms, and riddles: 'without my Tea I would be odd', he said constantly. Todd was an accomplished astronomer and physical scientist, the only one in the colony outside

[10] W A Verco and F S Hone, 'Obituary: Alfred Austin Lendon', *Medical Journal of Australia*, 26 October 1935, pp. 607–9; A A Lendon, *Clinical Lectures on Hydatid Disease of the Lungs* (London: Bailliere, Tindall and Cox, 1902); papers of Dr A A Lendon, State Library of South Australia, Adelaide, PRG 128.

[11] V A Edgeloe, *The Medical School of The University of Adelaide: A Brief History from an Administrative Viewpoint* (Adelaide: University Medical Faculty and Alumni Association Medical Chapter, 1991), p. 6.

[12] A D Hunt, 'The Bible Christians in South Australia', *Journal of the Historical Society of South Australia*, 1982, 10:15–31.

[13] A J Hannan, *The Life of Chief Justice Way* (Sydney: Angus and Robertson, 1960).

Fig. 5.2 Charles Todd, Adelaide, 1886. (Courtesy: Mrs E Wells.)

the university. He and William would find pleasure in each other's company and together they would later pioneer radio in Australia. Alice Todd, his wife, impressed William at once: so 'open and unconventional'. She and her husband both have memorials associated with the Overland Telegraph Line: the wide but usually dry riverbed through central Australia is named the Todd River, and the springs of clear water that were discovered there during construction were named Alice Springs in his wife's honour (see Figure 5.2).[14]

What William does not mention, but what we can guess caught his eye, were the other members of the Todd family. The two sons, Charles Edward ('Charlie') and Hedley Lawrence, were in their twenties, beginning medical and business careers respectively. In the future they would both consult the young professor on the new X-rays and on the electrification of the city. But most of all there were

[14] G M Caroe, *William Henry Bragg 1862–1942: Man and Scientist* (Cambridge: CUP, 1978), ch. 3.

Fig. 5.3 Gwendoline Todd, Adelaide, 1885, just a few months before she met William Bragg. (Courtesy: Mrs E Wells.)

four daughters: Elizabeth ('Lizzie', about 30 years old), Maude (22), Gwendoline ('Gwen' or 'Nina', only 16), and Lorna (about 10). Their irrepressible chatter delighted William most. It was a revelation to a man previously taught to weigh every word he uttered. After Market Harborough his life had been devoid of female companionship. Both King William's College and Cambridge University were male-dominated institutions, and although Cambridge had some female students, they were not members of the university and there is no evidence that William formed even a casual female friendship there. Now, suddenly, he was in the midst of a family whose members were an integral part of Adelaide young society, prominent at balls and parties, and full of fun and spontaneity. They nicknamed him 'The Fressor'.[15] As the two men rode back to the York Hotel, William's head was surely spinning; it had been an incredible day (see Figure 5.3).

[15] Ibid.

The following day Professor Bragg walked to the university on North Terrace and entered the front door of its main building. The north side of North Terrace housed Adelaide's cultural heart. Starting in the west and proceeding eastwards one passed the railway station, the original Legislative Council Chambers in brick and stone, and the construction site of a new Parliament House in marble. Crossing King William Road one then met Government House, modest but set in beautiful grounds, the Palladian South Australian Institute building, the Romanesque Public Library, Museum and Art Gallery, and then the University. Further on were the new Jubilee Exhibition Building, vacant land, Frome Road, the Adelaide Hospital, and finally the Botanic Gardens, small but attractive. My generation walked it many times during the 1950s: the railway station, for trains not gambling as now, the library, museum and art gallery for their treasures, the exhibition building (before it was demolished) to sit for public examinations, the botanic gardens for family gatherings, and the university, when it still had its post-WWII optimism.[16]

The universities in Australia were founded under a wide variety of circumstances, Sydney first in 1850, then Melbourne in 1853, Adelaide in 1874, and others later. No foundation was stranger than that of the University of Adelaide. Early in 1872 the Baptist, Congregational, and Presbyterian churches combined to establish Union College, primarily to train young men for the ministry but also to educate lay students. Rev. James Jefferis played a major role in devising the curriculum, which in the first two years of general study included classics, English literature, philosophy, mathematics, and natural sciences. It seemed likely that the copper magnate, Walter Hughes, would donate the money required to secure the college financially, but when he and Jefferis conferred they decided to suggest that Hughes' gift of £20,000 be used to establish a university. In turn, the Union College Council, the other religious denominations, and finally the colonial government, after long negotiations and a lengthy parliamentary debate, agreed. When the legislation received the Governor's assent on 6 November 1874 it was accompanied by another donation of a matching amount from the wealthy pastoralist Thomas Elder and by assurances of modest government financial support. The new university took its first students in March 1876, and in 1880 legislation passed the South Australian parliament to allow the university to confer degrees on women and to grant degrees in science. Late in 1885 Edith Dornwell became the university's first woman graduate and its first graduate in science.[17]

In the earliest years of the university, William's predecessor, Horace Lamb, voluntarily taught physics as well as mathematics, purchased an extensive suite of apparatus to illustrate his lectures, and sought a laboratory where his

[16] A separate art gallery was built in 1900; the main railway station building has been converted into a casino.

[17] W G K Duncan and R A Leonard, *The University of Adelaide* (Adelaide: Rigby, 1973), ch. 1; W Phillips, *James Jefferis: Prophet of Federation* (Melbourne: Australian Scholarly Publishing, 1993), pp. 92–106. Financially poor, Union College reverted to its original intention of training men for the ministry but lasted only a few more years.

students could repeat and extend the experiments.[18] The temporary accommodation was inadequate and the University Council quickly developed plans for its own building on space granted to it by the colonial government. The land was not ideal, but there were compensations in terms of its location and the vacant land that sloped down towards the River Torrens from the North Terrace frontage. The process by which the Gothic Revival building was designed and built, involving Vice-Chancellor Samuel Way, is not altogether transparent. However, by the time William arrived there was a spacious office for him, as well as a nearby group of four interconnecting rooms, ideally planned by Lamb for teaching mathematics and physics. Both the tiered lecture room, which followed the slope of the land, and the stabilized physical laboratory, with its adjoining optical room offering window access to the sunlight, had easy access to an apparatus storage and preparation room between them. They shared the ground floor with the other sciences, while the library and the arts teaching spaces were on the first floor. A rear basement was unoccupied but would later be ideal for storage, a mechanical workshop, and a research laboratory.[19] This was the period when systematic laboratory training was coming to be seen for the first time as essential for future physicists. Adelaide already had space and facilities, thanks to Lamb, and William had been trained by one of the practitioners of the new scheme, Richard Glazebrook. In comparison with circumstances elsewhere Adelaide and Bragg were especially fortunate, although William seems to have been unaware of how lucky he was to have been preceded by Lamb and the facilities he had acquired.[20]

William was immediately brought face-to-face with an institution still struggling to establish itself and with the magnitude of his new responsibilities. An account of the University of Melbourne also applies to Adelaide at this time: 'The basic weakness of the university was neither shortage of money nor conservatism of thought [although they were significant], but rather a shortage of students who wanted to study and who could afford to study. The university capped the pyramid of education, but the base of that pyramid was weak'.[21] Secondary education was widely available only to those families that could afford the fees of the private colleges. In addition, even in these schools the number of students pursuing their study to the top of the school and satisfying the matriculation examination with the intention of tertiary study was very small. In 1883 Richard Twopeny commented, 'As for Adelaide University, it is bound either to federate with Melbourne...or to drag on in extravagant

[18] J G Jenkin and R W Home, 'Horace Lamb and early physics teaching in Australia', *Historical Records of Australian Science*, 1995, 10:349–80 and references therein.

[19] Ibid.

[20] The literature on early physics laboratories is extensive; see, for example, R Sviedrys, 'The rise of physics laboratories in Britain', *Historical Studies in the Physical Sciences,* 1976, 7:405–36; G Gooday, 'Precision measurement and the genesis of physics teaching laboratories in Victorian Britain', *British Journal of the History of Science*, 1990, 23:25–51; F A J L James (ed.), *The Development of the Laboratory* (London: Macmillan, 1989).

[21] G Blainey, *A Centenary History of the University of Melbourne* (Melbourne: Melbourne University Press, 1957), p. 24.

grandeur. In five years of existence it has conferred five degrees at a cost of £50,000'.[22] Even in 1886 the numbers of matriculated students beginning degrees that might have included mathematics or physics were: one for BSc, ten for BA, and three for combined BA and BSc.[23]

In Australia, bachelor degrees in arts and science are normally of three years duration, while honours requires one further year of study. Initially William was responsible for the teaching in pure mathematics, applied mathematics, physics and practical physics at *all* levels. Much of the secondary-school public examining in these subjects—setting the papers and marking the students' answers—was also his responsibility. Fortunately, in his first year there were no third-year or honours students in mathematics or physics. Thereafter William devised arrangements whereby the small number of students could be combined into a smaller number of classes; for example, second- and third-year physics students in one class, with the content alternating from one year to the next. In addition there were forty-eight evening lectures to be given to a class of ten students in mathematics: men and women who were employed during the day and who sought to further their education in the evenings. And as if this load was not enough, William also gave lectures on acoustics to second-year music students. This course, in which he took a particular delight, was based upon his studies with Glazebrook, and he filled his lectures with demonstrations and analogies. Of all the lectures he gave in Adelaide that first year, he kept only his notes on acoustics.[24] As for examinations, by the end of 1886 William had set and marked twenty-nine major papers: seven in March just days after his arrival (for supplementary examinations), ten mathematics and physics papers for BA and BSc students near the end of the academic year in November, three scholarship papers, and nine papers for the Junior and Matriculation public examinations in December. One can readily picture the long hours he spent pouring over Lamb's syllabuses and previous examination papers.

During this first year William also wrote to the Council of the University (its senior governing body) on three occasions: first to ask for lengths of rubber tubing for the Physical Laboratory, second to point out that 'in the mathematical lecture room there are no desks or tables on which students may take notes during lecture', and third to request the purchase of seventeen books for the library.[25] Later the same year he had returned only six of the forty-seven textbooks he had borrowed from the library.[26] Preparation for his lectures and other basic matters of teaching filled his waking hours. In October he was elected Dean (and Chairman) of the Professorial Board.

[22] R E N Twopeny, *Town Life in Australia* (Harmondsworth: Penguin Colonial Facsimilies, 1973), p. 145 (original edition 1883).

[23] Such details have been gleaned primarily from the annual *Adelaide University Calendar* (later the *Calendar of The University of Adelaide*).

[24] W H Bragg, Adelaide lecture notes, RI MS WHB 31A.

[25] Letters W H Bragg to Council, 14 May, 16 November, and 15 December 1886, UAA, S200, dockets 171/1886, 447/1886, and 506/1886 respectively.

[26] List in possession of Barr Smith Library, University of Adelaide.

Some perspective on this extraordinary introduction to academic life can be obtained by comparing it with that of Richard Threlfall at the University of Sydney. Both Bragg and Threlfall were chosen by English selection committees and both arrived in Australia to chairs of physics in 1886. Threlfall was twenty-five years old, Bragg twenty-three. Their personalities were quite different, however. Threlfall was gregarious and outgoing, with a strong sense of self-confidence and self-sufficiency. Bragg was humble, private, and self-contained. At Cambridge Bragg had followed the traditional path through the Mathematical Tripos, while Threlfall had enrolled in the less fashionable Natural Sciences Tripos, had achieved first-class honours in both physics and chemistry, and had studied mathematics privately. As an undergraduate Threlfall was prominent in student life as a persuasive orator and a powerful rugby player. After graduation he gained valuable teaching experience as a demonstrator in the Cavendish Laboratory and a college lecturer in physics. His superb laboratory skills found expression in research projects, where J J Thomson rated him 'one of the best experimenters I ever met'.[27]

While Bragg was struggling to find his feet, Threlfall was already off and running. Before leaving England and unknown to the Sydney University Senate, he had purchased £2,400-worth of apparatus and engaged the services of a craftsman, who managed the laboratory efficiently and who had great workshop skills. Threlfall successfully petitioned the university for the erection of an elaborate new Physical Laboratory building. A year later he persuaded the Senate to give him a demonstrator, and he also began a varied program of research. As a result of moving more and more into industrial work, however, Threlfall did not subsequently participate in the development of the 'new' physics and, after his return to England in 1898, he devoted himself to industrial studies.[28] Bragg matured much more slowly, both personally and scientifically. Indeed, the remoteness of Adelaide would be crucial to his development, for it allowed him, slowly and methodically, to mature, to become self-confident and self-reliant in teaching, research, and community relations, and to develop his own ideas and programmes. Threlfall may have said many years later—albeit light-heartedly and without genuine regret—that going to Sydney was 'the greatest mistake of my life';[29] for Bragg 'going to Australia was like sunshine and fresh invigorating air'. The nature of the experience of working at a colonial outpost depended very much on the person concerned.

Still in that first year in Australia, William found time for recreational activities. The game of lacrosse was first introduced in South Australia in 1885, and in the winter of 1886 William joined his Cambridge team-mate, Talbot Smith, in the Adelaide team. *The Adelaide Observer* reported that William rapidly

[27] R W Home, 'First physicist in Australia: Richard Threlfall at the University of Sydney, 1886–1898', *Historical Records of Australian Science*, 1986, 6:333–57, and reference therein.

[28] Ibid.

[29] Recalled in Sir Harold Hartley, 'Sir Richard Threlfall, GBE, FRS', typescript in University of Sydney Archives, Threlfall papers, p. 3.

Fig. 5.4 Adelaide lacrosse team at its Victoria Park Racecourse ground, William Bragg standing third from left, mid-1886. (Courtesy: State Library of South Australia, SLSA: B 9516.)

established himself as 'without doubt, the finest all-round player we have'.[30] In future years he would be the central figure in the expansion of the competition, while in the summer there were games of tennis on the university court and elsewhere.[31] In October William took the male lead in a comic drama in two acts entitled 'The Jacobite', presented in the Torrens Park Theatre, a magnificent auditorium built by Robert and Joanna Barr Smith at their mansion at Mitcham, in the Adelaide foothills. Barr Smith's company, in partnership with Thomas Elder, had pioneered much pastoral settlement in South Australia and he was a rich man. His philanthropy became legendary, the university not the least of his beneficiaries. Mr & Mrs Barr Smith were lavish and charming hosts and the theatre became the venue for countless entertainments.[32] William's participation in at least one of these is a reminder of his love of theatricals, an indication of his immediate acceptance into the highest level of Adelaide society, and a crucial pointer to his future. Seventeen years later Barr Smith would

[30] J G Jenkin, 'William Bragg and lacrosse in Adelaide, 1885–1895', *The Australian Physicist*, 1980, 17:75–8, 76 (contact the author for printing errors in this paper). Men's lacrosse is a team sport played on a large outdoor field by two teams of ten players, each of whom uses a netted stick (the crosse) to pass and catch a rubber ball in order to score by throwing the ball into the opponent's netted goal. The game was invented by Native North Americans and developed by the French in Canada, where it is now the national summer sport (*Wikipedia* Internet encyclopedia).

[31] J G Jenkin, 'William Bragg in Adelaide: tennis too!', *The Australian Physicist*, 1981, 18:69–70, 131.

[32] J Brown and B Mullins, *Town Life in Pioneer South Australia* (Adelaide: Rigby, 1980), pp. 174–86; J G Jenkin, *The Bragg Family in Adelaide: A Pictorial Celebration* (Adelaide: Adelaide University Foundation, 1986), pp. 14–15; *Scotch College Magazine*, Scotch College, Adelaide, 1981.

provide the money with which William purchased his first radium sample and thereby began his extraordinary research career (see Figure 5.4).[33]

Socially William was also busy. Lendon recalled that he and William attended a Mayoral Ball, entertained freely on the balcony of their house, and together formed a whist club that met regularly at their home on Saturday evenings.[34] There are also the diaries of Chief Justice Way, still single at age fifty but living in his splendid mansion 'Montefiore', which contained large dining and reception rooms, a substantial library, and a large art collection. One of the city's best-known houses, Way used it extensively for entertaining, for which he was renowned. On 2 March 1886, three days after William's arrival and after his own full day in court, Way entertained Professor Bragg to dinner, with his brother (Edward Way), the University Registrar (Walter Tyas), the Professor of Chemistry (Edward Rennie), and three others. Way invited William on four other occasions in March, and then, on 30 May, he hosted a luncheon party for ten, including 'Mrs F. and two daughters, Drs Lendon, Giles & Todd, Miss Todd & Prof. Bragg'.[35] William Bragg and Gwendoline Todd were out together, officially, under the watchful eye of her brother.

There are two undated letters from 'Gwendoline Todd' to 'Professor Bragg', written in such a young hand compared with her later style that they must date from this time. 'Thank you so much for those lovely flowers you brought me; it is almost worth having the whooping cough to get them', she wrote in the first; and in the second, 'Thank you so much for the lovely pair of gloves you sent me. It was awfully good of you to remember my birthday. Although "my manner of spelling is far past all telling" I hope you won't find many mistakes in this letter, & if you do I hope you won't send it back corrected'.[36] William was entering new and uncharted territory, but so too was Gwen, socially mature beyond her years but a little concerned about her lack of formal education in comparison with her professorial beau.

What a year it had been! William was so busy that he hardly noticed the world around him, but he could not have missed the severe downturn in the local economy. The newspapers were full of it.[37] Ron Gibbs has called South Australia in 1886–87 'a colony in crisis'.[38] First there had been a series of bad years, particularly for farmers who had pushed too far into regions where average rainfall was too low to sustain wheat farming. When drought and rust arrived many settlers simply walked off their land. Declining copper prices depressed even the largest mines. In Adelaide 'a fire of speculation had consumed the attention of city interests. The fuel was cheap money'. On 24

[33] See chapter 11.

[34] *Autobiography , Dr A A Lendon*, State Library of South Australia, Adelaide, PRG 128/13/1, pp. 30–4.

[35] *Diaries, Samuel James Way*, State Library of South Australia, Adelaide, PRG 30/1.

[36] Bragg (Adrian) papers; William's daughter agrees with the likely dates, n. 14, p. 34.

[37] It is clear from William's collection of newspaper cuttings that he read the Adelaide newspapers very regularly, see RI MS WHB 39.

[38] R M Gibbs, 'A colony in crisis: South Australia 1886–1887', address to the Historical Society of South Australia, 13 October 1989, text in possession of the author.

February 1886 the Commercial Bank of South Australia failed and closed its doors; 'Embezzlement became a commonplace in the colony's general crisis', which 'continued against an unhappy background of unemployment and argument about its origins'. Then late in the season good rains fell, reviving dying crops and watering pastoral districts. Finally gold was discovered on the Teetulpa run. It sparked a rush and, although the gold quickly ran out, it boosted confidence and the economy gradually recovered.[39] For William it would mean that government and university funds would be tight. It has always been so in South Australia.

In January 1887, during the long summer vacation, William visited Melbourne and Sydney. Young, moneyed, and energetic, he was keen to explore his new homeland. In company with Chief Justice Way and Professor Rennie, he travelled the 500 miles by train to Melbourne, where he was able to use a letter of introduction from Edward Routh to Edward Nanson, the Melbourne Professor of Mathematics. Nanson was a Cambridge Mathematical Tripos graduate (Second Wrangler and Second Smith's Prizeman, 1873), a gifted pure mathematician but 'intellectually lonely in his theoretical interests'. He was delighted to hear recent news of Cambridge and its mathematics.[40] William then sailed to Sydney by steamer and was welcomed by Richard Threlfall.[41] Two months earlier, after several years of discussion, a preliminary meeting to establish an Australasian Association for the Advancement of Science (AAAS) had been held in Sydney, of which William was anxious to hear a first-hand report. It was to be based very heavily on the British Association. In the years ahead the regular meetings of the AAAS would provide him with invaluable opportunities to gather important information and to grow both professionally and personally.[42] William also took the opportunity to discuss apparatus with Threlfall: what could be purchased and what would have to be made, and therefore what workshop facilities he would need. Since Adelaide had little money to buy apparatus and neither workshop nor mechanic, William would have to become proficient in yet another skill.

On his return to Adelaide he wrote to the University Council requesting the following apparatus for the physical and chemical laboratories: a chronometer at £35 ('there is at present no apparatus whatever for the measurement of time'), a heliostat at £5 ('there are good optical instruments in the laboratory, and there is plenty of sunshine, but without a proper heliostat the sun's rays can not be brought

[39] Ibid.

[40] *Register*, 4 January 1887, p. 5; letter E Routh to W H Bragg, 22 December 1885, Bragg (Adrian) papers; G C Fendley, 'Nanson, Edward John (1850–1936)', in B Nairn and G Serle (eds), *Australian Dictionary of Biography* (Melbourne: MUP, 1986), vol. 10, p. 663.

[41] Letter W H Bragg to G Bragg, 5 January 1890, Bragg (Adrian) papers, mentions this earlier trip.

[42] For the foundation of the AAAS, see H C Russell, 'President's Address', in A Liversidge and R Etheridge (eds), *Report of the First Meeting of the Australasian Association for the Advancement of Science, held at Sydney, 1888* (Sydney: AAAS, 1889), pp. 1–21; for the importance of the AAAS to Bragg, see R W Home, 'The problem of intellectual isolation in scientific life: W H Bragg and the Australian scientific community 1886–1909', *Historical Records of Australian Science*, 1984, 6:19–30.

to bear on the instruments'), a lathe at £25 ('with a lathe, simple apparatus can be constructed and repairs effected'), bellows with table for glass-blowing at £5, and mercury distillation equipment at £5, total £75.[43] After the request was referred to the Science Faculty and the University Finance Committee and further enquiries were made as to alternatives, William's original request was granted in full.[44]

As to William's use of the new lathe, we have his son's recollection: 'He apprenticed himself to a firm of instrument makers, learnt to work the lathe, and made his class apparatus. I think that this experience must have given him his love of apparatus and his skill in designing it, at which he was supreme'.[45] The firm concerned was Edwin Sawtell's optical and watch-making business, which had premises at Port Adelaide and in Rundle Street, the principal shopping street in the city and close to the university. Edwin Sawtell was himself an expert craftsman, and his business manufactured, repaired, and was an agent for numerous items of equipment—nautical, optical, meteorological, and surveying—as well as watches, spectacles, charts, and books. It was here, in the firm's comprehensive workshop, that William met Arthur Rogers, the foreman instrument-maker whom he would soon lure to the university (see Figure 5.5).[46]

Uncle William wrote regularly to his nephew from England, encouraging him ('Your claim for money for your department is a good sign'), warning against over-exertion ('Don't overwork yourself; 10 or 11 hours is too much strain... 8 is quite enough in your climate too'), and offering extensive financial advice ('I wish you a continuance of the prosperity unfolding in your first balance-sheet'). William's surviving brother, Jimmy, took his inheritance from his father's estate in 1888 but William left his share with Uncle William, who husbanded it and continued to advise his nephew on financial matters.[47]

It was liberating to 'break away from being a subject to be now my own master', but during his second year in Adelaide (1887) William came to realize that his load was unsustainable. In July he recorded the details of his weekly teaching commitments: in the three undergraduate years, seven hours of mathematics and ten hours of physics and practical physics, six hours of evening classes and one hour for music students, a total of twenty-four hours per week. For the academic year of twenty-eight weeks, this amounted to 672 contact hours per year, 168 in the evenings.[48] Even by the standards of the day this was an extraordinary load, made all the more remarkable by the fact that William did not have a single academic colleague to assist with student difficulties or the twenty-one university examinations

[43] Letter W H Bragg to University Council, 24 February 1887, UAA, S200, docket 80/1887. In the absence of electric light, a heliostat was a device that, when placed close to a window, produced a narrow beam of sunlight in a constant direction by following the sun mechanically and reflecting its light into the room. Lamb had purchased a simple heliostat that was apparently unsatisfactory or lost (see n. 18).

[44] Council Minutes, meeting of 29 April 1887, UAA, S18, vol. IV, p. 140.

[45] Sir Lawrence Bragg, 'William Henry Bragg', *New Scientist*, 1960, 7:718–20, 718.

[46] 'Edwin Sawtell', in G E Loyan, *Notable South Australians* (Adelaide: author, 1885), pp. 223–4; *Sands & McDougall's South Australian Directory* annually.

[47] Letters W B Bragg to W H Bragg, 13 April 1887 and 5 May 1887, Bragg (Adrian) papers.

[48] Letter W H Bragg to University Council, 27 July 1887, UAA, S200, docket 290/1887.

Fig. 5.5 Arthur Rogers, photo celebrating his engagement to Anita Sheeran, Christmas 1894. (Courtesy: Mr E J Rogers.)

involved, and only one part-time laboratory assistant to help build, prepare, and supervise the lecture demonstrations and student laboratory experiments. It was said that William was initially an unimpressive lecturer, being too careful and too mathematical.[49] That he later became renowned as a lecturer without peer may owe something to the level of practice he had during his early years in Adelaide![50]

This commitment in no way reduced his other duties. In March he set and marked seven Matriculation examination papers, and in November another seven. He was also Chairman of the Board of Examiners for all the public examinations. In mid-year he investigated the possibility and cost of introducing electric lighting in the university, and accordingly wrote to the

[49] Caroe, n. 14, p. 31.
[50] J G Jenkin, 'W H Bragg and the public image of science in Australia', *Search* (Australia), 1987, 18:34–7.

Registrar reporting his findings.[51] His plan was based upon a gas engine that would drive a dynamo that would then power lights throughout the building, directly and through batteries that had been charged at off-peak times. Initial cost was estimated at £505, with recurrent costs of about £100 per year. The matter was immediately dropped!

On 21 June the Adelaide Jubilee International Exhibition opened in a purpose-built building on North Terrace, next to the university: 'the longest running spectacle in the history of South Australia'.[52] Before it closed early in 1888 it attracted 750,000 visitors when the population of the colony was only 320,000. It was based upon the London Crystal Palace Exhibition of 1851 as well as exhibitions in Melbourne and Sydney, but was delayed by strong opposition. It only proceeded when private and commercial donations supplemented the government funding, and was held as a joint celebration of the State's fiftieth birthday and Queen Victoria's jubilee. William paid it little attention. He was a late addition to the large Education and Science Sectional Committee and a member of the small 'Jury XV: Scientific Instruments', chaired by Charles Todd. The jury made several awards and especially commended 'the large collection of... apparatus exhibited by C Todd... No award is given in this instance, as it would be in contravention of Rule IV [prohibiting self-selection]'.[53] Amongst William's many surviving award medals there is a large one, in its own box, reading, 'For Service: Adelaide Jubilee International Exhibition, MDCCCLXXXVII', with 'W H Bragg B.A.' printed in gold letters inside the box.[54]

William pleaded with the University Council: 'I beg respectfully to call your attention to the large increase in the duties which devolve upon me as Professor of Mathematics, and to my need of assistance to enable me to fulfil them satisfactorily... Next year at least one new class must be started in accordance with the University regulations. These lectures are so many that I cannot make them fit in with the lectures of the other professors... [and] the strain of so much teaching is very heavy... I would rather suggest that when it is possible an assistant lecturer in mathematics be appointed'.[55] The matter was referred to the Education Committee, which recommended in August that, 'for the sake of the students as well as Professor Bragg, it is desirable that help should be given him next year if the funds will permit'.[56] It is notable that it was the mathematics for which William sought assistance; he was enjoying the physics and was happy to continue with all its teaching, including the practical classes. Council adopted the recommendation and in December

[51] Letter W H Bragg to Registrar, undated, UAA, S200, docket 362/1887.

[52] C McKeough and N Etherington, 'Jubilee 50', *Journal of the Historical Society of South Australia*, 1984, 12:3–21.

[53] *Adelaide Jubilee International Exhibition: Reports and Lists of Awards* (Adelaide: Exhibition, 1889), pp. 9–13, 43–4.

[54] RI MS WHB 25.

[55] See n. 48.

[56] Education Committee, Report to Council, 12/1887, 19 August 1887; Council Minutes, meeting of 26 August 1887, UAA, S18, vol. IV, p. 158.

William wrote to the Chancellor, urging that it be carried out. He proposed that a salary of £300 a year be offered, £100 from the Evening Class Fund and £100 from the University chest; 'the other £100 I will provide myself for the first two years, if the Council will then relieve me of that duty'.[57] It was a generous and astute offer.

There were six excellent applicants when the position was advertised shortly thereafter: four had British degrees, one an Irish BA, and one, Robert Chapman, had recent arts (mathematics and physics) and engineering degrees from the University of Melbourne.[58] The Education Committee discussed them fully in January 1888, and 'ultimately Professor Bragg, who was about to start for Tasmania, was desired *en route* to see one candidate in Melbourne and one in Tasmania; and to report which of the two he considered the better fitted for the lectureship, the Committee to recommend the gentleman so selected to the Council for the appointment'.[59] According to Alice Thomson, William's great-granddaughter, the circumstances of William's trip to Tasmania were as follows: 'Todd obviously considered Gwenny the beauty of the family, far too young to marry... [and] decided to pack his daughter off to Tasmania to keep her out of the way of the young Englishman, with her brother Charlie as chaperone. But Alice, perhaps remembering her own engagement, had other ideas. When Professor Bragg came to dinner and explained he might like to take a holiday in Tasmania, she handed him Gwenny's forgotten blue sash and explained that he would be doing her a great service if he would deliver it to her daughter'![60]

The interviews duly took place in Melbourne. After the second interview in Tasmania William wrote to the Registrar: 'I have chosen Chapman as assistant lecturer: he knew a great deal more than the other man, was energetic and strong in appearance, whilst the other was of the scholastic, weak-eyed type. I think Chapman will do very well. By the way, he is an oarsman, has rowed 6 for Trinity against Ormond [Melbourne university colleges]. Will you please send him a Calendar as soon as it comes out? Don't address any more letters to me here [Hobart], as I am coming home'.[61] The selection process would be unacceptable today, but the Education Committee and the Council speedily adopted William's recommendation.[62] It was an inspired appointment, for Chapman remained at the university for the next fifty years, serving it with dedication and distinction in a wide variety of roles, and retired in 1937 as Emeritus Professor Sir Robert

[57] Letter W H Bragg to Chancellor, 9 December 1887, UAA, S200, docket 511/1887.

[58] Letter of application, R W Chapman to University Council, 12 January 1888, with printed list of applicants, loose in Education Committee minute book, UAA, S23, vol. II.

[59] Education Committee, Report to Council, 1/188, 17 January 1888; Council Minutes, meeting of 27 January 1888, UAA, S18, vol. IV, p. 202.

[60] A Thomson, *The Singing Line* (London: Chatto and Windus, 1999), p. 255.

[61] Letter W H Bragg to Registrar, 1 February 1888, UAA, S200, docket 60/1888; letter Chapman to W H Bragg, 12 July 1937, RI MS WHB 2C/10: 'It is now 50 years since we first met in Melbourne and you offered me the appointment as your lecturer in Adelaide'.

[62] Education Committee, Report to Council, 4/1888, 10 February 1888; Council Minutes, meeting of 24 February 1888, UAA, S18, vol. IV, p. 211.

Chapman, having been Professor of Engineering from 1907 to 1937.[63] The Council took advantage of William's offer to pay part of Chapman's salary for two years, but then assumed total responsibility for it.[64]

William was now free to enjoy a holiday in Tasmania with Gwen and Charlie Todd. He proposed to Gwen and she accepted, subject to the approval of her parents. William wrote to Lady Todd and Charlie telegraphed saying, 'Professor Bragg wants to be engaged to Gwen'. The answer came back, 'Say everything kind to both'. Gwen later told her daughter that she had not wanted to be engaged so young, 'but she knew she'd never find anyone else so good'.[65] Letters of congratulation and good wishes arrived for Gwendoline, a few of which survive. Alfred Lendon expressed feigned regret that *his* permission had not been sought, a friend thought Gwen 'a most fortunate girl', while Uncle William and Fanny Addison wrote from Market Harborough, regretting that distance 'prevents our making your immediate acquaintance'.[66]

Gwendoline Todd, just eighteen years old, had grown up 'in happy freedom, avoiding education as far as possible (or so she always said); but she had one gift that she enjoyed enormously—painting'.[67] In 1886–7 and 1887–8 she attended the School of Design and did very well: in the first year she exhibited several pieces at the Adelaide Jubilee International Exhibition, while in the second year she received first grade certificates for geometry, freehand, model, and for the examination overall.[68] In view of her excellence as a painter, particularly in watercolours, Gwendoline presumably also attended the Gallery Art School (or School of Painting) at this time.[69] On 1 March 1887 the South Australian Post Office introduced postal notes for the first time, in seven denominations between one shilling and one pound. 'The notes', Charles Todd later reported, 'were designed by my daughter, Mrs Bragg'.[70] Gwendoline's love of painting continued throughout her life and she passed on her love and skill to many members of her family. She taught William to sketch and paint and he became quite proficient. They spent happy hours together in the countryside, painting landscapes.[71]

[63] R J Bridgland, 'Chapman, Sir Robert William (1866–1942)', in B Nairn and G Serle (eds), *Australian Dictionary of Biography* (Melbourne: MUP, 1979), vol. 7, pp. 613–15.

[64] Letter W H Bragg to Council, 2 December 1889, UAA, S200, docket 496/1889, expressing gratitude for being relieved of the duty.

[65] Caroe, n. 14, p. 34.

[66] Letters A A Lendon to G Todd, 31 January 1888; [friend, name unreadable] to G Todd, 13 February 1885; W B Bragg to G Todd, 22 March 1888; F Addison to G Todd, 22 March 1888: all RI MS WHB 37A/1.

[67] G Caroe, 'Bragg, Gwendoline (1869–1929)', in R Biven (ed.) *Some Forgotten... Some Remembered: Women Artists of South Australia* (Adelaide: author, 1976), no page numbers.

[68] *Report of the Board of Governors of the Public Library, Museum and Art Gallery of South Australia for 1886–7*, South Australian Parliamentary Papers, no. 83/1887, pp. 23–5; ibid., no. 83/1888, pp. 30–7.

[69] G R LeDuff, 'Adult education and the Institute movement in South Australia, 1836–1890', unpublished M.Ed. thesis, School of Education, Flinders University of South Australia, 1980; N Cato, 'Art in South Australia', *Australian Letters*, 1960, 2:67–78.

[70] C Todd, *Report of the Post Office, Telegraph and Observatory Departments*, South Australian Parliamentary Papers, no. 128/1896, p. 31.

[71] Caroe, n. 14, pp. 39–40.

6
Consolidation and marriage

After a difficult but stabilized childhood and youth, William had rushed to apply for the Adelaide professorship. When successful, he rushed to prepare for the journey and, immediately on arrival, began teaching, for which he was under-prepared and unpractised. Now he was rushing into engagement and marriage. He was a man in a hurry, which was not his natural style. His career would show that he did best when he considered new ventures carefully and prepared for them methodically, an approach he thus came to prefer. It was not surprising, therefore, that in 1888 and after two years of frenetic change, William suffered a period of uncertainty and doubt. It revolved around his relationship with his fiancé, Gwendoline Todd.

In early March 1888, while Gwen was recovering from a bad cold and holidaying at Port Elliot, a seaside resort south of Adelaide, William had returned to Adelaide for the start of university teaching. They wrote to each other almost daily. William received congratulations from his colleagues on his engagement and noted that Chapman was settling in well. He was examining students for possible entry to the university and wrote to Gwen, 'Miss Kirby was pretty bad, but I let her pass because she might just as well have a chance... There are not many entries for Arts and Science courses so far... a good many for the Medical, including two girls... I am not worrying... I do not understand myself when I do worry: I don't know what it means or of what it is the outcome: but I know I must, and mean to, conquer it, and when I have done so I shall have bettered my nature once and for all... you seem to understand... If I have work to do, that foolish introspection is driven away'.[1] The next day he wrote again, 'I saw that I was, and have been, acting a rather ignoble and selfish part: that I had asked you to marry me, that you had consented, and that I had thereupon sat down and expected to be made happy: that my own selfishness was the cause of my trouble'.[2]

At other times William was more buoyant. He bought a paint box and was looking forward intensely to the pleasure of using it, he was playing tennis socially, and his first lecture had 'a very decent number of freshmen and

[1] W H Bragg to G Todd, 27–28 February 1888, RI MS WLB 37A/1.

[2] W H Bragg to G Todd, 28–29 March 1888, ibid. The mention of a visit by 'Sheppard' in these letters does not refer to William's fellow mathematics student at Cambridge.

freshwomen...Now goodbye for the present...I love you well. Yours Will'.[3] When he sent the box to Port Elliot, Gwen thanked him for the 'sweet little paint box', and she used all the colours except the emerald green.[4] William lobbied for the election of a local judge to the University Council, and he was a member of the university tennis team in a friendly match. He had many lectures to give: 'I take a great interest in them, for I want the University to be popular'.[5] Professor Bragg had already hitched his star to the University of Adelaide (see Figure 6.1).

But then the unease returned: 'I have had rather a fright all day with fears and fancies, but I'm not afraid, and we'll conquer them in the end...I relieve myself by writing to you'.[6] There is no evidence as to what these 'fears and fancies' were, but since they faded with time, perhaps they were simply self-doubt after two tumultuous years in Adelaide. Gwen replied with patience, love, and a maturity beyond her eighteen years: 'I am going to give you a good lecture about your little fears and worries...try not to think of your own feelings so much...God has been very good to us, dear boy; don't let us spoil our happy contented time by wondering whether it should not be brighter'; and 'You will be here tomorrow. I got your letter. I am not going to answer it but will wait until I see you and we can talk it all over'.[7] And again, 'I must just write to you a bit tonight to say how glad I am that you are better'.[8]

In August/September William attended the first meeting of the AAAS in Sydney. The things he saw there and the people he met were important for his development as a scientist and were harbingers of things to come. 'There are about 700 members now and there is not the least doubt but that the affair is going to be a great success', he reported to Gwen. 'Over 100 papers have been handed in to be read. There are lots of excursions arranged...I have put my name down for two...Do you know, I find I have a reputation for skill in manipulating soap bubbles and films...I have been asked to give advice and a helping hand to a man who is going to show them at the conversazione'.[9] This was one area of physics that William had studied with Glazebrook, and news of his expertise had apparently preceded him. The elastic surface properties of liquids and the coloured components of white light were becoming popular topics for public demonstrations of physics, since they could be easily and elegantly illustrated by the use of soap bubbles and liquid films. In the same letter William preached to Gwen about the need for her to be peaceful and understanding, but then regretted it when he resumed the letter next day: 'What an old preacher I am! Bear with me, Gwenie dear, if I talk too much'. He then continued:

[3] W H Bragg to G Todd, 15 March 1888, ibid.
[4] G Todd to W H Bragg, 18 March 1888, Bragg (Adrian) papers.
[5] W H Bragg to G Todd, 16 March 1888, RI MS WLB 37A/1.
[6] W H Bragg to G Todd, 18 March 1888, ibid.
[7] G Todd to W H Bragg, 20 and 22 March 1888, Bragg (Adrian) papers.
[8] G Todd to W H Bragg, n.d., ibid.
[9] W H Bragg to G Todd, 28–30 August 1888, RI MS WLB 37A/1.

Fig. 6.2 The young professor, William Bragg, Adelaide, *circa* 1888. (Courtesy: Mrs E Wells.)

We had our first meeting of the whole Association last night in the University Great Hall... afterwards a few of us went to Threlfall's new laboratory. Oh, Gwen, it *is* a fine place: mine looks so small to it. You know my workshop with my one little lathe? Well, he has a room full of all sorts of machinery, driven by a gas-engine, and the whirr and the wheels and the belting make you think you're in a locomotive workshop or something of the kind. Then the lecture room and the laboratories and so on: I feel like the Queen of Sheba when she went to see Solomon. I admire the knowledge and genius of the man who can plan such a place and by sheer personal force and insistence get it built. Well, I must just learn as much as I can, and I think I can get a lot of assistance. I think this Association is going to do us a lot of good, especially such as, like me, are willing to work but don't quite know where to begin. Contact with other and more experienced workers will start us off on the right track. I'll be able to tell you more after today...

> And now I've got rather a sad bit of news to tell you: I shan't be home till two days later than I expected... There is so much going on next week: and so many opportunities of meeting men will occur then that I think I ought to stay till Thursday. I find it such a great advantage to me to meet these men and for a time to live in the atmosphere they create, that I do not think it right to miss the chance given me.

Gwen replied, 'I am glad you think it will be a good help to you in your work going to this Congress... It was just the sort of help you want to nerve you on, wasn't it'.[10] 'I don't know how many times I have read your letter', William responded. 'Just now I have the thought of you, which is sweet... Our meetings have been great successes... but the great advantage of the Association is, I find, the meeting with the other men in the colonies who are engaged in work similar to mine'.[11] He then outlined the various discussions he had had. William was indeed being 'nerved' for the many tasks that lay ahead, although he was apparently too reserved to approach the leading physical scientist such as State government astronomers Robert Ellery and Henry Russell or freelance physical chemist William Sutherland.

Late in the year, under the pressures of setting and marking examinations and of preparing a commemoration (graduation ceremony) oration, William's equilibrium was again upset, particularly when Gwen and her sister Maude went to Warland's Hotel at Port Victor, south of Adelaide. Gwen was again unwell, and William paid for congenial, seaside accommodation. 'Exams are thick', he wrote, 'I need rest, with novels and tennis and painting... I'm rather grumpy about I don't exactly know what, but I'm all right... I really am very tired... I had a very quiet little happy time with the mother [Alice Todd] last night and I enjoyed myself very much... I think my emotional faculties are rather played out'.[12] Worried about the tone of this letter, William wrote again next day, but it was even more dismal and indecisive: 'I just wanted to get a little encouragement to go down [to Port Victor]. I came over here [to the Observatory] to see the mother, but she had gone... Then I went back home with hope that Dr Lendon would say I had better go down, but he was out too... and I was rather in a stew for a bit... I'd like you to put your arms around my neck and kiss me and make me well'.[13] Gwen replied, 'I can not imagine what it is that makes you get in such a stew... the only thing to do is to have patience and it will all come right... You mustn't be always trying to love me very hard... just leave it alone... it will take care of itself... Lots of love, Gwen'.[14]

William responded with local news and concluded 'I shall go round to the Observatory again: I like being there best'.[15] At the Observatory he had found

[10] G Todd to W H Bragg, n.d., Bragg (Adrian) papers.
[11] W H Bragg to G Todd, 30 August–3 September 1888, RI MS WLB 37A/1.
[12] W H Bragg to G Todd, 10 November 1888, ibid.
[13] W H Bragg to G Todd, 11 November 1888, ibid.
[14] G Todd to W H Bragg, n.d., Bragg (Adrian) papers.
[15] W H Bragg to G Todd, 13 November 1888, RI MS WLB 37A/1.

a mother figure to love, as indicated by Gwen's letter to her mother at this time: 'Thank you for talking to that young man of mine and cheering him up as only you know how, and giving him a good kiss into the bargain'.[16] The fits of depression, William felt, came when he needed to rest but work prevented it. When work was finished he was much more cheerful, thinking ahead to possible houses to rent after their marriage, sympathizing with his students doing examinations in the very hot weather, and ending 'I am much better... I think I have been ill for a long time and I am getting to be myself again'.[17] For Gwen, who handled his melancholy extraordinarily well, 'it's awful nice to get your letters just before going to bed, especially when they are such sweet, cheerful letters as they are nowadays'.[18]

Education was a topic of considerable public interest and comment in Adelaide during these years. *The South Australian Register* newspaper was typical. At a time when illustrations were used infrequently, the broadsheet newspapers were packed with text. Important issues, speeches, and opinions were reported at length, and the newspapers were feisty in their views. They could be strongly, if not always consistently, critical of governments and organizations. They were keenly aware of developments interstate and overseas, particularly in Britain, and pushed for change and innovation locally. Education was one issue that attracted their constant attention.

In 1885, the year before William's arrival, the Register sponsored a vigorous debate about education by publishing two long articles by a recent arrival in the colony: 'Competitive examinations injurious to mind and body' and 'Reforms needed in the Adelaide University'.[19] The first, regarding public examinations and particularly that for matriculation to the university, criticized the existing system saying, 'The origin of the mischief lies in the Adelaide University's curriculum being so oppressive'. The second article said: 'We pronounce the Adelaide curriculum for B.A. more unbending and making less allowance for mental idiosyncrasies than the London University itself... In the cruel process of forcing the square men into round holes, few students are there who have not been pinched by the mould... giving the most unsatisfactory and shapeless results'. The *Register* supported these views editorially: 'The schools must put down their foot and enter an emphatic protest against the continuance of a system that tends to dwarf their energies, to minimise their capacities for good, and to turn them into mere grinding machines for the manufacture of question-answering puppets'.[20]

Letters to the newspaper followed, most using pseudonyms and bemoaning the elitism and restrictiveness of the system introduced by the new university, which, they claimed, ignored the fact that 'the greatest men the world ever

[16] G Todd to Alice Todd, n.d., Bragg (Adrian) papers.
[17] W H Bragg to G Todd, 22 November 1888, RI MS WLB 37A/1.
[18] G Todd to W H Bragg, n.d., Bragg (Adrian) papers.
[19] *Register*, 9 April 1885, p. 6 and 19 May 1885, p. 6 respectively.
[20] *Register*, 25 April 1885, p. 4.

knew are self-made men'.[21] Finally, however, a spokesman came to the defence of the university. Frederick Chapple, London University arts and science graduate, headmaster of a leading denominational secondary school, Prince Alfred College, and Warden of the University Senate,[22] responded strongly to a *Register* editorial, insisting that schools were improving, that the standards of the matriculation examination and of the university's degrees were not too high, giving several persuasive examples, and finally writing: 'That which is most distressing to me... is the general tone, the underlying assumption that in these schools we have no high aims and no noble purposes. I deeply regret that you should have come to this conclusion. Every profession has a generous purpose. We repudiate indignantly the charge of being hucksters and tricksters. No man or woman worthy of the name of teacher can be with opening minds, fresh young lives, hopes, and possibilities without learning to love them and live for them'.[23]

The *Register* cooled its rhetoric, wanted to give more power to the teachers to set the curriculum, and pleaded for amendment of the traditional classical education to include some science, since 'this scientific education is often practical in its fruits'.[24] The newspaper had a regular column entitled 'Science of the Day', which told its readers, for example, of Pasteur's research on rabies, of the undergrounding of wires for the telegraph, telephone, and electricity, of the practice of cremation, of bacteria in drinking water, and of possible rearrangement of the calendar. In reviewing the commemoration ceremony at the university in December, the newspaper applauded the address of the Professor of Classics, David Kelly, and suggested it 'deserves careful study as a powerful—to our minds convincing—apology for the study of the classics'.[25]

The university made no other response to the many attacks, but it did make substantial changes to the regulations for the public examinations, re-organizing, clarifying, and consolidating the procedures for Preliminary, Junior, and Senior (matriculation) public examinations, without altering its basic philosophy towards them.[26] Later it added yet another, the Higher Public Examination, for working people who wanted to attempt one or more university subjects or for students who wanted to enhance their maturity and preparation for university study by remaining at school for one more year and undertaking advanced study.[27] The new professor of mathematics and physics would have to tread cautiously in this tense environment. Indeed, in the years ahead William's personal profile would become central to the public's perception of the university and its programmes.

[21] *Register*, 26 May 1885, p. 5.
[22] R M Gibbs, *A History of Prince Alfred College* (Adelaide: Peacock, 1984), p. 362.
[23] *Register*, 29 May 1885, p. 7.
[24] *Register*, 25 August 1885, p. 4.
[25] *Register*, 17 December 1885, p. 4.
[26] *Register*, 21 September 1886, p. 7.
[27] *Register*, 5 December 1888, p. 5.

Always supportive of practical education, the *Register* noted the introduction of technical education in New South Wales and pointed out that, as manual work decreased and mechanical devices were introduced, the artisan would need a level of training in excess of the traditional apprenticeship.[28] A motion of the British House of Commons, 'that our national system of education should be widened so as to bring manual training and the teaching of the natural sciences and technical instruction within the reach of the working classes throughout the country', was quoted with approval.[29] At the 1886 Adelaide Commemoration, the Chancellor noted the substantial legislative changes made during the year: the provision of teaching to enable medical students to complete their studies in Adelaide, the addition of further courses to the BA and BSc degrees, and the creation of separate Faculties of Arts and Science and of a Board of Musical Studies. He thanked the professors for their labours in this direction, but none matched William Bragg in their efforts.[30] He had arrived at a time when the university had 'done with its days of small things' and was about to be tested on the world stage.[31]

As William consolidated his teaching to both day and evening students, the *Register* encouraged a broader cross-section of students to complete their secondary education and proceed to both ordinary and honours degrees at the university, in medicine, arts, science, and music. William contributed to this encouragement by participating in two public conversaziones: at the university, where he demonstrated 'the mutual influence of sounds and jets' to hundreds of visitors,[32] and at Charles Todd's Observatory for the South Australian Electrical Society, where 'much attention was paid to the forming of sand figures on discs that were made to vibrate from different points, the object of the experiment being to show the laws which govern sound and vibration'.[33] These demonstrations concerned the physics of sound and music and were the product of William's studies in the Cavendish Laboratory. Since sound is invisible, jets of lighted gas were used to show the vibrations of the air, and small particles of sand were added to make the vibration of the source of the sound clear. Thus, on a vibrating plate the sand gathered on the lines of minimum vibration and was thrown away from the areas of maximum motion, thereby making the patterns of vibration visible.

The year 1888 was the centenary of the founding of the Australian colony at Sydney Cove, but at the University of Adelaide much of 1888 was dominated by events surrounding the tenure of Vaughan Boulger, the Hughes Professor of English Language and Literature and of Mental and Moral Philosophy.[34] William chose not to be involved but was drawn in briefly and inadvert-

[28] *Register*, 15 May 1886, pp. 4–5 and 7 June 1886, p. 4.
[29] *Register*, 15 June 1886, p. 6.
[30] *Register*, 23 December 1886, p. 6.
[31] *Register*, 26 February 1887, p. 4.
[32] *Register*, 31 August 1887, p. 6.
[33] *Register*, 31 November 1888, p. 5.
[34] There were regular items in the *Register* and *Advertiser* newspapers from January through October 1888.

ently when his opinion was sought at one of the many meetings. Even then he was guided by Edward Rennie, the Professor of Chemistry, who acted as their joint spokesman, saying that he and Bragg were of the view that, in the absence of life tenure, termination at six-month's notice was a more permanent arrangement than reappointment every five years.[35] In truth the matter had never been resolved in the early days of the university, which was now making policy on the run. In addition, poor communication between the several parties aggravated the situation. The university would come to rely heavily on the personal and public qualities of Professor William Bragg, as he matured and increasingly adopted a leadership role in this strained environment.

In the wider educational sector, 1888 was notable for the formal establishment of the South Australian School of Mines and Industries.[36] During the first half of the nineteenth century Britain had dominated the industrializing world, but in the second half its rate of industrial progress was outstripped by those of Germany and America. Following the depression of the 1870s there was increasing analysis of Britain's relative industrial decline, and the lack of an organized system of technical education was seen as one significant factor.[37] Australian governments, constantly attuned to British thinking, were prompted to consider the matter in their own environments. In South Australia the colonial government appointed a Board in 1886 to inquire into and report upon the best means of developing a general system of technical education. Its Report in 1888 recommended alterations to the primary school curriculum, instruction in mineralogy, agriculture, and science for trainee teachers, secondary education in natural and applied sciences, and the establishment by the government of a School of Mines and Industries, the suggested characteristics of which were outlined, including the use of university classes but management by a separate Council.[38]

In order to begin as soon as possible the new School rented space in the Exhibition Building next to the university, and the Chamber of Manufactures handed over its existing museum and library to the School. It was hoped that 'the Professors of the University might lecture to the coming pupils of the School of Mines and Industries on elementary mathematics, physics, chemistry, geology, and mineralogy'.[39] How William received this news we can only guess; anticipating more difficulties no doubt, but also hoping for more acceptable class sizes if the School students joined his university classes. The School opened in March 1889 with 375 young students, and in the same year Langdon Bonython,

[35] *Advertiser*, 29 March 1888, 31 March 1888, 2 April 1888.

[36] A Aeuckens, *The People's University, 1889–1989* (Adelaide: S A Institute of Technology, 1989); D Green, *An Age of Technology, 1889–1964* (Adelaide: S A Institute of Technology, 1964).

[37] G Roderick and M Stephens, *Where Did We Go Wrong?: Industrial Performance, Education and the Economy in Victorian Britain* (Barcombe: Falmer, 1981).

[38] 'Report of the Board appointed by the Government to inquire into and report upon the best means of developing a general system of Technical (and Agricultural) Education in the Province', South Australian Parliamentary Papers, no. 33, 1888.

[39] *Register*, 3 December 1888, p. 4.

a local newspaper proprietor and philanthropist, assumed the Council chairmanship and presidency that he would occupy for the next fifty years.

As to his sporting interests, William participated in the three games of lacrosse played between the only two teams (Adelaide and Noarlunga, a country town south of Adelaide) during the winter of 1886.[40] It was in 1887 that he came to dominate the game. It was clear that, if the game was to progress after its initial year, it would need to attract more players and therefore more teams. In a large pre-season piece in the Observer newspaper, 'Facer' first reviewed the history of the game and then reported: 'We have three clubs—the Adelaide (and pioneer), the Noarlunga, and North Adelaide. The last-named has been formed this year, and will begin active work this day week on the North Park Lands... Professor Bragg, who, without doubt, is the finest all-round player we have, has left the Adelaides and cast in his lot with representatives of the northern portion of the city, whose Captain he has been elected'.

Early in May 1887 the Adelaide Club again obtained permission to play its home games in the centre of the Old (Victoria Park) Racecourse in the East Park Lands, while the North Adelaide team was allocated an allotment of the North Park Lands near the Buckingham Arms Hotel. The new season began on Saturday 28 May. Facer reported extensively and enthusiastically:

> The city clubs initiated a series of matches between themselves last Saturday, and it was a most judicious arrangement to play on the Old Racecourse, because hundreds of people were attracted thither by the races and consequently the game was splendidly advertised... The ground was in fairly good order and the play on the part of the new chums—that is the North Adelaides—was surprisingly good... Professor Bragg is a capital man for any team because he knows how to manipulate the crosse and ball, and therefore by scrutinizing his movements his subordinates may learn a good deal. Last season, however, the Professor showed considerably better form than on Saturday. A year ago if he got the ball it was generally regarded as a certainty that he would pass almost any number of opponents and have his throw, but on Saturday, possibly from lack of practice, he was very unfortunate in this respect, and the Adelaides constantly succeeded in baffling him, much to their credit, for it is no easy matter to wreck such a skilful exponent of the game.

In view of the manifold duties that William had accepted during his first year in Adelaide, and the major changes taking place in his personal life, his lack of form on the sporting field is surely understandable. Late in June the game in South Australia received its biggest boost with the visit to Adelaide of the South Melbourne Club. The bad weather and consequent poor condition of the Adelaide Oval could not dampen the enthusiasm with which the contests were received. William played in two matches, representing North Adelaide

[40] Details of Bragg's participation in lacrosse and references to the quotations from the *Observer* newspaper are contained in J G Jenkin, 'William Bragg and lacrosse in Adelaide, 1885–1895', *The Australian Physicist,* 1980, 17:75–8 (contact the author for printing errors in this paper).

in a losing encounter and then joining a combined South Australian team in a win at the Oval. From the extensive game reports it is clear that he had become something of a father figure or guiding spirit in the play, for it was reported that 'Davis, Michie...and Anderson are perhaps the most prominent men in the field [for South Melbourne]. Anderson is particularly dexterous, but on Monday Professor Bragg kept him well in hand, and the youthful exponent did not have so much liberty as at the first day's engagement'. At the end of the season Facer acknowledged that 'the North Adelaide Club was formed chiefly through the exertions of Professor Bragg', and recorded that the club had, in fact, won the competition with five wins, two losses and one drawn game.

The expansion of the game of lacrosse in the colony continued in 1888 with the formation of four new clubs, and William's contribution also changed markedly. His absence from the South Australian team was lamented, he withdrew from the Captaincy of North Adelaide, and Facer observed at the end of the season that 'Bragg has not played much'. In 1889 William's allegiance changed for the last time. His recreational interests joined his professional obligations in the University of Adelaide: 'The formation of the University Club has involved the withdrawal of five prominent men from the North Adelaides [including] Professor Bragg'. William played a little, but the team had a disastrous first year, winning only one of nine games played. In 1890, however, the university team won the competition, and it continued to do well in following years. William's contribution to the foundation of lacrosse in South Australia had been pivotal, but he now had other demanding duties and responsibilities.

After three years of quiet establishment, Professor Bragg now publicly announced his arrival as a significant contributor to education in his new home town. He accepted an invitation to deliver the commemoration oration at the university's 1888 graduation ceremony late in December and spent many hours composing it.[41] As he told Gwendoline early in November, 'I read [Chapman] a bit of the oration: he said I had put it very well, and that I was quite right. He said that I should be criticised a good deal: but of that I shall be glad rather than otherwise, as it would mean that I had really got contrary opinions to fight against'.[42] Further preparation and polishing of the text was a regular element of William's daily letters to Gwen while she remained at Port Victor. The address not only received extensive editorial coverage in the local newspapers but was also printed in full in 180 column centimetres of small and closely-spaced type:[43]

> There is nowadays a very general complaint that the education...is not practical enough...It is my wish today to help, if I can, in the enquiry as to how far this complaint is true, in making clear the nature of the defect in our manner of education, and in the determination of the best means of removing it...I must speak to you as one who has lately been a student, who is familiar with the anxieties and doubts as to his future occupation which

[41] Council Minutes, meeting of 29 June 1888, UAA, S18, vol. IV, p. 228.
[42] W H Bragg to G Todd, 11 November 1888, RI MS WLB 37A/1.
[43] *Register*, 20 December 1888, pp. 4–7; *Advertiser*, 20 December 1888, p. 4ff.

> often beset the student nearing the end of his term of education...I have felt then, and still do feel, that there is ground for complaint...[When] a boy leaves school...there is still some further training for him to undergo before he can obtain a living...But nothing stands still nowadays...if a man is to succeed he must do far more than learn certain rules thoroughly; he must understand the principles on which the rules are founded...It seems to me that from all that has been said, and from all that may in many ways be observed, there is one obvious practical conclusion to be drawn: it is, that improvement is desirable not so much in the nature of the subjects that we teach as in the way in which we teach them...We teach facts and rules instead of encouraging the student to discover the facts and rules for himself; there is too much teaching and too little educating...We ought to do our best, then, to educate in our boys faculties for observation, of reasoning from observation, and of applying the results of the reasoning; to train them to depend, in dealing with fresh facts, on their own intelligence, not on rules imperfectly understood...
>
> Let me, for example, try to suggest how, with this end in view, we might modify our manner of teaching mathematics and physics...The study of geometry is valuable on several grounds...geometry teaches the science of measuring [and] is also of the greatest value as a mental discipline...
>
> Every year I have answers from book-taught candidates which show quite as much practical ignorance of physics. The right way to teach the subject is that in which the principal part of the instruction consists of experimental work performed by the students themselves; and always this principle must be put first—that the function of the teacher in not to communicate rules but to direct enquiry...
>
> To sum up what I have said, our system of education will be greatly improved if more attention be paid to the teaching of our students to observe, to reason from observation, and to design...[and] whilst I am trying to show how our manner of teaching might be improved, I have not the wish...to find fault with the teachers. Teachers labour under two great disadvantages...there is no place where they may learn how to teach...[and] when they find out for themselves, the strong compulsion of routine often prevents their making [the desirable changes].

The *Register* and the *Advertiser* differed in their editorial reactions to William's address. The Register thought 'He is manifestly right...when he complains that "there is too much teaching and too little educating", [but] he leaves us without much definite guidance...He tells us that the function of the teacher is to direct enquiry...One who will make us understand how to do these things is the kind of person we want...but at all events we can patiently wait for him'.[44] On the other hand, after summarizing the address, the *Advertiser* found that, 'Professor Bragg attempts a difficult task. He suggests modifications in the conventional plan of teaching mathematics and physics, which would not only confer additional interest on those subjects, but add

[44] *Register*, 20 December 1888, p. 4.

to their usefulness as disciplinary studies'.[45] For some of his listeners/readers, William may indeed have been going over existing ground, but the debate continues without resolution even today, more than one hundred years later, and William was dipping his toe into water that he would warm with effective actions and substantial wisdom, soon in South Australia and later in Britain.

At the same ceremony, the Vice-Chancellor (Archdeacon Farr) saw the last of his seven children, Clinton Coleridge Farr, receive the only BSc degree of the day. 'Cole' was the first of a thin but regular line of Bragg graduates who would go on to later success in mathematics or physics.[46] The first woman BA graduate was cheered, as was Mr T M Burgess, who was Adelaide's first 'double-first', in classics and mathematics.[47] Letters from two leading Adelaide school teachers applauded William's commemoration address.[48] During his well-deserved summer holiday that followed, William experienced the extremes of Adelaide weather: on Christmas Day 1888, the Adelaide temperature rose to 107.5 °F (42 °C), while two inches of rain (5 cm) fell a week later on New Year's Day.[49]

In addition to these more public activities, William had also been active within the university. In 1887 he had asked the Council for further items of equipment, and in 1888 he was granted further funds to repair the oxygen holder for the limelight and to purchase a range of materials for his rudimentary workshop: rubber tubing, glass rod and tubing, glass containers, copper and German-silver wires, porous pots to make batteries, and sheet, rod, and tubing of ebonite and brass.[50] Lamb may have purchased a very useful set of apparatus, but now it needed to be repaired, improved, and supplemented by new items that could not be purchased. William's growing commitment to teaching at all levels was confirmed by an initiative he took early in 1889. On 8 March he wrote again to the University Council, which referred the matter to its Education Committee and the Faculty of Science.[51] William wrote:[52]

> The directors of the [Teachers'] Training College having expressed a desire that all their students should be taught Physics and Practical Physics, I have the honour to request your consent to a scheme by which this instruction may be given them by the Physics Department of the University. I propose:
>
> 1. That two classes in Elementary Physics and Practical Physics be Testablished, the one a day class, the other an evening class...

[45] *Advertiser*, 20 December 1888, p. 4.

[46] J G Jenkin and R W Home, 'Farr, Clinton Coleridge (1866–1943)', in C Cunneen (ed.), *Australian Dictionary of Biography: Supplement 1580–1980* (Melbourne, MUP, 2005), pp. 123–4.

[47] *Register Supplement*, 24 December 1888.

[48] *Register*, 28 December 1888, p. 7 and 29 December 1888, p. 7 respectively.

[49] *Register*, 15 April 1889, p. 4.

[50] Letters W H Bragg to Finance Committee, 23 February 1888, 22 May 1888, and n.d., UAA, S200, dockets 98/1888, 202/1888, and 354/1888 respectively.

[51] Council Minutes, meeting of 8 March 1889, UAA, S18, vol. IV, p. 283; Education Committee Minutes, meeting of 14 March 1889, UAA, S23, vol. II, p. 108.

[52] Letter W H Bragg to University Council, 8 March 1889, UAA, S200, docket 77/1889.

2. That half the students of the Training College attend the day class, half the evening class. That these classes be also the classes... for the first year of the School of Mines. That the class in the evening take the place of the old evening class, than which, I believe, it will be more efficient.
3. That the Training College pay the University for this instruction an annual fee of about £2.10.0 a student.

The extent of William's important contribution to teacher training in South Australia will be examined in later chapters, but one note is desirable here. From the beginning, the preparation of teachers in the colony had narrow utilitarian aims. Most first underwent a four-year apprenticeship as pupil-teachers in the classroom, and general studies and professional training were then squeezed into a single year at the Training College that had begun in 1876 in its own building in the city. 'The typical product of this training system was no more than an efficient craftsman, who obligingly brought the bulk of his young charges to the lowest common denominator of standards of schooling as prescribed by the [State] Education Department... By the end of the century the narrow craft training had clearly become an anachronism'.[53]

The same meeting of the University Council considered a second letter from Professor Bragg, in which he petitioned the Council for an increase in Mr Chapman's salary and for the appointments of a practical mechanic and a 'boy' for the physical laboratory, since the sole university laboratory assistant, already very busy, was now overloaded with work for the Medical School.[54] These items were referred to the faculties, and when they returned to the Council with support they were accompanied by a statement of probable income and expenditure that William had drawn up.[55] This showed that the cost of a mechanic and an increase for Mr Chapman would be £228.10.0 p.a., but that this would be more than offset by an increase in revenue of £357.0.0 p.a. from Training College and School of Mines students. The Council adopted all the proposals.[56] The School of Mines classes began on 13 March 1889, with the hope that the university would reduce the fees for its students,[57] and the schoolteachers at the Training College made the same request. The university agreed to reductions for some courses (physics and chemistry) but not for others (mathematics).[58]

[53] B K Hyams, 'University and Teachers' College: half a century of the academic nexus in South Australia', *Australian and New Zealand History of Education Society Journal*, 1972, 1:34–44, 34; see also id., 'The teacher in South Australia in the second half of nineteenth century', *Australian Journal of Education*, 1971, 15:278–94.

[54] Letter W H Bragg to University Council, 8 March 1889, UAA, S200, docket 78/1889, now lost.

[55] W H Bragg, 'Ways and means in regard to the proposed appointment of a skilled mechanic to the Physical Laboratory, and as to increase of Mr Chapman's salary', in Reports of the Faculties of Arts and Sciences (1887–99), UAA, S144.

[56] Council Minutes, meeting of 15 March 1889, UAA, S18, vol. IV, p. 286.

[57] *Register*, 12 March 1889, p. 4.

[58] Education Committee Minutes, meetings of 4 April 1889 and 30 May 1889, UAA, S23, vol. II, pp. 112ff, 123.

William now approached Arthur Rogers, whom he had met at Sawtell's. Arthur Lionel Rogers was born in London in 1861, the son of a veterinary surgeon of Brompton Road, South Kensington. Like his brother before him, he attended Berkhamsted School in 1871 but was forced to leave after only one year on the death of his father. In 1872 the opticians and scientific instrument makers Tinsley and Spiller opened for business in the premises next door to the Rogers' home, and Arthur struck up an acquaintance with workers in the firm. He had a natural inclination and ability in using his hands, in drawing and art, and in constructing mechanical contrivances in wood and other materials.[59] The presence of the firm seems to have determined Arthur's career choice. Certificates survive showing that he studied physics and drawing at the Science and Art Department of the Committee of the Council on Education at South Kensington in 1874–75.[60] He may have attended the previous and following years also, for on 1 July 1876 he entered into a six-year apprenticeship with Samuel Tisley and George Spiller, 'to learn the Art, Trade or Business of a Philosophical Instrument Maker'.[61] During his early years there Arthur received some instruction from William Pye, who was a senior member of the staff. Rogers proved a proficient apprentice, but he left after four years (in May 1888) to join Siemens Brothers in London. At Siemens Arthur gained experience working with some of the most up-to-date electrical technology in the company's instrument and gutta-percha electrical departments, but again he left unexpectedly. Some time earlier he had visited William Pye in Cambridge, where Pye was now foreman of the Cambridge Scientific Instrument Company, and perhaps Arthur was enquiring about possible employment as well as visiting his old mentor.[62] The reason for the truncation of Arthur's apprenticeship now became clear, however. Regularly bothered by ill health during his early life, Arthur had been advised to emigrate to a warmer climate, and he chose Australia. Armed with good references from Siemens, he soon found work with Sawtell's scientific instrument business in Adelaide. In Sawtell's workshop Arthur had 'very varied experience both practically and commercially in almost every instrument included under the term "scientific"... as well as every kind of electrical instrument... He is sincere, sober and truthful as a man, and as a workman he is thorough and English'.[63] In due time Arthur became foreman of the shop. It was here that William met Rogers, when William came to learn some workshop skills under Arthur's expert guidance.

[59] Copies of family papers in the author's possession; J Holland, 'Arthur Lionel Rogers (1861–1939)', *The Old Berkhamstedian*, 2002, pp. 52–4.

[60] I Inkster (ed.), *The Steam Intellect Societies* (Nottingham: University Department of Adult Education, 1985), pp. 28–29 notes: 'the creation of the Department of Science and Art in 1853... made government funds available for technical education for the first time'; certificates in the possession of Mrs S Timbury, Sydney.

[61] Indenture certificate in the possession of Mr E J Rogers, Melbourne.

[62] M J G Cattermole and A F Wolfe, *Horace Darwin's Shop: A History of the Cambridge Scientific Instrument Company 1878 to 1968* (Bristol: Hilger, 1987), p. 22.

[63] Reference for Arthur Rogers by C Sawtell, 29 March 1889, in possession of Mr E J Rogers, Melbourne.

Having received the Council's permission to appoint a mechanic, William wrote to Rogers saying, 'I am glad to give you the vacant place in the Physical Laboratory: and I shall expect you on Monday next. I do not think you need have any fear as to the permanency of the position'.[64] On 29 March William wrote to the Council, reporting that 'In accordance with the permission you have kindly granted me, I have engaged... a skilled mechanic, A L Rogers, late foreman at Mr Sawtell's establishment. I think he is just the man wanted for the place'.[65] William was delighted to have this expert mechanic join him at the university. It was his second inspired appointment, for Rogers would become his right-hand-man in every initiative he took: in teaching, public lecturing, research, and indeed in a number of family projects. While William was always the professor, Rogers soon became his very close colleague and collaborator.

Two more events of major importance for William occurred during the first half of 1889. The first was significant recognition of his future father-in-law, Charles Todd. Early in April it was announced that Todd was to be elected a Fellow of the Royal Society of London. In addition to the transcontinental telegraph connection, Todd was responsible for the connection of Adelaide to Melbourne and Perth. As Government Astronomer he had corrected the position of one of the State's boundaries, undertaken several important astronomical observations, and pioneered widespread meteorological recording, the publication of weather maps, and weather forecasting. He had also taken a prominent part in the government of public institutions of learning in South Australia, and he was beloved by his employees.[66] Todd was nominated unsuccessfully in 1874 and 1875 and then again in 1887 and 1888 before success in 1889. As Rod Home has pointed out, standards were not noticeably relaxed for overseas candidates and it was very rare for a candidate to be elected to the fellowship on first nomination.[67] There were several criteria on which Todd's case was based: 'having made discoveries in pure or applied science... as the author of scientific treatises or memoirs... distinguished for his knowledge of some branch of science... a distinguished engineer [and] one who is attached to science and anxious to promote its progress'.[68] The award, formalized on 6 June 1889, gave Todd, his family, and the citizens of South Australia much pleasure.[69]

[64] Letter W H Bragg to A L Rogers, 19 March 1889, Bragg (Adrian) papers.

[65] Letter, W H Bragg to Council, 29 March 1889, UAA, S200, docket 108/1889.

[66] G W Symes, 'Todd, Sir Charles (1826-1910)', in G Serle and R Ward (eds), *Australian Dictionary of Biography* (Melbourne: MUP, 1976), vol. 6, pp. 280–2.

[67] R W Home, 'The Royal Society and the Empire: the colonial and commonwealth fellowship; Part 2: After 1847', *Notes and Records of the Royal Society of London*, 2003, 57: 47–84.

[68] Royal Society of London, Certificate of a Candidate for Election: Charles Todd, CMG, MA Cambridge, FRAS, etc.

[69] A copy of the personal letter of congratulation to Todd from the Governor of South Australia, the Earl of Kintore, appeared in *Register*, 19 April 1889, p. 5.

The second event represented the culmination of William's early years in Adelaide; an event that seemed to cement his commitment to his new home and, in turn, its acceptance of him. The *Register* announced:[70]

MARRIAGES

> BRAGG–TODD: On the 1st of June, at St Luke's Church, Adelaide, by the Venerable Archdeacon Farr, LL.D., William Henry Bragg to Gwendoline, third daughter of Charles Todd.

The celebrant and location are interesting. George Henry Farr had been born in London and educated at Christ's Hospital, where he won a gold medal for mathematics and was Senior Grecian. Entering Pembroke College, Cambridge, on a scholarship, he graduated BA in classics and law, but on the death of his mother returned to Cambridge to study theology. He was ordained a priest in the Church of England in 1845 and married Julia Ord in February 1846. Serving in southwest England, he became dissatisfied with aspects of the established church, while his wife's health deteriorated in the damp climate. They decided to emigrate to a drier location, a change made possible by Farr's appointment as the second Headmaster of St Peter's College in Adelaide. They arrived in June 1854. Farr, with the able assistance of his wife, lifted St Peter's out of its unpromising beginning to be, on one assessment at his retirement in 1879, the pre-eminent educational institution in South Australia. He continued to be active after retirement, as a canon of St Peter's Cathedral, Adelaide, a priest of several parishes, notably St Luke's, Adelaide, and as a member of the Board of the Public Library, Museum, and Art Gallery. A member of the University Council from its inception, Farr also increased his commitment as Warden of the Senate from 1880 to 1882 and Vice-Chancellor from 1887 until 1893.[71]

The Todd family were loyal members of Stow Congregational Church in Adelaide, but William's membership of the Church of England prevailed when Farr was invited to officiate at the wedding in his parish church of St Luke's Mission on Whitmore Square. It was not a particularly elegant venue. A recent article has given an indication of the nature of the area:[72]

> The West End of Adelaide has been a contentious place for much of the time since European settlement... [it] had been settled early. It was the closest point within the city mile to the Holdfast Bay [Glenelg] Road, the bullock track from the Port and, after 1856, the railway. It was also close to the water supply—the River Torrens... Although there were

[70] *Register*, 6 June 1889, p. 4.

[71] J S Dunkerley, 'Farr, George Henry (1819–1904)', in D. Pike (ed), *Australian Dictionary of Biography* (Melbourne: MUP, 1972), vol. 4, pp. 155–6; J. Tregenza, *Collegiate School of St Peter, Adelaide: The Foundation Years 1847–1878* (Adelaide: St Peter's College, 1996).

[72] L Hammond, 'Class and control: social reform in the West End of Adelaide in the early twentieth century', *Journal of the Historical Society of South Australia*, 2003, 31:5–17, which gives references for the embedded quotations.

Fig. 6.2 St Luke's Church, Adelaide, photo *circa* 1985. (Courtesy: Mr R M Gibbs).

> some substantial dwellings, mainly on the fringes of the West End, by the 1880s workshops, warehouses and factories had encroached on the residential area. This resulted in an 'environment of noise, smell and pollution'...The first sewerage system drained directly into the Park Lands, and until deep drainage was completed in 1885, a 'mass of corruption... literally festered the surface of our city to a considerable depth, evolving gases fatal to health'...The West End had a certain reputation amongst 'respectable' folk. Some considered it 'morally the lowest part of the city', and [St Luke's Parish Paper] claimed that 'every form of evil is rampant in the West End'...The major Christian denominations were active in the area, several of their 'Missions' intent on moral reform, as well as providing some material assistance to the poor.

St Luke's Mission on Whitmore Square was an evangelical Anglican presence. In the early 1850s a prefabricated iron church was ordered from England for the site, but its likely cost caused the local committee to begin a cheaper structure of local bluestone. After much indecision, a new church of bluestone, timber, and some ironwork was consecrated in February 1856. It had a large nave, two aisles with clerestory above, attenuated columns, and Gothic arches. In 1889 the chancel was modest, but there was a new pipe organ and the church had been renovated and redecorated.[73] Here Archdeacon Farr had come, while Vice-Chancellor, for eleven years at the end of his ministry, and here Professor William Henry Bragg, aged twenty-six, and Miss Gwendoline Todd, aged nineteen, came to be married by him on 1 June 1889. Charles Todd ('Postmaster General') and Alfred Lendon ('MD' and William's best man)

[73] S Marsden, P Stark, and P Summerling (eds), *Heritage of the City of Adelaide: An Illustrated Guide* (Adelaide: Corporation of the City of Adelaide, 1990), pp. 197–8.

Fig. 6.3 Semi-detached house (R) on Lefevre Terrace, North Adelaide, rented by William and Gwendoline Bragg after their marriage and in which their two sons were born, photo *circa* 1985. (Courtesy: Mr R M Gibbs.)

signed the marriage certificate, and we might guess that Gwen's sisters were prominent, but no other details of the ceremony or the guest-list survive.

The young couple honeymooned in the Adelaide hills, at Gumeracha. Laid out in about 1845, Gumeracha remains a small township, winding along the valley of the Torrens River and serving the local farming and wine-growing area. In 1889 it was tiny, isolated, and quiet, the cottages were adorned with roses, and the surrounding paddocks were separated by fences of yellow and green English gorse.[74] Why the couple chose it is unclear. They stayed at the District Hotel and perhaps they just wanted to escape from the city and be by themselves. A letter of congratulation was sent to 'Mrs Bragg' at Gumeracha from seven members of her father's Telegraph Department, and her mother wrote: 'My own dear Gwen, Maude and Rosie have gone over to your house to work…Tell my son I will answer his letter to me tonight but, as he tells me he reads all your letters, he will, I hope, not feel neglected…I am so thankful you are both so well…Willie dear, I am obliged to you for bringing the roses to my dear one's cheeks & I am glad you have lost your white cheeks…With much love to you both. Your loving mother, Alice Gillam Todd'.[75] Gwen wrote to a sister saying, 'How nice everyone is', and 'it's something awful, we are always having to calm one another'.[76]

[74] D Whitelock, *Adelaide, from Colony to Jubilee: A Sense of Difference* (Adelaide: Savvas, 1985), pp. 260–1.

[75] Letter Alice Todd to Mrs W H Bragg, District Hotel, Gumeracha, 10 June 1889, RI MS WLB 37A/1.

[76] G M Caroe, *William Henry Bragg, 1862–1942: Man and Scientist* (Cambridge: CUP, 1978), p. 35.

William had earlier found a father in his Uncle William; now he had a mother and a wife. They were very happy (see Figures 6.2 and 6.3).

The young couple had decided to rent a large and elegant house in North Adelaide. It was owned by the well-known Adelaide colonial jeweller and silversmith, Henry Steiner, and he had lived in it himself until recently. Steiner had been born in Germany, had immigrated to South Australia in 1858, and had established his jewellery and silver business in the city.[77] So successful had this been that Steiner was able to finance a high-quality, speculative housing development on Lefevre Terrace, North Adelaide. Four pairs of semi-detached houses survive from the development, one of the first pair of which William and his growing family rented from mid-1890 until early 1898. Lefevre Terrace provided a delightful location, running north/south and with impressive houses on the western side that looked out over the Park Lands to the Adelaide hills in the east. William rented the house on the corner of Lefevre Terrace and Tynte Street: a two-storey home of local bluestone with stuccoed enrichments, in high Victorian Italianate style. Architecturally this well-preserved and renovated building is now regarded as important 'due to its design and detailing... with significant cast-iron fencing to Lefevre Terrace'.[78] Tragedy had beset Steiner as his housing development was nearing completion, however, for his wife and children died in the typhoid outbreak of 1883. He sold his jewellery business and returned to Germany in 1884. Arriving again in Adelaide in 1887, he left in 1889 and returned permanently to Germany. His personal residence thus became available for rent just when William and his bride were searching for a suitable home. Professor and Mrs Bragg moved in immediately on their return from Gumeracha.

[77] 'Henry Steiner (1835–1914)', in H J Gibbney and A G Smith (eds), A *Biographical Register, 1788–1939* (Canberra: Australian Dictionary of Biography, 1987), vol. II, p. 278.

[78] Marsden, Stark, and Summerling, n. 73, pp. 353–5.

7
Growth and maturity

The next six years, from mid-1889 to mid-1895, was a time of growth and maturation for Professor William Bragg, personally and professionally; a time when the unsettled young scholar became a settled family and professional man. During this period important public issues arose that the new Bragg family must have noted. In 1893 the great Kalgoorlie gold fields were discovered in Western Australia, and in 1894 the South Australian parliament became the first in Australia to grant women the vote. It was also during this time that the movement for the creation of an Australian nation—a federation of the six separate British colonies—gathered momentum, culminating in the birth of the Commonwealth of Australia on 1 January 1901. In this movement South Australians were to play a role out of all proportion to the modest size of their population.[1] Public and domestic life was also changed by the introduction of electricity, a topic that William had to master for teaching to university and School of Mines students and in order to answer questions that university officers and others directed to him.

Economic conditions in South Australia continued to be poor during the 1890s, and in 1893 the colony plunged into one of the most serious economic crises in its history. There was widespread unemployment in both urban and rural settlements, bitter disputes between employers and employees, and no system of social security to alleviate poverty and starvation. Banks were regularly under pressure and several closed. The firm of Elder Smith and Co. made its first and only loss.[2] There were outbreaks of typhoid fever, diphtheria, and influenza that caused concern; wide-ranging immunisation and antibiotics were far in the future. The lack of water has always been a serious problem in Adelaide, and in South Australia more generally. In 1891 the local branch of the Geographical Society urged the government to undertake experiments for the artificial production of rain and Charles Todd offered his support, outlining

[1] P A Howell, *South Australia and Federation* (Adelaide: Wakefield, 2002).

[2] R M Gibbs, *A History of South Australia* (Adelaide: Southern Heritage, 1984), pp. 154–6; *Election 1893: Come Out 1983* (Adelaide: Constitutional Museum, 1983), 2 vols; also see G Davison, J W McCarty and A McLeary (eds), *Australians 1888* (Sydney: Fairfax, Syme and Weldon Associates, 1987), ch. 11.

the atmospheric conditions likely to produce success.[3] Neither early trials nor those of more recent times have been successful, however.[4]

On the other hand, William's educational initiatives were already bearing fruit. In an editorial arising from the annual report of the Minister for Education for 1889, the *Register* made special mention of 'the introduction of scientific or technical training into the schools', and of the 'large number of teachers now taking pains to acquire a knowledge of science under the tuition of Professor Bragg'.[5] Not all William's suggestions were so readily accepted, however. Regarding the teaching of arithmetic, the *Register* contended that 'young children should not be required to do sums necessitating the exercise of reasoning powers', but rather that 'to drill a boy in what may be termed the dry and dreary first processes of rules...is the surest method of securing that he be ultimately a good arithmetician'.[6] Nevertheless, the newspaper noted with pleasure that 'reports of the proceedings at the different schools cannot fail to notice...the increasingly marked effect that the University is having...The past year...has seen an appreciable accession to their roll of students...[and] the number of students sent up to the University has been unusually large'.[7] The end of 1891 was marked by the introduction of free primary school education for all South Australian children.[8]

The Art Gallery of South Australia, the School of Design, and the School of Painting occupied rooms in the Jubilee Exhibition Building, and their students were obtaining commendable results in the South Kensington Science and Art examinations in London.[9] After her marriage Gwendoline Bragg continued to enjoy painting and other art forms and, at the 1891 prize-giving of the School of Design, she was awarded three of the prizes for members of the Everyday Art Club: for a landscape, a coastal view, and a still-life study.[10] Indeed, women were seeking formal education in increasing numbers: 'it is especially noticeable that the proportion of girl candidates is largely increasing, and it is clear that among the rising generation the monopoly of University study formerly enjoyed by the sterner sex will be more and more extensively challenged', the *Register* noted.[11]

The importance attached to education and to teachers and teaching at this time can be gauged from the following newspaper item: 'The first meeting for the new year of the South Australian Teachers' Association was held in the office of the Minister of Education on Saturday evening, April 27 [1894]...There was

[3] *Register*, 28 November 1891, p. 5.

[4] R W Home, 'Rainmaking in CSIRO: The science and politics of climate modification', in T Sharratt, T Griffiths and L Robin (eds), *A Change in the Weather: Climate and Culture in Australia* (Sydney: National Museum of Australia Press, 2005), pp. 66–79.

[5] *Register*, 6 June 1890, p. 4.

[6] *Register*, 9 June 1890, p. 4.

[7] *Register*, 19 December 1891, p. 4.

[8] *Register*, 31 December 1891, p. 4.

[9] *Register*, 9 October 1891, p. 4.

[10] *Register*, 21 November 1891, p. 6.

[11] *Register*, 19 September 1894, p. 4.

a large attendance... After the meeting the teachers and several representative gentlemen were entertained at luncheon at Beach's Rooms, Hindley-street, by the President (Mr J L Bonython). Mr Bonython... was supported by the Premier, the Minister of Education, the Mayor of Adelaide, Professors Mitchell, Tate, Rennie and Bragg, and the Inspector-General of Schools... Professor Bragg spoke on behalf of the Teachers' Guild, of which he is President'.[12] In June William chaired a meeting of the Guild, at which Professor William Mitchell discussed 'The principles of education'. Mitchell was the new Professor of English Language and Literature and Mental and Moral Philosophy, and during the next fifty-four years he would become the university's greatest servant: Professor 1894 to 1922, Vice-Chancellor 1916 to 1942, and Chancellor 1942 to 1948.[13] He succeeded Edward Boulger, who had finally left the university after the Council accused him of being unable to attend to his duties.[14]

At a personal level William's life now had an additional, family focus. On 31 March 1890 William Lawrence Bragg was born at the family home on Lefevre Terrace, North Adelaide, and he was baptised on 8 June by Vice-Chancellor Farr in the Whitmore Square church where William and Gwendoline had been married. In naming their first-born, the couple chose a name from each of their families. William followed the Bragg tradition of naming his first son 'William', perhaps with Uncle William particularly in mind, and it seems Gwendoline chose her brother's name that most appealed to her.[15] A little less than three years later, Robert Charles Bragg was also born at home, on 25 November 1892, with 'Robert' commemorating William's father and 'Charles' honouring Charles Todd. He was christened by Archdeacon Farr in St Luke's Church on 30 April 1893.[16] William Lawrence soon became 'Willie' (or occasionally 'Bill') to family and friends, while Robert Charles rapidly became 'Bob'. Willie had vivid memories of his earliest years in Adelaide:[17]

> At the time of my birth, on 30th [*sic*] March 1890, our family lived in a semi-detached house in Lefevre Terrace in North Adelaide, and we continued to live there till the visit to England in 1898. I can remember the house and garden vividly, though when I saw it again in 1960 I realised how much smaller it was than I had imagined it to be. It faced over the 'Park Lands'... A low fence separated the two gardens in front of the semi-detached houses. There was sometimes friction with the widowed lady who lived next door, who used to snip the stems of our nasturtiums

[12] *Register*, 29 April 1894, p. 6.

[13] V A Edgeloe, 'Mitchell, Sir William (1861–1962)', in B Nairn and G Serle (eds), *Australian Dictionary of Biography* (Melbourne: MUP, 1986), vol. 10, pp. 535–7.

[14] W G K Duncan and R A Leonard, *The University of Adelaide, 1874–1974* (Adelaide: Rigby, 1973), pp. 21–2.

[15] Her younger brother was Hedley Lawrence Todd; the clerk recorded W L Bragg's birth incorrectly and wrote 'William Lawrance Bragg' on the birth certificate.

[16] Birth certificates from the Office of the Principal Registrar of Births, Deaths and Marriages, Adelaide; 'Baptismal Records of St Luke's Church Adelaide', State Records of South Australia, Adelaide, SRG 94, Series 1, reel 22, pp. 295, 372.

[17] W L Bragg, Autobiographical notes, pp. 1–3.

if they wandered through to her side of the fence, even though they came back to our side; but there must have been amicable arrangements also, because I remember my mother telling me she used to share with our neighbour in buying half a lamb, a whole lamb at that time costing seven shillings. Next door again lived the Gills. Harry Gill was the leading artist in Adelaide and my mother attended his classes. His boy, Eric Gill, was just my age and my great crony.

I can remember my brother Bob as a baby in long clothes in the charge of the monthly nurse. I would be about two-and-a-half years old at the time. My mother had been very ill indeed at the time of his birth, and it was some time before I was allowed to see her. It is related in the family that my first words to her were, 'Mummy, do you know that I have got a baby brother?' While in bed in her convalescence she used to tell me stories, and I can remember my fury and tears when a visitor interrupted one. Bob and I were wheeled out in the pram together. I dreaded these excursions because the larrikins, the rough boys of the neighbourhood, would shout gibes on seeing so large a boy in a pram. The indignity was heightened by my mother's artistic taste, which led to our long hair being done in sausage curls formed around the nurse's wet finger. We later had, for best wear, blue tunics with red belts and broad-brimmed straw hats, when I felt my dignity demanded trousers and coat like other small boys of my age.

The cook and housemaid were called Tilly and Naomi. I had a great affection for them. They afterwards ran a boarding-house together and I used to visit them from time to time when I was much older. I cannot recall the early series of nursemaids, except one who had to be sent packing because she had nits, which she passed on to us.

But very soon Charlotte [Schlegel] must have arrived on the scene. Charlotte became a family institution, staying with us for nearly thirty years. She came from that part of Denmark acquired by Prussia in the war of 1867 [*sic*, 1864], and she remembered making bandages for the wounded as a girl...Looking back, I realise that Charlotte was not the right person to be a nurse; she was neurotic and fierce...I do not think my brother Bob was much affected by her—even as a child he had considerable calm self-confidence—but I was very impressionable and unsure of myself, and I am certain that Charlotte was very much the wrong person for me. She had a passion for our clothes always being smart, and sternly repressed any game devised by Eric Gill and myself by which we could conceivably get dirty, a severe restriction for small boys and very daunting to initiative. Our playground was the gravelled 'back-yard' so characteristic of Australian houses, with outhouses along one side and a huge wood-pile in one corner. Domestic fires were all fed with wooden logs, and the quantity required for a household was immense. Flowers and shrubs were reserved for the front garden, which had to be watered every evening in summer. Our house was a corner house, with an extension of the back-yard running between the house and the side street. Here was a tree in which some previous tenant had built a platform reached by steps, which overlooked the street and which was a great joy to us. My playtime was divided between our back-yard and that of the Gills [see Figure 7.1].

Fig. 7.1 The Todd family at the Adelaide Observatory during a visit by Lizzie Squires (née Todd), 1897 (refer to family tree, Figure 0.2). Back row (L to R): Elsie, William, Maude, Elsie's sister (Mrs Tower), Hedley, Lorna; centre row: Alice, Gwendoline, Charlie, Lizzie, Jessie, Charles; front row: Willie, Tower daughters, Bob (note boys' clothes). (Courtesy: State Library of South Australia, SLSA: B 28760.)

Willie's 'fury and tears' and his troubles with Charlotte are indicative of a sensitive child, who found some difficulty in riding the various bumps of childhood and who was sometimes treated inappropriately because his intellectual abilities were far in advance of his age. In addition, his mother was apparently insensitive to the difficulties with Charlotte, and Gwendoline even tried to foist her on to Willie and his new wife immediately after they were married many years later. Willie's childhood was sheltered and he became independent and self-reliant. His isolation continued until, at age twenty, he left the family home in England for study at Cambridge University, found a group of close friends, and began to take part in the challenging activities that need to be part of every young person's personal development. Lefevre Terrace had imposing homes and prosperous residents, but there were other sections of North Adelaide nearby that were occupied by unskilled, working-class families and where unemployment was high.[18] The assertive children from these families—the 'larrikins'—quickly became a source of irritation and sometimes fear for the sensitive child, until the family left North Adelaide in 1898 for a year in England and for a new home in the centre of the city.

[18] Davison et al., n. 2, p. 214.

William's love of sport continued to call him after his marriage. The location of his new family home enabled him to take up a new sport in a way that was less demanding on his time, golf. The Adelaide Golf Club was founded in 1870, but by 1876 it had lapsed because of lack of support. It was re-formed in August 1892 and obtained permission from the Adelaide City Council to play on the North-east Park Lands; as it happened, directly across Lefevre Terrace from the Bragg's home.[19] William watched with increasing interest as the members struggled to lay out a nine-hole course on the hard and waterless ground. Willie and Eric Gill sometimes played there, although this could bring a scolding from Charlotte, since the ground was dusty in summer and muddy in winter. Initially the course was composed of narrow fairways of natural grass and weeds, cropped by grazing cows, while the tiny greens were cut by scythe when necessary. Distances were short, but the nature of the course, the wooden clubs with beech heads and ash shafts, and the gutta-percha balls made low scoring difficult.[20] The game attracted those in the upper classes, with status and wealth, and was exclusive to them.[21]

William joined the club in April 1893 and again demonstrated the characteristic that was a feature of every activity he undertook, whether in teaching, research or recreation: 'Professor Bragg's golf is the result of an infinite capacity for taking pains, as during all his golfing career he has set himself to master individual shots by constant daily practice'.[22] With the course outside his front door, it was possible for William to spend an hour or two in the evenings to improve his game. He quickly reduced his handicap from thirteen to one. A ladies' club was formed in 1893 and was given permission to play on the same course.

In January 1894 William was elected Honorary Secretary/Treasurer of the club for the ensuing year. The neat and economical style of his hand-written minutes stands out in contrast to the heavy and flamboyant hands of his predecessor and successor. The same year Mr Ayers and Professor Bragg were appointed a sub-committee to superintend the preparation of the greens, and William's year in office was not an easy one. There was much dissatisfaction with the greensman, and J R Baker, the gifted but erratic tennis player that Bragg had partnered in doubles to good effect, now frequently complained about the golf course and the committee.[23] Some members were unhappy about the presence of lady players, and the committee decided, on the motion

[19] Letter J M Gordon to Adelaide Town Clerk, 9 August 1892, Adelaide City Council Archives, ACC docket 2176/1892; Digest of Proceedings of ACC, S35, 1891–2, p. 211.

[20] M Ridgeway, *South Australian Golf, 1869–1970* (Adelaide: South Australian Ladies' Golf Union, n.d.).

[21] Royal Adelaide Golf Club, *Minute Book, 1892–1896, 1896–1902, 1902–1907, 1907–1915* (for committee and general meetings and including printed notices, financial accounts, annual reports, newspaper cuttings etc.); much of the information on Bragg's golf has been gleaned from this source and was published previously as J G Jenkin, 'William Bragg in Adelaide: and finally golf', *The Australian Physicist*, 1986, 23:138–40. See also J A Daley, *Elysian Fields: Sport, Class and Community in Colonial South Australia, 1836–1890* (Adelaide: author, 1982), pp. 134–5.

[22] *The Critic*, Adelaide, 16 October 1907.

[23] J G Jenkin, 'William Bragg in Adelaide: tennis too!', *The Australian Physicist*, 1981, 18:69–70, 131.

of the Secretary, 'that in consequence of the danger resulting from members of the club going the reverse way round on days on which the ladies were allowed to play, members should go the ordinary way on Mondays and Tuesdays'. Despite the difficulties, the weekly handicap match offered William the satisfying relaxation he sought. The official course consisted of two rounds of the parkland links, and in 1894 only Dr Swift (scratch) had a lower handicap than Professor Bragg (one) for the Browne Trophy competition. Winning net scores were usually close to one hundred. The future prospects of the course were uncertain, however, and in December 1895 it was decided to amalgamate with the Glenelg Golf Club while retaining the identity of the Adelaide club and its course for practice. A convenient train service ran from the city to the seaside suburb and William's handicap dropped to scratch early in 1896.[24]

Decades later, after the family had moved permanently to England, William Bragg became renowned as a public speaker on science, with a special ability to explain complex scientific ideas to lay audiences. One reason for this was his use of familiar examples and analogies in meticulously prepared addresses. Thus, during his six Christmas Lectures at the Royal Institution in London in 1923–24, 'Concerning the nature of things', William used many sporting analogies. In this notable series on recent discoveries in science he examined the atomic nature of things: gases, liquids, diamond, ice and snow, and metals. In discussing the cooling of a gas in a cylinder fitted with a piston that is progressively withdrawn, William noted: 'If we have played cricket, we know that when we want to catch a ball we must draw our hands back as the ball begins to touch them... So also when a lacrosse player catches a ball, he draws his crosse downwards when the ball first enters it... A tennis raquet can be used to catch a tennis ball in the same way'.[25] And later in the same lecture: 'A very pretty example of the laws of the dynamics of the air is to be found in the swerve of a spinning ball... We see it and make use of it in nearly every game, though perhaps the golf ball shows it most because its speed is greatest... Heavy balls swerve less than light balls going at the same speed... the swerve that the pitcher can give in baseball is a marvellous spectacle'.

In a final note on making models of atomic structures, William suggested that, 'Balls representing the atoms may be made of hard dentists' wax, which softens in boiling water and can then be pressed into proper shape in metal moulds made for the purpose, just as we used to remake our golf-balls in the old days'.[26] Similarly, in one of the finest yet most humble accounts of the frustrations and elations of postgraduate research—of which he was as yet unaware but of which he would become a leading British authority—William proposed, during a prize-giving address at the Sir John Cass Technical Institute on 30 January 1924, that 'Research is rather like playing against bogey at golf: Nature never has any weakness of which advantage may be taken; there is no

[24] Royal Adelaide Golf Club, *Minute Book*, 1892–1896.
[25] Sir William Bragg, *Concerning the Nature of Things* (London: Bell & Sons, 1925), pp. 55, 70–3.
[26] Ibid., pp. 73, 232.

hole to be won by bad play because our opponent plays worse. Yet research is very human, for the researcher finds himself one of a company who have in their turn striven and denied themselves, very happily; and have handed on their experience to those who take up the quest where they have left it'.[27]

Soon after their marriage the couple began a regular series of Christmas holidays with their children at the seaside. For the first three summers, 1889–90, 1890–91, and 1891–92, they stayed at Port Willunga, another small village on the eastern shore of Gulf St Vincent.[28] The sand is white and wide on the beaches south of Adelaide, while the sea is calm and sheltered from the breakers of the Southern Ocean, and cool and refreshing in the severe summer heat. These beaches have always been popular, and the Adelaide suburbs have now extended to embrace many of them. In the following five years, holidays were spent at Port Elliot, on the southern coast, where the sea can be rough but still pleasant and where Gwendoline had convalesced before her marriage.[29]

Gwendoline continued to pursue her favourite pastime, painting, especially while on holiday. She and William were each represented in a collection of twenty-one works by South Australian artists selected for a display in Melbourne by the Royal Anglo-Australian Society of Artists,[30] and the couple exhibited regularly at the annual exhibitions of the South Australian Society of Arts, where Gwendoline's work was reviewed favourably. Landscapes predominated, such as 'Green Bay, Port Elliot', and 'Lake Sorrell, Tasmania' by Gwen in 1893; but more intimate subjects appeared occasionally, 'A portrait' by Gwen in 1894, for example.[31] The reviewer for the *Register* commented that Gwen's pictures had 'solidity and boldness of outline, and their colouring is admirable';[32] while *Quiz and the Lantern*—a weekly satirical, social and sporting journal—thought Professor Bragg 'not happy in his sea effects, but his "Aloes" are well drawn, and the scene is characteristic and natural'.[33]

At the university

William took the horse-tram along O'Connell Street, King William Road and over the River Torrens to the university each day. The pure and applied mathematics courses that he presented to his undergraduate students contained nothing radical or unexpected. He was guided by his own studies and the courses that Horace Lamb left behind, and any tendency Bragg had to innovate

[27] Sir William Bragg, 'Research work and its applications', *Nature*, 1924, 113:311–312, being the text of an address at the Sir John Cass Technical Institute on 30 January 1924.

[28] W H Bragg, 'Christmas Holidays in Australia', a hand-written list, Bragg (Adrian) papers.

[29] L Abell, 'Holidays and health in nineteenth and early twentieth century South Australia', *Journal of the Historical Society of South Australia*, 1994, 22:82–97.

[30] *Register*, 20 February 1892, p. 5.

[31] *Royal South Australian Society of Arts, Exhibition Catalogues, 1857–1920*, State Library of South Australia, Adelaide, SRG 20.

[32] *Register*, 15 June 1894, p. 7.

[33] *Quiz and the Lantern*, 20 June 1895, p. 5.

was snuffed out by the limited knowledge and small number of his Adelaide students. He settled for a reduced version of the Mathematical Tripos he had completed at Cambridge. The subject descriptions that appeared in the university *Calendar* were very limited, allowing maximum flexibility in dealing with the particular students each year. In 1890, Elementary Pure Mathematics at first year involved the geometry of the straight line and circle, elementary solid geometry, and the elements of algebra and trigonometry, while 'honours' added elements of statics, dynamics, and hydrostatics. At second year, Pure Mathematics listed only algebra and trigonometry, Applied Mathematics involved elements of statics, dynamics, and hydrostatics, and 'honours' added elementary analytical conics, elementary differential and integral calculus, elementary spherical trigonometry, and astronomy.[34]

The 1890 *Calendar* also reported on the examinations for 1889. William was still setting and marking most of the university mathematics papers, but Chapman relieved him of some of the public (school) examining, including matriculation, where criticisms of the difficulty of the examination papers continued.[35] The number of university students who completed their whole year of study satisfactorily was still tiny: two at both first- and second-year levels of the BA, five at both levels of the BSc, and one and two respectively at third-year, while twelve evening students passed the introductory mathematics subject.[36]

By 1900 both the standard and the extent of syllabus information had increased, and Chapman was taking a larger fraction of the teaching and examining. 'The most elementary portions of analytical geometry in two dimensions, and of the infinitesimal calculus' had been added to Pure Mathematics at first year; Applied Mathematics involved statics, dynamics, and hydrostatics, treated with the aid of the elements of analytical geometry and of the infinitesimal calculus. Pure Mathematics II was restricted to analytical geometry of two dimensions and infinitesimal calculus, and two other subjects appeared under the heading 'Mathematics for the Ordinary Degree'. 'Applied Mechanics' involved: the testing of materials (iron and steel under stress and the properties of colonial timbers); calculations regarding the strength of rods, ropes, chains, struts, columns, and beams; computations regarding structures such as roofs and bridges; the strength of boilers and pipes; calculations with special application to mining; and the theory of the steam engine. This subject was clearly presented especially for senior students of the School of Mines and Industries. 'Spherical Trigonometry and Astronomy' was the final mathematics subject available and included use of the telescope, elementary computations, and descriptive astronomy. The number of evening and 'non-graduating' students had risen markedly: from the School of Mines, from the Education Department, and from the Pharmacy Society.[37]

[34] *Adelaide University Calendar for the Year 1890* (Adelaide: Thomas, 1890).
[35] *Register*, 5 December 1890, p. 6, 8 December 1890, p. 7, and 10 December 1890, p. 4.
[36] *Calendar*, n. 34, individual subject results are not given.
[37] *Calendar of The University of Adelaide for the Year 1900* (Adelaide: Thomas, 1900).

The English system of external examiners was employed at Adelaide, where a fellow academic in the same discipline but from another university set examination papers or commented on the papers set by local staff. This ensured that the standard of the courses did not stray seriously from that at other institutions. In 1889, for example, Professor Nanson of Melbourne and Mr Newham of the University of Sydney were approached to fulfil this role in mathematics,[38] and later *Calendars* show that others were additional examiners. William was also consulted from time to time when mathematics problems arose outside the university. Captain Patrick Weir, a tug owner/master at Port Adelaide, had noted to local naval authorities that a British Admiralty book on gunnery included a number of errors. Disbelieving, they consulted Professor Bragg, who confirmed the errors. Furthermore, a diagram invented by Weir for the use of navigators had been approved and adopted by the Admiralty: 'So South Australia not only helps to navigate British "men of war", but puts the Britisher right in his sums', *Quiz and the Lantern* boasted.[39]

The undergraduate physics programme for 1890 prompts remarks similar to those for mathematics. For the BSc degree, first-year Elementary Physics included the expected classical components: the first principles of mechanics, hydrostatics, heat, light, sound, electricity, and magnetism.[40] The early introduction of electricity and magnetism (probably electro- and magneto-statics) is notable. At the second and third-year levels, a substantial amount of detail was given, perhaps reflecting William's increasing commitment to physics. For the second-year he specified: sound (general theory of waves and vibrations, Lissajous' and other methods of studying vibrations, waves in solids, vibrations in columns of air and of strings, resonance, analysis of sounds, quality, interference, beats), geometrical optics, heat, electricity, and magnetism. For third-year physics, including practical physics, the syllabus was the second-year topics treated more fully and more mathematically, with the following additions: mechanics (moment of inertia, Kater's pendulum, motion of liquids and gases, Torricelli's theorem), properties of matter (elasticity, viscosity, capillarity, diffusion), sound (vibration of bars and plates, consonance and dissonance, combined tones), optics (velocity of light, spherical aberration . . . colour, reflection and refraction, interference, diffraction . . .), heat, electricity, and magnetism. In October 1891 the Council adopted a recommendation from the faculties of science and medicine 'that work, more especially practical work, done by students during the year, shall be taken into consideration at the Annual Ordinary Examinations'.[41] Bragg and Chapman carried the examining load, and Professor Thomas Lyle and Ernest Love of Melbourne University were external examiners.

[38] See, for example, letter W H Bragg to Education Committee, 7 November 1889, UAA, S200, docket 457/1889, regarding an external examiner for third-year mathematics in 1889.

[39] *Quiz and the Lantern*, 25 April 1895, p. 7.

[40] *Calendar*, n. 34.

[41] Council Minutes, meeting of 30 October 1891, UAA, S18, vol. V, p. 113.

Medical students constitute the main source of contemporary student evaluation of William's teaching. They were required to undertake physics as part of their first year of study. Such 'service' courses are traditionally unattractive to students and teaching staff alike, and favourable student comment can be regarded as especially revealing. In a letter to his grandmother in 1897, first-year medical student Elliott Brummitt wrote, 'Professors Rennie and Bragg know how to explain things nicely—they take us in Chemistry and Physics';[42] while prominent Adelaide medical identity Sir Henry Newland recalled, 'The Professorial staff of the Medical Faculty consisted of Professors Watson, Stirling, Bragg, Rennie and Tate. As lecturers, Bragg and Rennie were outstanding'.[43] Emeritus Professor J B Cleland had more intimate memories of his student days: 'Professor Bragg lectured to us in Physics, assisted by a young Mr Chapman. The former seemed quietly settled in South Australia and contented with his lot. As a medical student I came in very slight contact with him, but my wife, a few years later, taking the course in Science, came to know him much more intimately. In any little mathematical difficulty she had, he was always so nice and self-deprecatory that in the end the student felt she had solved the problem herself. On one occasion she covered two foolscap sheets in working out a complicated problem. Professor Bragg annotated the result "well-earned" but gave an alternative solution in some half-dozen lines'.[44]

As Arthur Rogers established his new workshop in the basement of the university building, William's requests for a range of accessories and materials increased. Two lists from 1889 include items of glassware, metals in wire, rod, sheet, and tube forms, wood, screws, and hand tools. The Optics Room was also put into full operation, with the purchase of four-dozen lenses, one-dozen prisms, and mirrors.[45] Following Rogers' arrival William also took on an apprentice named Horace Woolcock. Although no wages were normally paid for the duration of the five-year apprenticeship, William asked the Finance Committee to consider five shillings per week in Horace's third year, ten in his fourth, and fifteen in the fifth.[46] The Finance Committee agreed to consider the proposal favourably, but not until the time came, although William thought this 'as good as a promise'.[47] In 1892, however, the Finance Committee agreed to pay only five shillings per week for the final three years of the apprenticeship, and Horace's father, Rev. R. Woolcock, wrote to the committee to express his disappointment. William also wrote to the committee saying, 'I am

[42] Letter E A Brummitt to grandmother, 8 August 1897, privately held in the family.

[43] Sir Henry Newland, 'My student days', *University of Adelaide Gazette*, 1952, 1:26–7.

[44] J B Cleland, 'An undergraduate of the nineties', *Adelaide University Graduates' Union Gazette*, 1959, 2: no page numbers.

[45] Letter W H Bragg to Finance Committee, 29 November 1889, and W H Bragg, 'Estimate of expenses for the Physical Laboratory for October [1889]', UAA, S200, dockets 491/1889 and 390/1889 respectively.

[46] Letter W H Bragg to Finance Committee, 28 November 1889, UAA, S200, docket 492/1889.

[47] Letter W H Bragg to Woolcock's father, 30 November 1889, attached to letter R Woolcock to Finance Committee, 25 May 1892, UAA, S200, docket 218/1892.

afraid the man is in very poor circumstances'.[48] After interviewing both men, the Finance Committee and the University Council agreed that the original suggestions should be paid and that 'Professor Bragg be requested in future to at once report the proposed engagement of lads'.[49] In 1894 another young man was also taken on 'to receive the benefit of Mr Rogers' instruction... under the direction of Professor Bragg as well as Mr Rogers'.[50]

In the university building the library stretched across the front of the upper storey and occupied considerable space. It was the venue for numerous committee meetings and ceremonial functions such as the annual commemoration, but conditions were poor. It was cold and damp in winter and could be unbearably hot in summer. Horace Lamb generously acted as the library's honorary buying agent in Britain, but the number of books was modest because of the university's limited financial resources. The library needed a benefactor, and it ultimately found one in the person of Robert Barr Smith. Like his fellow Scots, Hughes and Elder, who had helped to found the university, Barr Smith became interested in the university in the 1890s, joined its Council, and in 1892 gave £1,000 for the library. In following years he donated a total of £8,000 for the same purpose, and in 1899 the library was officially named The Barr Smith Library. In the 1920s his son, Tom Barr Smith, donated nearly £35,000 for the construction of a magnificent new building to house the library, which still stands, carrying the family name.[51] The State Library of South Australia, on North Terrace near the university, was also expanding its book collection as much as its limited finances and space would allow, 'especially augmenting the division devoted to mathematics and physics, in doing which they have taken advantage of the services of Professor Bragg'.[52]

The financial state of the university had never been prosperous, or even particularly healthy, and things came to a head early in 1890 when the Council considered the Finance Committee's estimates for the next year. It requested the committee to reduce expenses so as to balance the budget, and also appointed a Special Committee to consider changes to student fees.[53] The latter, including Professor Bragg, met in April and June and recommended significant increases in the fees for many of the subjects for the bachelor degrees of medicine, surgery, science, and music.[54] The Council agreed and, no doubt to its relief, there appears to have been little public comment.

[48] Letter W H Bragg to Finance Committee, 26 May 1892, UAA, S200, docket 221/1892.

[49] Finance Committee Report to Council, 28 July 1892, UAA, S145, vol. 2, report 17/1892; Council Minutes, meeting of 29 July 1892, UAA, S18, vol. V, p. 160.

[50] Education Committee Report to University Council, UAA, S140, no. 30/1894; Council Minutes, meeting of 30 November 1894, UAA, S18, vol. V, p. 324.

[51] W G K Duncan and R A Leonard, *The University of Adelaide, 1874–1974* (Adelaide: Rigby, 1973), ch. 12.

[52] *Register*, 13 October 1892, p. 4.

[53] Council Minutes, meeting of 28 March 1890, UAA, S18, vol. IV, p. 368.

[54] Special (Fees) Committee Minutes, meetings of 22 April and 24 June 1890, UAA, S22, vol. I, pp. 19–20, 23–24; Council Minutes, meeting of 27 June 1890, UAA, S18, vol. V, p. 13.

The founding, legislative Act of the University of Adelaide gave power to confer degrees in arts, science, law, medicine, and music, although when the university opened in 1876 the only degree offered was the Bachelor of Arts. Degrees in science, law, and medicine followed in 1881, 1883, and 1885 respectively. Under the enthusiastic leadership of the Governor and University Visitor, Sir William Robinson, and further financial support from Sir Thomas Elder, the first chair of music in Australia was established in 1884 and filled in 1885 by Joshua Ives. A course of study leading to the degree of Bachelor of Music (MusBac) was begun the same year. It was this course to which William contributed his 'acoustics' subject from the time of his arrival in Adelaide. Since a chair of music had been initially assured for only five years, the University Council declined to establish a Faculty of Music and instead created a Board of Musical Studies.[55] William was a member of the Board from its inception in 1886 until 1903.

William's attendance at meetings of the Board was somewhat irregular, and he was absent on 25 March 1890 when the Board decided to recommend the colour magenta for the hood accompanying the gown for the MusBac degree.[56] When the matter came to the University Council three days later, however,[57] it was accompanied by a letter of protest from Professor Bragg saying, 'Magenta is a colour which destroys the effect of nearly all colours with which it may be brought into juxtaposition. In place of a magenta and black hood I beg to submit rough sketches of other hoods from which a choice might be made if any of them seem suitable to you'.[58] The matter was referred back to the Board, which then recommended the colour rose-pink.[59] The Council, still dissatisfied, referred the whole question of the shape and colour of hoods for the various degrees to the professors.[60] The Special Committee met and decided to seek the views of the Master of the School of Design, Mr H P Gill, with Professor Bragg as intermediary.[61] The Committee did not reconvene until the following September, with William in the chair, when it decided to recommend as follows: that hoods be of the same shape and cut as those of Cambridge University, that each faculty be distinguished by a separate colour, that bachelors' hoods be of black stuff, lined to a width of six inches with coloured silk, that masters' hoods be black, entirely lined with silk, and that doctors' hoods be entirely silk, inside and out. The colours chosen were Arts—grey, Science—yellow, Law—blue, Medicine—rose, Music—green, of precise colour numbers.[62] The

[55] V A Edgeloe, *The Language of Human Feeling: A Brief History of Music in The University of Adelaide* (Adelaide: Friends of the Elder Conservatorium, 1985).
[56] Board of Musical Studies Minutes, meeting of 25 March 1890, UAA, S129, vol. I, p. 61.
[57] Council Minutes, meeting of 28 March 1890, UAA, S18, vol. IV, pp. 373–4.
[58] Letter W H Bragg to University Council, 28 March 1890, UAA, S200, docket 137/1890.
[59] Board of Musical Studies Minutes, meeting of 11 April 1890, UAA, S129, vol. I, pp. 62–3.
[60] Council Minutes, meeting of 25 April 1890, UAA, S18, vol. IV, p. 379.
[61] Special (Hoods) Committee Minutes, meeting of 28 October 1890, UAA, S22, vol. I, pp. 27–8.
[62] Ibid., meeting of 14 September 1891, pp. 29–30.

recommendations were accepted and implemented.[63] Indeed, my own BSc hood from 1959 fits these specifications exactly, and William's suggestions—for he was largely responsible for them—have survived the intervening century and more.

In the earliest years of the new university building, the handful of students were content with a small, unfurnished Students' Room, but by 1889 it was quite inadequate. Angry students significantly damaged the room on 31 July and the university appointed a Board of Discipline to deal with the matter.[64] It recommended that the room be repaired and improved and that the students be excluded until they had paid for the repairs and furnished the Board with the names of a committee who would be responsible for any future damage.[65] Since the students were unable or unwilling to meet these financial requirements, the room remained closed for two years. Following a growing custom, born of relief following the annual and all-important examination period, students also disrupted the commemoration ceremony in December 1890. The Council again referred the matter to a Board of Discipline, this time including Professor Bragg, to report and recommend action.[66] This time the Board was more thoughtful, saying it had no knowledge of the participants and urging the staff to use their personal influence to prevent future disturbances.[67] Encouraged by this development, seventeen students wrote to the Council asking for the Students' Room to be reopened.[68] This too was referred to the Board, which informed the students that it was inclined to comply with their request provided they made an acceptable proposal for safeguarding the room and its furniture.[69] The students readily accepted these suggestions and the Students' Room was reopened.[70] In these negotiations Professor Bragg increasingly took the lead, forging an excellent relationship with the student body. It would be an invaluable asset in the future.

In the wider public arena William gave an increasingly varied range of illustrated lectures and became a well-known spokesman for science. In a conversazione at the university in May 1889 he 'demonstrated the polarisation of [light from] the sky',[71] and in May 1890 he addressed a monthly meeting of the Field Naturalists' Club on the topic 'Some of the effects of plant extracts on light'.[72] In 1891 he gave three further popular lectures: the first

[63] Council Minutes, meetings of 9 October and 30 October 1891, UAA, S18, vol. V, pp. 109 and 114 respectively.

[64] The best account of these matters is to be found in the history of the Adelaide University Union: M M Finnis, *The Lower Level* (Adelaide: Adelaide University Union, 1975), chapter II, where the major source of information is the *Medical Students' Society Review*.

[65] Board of Discipline Minutes, meetings of 5 and 8 August 1889, UAA, S208, vol. I, pp. 4–8.

[66] Council Minutes, meeting of 22 December 1890, UAA, S18, vol. V, p. 62.

[67] Board of Discipline Minutes, meeting of 21 March 1891, UAA, S208, vol. I, p. 12.

[68] Council Minutes, meeting of 26 March 1891, UAA, S18, vol. V, p. 73.

[69] Board of Discipline Minutes, meeting of 17 April 1891, UAA, S208, vol. I, pp. 14–15.

[70] Board of Discipline Minutes, meeting of 23 April 1891, UAA, S208, vol. I, pp. 15–16; Council Minutes, meeting of 24 April 1891, UAA, S18, vol. V, p. 82.

[71] *Register*,27 May 1889, p. 6.

[72] *Register*, 21 May 1890, p. 7.

during an industrial exhibition was entitled 'Capillarity', the second to the Teachers' Guild and Collegiate Schools' Association addressed 'The teaching of elementary practical physics', and the third to the Boys' Field Club concerned 'The tin whistle'. The first drew heavily on the work of Charles Boys on soap bubbles,[73] the second was designed to encourage the teaching of physics in secondary schools and contained practical suggestions for establishing a physical laboratory,[74] while in the third 'Professor Bragg...entered into a lucid explanation as to how sound was produced, demonstrating his points by apt illustrations'.[75] In November 1892 William spoke to the South Australian Photographic Society about the 'simpler rules of optics',[76] while in May 1894 he addressed the Chamber of Manufactures on the elementary principles of the 'Electric transmission of power', when the lecture had to be moved from its usual location to the university to accommodate the large audience.[77] Less than ten years after his arrival in Adelaide, William had grown in maturity and confidence and become a household name in South Australia.

There had been passing references to the possibility of electric lighting in Adelaide for many years, but the discussion became more serious in the 1880s, when Charles Todd experimented with it from the top of the Post Office tower, when the South Australian Electric Light Act passed the local parliament but left supply to small private plants acting under contract, and when numerous dynamo electric machines were displayed at the Adelaide Jubilee Exhibition.[78] Development was rapid during the 1890s, and by the end of the century a number of Adelaide's streets had electric lighting. With these developments came a need for electrical engineers, not only with the practical skills of the earlier telegraph men or electricians but also with knowledge of the new theoretical developments of Maxwell and his followers. In Britain a fierce debate was raging between the electricians (telegraph engineers led by Todd's colleague William Preece of the British Post Office) and the Maxwellian scientists, a debate won by the latter and the emergence of college-trained electrical engineers.[79]

In Australia, Adelaide tackled the problem first and with foresight. The Council of the School of Mines planned to introduce electrical engineering at an early stage but, apparently aware of the British problem, soon wrote to the university saying, 'the Council of the School of Mines and Industries proposes next term to re-organise the Electrical Engineering class, [and] it is the desire of the Council that the students should have the advantage of tuition by one

[73] *Observer*, 30 May 1891, p. 33; C V Boys, *Soap-Bubbles and the Forces Which Mould Them* (London: Society for the Promotion of Christian Knowledge, 1890ff), also (London: Heinemann, 1960) and (New York: Dover, 1959ff).

[74] *Register*, 13 August 1891, p. 6.

[75] *Register*, 24 August 1891, p. 4.

[76] *Register*, 11 November 1892, p. 5.

[77] *Register*, 26 May 1894, p. 7.

[78] D Wakelin, *Fifty Years of Progress, 1896–1946: Being a History of The Adelaide Electric Supply Company Limited* (Adelaide: AESC, 1946).

[79] See, for example, B J Hunt, 'Practice vs. Theory: the British electrical debate, 1888–1891', Isis, 1983, 74:341–355.

of the University Professors... one evening or two evenings per week'.[80] The matter was referred to the university's Education Committee, with power to confer with the Faculty of Science, Professor Bragg, and the School of Mines in order to develop 'a scheme'.[81] The task immediately fell to William, who reported as follows:[82]

> In the first place, if such a class is to be started in the colony, I consider it advisable that the University should have it: we should not lose an opportunity of establishing the University's authority and influence in such matters.
>
> In the second place... it is probable that a special evening class which we are holding this year will not be required again... this will next year leave us the necessary time.
>
> In the next place... I think it would be well if the class met for two hours on one evening in each week throughout the University year; part of this time should be occupied by a short lecture, part by practical work. It should be understood that we are going to teach the *principles* of Electrical Engineering to those who are, or are about to be, engaged in electrical work, but that it is not our object to train—say—bellfitters or telegraph operators.

William's letter concluded with an appeal for funds to purchase a small gas engine and dynamo and a set of accumulators or batteries; for, while 'we might use the School of Mines dynamo to illustrate our teaching, it would be much better to have our own'. We might wonder at William's ability to lead such an initiative, given his earlier lack of knowledge and his nervousness regarding the physics of electricity and magnetism, but these had recently been overcome, as the next chapter will show. The matter then travelled around the university committee system until a meeting in February 1891 between leading members of the Education and Finance Committees, the School of Mines, and Professor Bragg, at which it was agreed that 'a class on Electrical Engineering be established... on the understanding that the class should not involve the University in any expense for apparatus and on the condition that the School of Mines permit the use of their dynamo'. The University Council adopted the recommendation.[83] William was not to be put off so easily, however, regarding a university gas engine and dynamo. He wrote to the Finance Committee in April 1892 to request a second-hand system, and it then persuaded the Council to buy a new one![84]

[80] Letter Assistant Director of School to University Registrar, 19 August 1890, UAA, S200, docket 284/1890. The School Council had appointed an instructor in electrical engineering and authorised expenditure on a dynamo, lamps, cable etc. to light its building in place of gas: School Council MInutes, meeting of 23 September 1889, vol. 1, p. 107.

[81] Council Minutes, meeting of 29 August 1890, UAA, S18, vol. V, p. 22.

[82] Letter W H Bragg to Education Committee, 10 October 1890, UAA, S200, docket 346/1890.

[83] Education Committee Minutes, meeting of 13 February 1891, UAA, S23, vol. III, pp. 20–21; Council Minutes, meeting of 27 February 1891, UAA, S18, vol. V, p. 70.

[84] Council Minutes, meeting of 29 April 1892, UAA, S18, vol. V, p. 148.

The new course was one option for graduation with a Diploma of Associate of the School and involved three years of study: in the first year, preliminary mathematics, physics and chemistry, wood or metal work, and drawing; the second year, drawing, wood or metal work, applied mathematics and mechanics, and advanced physics; while the third year was entirely devoted to electrical engineering and related topics.[85] At the end of the first year of the new course (1891) Mr Chapman reported: 'Seven students from the School of Mines attended this class, which was held at the University on Friday evenings from 7 till 9, one hour being devoted to lecture and the remaining hour to practical work in the laboratory. The students being mostly beginners at the subject, we went through a general course of electricity in the first part of the year, and afterwards passed on to consider its most important practical applications. Four students entered for the final examination, three of whom passed'.[86]

In response to student requests and when their number justified it,[87] a more advanced course was given from 1894 onwards;[88] but William never taught any of the classes, although he retained overall supervision of the programme. He was admitted as an Associate of the British Institution of Electrical Engineers in January 1893 on the sponsorship of Charles Todd, William Preece, and others, and then transferred to full membership in March 1894, supported by Preece and Professor John Hopkinson of King's College, London.[89] Many years later William was made an honorary member of the Institution, gave two of its Kelvin Lectures (in 1921 and 1935), spoke at several of its annual dinners, and in 1936 was awarded its Faraday Medal.[90]

William's association with the School of Mines and Industries became closer and permanent in September 1890 with his election as government representative on the Council of the School, in company with his father-in-law.[91] Soon thereafter, early in 1891, William was appointed to represent the School in discussions to consider overlap in the courses of the School of Design and the School of Mines and Industries.[92] Furthermore, in May 1892 he and Todd were appointed a Special Committee to report and then act upon the safety of the School's existing electric lighting installation, and late in the year extensive improvements were made with the assistance of Hedley Todd and his Brush Electrical Engineering Co. Ltd

[85] *The South Australian School of Mines and Industries and Technological Museum, Annual Report, 1890* (Adelaide: School of Mines, 1891), pp. 54–55 (hereafter *School of Mines Annual Report*).

[86] *School of Mines Annual Report, 1891*, pp. 29–30.

[87] Letter W H Bragg to University Council, 29 March 1894, UAA, S200, docket 183/1894; letter H A Pilgrim and others to Registrar, 10 April 1895, ibid., docket 214/1895 and Bragg annotations thereto.

[88] 'Examinations' in *School of Mines Annual Report*, 1897 and 1900.

[89] Papers in possession of Institution of Electrical Engineers (IEE), London, and correspondence between IEE Archivist and the author, October–November 1984.

[90] Ibid.; R Appleyard, *The History of the Institution of Electrical Engineers, 1871–1931* (London: Institution, 1939), Appendices 2 and 3.

[91] Council Minutes, meeting of 26 September 1890, UAA, S18, vol. V, p. 28; School Council Minutes, meeting of 6 October 1890, vol. 1, p. 202. Stirling's poor attendance and resignation implies disinterest in the School's activities.

[92] School Council Minutes, meetings of 14 and 27 April 1891, vol. 1, pp. 251 and 253 respectively.

agency.[93] Early in 1893 William joined the School's Education Committee. He had been concerned that 'nearly all the students have been hampered by an inadequate knowledge of arithmetic and elementary algebra',[94] and now he urged the rejection of exemptions for passes in public examination mathematics as well as maintenance of both pure mathematics and advanced mathematics in a re-arrangement of the Mechanical Engineering course.[95]

Charles Todd was the only scientist whose work and public service gave him more exposure than William Bragg. The newspapers were full of his personal contributions and news of his many responsibilities. The telegraph and telephone systems were constantly expanding, and he was a member of numerous committees. He reported regularly to the newspapers on interesting astronomical events, and in 1892 he inaugurated the Astronomical Section of the Royal Society of South Australia. Todd was a central player in Australian meteorology, and he played a leading part in Australia-wide discussions regarding postal services and in the introduction of electricity in South Australia. The various services for which he was responsible were not always as reliable or expanding as fast as some critics desired, but he supported his staff unflinchingly and was very widely loved and respected.[96] On 6 December 1891 Charles Todd celebrated fifty years of imperial and colonial service, having begun at the Royal Greenwich Observatory on that date in 1841.[97] He was knighted (K C M G) in June 1893, with much family satisfaction and public celebration.[98]

[93] School Council Minutes, meetings of 9 May, 23 May, and 14 November 1892, vol. 2, pp. 27, 32, and 76–77 respectively.

[94] *School of Mines Annual Report, 1889*, p. 18.

[95] School Council Minutes, meetings of 27 March and 12 June 1893, vol. 2, pp. 108 and 128 respectively.

[96] *Register, passim*; G W Symes, "Todd, Sir Charles (1826–1910)", in G Serle and R Ward (eds), *Australian Dictionary of Biography* (Melbourne: MUP, 1976), vol. 6, pp. 280–2.

[97] *Register*, 5–8 December 1891, *passim*.

[98] *Register*, 5 June 1892, pp. 4,7.

8
Towards research

Research is now such a common and essential ingredient of academic life that it is salutary to be reminded that it was rare in many universities in the late nineteenth century. In Australia the teaching load of the professors, and their other educational and public roles, left them scant time for research. Only a few, such as Richard Threlfall in Sydney, made deliberate provision for a major research programme in science. Many years later William recalled without embarrassment that, 'For seventeen years I worked steadily in Adelaide... It never entered my head that I should do any research work',[1] and his daughter, son, and other scholars have accepted this judgement. Indeed, his daughter recalled a comment by Sir George Thomson: 'the great question was "why did he come to research so late?"'.[2]

As earlier, however, William's self-effacing personality and excessive modesty misled his readers. He had studied for nearly a year in the Cavendish Laboratory, he had attended the first meeting of the AAAS in Sydney in 1888 and expressed his pleasure that the Association would be a guide for those who 'like me, are willing to work, but don't quite know where to begin', and he was very conscious of Threlfall's initiatives in Sydney. Although he did not begin a major research programme until 1903, William was contemplating the possibility of research from the time of his arrival in Adelaide. He made initial steps in that direction from 1890 onwards, when he began to wrestle seriously with physics problems from his teaching, whose solutions could not be found in the standard textbooks.

Because he had not studied electricity or magnetism at Cambridge he devoted a good deal of time to this area in his early years in Adelaide. Threlfall, on the other hand, had studied the subject as part of his natural sciences degree, and subsequently it became a major part of his research and publication programme in Sydney.[3] At the second AAAS meeting, held in Melbourne

[1] W H Bragg, 'In the days of my youth', *T P's & Cassell's Weekly*, 3 April 1926, p. 834 (copy at RI MS WHB 39/2).

[2] G M Caroe, *William Henry Bragg 1862–1942: Man and Scientist* (Cambridge: CUP, 1978), p. 2; J L Heilbron, 'The scattering of α and β particles and Rutherford's atom', *Archive for History of Exact Sciences*, 1968, 4:247–307, 256.

[3] R W Home, 'First physicist in Australia: Richard Threlfall at the University of Sydney, 1886–1898', *Historical Records of Australian Science*, 1986, 6:333–57.

in January 1890, Threlfall delivered the Presidential Address for Section A on 'The present state of electrical knowledge', where he presented 'a sketch of Maxwell's theory, because it has recently received ... a great deal of striking confirmation'.[4] About 850 people had attended the inaugural AAAS meeting in Sydney in 1888; 1160 came to the Melbourne congress.[5] William attended, and during a conference session he wrote to his pregnant wife, who was at Port Willunga on holiday with the Todd family: 'I got your letter from the Post Office this morning. I was wondering all Sunday and Monday whether you would be able to get a letter to me: I knew I had left you no address ... Gwenny, how can I write what I want to say when that blessed idiot will go on talking at the top of his voice, 19 to the dozen, on the ravages of red rust? There, he's shut up! ... I say, Threlfall is marrying one of the very party who went up the Hawkesbury River last year! ... There was a meeting at the Townhall last night: Governor presided. Baron von Mueller read a magnificent presidential address. I was on the platform'.[6] William was missing Gwendoline's support and encouragement. Threlfall soon requested the University of Sydney to provide a house for him and his new wife in the university grounds; William quietly rented a house some distance from the University of Adelaide.[7]

Threlfall offered a point of reference and a source of professional advice and security regarding electricity and magnetism, which William quickly took up. Soon after the 1890 meeting he began a large foolscap notebook in which he entered a long series of notes on electromagnetism,[8] and in June he began a series of letters to Threfall, seeking his guidance.[9] The notebook contains extensive references to introductory textbooks by Gray, Mascart and Joubert, and William Thomson, but a few only to Maxwell's difficult work.[10] Of particular interest are William's several forays into experimental work, as outlined in the notebook. They mark the start of his recorded career as an experimental physicist and are entitled: 'To find the specific resistance of some mica belonging to the Brush Co.', 'Comparison of EMF of Latimer-Clark [cell] with copper deposit [cell]', and 'Determination of capacity of condenser by Wheatstone's

[4] R Threlfall, 'The present state of electrical knowledge: Presidential address in Section A', in W B Spencer (ed.), *Report of the Second Meeting of the Australasian Association for the Advancement of Science, held at Melbourne, 1890* (Melbourne: AAAS, 1890), pp. 27–54.

[5] 'Appendix 5: Attendances at AAAS (ANZAAS) congresses', in R Macleod (ed.), *The Commonwealth of Science* (Melbourne: OUP, 1988), p. 377.

[6] Letter W H Bragg to Gwendoline Bragg, 5 January 1890, Bragg (Adrian) papers; letter W H Bragg to Gwendoline Bragg, 8 January 1890, RI MS WLB 37A/1. Ferdinand von Mueller was an important botanist and explorer who, over many years, led and developed the Botanic Gardens and its Herbarium in Melbourne, Australia.

[7] Home, n. 3, p. 338.

[8] W H Bragg, untitled notebook, RI MS WHB 38/12.

[9] Letters W H Bragg to R Threlfall, RI MS WHB 6C; apparently the Threlfall to Bragg half of the correspondence has not survived.

[10] A Gray, *The Theory and Practice of Absolute Measurements in Electricity and Magnetism* (London: Macmillan, 1888); E Mascart and J Joubert, *A Treatise on Electricity and Magnetism*, tr. E Atkinson, 2 vols (London: De La Rue, 1883); Sir William Thomson, *Reprints of Papers on Electrostatics and Magnetism* (London: Macmillan, 1872); J C Maxwell, *A Treatise on Electricity and Magnetism* (Oxford: OUP, 1873), 2 vols.

bridge'. The first, in June 1892, related to the fact that Hedley Lawrence Todd had recently become the local agent for the London-based Brush Electrical Engineering Co. Ltd.[11]

In his letter of 10 June 1890 to Threlfall, William wrote: 'I have been very interested just lately in deducing the elementary electrostatic theorems from the "elastic medium" theory of Maxwell, of which Oliver Lodge gives the mechanical analogy...The extent of my reading is very limited, and I wish you would tell me if you have seen the thing put as on the accompanying slip of paper....The advantages of the method, as a teaching one, are that it gives clear mental pictures, but it also puts some things in a truer way than usual....I haven't seen the ideas mathematically expressed, so if you haven't seen them either I might make a bit of a paper for the Association [AAAS] out of them....My first born was christened on Sunday last amidst much rejoicing. He is a sturdy little chap'.[12] William did, indeed, prepare and deliver such a paper to the January 1891 meeting of AAAS, held in Christchurch, New Zealand;[13] and, following advice from Lodge not to publish in *The Electrical Review*,[14] it appeared in the *Philosophical Magazine* and was later noted by Whittaker.[15] The British preference for 'clear mental pictures' is evident, as is the error of the suggestion that William did not contemplate research until he was more than forty years old: he was twenty-eight in January 1891.

Threfall apparently wrote a number of notes to William during 1890–91, for on 10 October 1891 William replied to him at length saying, 'Thank you for your notes. I was delighted to have them. The question has worried me for a year & more, and I wish I had written to you sooner: but I was rather afraid of expressing my ignorance. Now my difficulties are clearing away'.[16] In a letter that quickly followed a reply from Threlfall, William noted, 'I have not read all Maxwell yet...I did not read any Electricity at Cambridge'; and again, 'Now as to dimensions. Of course Maxwell put down B and H as having the same dimensions, but then he left out μ. Putting it back, I think it makes it all right'.[17] Now even Maxwell's work had lost its dread.

William was elected President of Section A for the January 1892 Hobart meeting of the AAAS, and he confidently presented his new understanding of electromagnetism by drawing wide-ranging analogies between electrostatics, current electricity, magnetism, heat, hydrokinetics, and mechanics (twist and

[11] Untitled notebook, n. 8, pp. 115–19, 133–43.

[12] Letter W H Bragg to R Threlfall, 10 June 1890, RI MS WHB 6C/18 and 19; the Lodge reference is O Lodge, *Modern Views of Electricity* (London: Macmillan, 1889).

[13] W H Bragg, 'The "elastic medium" method of treating electrostatic theorems', in Sir James Hector (ed.), *Report of the Third Meeting of the Australasian Association for the Advancement of Science, held at Christchurch, 1891* (Wellington: AAAS, 1891), pp. 57–71.

[14] Letter O Lodge to W H Bragg, 5 March 1891, RI MS WHB 4A/25.

[15] W H Bragg, 'The "elastic medium" method of treating electrostatic theorems', *Philosophical Magazine*, 1892, 34:18–35; Sir Edmund Whittaker, *A History of the Theories of Aether and Electricity* (London: Nelson, 1951), vol. 1, p. 272.

[16] Letter W H Bragg to R Threlfall, 10 October 1891, RI MS WHB 6C/20.

[17] Letters W H Bragg to R Threlfall, 23 November 1891 and 3 December 1891, RI MS WHB 6C/21 and 22 respectively.

spin). 'I believe it is most important that every physical student should examine this analogy', he wrote, 'because... we shall be rewarded for our pains by finding ourselves able to make a fresh start—a further advance into regions as yet unknown'.[18] He had earlier tried out his ideas in a letter to Threlfall, and a brief abstract of the paper appeared in the journal *Nature*.[19] These suggestions were developed further in papers to the Royal Society of South Australia and to the January 1895 Brisbane meeting of the AAAS.[20]

Although William was a little behind the latest developments of electromagnetism in Europe, it was a confused area of study and his work was a scholarly inquiry for the benefit of himself, his students, and his Australian colleagues.[21] As his daughter later observed, these early events 'illustrate one of his most striking characteristics; WHB could not rest until he had mastered some new idea completely, reduced it to a logical form which satisfied him, and expressed it in the simplest possible way'.[22]

The fifth AAAS congress was held in Adelaide during September 1893. Rennie and Bragg were joint secretaries, and they worked hard to maintain the scope and attendance of the earlier meetings. There were difficulties, however, not least in choosing September to avoid the hot January weather in Adelaide.[23] In surveying the early AAAS meetings, Roy MacLeod commented on the 1893 congress as follows: 'After two sea voyages in succession [to Christchurch and Hobart], there were fears that the Association might be deserted... In September 1893 the Association travelled hopefully to Adelaide... There took place its fifth and smallest meeting to date... The timing of the meeting conflicted with university terms, thus losing academic Victorians... and many from the other colonies'.[24] William did not present a paper, but at the meeting of the General Council of the Association, held at the end of the congress, Professor Bragg moved 'that a committee be appointed to report on... the thermodynamics of the voltaic cell', and Professor Kernot of Melbourne moved 'that the best thanks of the Association be offered to... Mrs Bragg and the ladies associated with her'.[25]

[18] W H Bragg, 'Mathematical analogies between various branches of physics', in A Morton (ed.), *Report of the Fourth Meeting of the Australasian Association for the Advancement of Science, held at Hobart, 1892* (Hobart: AAAS, 1893), pp. 31–47.

[19] Letter W H Bragg to R Threlfall, 20 December 1891, RI MS WHB 6C/23; 'Report of AAAS meeting...', *Nature*, 1892, 45:423.

[20] W H Bragg, 'The energy of the electromagnetic field', *Transactions of the Royal Society of South Australia*, 1892, 15:74–6; W.H. Bragg, 'The energy of the electromagnetic field', in J Shirley (ed.), *Report of the Sixth Meeting of the Australasian Association for the Advancement of Science, held at Brisbane, 1895* (Brisbane: AAAS, 1896), pp. 228–31.

[21] F Bevilacqua, *The Principle of Conservation of Energy and the History of Classical Electromagnetic Theory (1845–1903)*, Ph.D. dissertation, University of Cambridge, 1983.

[22] Caroe, n. 2, p. 31.

[23] *Register*, 25 September 1893, p. 4.

[24] R MacLeod, 'From imperial to national science', in MacLeod, *The Commonwealth of Science*, n. 5, ch. 2, p. 47.

[25] 'Extract of minutes of General Council meeting held on 2 October 1893', in R Tate, E H Rennie, and W H Bragg (eds), *Report of the Fifth Meeting of the Australasian Association for the Advancement of Science, held at Adelaide, 1893* (Adelaide: AAAS, 1894), pp. xxi–xxiii; both motions were carried.

Towards the end of 1890 the University of Adelaide received notification that some of the income resulting from the investment of funds remaining from the Great Exhibition of 1851 in London were to be used to establish a scheme of Science Research Scholarships.[26] These awards were intended, 'not to facilitate attendance on ordinary collegiate studies, but to enable students who have passed through a college curriculum and have given distinct evidence of capacity for original research, to continue the prosecution of science with the view of aiding its advance, or its application to the industries of the country'.[27] The universities of Sydney, Melbourne, Adelaide, and New Zealand were initially invited to recommend candidates every second year. The Adelaide University Council considered the offer, instructed the Registrar 'to write and accept the Scholarship for 1892 and to say that the University was taking steps to give effect to their offer', and referred it to its Education Committee for this purpose.[28]

The requirement that candidates should have demonstrated a capacity for original work placed an obligation upon Adelaide's science professors to begin (or enhance) a research programme. When William wrote in 1892 to the Commissioners for the Exhibition of 1851, recommending Bernard Allen for Adelaide's first award, he had to suggest, perhaps with some embarrassment, that 'although he [Allen] has not, during his undergraduate course, done any original work of consequence, he has done a considerable amount of practical work in Physics; and he indicates high promise of capacity for advancing science by original research'.[29] The Commissioners accepted this recommendation and Allen went to Sydney to work with Threlfall on precision measurements of the electrical properties of sulphur. Earlier William had sent another of his Adelaide students, Coleridge Farr, to Sydney to study electrical engineering under Threlfall's guidance, with the assistance of Adelaide University's Angas Engineering Scholarship.[30]

The family of George Fife Angas, originally of Newcastle-upon-Tyne, England, took an interest in South Australia from its beginning,[31] and his second son, John Howard Angas, gave generously to the University of Adelaide: £4,000 in 1878 to found the Angas Engineering Exhibition and Scholarship, and £6,000 in 1884 for the chair of chemistry.[32] The three-year exhibition was awarded annually for study leading to a B.Sc. degree, while the three-year

[26] Letter Agent-General for South Australia, London, to Registrar, 22 August 1890, UAA, S200, docket 13/1890.

[27] *Record of the Science Research Scholars of The Royal Commission for the Exhibition of 1851, 1891–1960* (London: The Commission, 1961), pp. 99–101.

[28] Council Minutes, meeting of 26 September 1890, UAA, S18, vol. V, p. 29.

[29] Letter W H Bragg to the Commissioners, 28 March 1892, Archives of the Royal Exhibition of 1851, Imperial College, London, in J B Allen file, no. 31.

[30] Home, n. 3, *passim*; the Commissioners are unlikely to have accepted the Allen nomination in later years.

[31] E J R Morgan, 'Angas, George Fife (1789–1879)' and 'Angas, George French (1822–1886)' in A G L Shaw and C M H Clark (eds), *Australian Dictionary of Biography* (Melbourne: MUP, 1966), vol. 1, pp. 15–18 and 18–19 respectively.

[32] S O'Neill, 'Angas, John Howard (1823–1904)', in N B Nairn, A G Serle and R B Ward (eds), *Australian Dictionary of Biography* (Melbourne: MUP, 1969), vol. 3, pp. 36–8.

scholarship was awarded every three years to a graduate of the university to study civil engineering in the United Kingdom.[33] Farr received the first exhibition in 1888 on the personal nomination of the donor, and then the scholarship in 1889. His time in London was cut short by illness, however, and on his recovery he was allowed to complete his scholarship in Sydney with Threlfall. In 1894 Farr taught the first class in advanced electrical engineering at the South Australian School of Mines, under William's guidance. Late in 1896 he received a lecturing appointment in New Zealand, and he spent the rest of his life and career there, rising to become Professor of Physics at Canterbury College, Christchurch. His most notable research achievement was a comprehensive magnetic survey of New Zealand, for which he was awarded Adelaide's first DSc degree[34] Allen's research work in Sydney was regarded highly by Threlfall, but it was also cut short by illness. He later held mathematics and physics appointments in Adelaide and then at the Technical School in Perth, Western Australia. In 1911 he completed an Adelaide BA degree with honours in mathematics under William, but it seems his health was never strong, and he contracted typhoid fever and died in Perth in March 1912.[35]

In April 1891 a group of undergraduate students founded the Adelaide University Scientific Society.[36] The objects of the society were to gather together those interested in science and to promote the study of both pure and applied science and 'especially of those which relate particularly to Australia'.[37] A year later the stated aims had been amended to include 'encouraging original research', and it was reported that 'the members have carried out the idea well... devoting themselves earnestly to the fascinating pursuit of scientific knowledge'.[38] The Society held its first annual conversazione on 30 May 1892, when 'a most instructive and entertaining programme was carried out in a manner delightful to the large number [nearly 400] of ladies and gentlemen attending... in the physical Laboratory were exhibited the electromagnet, Wimshurst machine, sound-plates, smoke-rings &c.'.[39]

The year 1892 also saw further agitation for scientific research in the Australian universities. In Adelaide the *Register* interviewed Professor Anderson Stuart of the University of Sydney, returning from a trip to Europe, and published a long editorial. It quoted Stuart's observation that 'more original

[33] See, for example, *Calendar of The University of Adelaide for the year 1900* (Adelaide: University of Adelaide, 1900), pp. 44–8, 144–6.

[34] J G Jenkin and R W Home, 'Farr, Clinton Coleridge (1866–1943)', in C. Cunneen (ed.), *Australian Dictionary of Biography: Supplement 1580–1980* (Melbourne: MUP, 2004), pp. 123–4.

[35] For Allen see J G Jenkin, 'Frederick Soddy's 1904 visit to Australia and the subsequent Soddy-Bragg correspondence', *Historical Records of Australian Science*, 1985, 6:153–69, 154–5.

[36] *Register*, 31 May 1892, p. 5.

[37] 'Rules of the Adelaide University Scientific Society', Adelaide, 1891, copy of leaflet in possession of author; letter Alex. Wyllie and fourteen other students to Registrar, 25 June 1891, UAA, S200, docket 266/1891.

[38] See n. 36.

[39] Ibid.

research must come from the Australian Universities', and added its own characteristic observation that 'it is a very serious and only too well-founded reproach against the Universities of Australasia that up to the present time they have contributed very little indeed to the sum total of human knowledge'.[40] In May, in an article reporting the return from Europe of Adelaide's Professor Edward Stirling, the *Register* noted the difficulties of keeping up with advances in science, thousands of miles from the centres of learning, and admitted that 'the system in Adelaide necessitates... the whole energies of the teacher being devoted to teaching, and very little time remains to him for anything like original work'.[41] This was certainly true of William Bragg. When Edward Rennie, the Angas Professor of Chemistry, prepared his graduation address late the same year, he included remarks that the newspaper also fulsomely reported: 'Professor Rennie's oration was in part a protest against the purely utilitarian and materialistic view of the function of the teacher of science... it may safely be asserted that there is no more insidious enemy to the happiness of mankind than the same spirit of opposition to everything which has not an immediate money-making result... There is for many reasons much cogency in Professor Rennie's appeal to students to apply themselves to original researches'.[42] William Bragg could not have missed all this rhetoric regarding the desirability of research.

At home, William's sons were maturing. Willie remembered his early childhood as follows:[43]

> I was sent, I suppose when five, to a convent school on the other side of North Adelaide; I used to walk both ways. I must have been a very conventional and timid small boy, because I remember once the butcher... chanced to pass when I was being subjected to some mild bullying by the larger boys and girls. Next day the whole school chanted 'Tell Tale Tit' at me; I had not breathed a word to anyone about the bullying—it was, of course, the butcher who had reported it to my parents, who had taken the matter up with the headmistress. How powerless the young are when dealt injustice! I could not possibly have explained. I remember, too, a fierce argument with one of the nuns about the way a mirror worked. She was, of course, right, but it perhaps showed an early interest in science.
>
> On Sundays we traditionally paid a visit to the grandparents at the Observatory. We went in a four-wheeler with two horses which had a rich smell, partly horse, partly cab, and I suspect partly driver... The Observatory was a wonderful place for small boys. It was a rambling two-story house on West Terrace, with deep latticed verandahs and balconies in front. There was a circular drive... in a lawn of buffalo grass planted with almond trees, was a cluster of buildings which housed the offices, the transit telescope, and other astronomical and telegraphic equipment. At the back of the house was another wide verandah and at one end of this

[40] *Register*, 23 March 1892, p. 4.
[41] *Register*, 5 May 1892, p. 6.
[42] *Register*, 15 December 1892, pp. 4, 6.
[43] W L Bragg, Autobiographical notes, pp. 3–6.

was a bathroom...The bath on the verandah was the centre of endless games. Stones and bricks made islands and harbours for our boats...

The outbuildings were fascinating because all sorts of junk which could be incorporated into our games had accumulated there...Another outhouse contained souvenirs our grandfather had brought back from the interior when he put up the overland line; in particular, gorgeous shells from the Tasman [sic, Arafura] Sea; there were boxes of old letters which we ransacked for stamps. The carriage gate at the bottom of the backyard, overhung by a large fig tree which had especially luscious figs, led on to the back drive, and beyond that was the large equatorial telescope in its dome, surrounded by various smaller buildings housing the meteorological instruments. We used to accompany grandfather when he went the round and made the readings on Sunday. The evaporation tank was part of the programme, but its readings must have been rather untrustworthy when we were there because we fished for the tadpoles it contained. The cellar under the main telescope building also contained a glorious collection of junk: insulated wire, battery elements and chemicals which, when we were older, we used for the electrical gadgets we made...

The Sunday lunch was presided over by our dear, placid, vague grandmother in her old lady's cap with lace frills...Knowing my love of it, she always provided for me a custard in a stemmed glass. After lunch my father and grandfather smoked their cigar of the week, and we were dispatched to play in the grounds. All has gone now; South Australia has ceased to have an Astronomer of its own, and the lovely old house as well as the astronomical buildings have all been demolished[44]...

Around the time of the Queen's birthday, many Aborigines within reach of Adelaide would arrive and camp in the Park Lands around the Observatory. In the nineties, there were still quite a number who came each year. A man received a blanket from the Government and a woman a pair of stockings. They could also get medical treatment...Women and men had strange voices, deep and guttural and yet liquid. These Aborigines were the last relics of the tribes which had lived in that area, and a few years later there were no more of them to claim the bounty.

Most of our holidays were spent at the sea, some seven miles away from Adelaide. There were several seaside settlements...each with a long wooden jetty...The journey to the sea was made in a horse tram from Adelaide to Henley Beach, and there were two great excitements on the way. At one point the road crossed a bridge over a stream bed, and how we craned out of the tram to look down, because sometimes one could glimpse WATER! It is hard to convey what a thrill this was; there were many stream beds in the country around Adelaide, worn deep into the ground, but they were quite dry except when there was torrential rain. I do not think my brother and I ever saw a running stream till we came to England. The other thrill came on the return journey. As we approached the plateau on which Adelaide is built, there was a fairly steep hill leading up to West Terrace. At the bottom of this hill was a tin shed, the habitat

[44] Adelaide High School now occupies the site.

> of a very large horse and a very small boy. When the tram hove in sight the boy trotted out on the horse, flung a cable to the driver which he cleverly hooked on to the moving tram, and our steeds, now augmented to three, were lashed into a hand-gallop which bore us triumphantly to the crest, where the extra horse and its postilion were released and returned to their post. There was a dash about the whole affair which appealed to us greatly.

Willie recalled the argument about the working of a mirror as an early indication of his interest in science, but his playing in and about the Observatory was surely more formative; and there was his father. Evidence has now emerged that the 'convent school' was that of the Dominican Sisters in North Adelaide. Founded in the 1840s at Stone in England, the pastoral work of the originating Dominican congregation was focused on schools, the sick, and the poor, while their spiritual work lay in the reintroduction of Roman Catholic devotions in England. Six of the sisters responded to a request from Adelaide to care for the sick, and they arrived in 1883. Their leader, Mother Rose Columba, was the daughter of a wealthy Protestant family, and her companions were similarly educated in the liberal arts. When general nursing was not possible they turned instead to teaching and needlework. Mother Columba reported back to Stone with brutal honesty: 'We must have a school to help us pay the rent and exist'.[45] Supported by the Adelaide community, the sisters built a chapel, a priory, and a successful college for girls and young women, from grade 1 to matriculation. Willie Bragg probably attended in 1896 and 1897, although only 1897 can be documented: 'Willie Bragg, £2.2.0' for each half of 1897.[46] Such schools commonly took boys in the youngest grades, and surviving evidence at the College testifies to the sisters' exceptional ability in music, painting, pottery, illumination, and needlework, skills that would have attracted Gwendoline. The local catholic parish was upper-middle-class and boasted gifts from several of the early South Australian governors, who were high Church of England adherents.[47] This was Willie's introduction to education.

The annual exhibition of the Society of Arts in 1896 included works by Gwendoline and William: 'The Old Threshing Floor by W H Bragg has good points in colour, but is formal in detail', while 'in G Bragg's Study by Lamplight the conflict of yellow gleams and flickering shadows is skilfully interpreted, but success has not been attained in all the faces of the family reading around the table'.[48] In September William gave a lecture to the Society on 'Colour',[49] while in the previous November he had been elected by the Society as its representative on the Board of Governors of the Public Library, Museum, and Art

[45] S Burley and K Teague, *Chapel, Cloister & Classroom: Reflections on the Dominican Sisters of North Adelaide* (Adelaide: St Dominic's Priory College, 1993), pp. 8–20, 14.

[46] Account Book, from 1893 to 1916, handwritten, with students' names and fees paid, St Dominic's Priory College Archives, North Adelaide.

[47] D O'Sullivan, *Dominican Sisters of North Adelaide: Their History and Spirituality, 1883–1983* (Adelaide: St Dominic's Priory, 1983), ch. 1.

[48] *Register*, 18 June 1896, p. 6.

[49] *Register*, 8 September 1896, p. 5.

Gallery of South Australia.[50] Sir Thomas Fowell Buxton, the new Governor of South Australia following the departure of the Earl of Kintore, arrived in Adelaide in October 1895, and Professor Bragg was amongst those received at his first reception, a levee at Government House.[51] As Dean of the Faculty of Arts, William presented the governor for the award of the degree M.A. *ad eundem gradum* at the university commemoration ceremony in December, and in July 1896 William and Gwendoline were among the guests at St Peter's Cathedral for the wedding of the governor's daughter.[52] Professor Bragg was now a major public intellectual and William and Gwendoline were prominent at most major events in Adelaide.

At the university the inadequacy of the facilities and accommodation for students was still causing resentment. A comprehensive solution was needed, and in April 1895 Rev. Canon F Slaney Poole, a prominent member of the Council, convened a public meeting to develop a plan to provide a room, a name, and a context for the emerging student and graduate body of the university. The basic aim of the proposed 'Adelaide University Union' was to coordinate the social, sporting, and intellectual activities previously organised by separate clubs and societies, a particular necessity in a non-residential university such as Adelaide.[53] In May the Council considered a letter from the Vice-Chancellor (John Hartley) and Professor Bragg, on behalf of the committee of the proposed Union, asking the Chancellor 'to mention to the Council the suggestion to build a room on the University grounds, which should be a home for the Union and a centre of social life for the students'.[54] The Council responded that it would consider the proposal favourably when a definite scheme was submitted.[55] From the beginning there was no suggestion that the university could—or the government would—fund such an undertaking.

William apologised for not attending the April and May meetings of the Union Committee but he was elected a Vice-President in his absence, and his signature on the letter to the Chancellor is revealing. His excellent relationship with staff, students, and graduates alike was bound to draw him into the project. At a meeting in June, 'Professor Bragg... presented his report on the Scheme of Building a room and... [it] was accepted and referred back to the Committee for the best means of carrying it into effect'.[56] Donations and loans were sought and the response was so satisfactory that Bragg, with the Vice-Chancellor, was able to write to the Council in April the following year saying, 'We have the honour... to ask your permission to erect... a building for the use of members of the Union. We enclose a plan of the proposed building, prepared

[50] *Register*, 8 November 1895, p. 5.

[51] *Register*, 1 November 1895, p. 6.

[52] *Register*, 23 July 1896, p. 4.

[53] W G K Duncan and R A Leonard, *The University of Adelaide, 1874–1974* (Adelaide: Rigby, 1973), p. 58; M M Finnis, *The Lower Level: A Discursive History of The Adelaide University Union* (Adelaide: Adelaide University Union, 1975), p. 10.

[54] Letter J A Hartley and W H Bragg to Chancellor, 30 May 1895, UAA, S200, docket 281/1895.

[55] Council Minutes, meeting of 31 May 1895, UAA, S18, vol. V, p. 370.

[56] Finnis, n. 53, p. 59; ch. III is devoted to '1897–1913: The First Union'.

by Mr Naish, the University architect. The site proposed is the old and deserted tennis court immediately behind the University... We have enough money to enable us to put up at once the main hall of the building. We propose to add the side rooms, the verandah, the porch, the roof panelling and the wainscoting as the growth of our building fund permits us'.[57]

The Council approved the request.[58] The proposed location was directly behind the main university building, which had received a small extension on its north-east corner in about 1885 for physiology teaching in connection with the new Medical School.[59] The State Governor laid the foundation stone on 5 August 1896.[60] The basic structure rose quickly, a rectangular stone building, 36 feet by 24 feet (11 m × 7.3 m), with a large window on each side of a central door and a pitched roof; a small and insignificant building, simple and functional, behind the ornate main building. The anticipated side rooms, verandah, and porch were never built, although this charming reminder of earlier days remained as the university book-room during my own undergraduate days and until 1972, when it was demolished to allow for major extensions.[61]

William was the leading activist throughout this building programme, and even after it was completed promised subscriptions had to be collected and the room furnished. Few financial details survive, but there is one record of the Union Committee being able to reduce its 'debt to Professor Bragg to a very appreciable extent as will be seen from the Treasurer's Balance Sheet'.[62] William was surely one of the original donors, and clearly he also provided significant additional funds as a loan. On his departure for leave in Britain during 1898, the minutes of the Union Committee record that 'the Committee made an effort...to show how we recognise his unceasing labours on our behalf, but...had to content ourselves with a written expression in the name of the members of the Union'.[63] In 1896 the boats, tennis, and lacrosse clubs of the university had amalgamated to form the Sports Association, and in 1897 it became an affiliated member of the Union. William had played a large part in the health and progress of the last two clubs. The inauguration of non-academic facilities in the University of Adelaide owed most to Professor William Bragg.[64]

Public or 'extension' lectures began at the University of Adelaide in 1877, during only its second year of teaching, but they had declined and were replaced in 1885 by a programme of regular evening courses.[65] From 1887 until 1897

[57] Letter W H Bragg and J Hartley to Council, 24 April 1896, UAA, S200, docket 192/1896.

[58] Council Minutes, meeting of 24 April 1896, UAA, S18, vol. VI, p. 29.

[59] In 1961 the original university building was named the Mitchell Building on the occasion of Sir William Mitchell's 100th birthday and in acknowledgement of his unique contribution to the institution.

[60] *Register*, 6 August 1896, p. 6.

[61] Finnis, n. 53, pp. 1, 61; there is also a photograph of the building in Finnis, opposite p. 66.

[62] Ibid., p. 63.

[63] Ibid., p. 64.

[64] Ibid., p. 61.

[65] 'Continuing education at The University of Adelaide, 1876–1983', a catalogue prepared by S Woodburn to accompany 'An exhibition of documents, photographs, newspaper cuttings and publication', Barr Smith Library, University of Adelaide, March 1984.

William conducted the first-year physics and practical physics evening classes, averaging 110 hours per year to an average class size of 26 students.[66] In 1888 William proposed a series of evening lectures on general subjects and obtained the support of the university's Education Committee, but the Council, remembering the earlier experience, did not adopt the proposal. In August 1895, however, in response to requests from a variety of city and country bodies, the Education Committee appointed an Extension Lectures Committee, including Professor Bragg, and the scheme was resurrected.[67]

Four series, of six lectures each, were presented by the professors in 1895: Rennie on 'The atmosphere', Bensley on 'Rome', Mitchell on 'English literature and philosophy from 1700 to 1750', and Bragg on 'Radiation'. William's syllabus was extensive, again involving topics he had studied with Glazebrook in Cambridge and that he had come to love for their popularity and ease of illustration by experiment and diagram.[68] His six lectures involved: the nature of radiation (the wave theory), the reflection and refraction of radiation (using light), use of a prism to form a spectrum (light, sun, stars), colour (absorption, mixtures, sensation), heat radiation, and chemical effects (photography, fluorescence, electric waves and their transmission and reflection, the 'coherer'). The 'electric waves' referred to the newly-discovered wireless or radio waves, which will be discussed later in this chapter. The substance of the lectures became the basis of William's 1931 Christmas lectures at the Royal Institution in London and of the fuller treatment that followed in his book *The Universe of Light*.[69]

In a second set of extension lectures in July-August 1896, William planned to speak about 'The elementary principles of the electric transmission of power', but a last-minute change to this plan occurred and an amended syllabus of four lectures was printed, entitled 'The electric discharge and its latest development—Röntgen rays'. The story of the discovery of X-rays is available very widely and need only be outlined here.[70] When two metal discs are sealed into a glass envelope or discharge tube and connected to a source of high voltage, a series of curious and colourful effects are produced as the air in the tube is progressively withdrawn. At very low pressure the glass itself glows, and it was known that this was caused by 'cathode rays' emanating from the negative disc or cathode. The German physicist Wilhelm Röntgen, Professor of Physics at the University of Würzburg, began studying cathode rays in 1895 and, repeating earlier work, he covered the tube with black cardboard to mask

[66] Successive annual *Adelaide University Calendars*.

[67] 'Continuing education', n. 65, p. 4; Council Minutes, meetings of 29 March, 31 May, and 30 August 1895, UAA, S18, vol. V, pp. 354, 368, and 381 respectively.

[68] 'University extension lectures: syllabi of lectures', advertising leaflets in Minutes of Special Committes, UAA, S22, vol. I; see also J G Jenkin. 'W H Bragg and the public image of science in Australia', *Search* (Australia), 1987, 18:34–7; D Knight, 'Getting science across', *British Journal for the History of Science*, 1996, 29:129–38.

[69] Sir William Bragg, *The Universe of Light* (London: Bell & Sons, 1933).

[70] See, for example, O Glasser, *Wilhelm Conrad Röntgen* (Springfield: Thomas, 1934); id., *Dr. W C Rontgen* (Springfield: Thomas, 1945).

the fluorescent glow of the tube. Late in the year he noticed that some crystals on a distant bench were glowing. Fascinated, Röntgen spent the next seven weeks investigating the phenomenon, and on 28 December 1895 he communicated his initial findings to the Physikalisch-medicinischen Gesellschaft of Würzburg, which immediately published them. A second communication followed in March 1896.[71]

Many of the characteristics of the new rays, widely called X-rays because their nature was unknown, were elucidated by Röntgen himself in his initial publications. Their most intriguing property was the ability to penetrate solid materials. Its most dramatic illustration occurred when a hand was placed between the tube and a fluorescent screen and the bones could be seen within the flesh. In addition, Röntgen suggested that the source of the X-rays was the area where the cathode rays struck the glass tube, that photographic plates responded to the rays but the human eye did not, that, unlike visible light, X-rays could not be reflected, refracted or polarised, that, unlike cathode rays, they could not be deflected by a magnet, and that the air became an electrical conductor when X-rays passed through it.[72] So what were these X-rays? It was a fascinating mystery and there were many suggestions.

Röntgen sent draft copies of his findings to a small group of fellow scientists and friends, *Die Presse* in Vienna published a long article on 5 January 1896, and accounts soon appeared in newspapers and periodicals around the world. For readers of English, the journal *Nature* published a translation of Röntgen's first communication on 23 January.[73] As William said many years later, 'no scientific discovery before or since that of Röntgen in 1895 has excited such immediate or universal interest'. The ability of X-rays to see through solid objects, and their consequent medical potential, aroused enormous public fascination and fear.[74]

In Australia the first brief press reports appeared on 31 January 1896 in Melbourne, Sydney, and Adelaide, and the first experiments were undertaken by Thomas Lyle at the University of Melbourne, by Walter Filmer, a Newcastle electrical engineer, and by Rev. James Slattery at St Stanislaus College in New South Wales.[75] Richard Threlfall, with his colleague James Pollock, conducted an experimental investigation of the nature of the new radiation in April and May 1896, and considered a series of possibilities; namely, that the radiation consists of a swarm of material particles, an 'aether' wind, aether vortices

[71] W C Röntgen, 'Ueber eine neue Art von Strahlen [On a new kind of ray]', *Sitzungsberichte der Wurzburger Physikalischen-Medicinischen Gesellschaft*, December 1895, pp. 132–41, and March 1896, pp. 11–19.

[72] Glasser, *Dr. W C Rontgen*, n. 70.

[73] Translation by A. Stanton in *Nature*, 1896, 53:274–6.

[74] Sir William Bragg, 'The early history of X-rays', as reported in *Nature*, 1929, 123:218; for initial reaction, see, for example, R Winau, 'The impact of Roentgen's discovery on medicine', in *Wilhelm Conrad Roentgen, 1845–1923* (Bonn: Inter Nationes, 1973), p. 49.

[75] For a detailed account of the Australian responses to the discoveries of X-rays and radioactivity, see H Hamersley, 'Radiation science and Australian medicine', *Historical Records of Australian Science*, 1982, 5:41–63; see also J P Trainor, *Salute to the X-ray Pioneers of Australia* (Sydney: Watson & Sons, 1946).

moving to or from the source, aether waves, electromagnetic waves of either very small wavelength or having a longitudinal component, or a phenomenon of a new order entirely. All the results were negative and the authors did not pursue the matter further.[76] As a result of this work, however, Threlfall was able to add a very useful appendix, entitled 'On the preparation of vacuum tubes for the production of Professor Röntgen's radiation', to the book he published from Sydney, one that gave evidence of his superb laboratory skills.[77] Oliver Lodge, the Professor of Physics at University College, Liverpool, summed up the nature of X-rays for most researchers when he said, as early as July 1896, that in all probability they were ordinary transverse ethereal waves, moving with the velocity of light, of various wavelengths down to 10^{-8} cm.[78] Threlfall and Pollock's paper, the problem it addressed, and the laboratory techniques required for the investigation stayed in William's mind and, later in Adelaide, he was to propose a very different model (see chapter 13).

In Adelaide the *Register* and the *Advertiser* carried editorials and articles about the new rays,[79] but it is unclear when local experiments were first attempted. Mr S Barbour, a local manufacturing chemist and photographic enthusiast, had been in England and America on holiday and had secured two examples of the necessary high-vacuum tubes. On his return to Adelaide he and an assistant obtained indicative X-ray photographs, but they lacked an adequate source of high voltage to obtain better results.[80] Meanwhile, William had been unsuccessful in his initial attempts at the university because Rogers had to make a tube, and the level of vacuum achieved was inadequate. However, on Friday evening, 29 May 1896, the two groups pooled their resources—William provided a high-voltage induction coil from his father-in-law's department and Barbour one of his tubes—and demonstrated the new technology to a group of doctors. Excellent photographs were achieved in five-minute exposures of William's hand, Alfred Lendon's foot, and Dr Swift's knee and wrist, clearly showing the separate bones and several defects.[81] Further experiments were made the following week, locating an embedded needle in a lady's hand and demonstrating X-rays to the Governor and his family.[82]

Public excitement grew with the publication on Saturday 6 June of a full-page newspaper account of recent events,[83] and the *Register* devoted an editorial to William's first upcoming public lecture on X-rays, which was also 'the means of helping forward considerably the funds of the Students' Union, to

[76] R Threlfall and J A Pollock, 'On some experiments with Röntgen's radiation', *Philosophical Magazine*, 1896, 42:453–63.

[77] R Threlfall, *On Laboratory Arts* (London: Macmillan, 1898), pp. 90–107.

[78] O Lodge, 'The surviving hypothesis concerning the X-rays', *The Electrician*, 1896, 37:370–3.

[79] See, for example, *Register*, 1, 15, and 21 February 1896, pp. 4, 5, and 5 respectively, ibid., 28, 29 May 1896, pp. 5, 5 respectively; *Advertiser*, 2 March 1896, pp. 3, 4.

[80] *Register*, 30 May 1896, p. 5.

[81] Ibid.; *Advertiser*, 30 May 1896, p. 4.

[82] *Register*, 2 June 1896, p. 5 and 5 June 1896, p. 5.

[83] *Supplement to the Adelaide Observer*, 6 June 1896, p. 3.

which the proceeds over and above the expenses are to be applied'.[84] So many people had to be turned away from the university library that William agreed to repeat the lecture in a large public hall.[85] His extension lectures on the physics of the phenomenon then followed, the first being devoted to the fundamentals of electricity, the second and third to the varied and colourful effects taking place within a discharge tube at different pressures. Tubes made by Rogers were used for the first time, and all the lectures drew high praise for their clarity and delightful illustrations.[86] The fourth and last lecture was concerned with Röntgen's work and discovery.[87] For the remainder of 1896 William lectured regularly on the new rays: to about 100 members of Our Boys Institute, to the North Adelaide Young Men's Society, to the teachers' conference, and a repetition of his four extension lectures at the Mount Barker Institute in the Adelaide hills.[88] These lectures were assisted by the arrival of new 'Röntgen-Crookes tubes' and a 'really good spark coil', the result of requests from Charles Todd to Sir William Preece in London and prompt payment by Uncle William, who was 'holding the Professor's money in England'.[89]

A related event now occurred that intimately involved William, his elder son Willie, and the new X-rays. It was the first time these three elements came together, but it would not be the last. In fifteen years' time they would again become associated, and in a way that would produce a Nobel Prize and set them both on the high road of outstanding research. But that was in the future; Willie's autobiographical notes explain:[90]

> When I was six I had an accident which might have had worse results if it had not been for the skill of my doctor uncle (Uncle Charlie Todd). We used to play in the afternoon in a square at the centre of North Adelaide [Wellington Square], and once when I was riding my tricycle Bob jumped on behind and upset me. The weight of both of us fell on my left elbow, which was smashed into numerous pieces. I remember well the walk home, with my arm feeling strangely stiff and the consultation round my bedside. The family doctor, Dr Lendon, thought the smash to be beyond repair and could only advise that the arm be allowed to set stiff in the most useful position. Uncle Charlie, however, determined to do better. Every few days I was put under ether, and the doctors flexed the arm backwards and forwards so as to coax a new joint to form. How I hated

[84] *Register*, 17 June 1896, p. 4.

[85] *Register*, 18 June 1896, p. 6.

[86] *Register*, 15, 22, and 29 July 1896, p. 7, 7, and 3 respectively.

[87] *Register*, 5 August 1896, p. 6.

[88] Unidentified Adelaide newspaper cuttings in RI MS WHB 39. William thought to refract X-rays in a sulphur prism, but he then realised that, even if they were waves, their characteristics would make the experiment unproductive: see letter W L Bragg to Sir Mark Oliphant, 13 October 1966, RI MS WLB 54A/27: 'I remember in the cupboard in his office an enormous prism of sulphur, which he used to point to as an object lesson to remind him to be modest—had tried to find the refractive index of x-rays with it!'

[89] Letter of gratitude, C Todd to W. Preece, 13 October 1896, Todd letter books, State Records of South Australia, Adelaide, GRG 31/1, vol. 4, pp. 243–246.

[90] W L Bragg, Autobiographical notes, pp. 6–7.

> these occasions! I would be quietly playing and hear the dreaded voices in the hall which announced their arrival, and start yelling at the top of my voice. The treatment was successful, and left me with a very useful left arm, although it is out of the straight and shortened. [Willie was fortunately right-handed]...
>
> Incidentally, I must have been one of the first patients to be X-rayed in South Australia. Very soon after Röntgen's announcement of the new radiation, my father set up a tube worked by an induction coil, and he took radiographs of the broken elbow. I was scared stiff by the fizzing sparks and smell of ozone, and could only be persuaded to submit to the exposure after my much calmer small brother, Bob, had his radiograph taken to set me an example.

The repair made to Willie's elbow was, indeed, timely and happily successful, although it almost certainly restricted the sorts of sports he would later be able to enjoy. The very different personalities of the two boys also show clearly in this extract.

For his 1897 extension lectures William's topic was 'Sound', by now an old favourite. The six lectures covered the nature and travel of sound, pitch and noise, scales and strings, air columns (organ pipes and resonance), other vibrations (bars, membranes, bells, phonograph, telephone, interference and beats), and consonance and dissonance.[91] He would return to the topic yet again at the Royal Institution in London in 1919, in a series of lectures entitled *The World of Sound*.[92]

Contrary to newspaper expectations, the Adelaide extension lecture programme was a major ongoing success, not least the lectures of Professor Bragg. In 1895, for example, after his second lecture the *Register* reported: 'Following up the first lecture, which dealt with a general explanation of radiation in the form of waves of light, the lecturer proceeded to demonstrate by diagrams on the blackboard and an electric-lighted lantern with screens, lenses, mirrors, prisms, and other ingenious appliances the conduct of waves of light when manipulated to strike the eye...The deftness and success with which the Professor carried out his experiments and the exceedingly simple method of unravelling the mysteries of a difficult branch of science did him infinite credit'.[93] Similar reports continued to attract large audiences to public lectures throughout the remainder of William's time in Australia. His university students also remembered his facility to simplify and clarify the most complex topics: 'He had a wonderful flair for illustrating knotty points by some homely simile, a facility which was later to win him great fame and popularity'.[94] His daughter was told that her father was 'an unimpressive lecturer to start with, he would be too careful, too mathematical',[95] but obituaries

[91] 'University extension lectures', n. 68.

[92] Sir William Bragg, *The World of Sound* (London: Bell & Sons, 1920).

[93] *Register*, 11 July 1895, p. 5; for its gloomy prediction see *Register*, 14 June 1895, p. 4.

[94] Reminiscences of F K Barton, Adelaide undergraduate student 1907–09, extract from typescript in possession of the author.

[95] Caroe, n. 2, p. 31.

and articles paint a very different picture. Andrade noted that, 'By 1930 Bragg had become...something of a national figure';[96] and the historian John Heilbron wrote, 'Bragg spoke and wrote with exemplary clarity. Between the wars he was literally the voice of science in England'.[97]

William's approach to public lecturing may be summarised as follows. First, his subjects were areas of physics accessible to the general community. Second, William mastered his subject before daring to speak about it publicly. A reporter who interviewed him in 1926 commented, 'His mastery is revealed in his ability to put his thoughts into the simplest language...Only masters of an art or science are able to make it look or sound easy'.[98] Third, William freely expressed his own personality in his lectures. Andrade spoke of 'the originality and personal qualities which he could bring to an apparently hackneyed subject', and of 'His warm, simple, persuasive utterance, his personal tone in lecturing, which made each member of the audience think that the remark was intended for him'.[99] Fourth, William used copious analogies, blackboard illustrations, lantern slides, experiments, and demonstrations, all of which took valuable time for preparation and practice. His technician, Arthur Rogers, provided crucial assistance, which William gratefully acknowledged. In 1892 and 1897 William wrote to the Finance Committee of the university, successfully supporting a salary increase for Rogers,[100] while in 1896 he wrote to say, 'I have the honour to ask that a sum of £1 be paid to A L Rogers out of the Extension Lectures Fund...I think he well deserves a little recognition of the way in which he has worked to make the Rontgen tubes and photographs a success'.[101]

Finally, William conveyed an ease and confidence of presentation that led many commentators to suggest that he had unusual gifts and ability in the area. The truth was that this illusion was created only after hours of preparation. More perceptive writers noted: 'Beforehand he appeared to feel some diffidence, and certainly the utterance that seemed to come so easily and so spontaneously was the fruit of more labour and thought than the audience often suspected',[102] 'He took infinite care in their preparation',[103] and 'The amount of work, the continual polishing that go to produce this sense of ease can hardly be gauged by others'.[104] His recipe was simple but profound: 'the value of a lecture is not to be measured by how much one manages to cram into an hour...it

[96] E N da C Andrade, 'William Henry Bragg, 1862–1942', *Obituary Notices of Fellows of the Royal Society of London*, 1942–44, 4:276–300, 289.

[97] J L Heilbron, *W H Bragg and W L Bragg: A Bibliography of Their Non-Technical Writings* (Berkeley: University of California Office for History of Science and Technology, 1978), p. ii.

[98] W Steed, 'Seeing the invisible: A talk with Sir William Bragg', *Review of Reviews*, 1926, 73:113–122, 113.

[99] Andrade, n. 96, pp. 285, 291 respectively.

[100] Letters W H Bragg to Finance Committee, 25 February 1892 and 27 August 1897, UAA, S200, dockets 91/1892 and 482/1897 respectively.

[101] Letter W H Bragg to Finance Committee, 23 September 1896, UAA, S200, docket 405/1896.

[102] Andrade, n. 96, p. 289.

[103] Caroe, n. 2, p. 98.

[104] Steed, n. 98, p. 113.

is to be measured by how much a listener can tell his wife about it at breakfast the next morning'.[105] All his skills in this area were developed in Adelaide although, as his daughter's later recollections showed, his shyness and modesty remained: 'One evening, opening his study door, I noticed how wearily he looked up from his papers. "Daddy, need you work so hard?" I cried, and he answered simply, "I must dear; I'm always afraid they'll find out how little I know"'.[106]

At the School of Mines, Bragg and Rennie found themselves very active in the deliberations of its Council, for a tension between the School and the university always simmered below the surface. In 1895 Rennie tendered his resignation, saying that he found himself too often opposed to the policies that the majority of Council members wished to pursue, and he declined overtures to reconsider his decision.[107] William would have to walk a tightrope as he carried the academic banner in a Council focused on very practical considerations and outcomes. In 1897 he drew attention to the many students who were too young and possessed an inadequate educational background to benefit from the School's programmes, particularly those entering the Associate Diploma courses. The Council's Education Committee recommended the introduction of a one-year bridging course, and the plan was adopted on the motion of William and a fellow Council member.[108] Later the same year William drew the Council's attention to new courses in mining engineering and metallurgy established by the university, and he proposed a conference between delegates of the two institutions.[109] A deputation from the School met representatives of the University Council in December, and potential conflict between the two bodies was avoided when it was agreed that the university course was an advanced programme and would not compete with the Associate Diploma course at the School.[110]

William continued to be concerned about teachers—their training, their teaching, and their general welfare. He wished always to assist and support them rather than criticise. He knew their accomplishments were modest, but he also recognised the difficulties under which they trained and worked and the crucial role they played in preparing students for the further study that was his prime concern. Names of the various teacher organisations changed over the years, but by 1895 there were three of note: the South Australian Public Teachers' Association, the major organisation for primary school teachers, and both the Collegiate Schools' Association and the South Australian Branch of the (British) Teachers' Guild for teachers in the private secondary schools. In 1897 Professor Bragg was President of both the latter bodies. In October

[105] I James, *Remarkable Physicists From Galileo to Yukawa* (Cambridge: CUP, 2004), p. 206.

[106] Caroe, n. 2, p. 109.

[107] School Council Minutes, meetings of 29 July and 12 August 1895, University of South Australia Archives, vol. 2, pp. 270 and 273 respectively.

[108] Ibid., meetings of 10 and 31 May, and 28 June 1897, vol. 3, pp. 124, 128, and 135 respectively.

[109] Ibid., meeting of 30 November 1897, vol. 3, p. 175.

[110] Ibid., meeting of 13 December 1897, vol. 3, pp. 181–2; A. Aeuckens, *The People's University, 1889–1989* (Adelaide: S A Institute of Technology, 1989), p. 30.

he chaired the annual meeting of the Guild, which involved a discussion, introduced by Professor Mitchell, on 'The advantages of having a Union of all teachers'. Members of the Association were also present.[111] This was perhaps an invitation for the two secondary-school groups to merge, or for the secondary teachers to join their primary colleagues, for the latter had formed the South Australian Teachers' Union the previous year and had mounted an impressive conference in September, 'the largest gathering of teachers... ever seen in the colony'.[112] William addressed this conference on the new Röntgen rays, and also the 1897 meeting on 'Telegraphy without wire'.[113]

In May 1897 Professor Mitchell submitted a paper to the University Council in which he proposed that the university should assume full responsibility for the training of teachers in the colony.[114] It was the university's view that teachers received a sound practical training but that it lacked breadth and intellectual development. A two-year course was proposed that involved subjects from the general arts and science courses and lectures in educational theory and that would provide a basis for the practical training. Since many trainee teachers lacked university entrance qualifications, it was proposed to introduce a bridging course at the university similar to that for the Senior Public (matriculation) Examination. William was a strong supporter of the suggestion and he was appointed to a sub-committee that quickly made some initial progress.[115] The proposal would advance further during his forthcoming absence on leave, and he would be heavily involved on his return.[116] For the present he persuaded the Council to provide an evening's entertainment at the university for the schoolteachers during their annual conference.[117]

In 1894 a major dispute involving the Adelaide Hospital, the colonial government, and the university as an affected bystander, arose and became infamously known as the 'Hospital Row'. William wisely avoided involvement, despite several of his colleagues and friends being heavily enmeshed.[118] When the government announced its intention to discontinue its annual grant to the Medical School, and it then closed in 1896, however, William must have noticed the loss of students from his classes. The students sought their training interstate and the School was not re-established until 1901.[119]

[111] *Register*, 27 October 1897, p. 10 and 3 December 1897, p. 7 respectively.

[112] *Register*, 29 September 1896, p. 5.

[113] *Register*, 22 September 1897, p. 6.

[114] Council Minutes, meeting of 28 May 1897, UAA, S18, vol. VI, p. 106.

[115] B K Hyams, 'State school teachers in South Australia, 1847–1950: A study of their training, employment and voluntary organization', unpublished Ph.D. thesis, Flinders University of South Australia, 1972, ch. 5.

[116] Education Committee Minutes, meetings of 11 June, 9 July, and 10 September 1897, UAA, S23, vol. IV, pp. 108, 112, and 119 respectively.

[117] *Register*, 31 August 1897, p. 7.

[118] R van den Hoorn and J Playford, 'The Adelaide Hospital Row', in D Jaensch, *The Flinders History of South Australia: Political History* (Adelaide: Wakefield, 1986), ch. 6 appendix, pp. 215–25.

[119] Council Minutes, meetings late 1895 through 1896, UAA, S18; *Register*, December 1895 through February 1896 and beyond.

Sir Thomas Elder was the youngest of four brothers born in Kirkcaldy, Scotland, all of who made important contributions to early South Australia. Living quietly and never married, Thomas was the most influential. He entered into a partnership that financed the Wallaroo and Moonta copper mines and the venture brought the partners great wealth. When two brothers retired, Thomas and Robert Barr Smith established Elder Smith & Co., which acquired huge pastoral holdings and rose to become one of the world's largest wool-selling firms. Elder died on 6 March 1897, and his estate was sworn at £615,570, with an additional £200,000 outside South Australia. His biographer recorded that, 'His philanthropy is everywhere evident in South Australia, not least at the University of Adelaide'. He had given £20,000 to endow the two initial science chairs at the university, and thereafter he gave, mostly in his will, £31,000 to the Medical School, £21,000 to the School of Music, and £26,000 for general university purposes. His will also included £25,000 for pictures for the Art Gallery, £18,000 for the Presbyterian, Anglican, and Methodist churches, £25,000 for working men's homes, and £16,000 for hospitals.[120]

The university was aware of Elder's intentions for, six days after his death, the Chancellor, who was going to England, sought permission to inquire at the Royal College of Music how the money for music might best be used.[121] The Professor of Music, Joshua Ives, also made inquiries in Europe and sent a letter to the Chancellor setting out detailed proposals for the use of the money, centred around a new conservatorium.[122] The letter, together with estimated costs and likely income and expenditure, was also considered at a meeting of the Board of Musical Studies, which set up a small sub-committee, including Ives and Bragg, to prepare a paper for the University Council.[123] The university architect was consulted, estimates prepared, and the resultant report was considered and approved by the Council on 5 November 1897. It was resolved to name the conservatorium, the new building, and the associated scholarships after the donor, and the Chancellor was empowered to negotiate with the State Premier on the land to be used.[124]

All facets of the plan went forward enthusiastically, including the building of Elder Hall. Designed in Florentine Gothic style and built of freestone, it was set back and placed between the university and the Exhibition Building.[125] The lawn in front later carried a statue of Sir Thomas Elder. William departed on study leave before the details were finalised and did not influence the acoustic design of the building, but on his return he would be asked to investigate the

[120] F. Gosse, 'Elder, William…, Alexander…, George…and Sir Thomas (1818–1897), businessman, pastoralist and public benefactor', in B Nairn, G Serle and R Ward (eds), *Australian Dictionary of Biography* (Melbourne: MUP, 1972), vol. 4, pp. 133–4.

[121] Council Minutes, meeting of 12 March 1897, UAA, S18, vol. VI, p. 190–2.

[122] Letter J. Ives to Chancellor, 23 September 1897, UAA, S200, docket 567/1897.

[123] Board of Musical Studies Minutes, meeting of 1 October 1897, UAA, S129, vol. I, pp. 184–6; the sub-committee and Board met frequently during October.

[124] Council Minutes, meeting of 5 November 1897, UAA, S18, vol. VI, pp. 139–41 (Board Report 10/1897).

[125] Duncan and Leonard, n. 53, ch. 4.

impossible echoes in the classrooms and the poor acoustics of the hall.[126] He also considered Elder's Art Gallery bequest in his role as Society of Arts representative on the Board of the Public Library, Museum, and Art Gallery.[127] A suggestion to government from the Board that the Elder gift necessitated a purpose-built Art Gallery was recognised and acted on at once.[128] It rose on the western side of the university building, so that the university would be sandwiched between the two new buildings made possible by the Elder bequest. William had an abiding interest in all three structures.

Despite these various activities and a growing family, William could not forget the research challenge sounding at the back of his mind, particularly when the science faculty had to admit that it was unable to nominate a candidate suitable for the Exhibition of 1851 Science Scholarship available in 1898.[129] One topic in the literature that caught William's eye (and that of his father-in-law, Charles Todd) was telegraphy without wires or wireless telegraphy—radio. Threlfall had repeated Heinrich Hertz's 1888 experimental confirmation of Maxwell's prediction of such waves in Sydney in 1888, but it was not until mid-1895 that William attempted the same. That winter he gave six extension lectures on 'Radiation' and, according to the advertising brochure, the last was addressed to: 'Electric waves, their production and detection. The "coherer" or "electric eye". Electric waves pass freely through non-conductors of electricity such as walls and doors, but are reflected by conductors such as sheets of metal'.[130] The *Register* was effusive in its praise of the lectures.[131]

It was at this time that the young Ernest Rutherford, on his way by sea from New Zealand to Cambridge as J J Thomson's first 'B.A. by research' student, stopped briefly at Adelaide. Eve recorded that Rutherford 'called Bragg from a dark-room where he was trying to get a Hertz oscillator to work and enthusiastically showed him the magnetic detector that he was taking to England'.[132] It was the start of a pivotal and lifelong friendship that would grow in Adelaide and blossom later in England (see Figure 8.1).

In mid-1897, while Adelaide newspapers carried reports of lectures and demonstrations of radio by Preece and Marconi in England, Charles Todd was frustrated by regular rupture of an underwater communications cable linking the Althorpe Island lighthouse with the mainland at the southern tip of Yorke Peninsula. He wondered if a radio link might be possible and consulted his son-in-law.[133] In August Rogers recorded in his diary that he was 'making Marconi's

[126] J G Jenkin, *The Bragg Family in Adelaide: A Pictorial Celebration* (Adelaide: Adelaide University Foundation, 1986), p. 39.

[127] *Register*, 20 September 1897, p. 3.

[128] 'Report of the Public Library, Museum and Art Gallery for year ending June 30th, 1897', South Australian Parliamentary Papers, 1896–1897, no. 94, p. 2.

[129] Council Minutes, meeting of 10 December 1897, UAA, S18, vol. VI, p. 153.

[130] See n. 68.

[131] *Register*, 4 July, 11 July, 18 July, 25 July, 1 August, and finally 8 August 1895.

[132] A S Eve, *Rutherford* (Cambridge: CUP, 1939), p. 13.

[133] J F Ross, *A History of Radio in South Australia, 1897–1977* (Adelaide: author, 1978), ch. 1.

Fig. 8.1 Ernest Rutherford, shortly after he visited William Bragg in Adelaide and arrived at Cambridge, late 1895. (Courtesy: Dr J A Campbell.)

app[aratus]'.[134] Some success was obtained, for on 21 September when William spoke to the Teachers' Union conference, 'Marconi's apparatus was shown in action, a bell responding merrily in the lecture hall to an impulse sent from a vibrator in quite a different part of the building'.[135] Referring to William's lecture, the *Register* editorialised on the 'New wonders of electricity', saying, 'Electric signalling seems to be on the eve of extraordinary developments'.[136]

In England the next year William used a letter of introduction from Todd to Preece, and he met and discussed with Marconi the latest developments in radio.[137] Rogers' diary for 6 March 1899 notes, 'Prof. Bragg returns to Adelaide from Europe', and on 23 March Rogers was 'working on wireless telegraphy'. The work continued through April and May, with 'messages sent a measured mile [1,600 m]' at Todd's observatory on Saturday 13 May. In June and July

[134] Annual personal diaries of A L Rogers, held by his family, entry for Friday 13 August 1897.
[135] See n. 113.
[136] *Register*, 28 September 1897, p. 4.
[137] Ross, n. 133, p. 17.

signals sent from the observatory's wireless hut by Rogers and his assistant were received clearly by William at Henley Beach, and return messages were later received at the observatory, a distance of about nine kilometres.[138] Todd's youngest daughter, Lorna, recalled many years later: 'I remember father asking me to send a wireless message to my brother-in-law to say that we were bringing down afternoon tea. We drove down in half an hour, and approaching the mast we could see the professor on top of the sandhill, waving his arms to let us know that he was expecting us'.[139]

In September William gave a course of three extension lectures on 'Wireless telegraphy' to overflowing audiences. 'If any fault could be found with the Professor', the *Register* suggested, 'that fault lay in the extreme simplicity of his language which made his subject so easy to understand that everyone carried away a clear impression of ether waves'.[140] Todd planned to experiment with transmission over water, but in correspondence with Preece he confessed that he, like Bragg, could give little time to the project because of regular work commitments. Furthermore, he estimated that the cost of a lighthouse link would be prohibitive because a skilled operator, properly housed, would be required at each end of the transmission.[141] Bragg and Todd were not the first in Australia to transmit a message by radio, and their project was abandoned early in 1900. Nevertheless, they were the first to demonstrate transmission and reply over a significant distance, and William had further widened his experience of experimental physics in a way that helped to prepare him for original research.

X-rays and radio had so dominated William's scientific thinking that it appears another important discovery escaped his notice, at least for the present time—natural radioactivity. I have found only one small notice in the Adelaide newspapers before 1900: 'We have received a copy of *Knowledge* for June. This magazine of science, literature and art continues... to maintain a high standard... The nature of the Röntgen rays is discussed by Mr J J Stewart, B.A., B.Sc., who mentions that a French savant, M Henri Bequerel, has recently discovered a form of radiation possessing characteristics intermediate between that of ordinary light and of the X-Rays of Röntgen'.[142] This, too, was a topic to which William would turn later.

A severe earthquake shook South Australia on Monday 10 May 1897, and Charles Todd's department received immediate telegraph reports from all parts of the colony.[143] There were several aftershocks, and on 17 May the *Register* printed a letter from Sir Charles saying, 'Professor Bragg and myself are collecting records of the recent earthquakes... and will be very glad to

[138] Rogers' diaries, n. 134.

[139] *Advertiser*, 15 September 1950, p. 4.

[140] *Register*, 15 September, 22 September, and 28 September 1899; the quotation is from an unidentified Australian press cutting in RI MS WHB 39.

[141] Letter books of Sir Charles Todd, State Records of South Australia, Adelaide, GRG 31/1, vol. 4.

[142] *Advertiser*, 8 July 1897, p. 4; Hamersley says this is true for all of Australia, n. 75.

[143] *Register*, 11 May 1897, pp. 4, 6.

receive from observers information on the following points'.[144] This again was something that time and circumstance would not allow William to pursue, but his elder son would take up the matter a decade later in Adelaide, in his first scientific presentation. In amongst all this hectic activity, the annual family holiday survived:[145]

> Sometimes, instead of having a holiday on the nearby shore of the St. Vincent Gulf, we went to Port Elliot on the ocean coast. I suppose the train journey was some sixty miles, but it seemed to take the best part of a day...A landmark on the way was a huge gum tree, from which a rectangular section of bark had been cut out by the Aborigines to make a canoe...The Bragg and Todd families combined to take a large part of a boarding house at Port Elliot. The water-colour painted by my mother in 1896 shows Grannie [Todd], Aunt Maude, Aunt Lorna and my father grouped around the table in the sitting-room of this boarding house. Port Elliot was a tiny township, just one main street with the station at one end and our boarding house at the other, seaward end.
>
> My father joined a few friends for a before-breakfast bathe each day and took me with him. Once he had me on his back in deeper water when a wave bowled us completely over, but he managed to retain a grip on my ankle. I am told the first thing I said, when able to speak again, was, 'Daddy, you shouldn't have done it!' One member of the party, as we returned home, used to regale us with delicious pears. My mother spent much time sketching, and my father was prompted to try his hand at it...My father also joined in making some sort of golf course in the rough ground in front of the hotel. He had a great interest in games of all sorts...
>
> My great friend and confidante on these holidays was my dear Aunt Lorna, the youngest of the aunts, who was then still in her teens. She looked after me, invented games for me, read Grimm's Fairy Tales to me and altogether constituted herself my guardian. We used to make excursions to Victor Harbour, a romantic place where Granite Island was connected to the shore by a long wooden jetty, along which plied a small horse-tram. The ocean rollers, falling on the granite cliffs, were always a thrilling sight. In the other direction we went to Middleton, a long sandy beach where the same rollers broke in a terrace of foam lines seven or more behind each other, and all the shells were stained blue by some chemical coming from the seaweed. I can remember well the deafening roar of that long series of rollers thundering on the beach.

Apart from the annual holiday, William was extraordinarily busy; even Saturday mornings were work-filled. He must have seemed a distant husband and father at times. Particularly while they were young and could accompany their mother to the Observatory, the Todd influence on the young boys, as well as on William and his wife, was pervasive. Sir Charles was particularly influential in regard to his use of physics and engineering knowledge and in his

[144] *Register*, 17 May 1897, p. 5.
[145] W L Bragg, Autobiographical notes, pp. 7–8.

love of adventure and his willingness to take risk, as in the introduction of new technologies such as the telephone, radio, and electricity and in the building of the Overland Telegraph Line. Willie was also close to his Aunt Lorna, and she later became the centre of the family network, world-wide.

With the boys reaching school age and with the assistance of several maids in the house, Gwendoline had ample time available. For a Todd daughter with an artistic temperament the social round called unerringly. The social and gossip columns show that Sir Charles, Lady Todd, and their children were represented at almost every social function of note in Adelaide, and Gwendoline maintained the family tradition. She reciprocated her many invitations by giving a musical 'at home' at the Lefevre Terrace residence, for example, when 'His Excellency the Governor and Captain Willington arrived for afternoon tea on their way from the Golf Grounds'.[146] While William was securing his family's future and unknowingly its fame, Gwendoline was building a social network that would remember the family with pride and affection long after it had left Adelaide and Australia.

[146] *The Critic*, Adelaide, September though November 1897, *passim*; quotation ibid., 9 October 1897, p. 16.

9
Leave-of-absence

William's predecessor, Horace Lamb, had first sought leave-of-absence from the University of Adelaide in 1883, and this had led the university to establish a procedure for 'study leave' for up to one year.[1] After twelve strenuous years of service and with several interrelated thoughts in mind, William wrote to the Council in July 1897:[2]

> I have the honour to apply for leave of absence during 1898. My reasons for making this request are:
>
> (1) that during the twelve years I have been here the science of Physics has made great strides and I feel that, to keep myself abreast of the times, I should study in England the advances both of the subject itself and the methods of teaching it;
>
> (2) that I would be grateful for an opportunity to see again my family in England.
>
> I have been endeavouring to make satisfactory arrangements for the taking of my work in my absence, supposing that you grant my request. I would propose that Mr Chapman take charge of my department, and that he be assisted by Mr J B Allen B.Sc.

There followed a ringing endorsement of Chapman's performance since his appointment and an outline of Allen's career, emphasizing his suitability as an assistant for Chapman. William also pleaded for the retention of 'the services of the present laboratory boy, Bromley, [who] at present receives no pay: but to induce him to remain here, I propose to pay him during next year 10/- [shillings] a week'.

William's reasons are revealing. He wanted to hear about the latest discoveries in physics, and there were several of great significance: electromagnetic waves and radio, X-rays, natural radioactivity, and the pregnant matter of cathode rays and the electron. It was not clear that he was fully aware of all these,

[1]For full details see J G Jenkin, 'The appointment of W H Bragg, F R S, to the University of Adelaide', *Notes and Records of the Royal Society of London*, 1985, 40:75–99. Since such leave later became an entitlement every seven years, it became known as Sabbatical Leave, later still Study Leave, and most recently OSP (Outside Studies Program).

[2]Letter W H Bragg to Council, 30 July 1897, UAA, S200, docket 422/1897.

but the general excitement had certainly communicated itself to Australia. Teaching was specifically mentioned, and it was clear that 'the advances... of the subject itself' were of great interest and might therefore offer an opportunity for research. In addition, William spoke with attractive honesty of his desire to see his English relatives again, and no doubt to show them his new Australian family. The Council granted his request and adopted the report of its Education Committee on the arrangements for undertaking Professor Bragg's work during his absence.[3]

When William's plans became known, he accepted other assignments that reflected his commitments to science and education beyond the university. Charles Todd asked him to consult Preece and Marconi regarding radio developments and to visit astronomical observatories for information on the spectroscopic study of the light from stars.[4] William had a commission from the local colonial government to enquire into teacher training and higher primary education in Britain.[5] Furthermore, at the meeting of the Council of the School of Mines and Industries at which William was granted twelve months leave-of-absence, he was also commissioned to report upon the desirability of purchasing such exhibits and apparatus as might be advantageous to the School, although later events showed that he viewed this brief more broadly.[6]

Gwendoline's eldest sister, Mrs Lizzie Squires, had arrived from England in May,[7] and her forthcoming return now offered the Bragg family a useful variety of travel arrangements. Gwen and William would depart first, travel by liner, and enjoy stopovers in Egypt and southern Europe, while the boys would remain in Adelaide and sail later and more directly with Aunt Lizzie and Charlotte. Professor and Mrs Bragg departed on the *Australia* on 15 December 1897.[8] Their daughter's biography of her father records that they sailed via the Indian Ocean to Colombo and then on to Aden, the Suez Canal, and overland to Cairo. They visited the pyramids and a steamer took them up the Nile. Gwendoline was constantly excited and regularly painted scenes in a tiny sketchbook. They visited tombs and temples and met a group of English soldiers going to war in the Sudan.[9] In a letter to Lamb, Samuel Way told of the Braggs' travel arrangements and standing in Adelaide: 'Bragg goes away for a year's leave of absence... and he and Mrs Bragg are going to have a trip up the Nile. They are sensible enough to travel to Port Said second class, so as to have more money to

[3]Council Minutes, meetings of 30 July and 27 August 1897, UAA, S18, vol. VI, pp. 118 and 121 respectively.

[4]Letter C Todd to Sir William Preece, 2 November 1898, Todd letter books, State Records of South Australia, Adelaide, GRG 31/1, vol. 4, p. 319.

[5]*Register*, 7 March 1899, p. 5. This item resulted from an interview with Bragg on his return to Adelaide, but a vigorous search in Adelaide for the report that William prepared has proved fruitless.

[6]School Council Minutes, meeting of 30 November 1897, University of South Australia Archives, vol. 3, p. 176.

[7]*Register*, 3 May 1897, p. 4.

[8]*Register*, 16 December 1897, p. 4.

[9]G M Caroe, *William Henry Bragg, 1862–1942: Man and Scientist* (Cambridge: CUP, 1978), pp. 41–2.

spend in Europe. It would be impossible to exaggerate the great work Bragg has done for the University. Besides his direct academical work he has had a most useful and healthy influence upon the students, and in developing the social side of our University life. He is sure to look you up'.[10] Leaving Egypt, the couple sailed to Italy, where they visited Naples and Pompeii, Rome and Florence. In Rome they enjoyed a 'jollification', as reported later in Adelaide:[11]

> Prosperous South Australia
>
> Professor Bragg was responsible for a little merriment at the meeting of the Council of the School of Mines and Industries on Monday afternoon by relating an anecdote. Whilst in Rome the professor attended a jollification, and on returning to his conveyance found the coachman studiously engaged at a book . . . The book proved to be a lexicon, and on one of the pages were the following significant statements: 'Are you leaving the country?'—'Yes'. 'Where are you going?'—'I am going to Adelaide, South Australia, where there is plenty of work, and where the people have plenty of clothes to wear and food to eat'. This complimentary reference to the state called forth a chorus of 'Hear, hear' from the members of the Council.

The two boys wrote to their parents regularly during these weeks. Willie wrote to his mother from Henley Beach:[12]

> I hope you are not very seasick. We were going on a little steam launch this morning, only Aunt Lorna said she thought that we would get a little seasick, and Roby would not go without Sharlit [*sic*] . . . I am going to write to Dad as well in this same letter. Dear dad, Grandfather and me and Rob are all going to play cricket if Rob will come, only Rob is just having his sleep now, but I think he is really awake don't you. I like my lessons very much, and I know lots of funny sums, but Rob mostly has holidays.
>
> [And the next day] Today morning we went in a boat that was lying on the beach and me and another little boy pushed it out in to the sea and waited till the tide came in, and then we poped in it and another big boy who was passing came and joined the fun and did most of the workery, and we pulled up the anchor and got the cricket stumps and pushed right away far out to sea . . . Good Bye. I still remain your loving son, William Bragg.

Charlotte wrote, in her broken English: 'Willie was going to write you a long letter but he is allways so bussy on the beach . . . the Children are very well . . . The Children are very good, they play so nice together; and poor Rob so soon he has sayed his prayes he is of too sleep, because he never sleeps in the morning now. Rob sends 100 kisses to you and the Professor; they will be so pleased to see you again. I hope you and the Professor enjoying yourself very much'.[13]

[10]Letter S Way to H Lamb, 8 November 1897, Way letter books, State Records of South Australia, Adelaide, PRG 30/5, vol. 4, pp. 330–1.

[11]*Observer*, 19 October 1901, p. 31.

[12]Letters W L Bragg to his parents, 3 and 4 January 1898, Bragg (Adrian) papers.

[13]Letter C Schlegel to G Bragg, 1 February 1898, Bragg (Adrian) papers.

Willie recalled his trip to England as follows:[14]

> In 1897, when I was seven years old and Bob four, the parents made the momentous decision that my father should take a year off and the whole family should spend it in England. I think the main object was Uncle William... it was a great blow to him when my father left England for the post at Adelaide University. My father felt he owed everything to Uncle William and must see him again as he was getting to be an old man... We were parked at the Observatory until we sailed for England. I can remember the departure of the parents from the Observatory, a very tearful farewell on my mother's part and a completely unemotional one on ours; the difference between an absence of a day or of some months meant nothing to us.
>
> Our P&O liner was called the 'Oceana'. She had four tall masts with yards and we used to set some sail, though more, I think, to steady the boat than to assist our progress. Colombo produced a great impression on me. The gentlemen with complete upper garb, but with only a kind of dish-cloth from the waist down, the punkahs and punkah boys in the shops, the naked children who ran after our carriage begging... These, and the heat, and the crowd of bargaining pedlars on board, were a great thrill.
>
> I can also remember the arrival at the Marseilles quay, when I was far more interested in watching intrepid Frenchmen sail canoes around the liner than in the parents we were joining.

While the differences should not be overstated, it is clear that Willie was an introverted and self-conscious child, while Bob was more easy-going and sociable, closer to his parents and therefore something of a favourite. Many years later Willie told his sister he found '*things* easier than people', whereas Bob was widely popular, 'with his mother's intuitive knowledge of how to deal with people'.[15]

The family travelled overland from Marseilles to England and then moved around Britain energetically. While their more important and interesting activities are clear, the following suggested chronology is circumstantial. The early months, April through June 1898, were spent visiting family in Market Harborough, gathering information on education, and enquiring about radio in London. Willie continued:[16]

> We first made a long stay with Uncle William at Market Harborough. Uncle William had created for himself a leading position in the town. He had built a house in the market square, which is now, I believe, a bank. It is a very ugly, Victorian, semi-detached house [called 'Catherwood House'], but the view over the market square was very attractive.[17] One market day a bullock got loose and ran into our hall, a great excitement!... Uncle William also owned a good deal of property in the town, including a brick works which we enjoyed visiting. I loved seeing the clay

[14]W L Bragg, Autobiographical notes, pp. 8–9; the party left Adelaide on 23 February 1898.
[15]Caroe, n. 9, p. 39.
[16]W L Bragg, Autobiographical notes, pp. 9–10.
[17]The house had two stories and high gables and had been named by Uncle William.

> oozing out as a long rod, and being cut into bricks by wires. There was a garden behind the house which seemed enormous to us, and at the far end stables and Uncle James's workshop. Uncle James was very gentle and kind and made toys, whistles, kites and endless other things for us; we were very attached to him... I remember that Bob and I demolished most of the rockery to make a fort in one corner of the garden under a large elder tree, and any annoyance felt by Uncle William was mastered by his amazement at our imagination and the ant-like perseverance with which we moved, inch by inch, stones almost our own weight. He did, however, take the precaution of keeping his beehives in the strawberry bed. He was rather a vulgar old man... but even we realised his powerful personality and stood in great awe of him.
>
> During our stay our parents and Uncle James went off for a bicycle tour in Wales, while we remained at Market Harborough in the care of Charlotte. My father had bought two magnificent Humber bicycles for my mother and himself, which were afterwards taken to Australia. Bicycles were still a comparative novelty. When my parents were married, bicycling parties were the order of the day... in his younger days he rode a penny-farthing bicycle and he had a scar on his forehead due to pitching over the handlebars. I remember my father's story of a Welsh ferry, where bicycles were so unknown that no provision was made for their tariff. The girl in charge deliberated and then said, 'I'll charge you as pigs'.

Willie also wrote to his parents from Market Harborough. The letters were an early attempt at script writing: 'Dear Mother, This is the last time I am wright to you. I must be very. Uncle James says so... I cannot wright to you both, so I thought I would write to you. Rob and Freda are just gone out in the garden to play... Good bye, Your afectionate son, William Bragg' (see Figure 9.1).[18]

William made enquiries about educational initiatives in Wales, and the *Register* reported that: 'Professor Bragg... has been devoting a good deal of attention to specially interesting features in the Welsh system of teaching. In the matter of the training of teachers he is of the opinion that South Australia has something to learn from the old country. He has gathered that the idea of more directly utilizing the Universities in the carrying on of this important work has many supporters in England'.[19] It seems likely that they next visited London. Willie again:[20]

> Then we paid a visit to Dad's cousins [Fanny and Willie Addison and their families]. My father was very fond of Fanny, who married a farmer called Kemp-Smith. We stayed with the Addisons in Croydon and history repeated itself. Will Addison, now a doctor, had taught his boys to box and set them on to us, when I am afraid the Braggs put up a poor resistance. We did not enjoy our stay. On the other hand I have pleasant memories of our stay in the country with the Kemp-Smiths. The daughter,

[18] Letter W L Bragg to G Bragg, June 1898, RI MS WLB 37A/2.
[19] *Register*, 2 August 1898, p. 5.
[20] W L Bragg, Autobiographical notes, pp. 11–12.

Fig. 9.1 Gwendoline Bragg and her two sons, Willie (L) and Bob (R), at Market Harborough, England, 1898. (Courtesy: Dr S L Bragg.)

> Freda, was about my age... During this stay my father started to tell me bed-time stories and they were always the same—about the properties of the atoms. We started with hydrogen and ran through a good part of the periodic table. Also I remember I had been given a large volume about Captain Cook's voyages, really a grown-up book. How little it matters to a child whether he understands all of a book; it is the atmosphere he likes. I ploughed through it from end to end. We did the sights of London... when staying in Croydon. I am glad I remember the Hansom cabs, with the clip-clop of all the horses on the wooden sets, and the adventure of riding behind the apron with the cabbie perched up above.

While in London, William met Sir William Preece using Charles Todd's introduction. He was a guest at Preece's home, 'Gothic Lodge' in Wimbledon, and was introduced to Marconi, who had been in London since 1896 and enjoyed Preece's patronage.[21] William was Preece's guest at a Royal Society dinner in London, where 'he met a large number of men justly regarded as eminent in science'.[22] Like much of his Adelaide experience, radio found expression in William's later years in Britain. In 1916, during the Great War and in a lecture to the Textile Institute on 'Physical science and its application to industry', William noted that 'no one can foretell what scientific research will enable us to do', and he used the invention of wireless telegraphy as his first example

[21] J F Ross, *A History of Radio in South Australia, 1897–1977* (Adelaide: author, 1978), ch. 1.
[22] *Register*, 23 January 1899, p. 6.

of this principle.[23] Likewise, during the Second World War, William wrote a small book entitled *The Story of Electromagnetism: Specially Written for the Air Training Corps and... all interested in radio*, the cover of which carried the following explanation: 'This luminous little book by one of our most eminent living physicists describes the great scientific discovery which lies at the very root of understanding how wireless works. It is an expanded version of a lecture given on several occasions to cadets belonging to the Air Training Corps'.[24]

The pull of Cambridge now became irresistible: there they found Todd-family conviviality with Lizzie and the Squires family, Trinity College and its fellows, J J Thomson and the Cavendish Laboratory, and the sights and sounds of William's student days. His son remembered:[25]

> I know we stayed for quite a time in Cambridge, boarding in a house on the corner where Station Road diverges from Hills Road. Our living-room had two pictures illustrating the broad road which leads to Hell and the narrow path which leads to Salvation... I can see them in my mind's eye now, with a lady encouraging a child to enter the wicket gate which led to the narrow path. I played with Stenie Squires in Uncle Charlie's garden, which was in another street some way from Vale House. Once a circus came to Cambridge and we were all to go to it. On the previous day I had some minor ailment, and it was judged that it would not be advisable for me to go to the circus; I was taken instead to a flower show on the same afternoon. This fine distinction caused me much grief. We had a holiday with the Squires at Hunstanton, and it was then that we heard of my grandmother's death. She was a dear old lady—I wish I had been older during her lifetime and known her better.

Fairs and circuses had been coming to Cambridge for centuries: Isaac Newton had purchased triangular glass prisms at the Sturbridge and Midsummer fairs and thereby explained the 'celebrated phenomena of colours'.[26] Alice Todd's unexpected death on 9 August saddened Gwendoline and William greatly, for she had been important to them both, and they were fortunate to be able to share their grief with Lizzie. The grief of Charles Todd and his two unmarried daughters, Maude and Lorna, was communicated to Cambridge in many letters.[27] Lady Todd had been in good health until an illness necessitated an operation, and she died within a month. The *Register* observed that she was 'beloved in philanthropic and social circles', and noted that she would be mourned particularly in West Adelaide, where 'her charitable deeds and acts of kindness' would not be forgotten by the poor.[28]

[23] W H Bragg, 'Physical science and its application to industry', *Journal of the Textile Institute*, 1916, 7:185–93.

[24] Sir William Bragg, *The Story of Electromagnetism* (London: Bell & Sons, 1941).

[25] W L Bragg, Autobiographical notes, pp. 10–11.

[26] R S Westfall, *Never at Rest: A Biography of Isaac Newton* (Cambridge: CUP, 1980), pp. 156–75.

[27] Bragg (Adrian) papers; Todd letter books, n. 4, *passim*.

[28] *Register*, 10 August 1898, p. 4.

Also in August, since he was already in England, William was asked to represent South Australia at the combined Fourth International Congress of Zoology and the International Congress on Physiology in Cambridge.[29] Later that month he went to Bristol for the sixty-eighth meeting of the British Association for the Advancement of Science. William was not listed as a member, but his attendance is recorded in the *Register* under its regular heading 'Anglo-Colonial Notes: Professor Bragg in September went to the meeting of the British Association at Bristol and found it extremely interesting. Not only were there many fine papers and instructive discussions to be heard, but there was the opportunity of seeing, and sometimes speaking to, many distinguished men formerly known to him only by name'.[30]

Reports were presented to the meeting on a variety of topics that would have interested Charles Todd—meteorological phenomena, magnetic observations, practical standards of electrical measurement, and seismological investigations—and William himself would have valued the report on 'The teaching of sciences in elementary schools'.[31] The presidential address by Sir William Crookes focused upon three topics: 'the important question of the supply of bread to the inhabitants of these Islands', ultraviolet spectroscopy, and 'to me the weightiest and the farthest reaching of all . . . psychic researches'.[32] In the papers of Section A—Mathematical and Physical Science, William found interesting papers on electricity and magnetism, mathematics, meteorology, and the recent solar-eclipse expedition. They were given by leaders of nineteenth-century experimental and mathematical physics in Britain: Oliver Lodge, Lord Kelvin, Hugh Callendar, Silvanus Thompson, and Sir George Stokes. There was nothing in Section A on radio, X-rays, or radioactivity, although these topics surely came up in informal discussions.[33] William did have 'one whole day with Sir William Crookes',[34] when X-rays and radio were presumably discussed, given Crookes' central role in gas-discharge and early X-ray research and his 1891 presidency of the Institution of Electrical Engineers.[35]

It was some time after these meetings that William took up a task recently entrusted to him by the University of Adelaide. On 27 September the Library Committee noted that Horace Lamb had asked to be relieved of his long-standing honorary role as its buying agent in England, and the committee resolved to call the attention of the Council to 'the obligation the University is under to Professor Lamb, in undertaking the commission in connection with the Barr Smith Library, and to the many services he has rendered to the University since

[29] *Register*, 30 August 1898, p. 5 and 3 October 1898, p. 5.

[30] *Register*, 23 January 1899, p. 6.

[31] *Report on the Sixty-eighth Meeting of the British Association for the Advancement of Science, held at Bristol in September 1898* (London: Murray, 1899).

[32] Ibid., pp. 3–33.

[33] Ibid., pp. xiv–xvi. There was, in fact, only one paper on these topics in the whole meeting: on the absorption of X-rays by chemical compounds, in Section B—Chemistry.

[34] *Register*, 7 March 1899, p. 5.

[35] See, for example, G W C Kaye, *X-rays* (London: Longmans Green, 1918), *passim*, and R Appleyard, *The History of the Institution of Electrical Engineers, 1871–1931* (London: Institution, 1939), p. 290.

his return to England'. Furthermore, it noted that, because Lamb's departure had been unacknowledged by the university, 'some substantial token of our gratitude, say a collection of books, a handsome bookcase or a piece of plate suitably inscribed, be presented to Professor Lamb and that Professor Bragg be empowered to act for the Council'. The Council agreed and authorized William to 'spend up to £100 in procuring some suitable present'.[36] He therefore travelled to Manchester to see Lamb and to present to him an oak bookcase, an oak bureau, an inscribed silver rose bowl, and four silver candlesticks that he had purchased for a total of £99.14.5, 'which will leave me a few shillings to return to you'.[37] The Lamb family generously presented the silver bowl to the University of Adelaide in 1966.[38] Lamb reciprocated with further assistance to Bragg and his university. The matter is best explained in the letter William wrote to the university's Finance Committee after his return to Adelaide:[39]

> I have the honour to report that, when I was in England, I received instructions through the Registrar to purchase books and apparatus for the new mining engineering school. The total cost was about £150: I am unable to give the sum exactly as the accounts are not yet all in. When I set about the work given me, I found that I could get large reductions by paying cash: in the case of the apparatus the saving was 10% to 15%. This reduction was obtained largely by ordering through the Engineering School of Cambridge, where the authorities were so kind as to get the things on their own account and hand them on to me.
>
> On consultation with Professor Lamb, it seemed a pity not to take advantage of the reduction I could get by paying ready money: and he offered to advance me the sum I wanted out of [Manchester] University funds at his disposal. Accordingly he gave me £150, account of which I hope to present to you in a few weeks, on receipt of all the invoices.
>
> We further arranged that I was to ask to have the money sent to him as soon as I could after my return. Unfortunately I have missed one meeting of your committee through illness, and I now take this opportunity of explaining the matter to you.

This letter takes us back to Cambridge, where William and his family returned to stay with the Squires for the last weeks of their time in England. Here William held detailed discussions with the staff of the Cavendish Laboratory about the teaching of mathematical and experimental physics and about the possibilities of research. 'The arrangements for teaching practical physics ... he found greatly improved since last he was there',[40] and with the guidance of the staff he ordered a range of new items for the Physical Laboratory in Adelaide. He also heard of the extraordinary advances made by J J Thomson, his staff, and

[36]Council Minutes, meeting of 30 September 1898, UAA, S18, vol. VI, p. 215, referring to Library Committee Report, 5/1898, 27 September 1898, UAA, S148, vol. II.

[37]Letter W H Bragg to Council, 28 June 1899, UAA, S200, docket 620/1899.

[38]R B Potts, 'Horace Lamb rose bowl', *Adelaide University Graduates' Union Gazette*, June 1966, pp. 11–12.

[39]Letter W H Bragg to Finance Committee, 27 April 1899, UAA, S200, docket 381/1899.

[40]*Register*, 23 January 1899, p. 6.

their research students in the last few years, which he found 'very stimulating'.[41] These included J J Thomson's own work on cathode rays, electric discharge in gases, and related topics, his work with Rutherford and other students on the passage of electricity through gases exposed to Röntgen rays, C T R Wilson's early research on the Wilson cloud chamber, and the researches of Townsend, McClelland, Skinner, Shaw, and others.[42] William must have heard of Becquerel's discovery of natural radioactivity and of Thomson's suggestion that cathode rays were particulate and negatively charged: the 'corpuscles' that became 'electrons'.[43] He was impressed and inspired by the level of research activity and success in Thomson's laboratory, and he surely recognized now, if not before, that the unexpected and unexplained results offered fertile ground for further study. The modest nature of the apparatus being used must have impressed him also, especially when compared to the elaborate building and apparatus that Threlfall had built to carry out his classical physics research in Sydney and that had overawed William earlier. The apparatus William purchased is again best revealed in the letter he wrote to the Finance Committee of the university after his return:[44]

> I have the honour to report to you that, whilst I was in England, I took upon myself to make some purchases of apparatus for the Physical Laboratory. I did so in the hope that you would overlook the irregularity of the expenditure: for I had special chances of noting points in which we were deficient in comparison with English laboratories, and of choosing personally the things needed to make the defects good. I bought as follows:
>
> a d'Arsonva galvanometer...lamp & scales...alloy for joining glass to brass & making tops of resistance boxes...glassware, helium & argon vacuum tubes etc....10 accumulators...aluminium...screw-plates...small goods...Zirconium cylinders for lantern...packing, insurance, freight, landing charges etc.... [total cost] £45.7.0.

Much of this was clearly destined for Rogers' workshop, and William added one other request to the letter. In Adelaide he had exhibited the effects and uses of X-rays using an induction coil lent by Sir Charles Todd. Ashamed to keep it any longer, he had sought to buy a coil for the university but funds were limited. He had obtained one by paying for it himself and it had proved invaluable. He would be grateful if the university would now buy the induction coil from him for its purchase price. The Finance Committee and the Council agreed.[45]

[41] Ibid.

[42] 'A list of memoirs containing an account of work done in the Cavendish Laboratory', in *A History of the Cavendish Laboratory, 1871–1910* (London: Longmans Green, 1910), particularly for the years 1894–8, pp. 297–302.

[43] E A Davis and I J Falconer, *J J Thomson and the Discovery of the Electron* (London: Taylor & Francis, 1997).

[44] W H Bragg to Finance Committee, 29 June 1899, UAA, S200, docket 538/1899.

[45] Council Minutes, meeting of 30 June 1899, UAA, S18, vol. VI, p. 275.

Some of the firms mentioned in William's list were London-based, and he had other reasons to journey south. On 10 November Todd wrote to William, apologizing for not writing sooner but explaining: 'I think you will forgive me. I have often been going to write—the flesh was willing but the spirit sad and weak... since the sad event [his wife's death]... You do not mention Christie's or Huggins' names. Have you met them? The Royal Observatory and Huggins' private observatory and laboratory would be well worth your seeing, and I want you to bring me some practical information as to stellar spectroscopic work and the best spectroscope for our Equatorial [telescope]. Turner of Oxford writes that he met you and that he and you were at Cambridge together as under-graduates. His would also be worth a visit and he is a nice fellow in every way. I want you to select me some really good modern astronomical [lantern] slides, especially, if they are to be got, some illustrating the beautiful phenomena at the recent, or some other, total solar eclipse... the Secretary of the Royal Astronomical Society would assist you to procure'.[46] During January 1899 William hastened south to London to buy the equipment he needed and to visit Greenwich, Sir William Huggins at his private observatory at Tulse Hill,[47] and the Royal Astronomical Society. He then turned to Oxford to see Herbert Turner.

Uncle William hoped they would stay in England 'for good', but this was not possible.[48] The family travelled again to Marseilles and, on 2 February 1899, embarked on the *Victoria* for the journey home. Willie had 'little recollection of our return voyage when I was eight, except for staying in the same hotel in Marseilles, where the chambermaid recognised Bob and me and said, "Bébés upgrown", which we thought very insulting'.[49] They arrived back in Adelaide late on Sunday 5 March, just in time for Professor Bragg to begin a new year of teaching.

Return to Adelaide

William's head was full of things to be done as a result of his year of leave, and there were domestic decisions to be made as well. Initially the family stayed at the Observatory, where there was ample room and where Sir Charles welcomed their companionship. William was now very settled and content in Adelaide, a feeling confirmed by his year away, and he and Gwendoline had decided to build a permanent home in the city. They purchased a vacant block of land on the corner of East Terrace and Carrington Street, with characteristics that mirrored those of their rented house in North Adelaide. The block looked out over the Park Lands and the Victoria Park racecourse to the Adelaide hills in the

[46]Letter C Todd to W H Bragg, 10 November 1898, Todd letter books, n. 4, pp. 321–4.
[47]Letter W Huggins to W H Bragg, 29 December 1898, RI MS WHB 3C/52, suggesting 17 to 24 January for Bragg's visit and inviting his wife.
[48]Letter W B Bragg to W H Bragg, no date, Bragg (Adrian) papers.
[49]W L Bragg, Autobiographical notes, p. 12.

distance, and there would be room for a large garden. The boys could play on the land where William had first played lacrosse.

With the assistance of the university architect, William designed a large, two-storey, brick house with Edwardian gables, in general appearance not unlike Uncle William's home in Market Harborough. When Gwendoline protested that she could not understand plans, William 'made a model of stiff paper, fastened by a clipping machine'.[50] Similarly, 'I remember my dear mother, whose mind worked in a strange way, urging [my father] to install the tank early so that it would be full of water when the house came to be built! You will remember the huge galvanized-iron tanks fed from the roof'.[51] On 9 September 1899 Sir Charles Todd laid a foundation stone in the brickwork of the front veranda, and the house was named 'Catherwood House', recalling William's youth and the family's recent visit to Market Harborough. There were library/study, reception, dining, and family rooms on the ground floor, together with the kitchen area, and bedrooms and servants' quarters upstairs. The house was built at a cost of £2,300 and with the aid of a loan of £1,300 from the Savings Bank of South Australia.[52] The family lived in rented accommodation on South Terrace, Adelaide, while the house was being built.

The boys' schooling was another matter to be decided and arranged. Bob was sent to Canterbury House, a preparatory school run by Rev. Slaney Poole in a two-storey house with iron lace-work on the eastern side of Dequetteville Terrace and close to the malting business of Barrett Brothers. It was a pleasant walk across the Park Lands from their new home on East Terrace.[53] Willie was sent further. In 1883 a private school had been opened in North Adelaide using the Christ Church Schoolroom for classes. At the end of 1891 it was sold to James Lindon and Edmund Heinemann, second and fourth masters respectively at an unhappy St Peter's College, who named it Queen's School in honour of Queen Victoria and who saw student numbers jump from 34 to 72 in their first year. They were talented teachers and well known.[54] They were members of the Adelaide Whist Club, along with other leaders in South Australian education and business, and must have been known to William.[55] To accommodate the increasing numbers, a new building was erected on Barton Terrace,

[50]Ibid.

[51]Letter W L Bragg to Sir Mark Oliphant, 13 October 1966, RI MS WLB 54A/27.

[52]Correspondence of present author with Archivist of the Corporation of the City of Adelaide and with the Savings Bank of South Australia (SBSA), including copy of the Certificate of Title; letter Accountant, SBSA, to W H Bragg, 14 December 1899, Bragg (Adrian) papers. The house later became the Public Schools Club, was altered and enlarged, was then under threat of sale and demolition, and finally, in 2004, was placed on the South Australian State Heritage Register, ensuring its preservation.

[53]'A man who remembers', *The Mail* newspaper, Adelaide, 4 April 1942; *South Australian Directory, 1902* (Adelaide: Sands and McDougall, 1902).

[54]B O'Connor, *Queen's College North Adelaide, 1883–1949* (Adelaide: Queen's Old Boys Association, 1999); for St Peter's College, see ch. 12.

[55]Photo of members of Adelaide Whist Club, 1895, State Library of South Australia, Adelaide (Mortlock Library), photo B3168 (with names), reproduced in C Thiele, *Grains of Mustard Seed* (Adelaide: Education Department of South Australia, 1975), opposite p. 59.

with four classrooms, a Big Schoolroom, a chemical laboratory, and playing fields on the Park Lands opposite the school. Within a few years, however, Lindon's health deteriorated and Heinemann left to join the Adelaide Stock Exchange. R G Jacomb Hood became Acting Headmaster in 1896 and, following Lindon's death, he purchased the school from Lindon's wife. Willie remembered:[56]

> I was sent to a preparatory school called Queen's on the far side of North Adelaide. This meant a walk to catch the [horse] tram, quite a long journey, and then a walk at the far end. The trams were double-decker, with an awning on top and canvas sheets which could be lashed along the sides of the upper storey in wet weather. There were no points; at a junction of routes the driver skilfully swung his pair of horses to one side so that the wheels should follow the right grooves. Sometimes they did not, and then we all had to help manhandle the tram on to the rails again.
>
> Both the walking sections of the route were fraught with danger. Returning from school in North Adelaide we were from time to time set upon by 'larrikins', so we avoided returning singly. Our best weapon of defence was our satchels, which we swung round our heads by the straps... At the other end, between tram and home, there was a boy with whom I had a vendetta... I think our feud started because I came on him once beating up his younger sister. Knight errantry prompted me to interfere, whereupon both brother and sister attacked me; a lesson in human reactions...
>
> Our headmaster was called Hood. He was a believer in corporal punishment... We used to be given sets of ten words to spell. We were allowed, if I remember rightly, to get two wrong but had a cut of the cane on the palm of the hand for each mistake beyond that number. I was fortunately a good speller and only once had a cut; I can still sense the numbness, which lasted for the rest of the morning.
>
> There was a fair amount of bullying... I was a misfit at school, being so very immature in some ways and so precocious in others. We had lunch in the boarding house attached to the school, and after lunch a scratch game of hockey was organised by the masters. Boys were asked if they wished to play and supplied with sticks. Now it would have done me the greatest good to have played in these games, but I could not do anything so decisive as announcing I wanted to play, and so moped rather miserably in the lunch hours... On the other hand I was precocious in lessons. We generally had more than one class in each room, and I remember being in the same room as a very senior class doing Euclid. Hood... pulled me, a very small boy, out of my class and made me explain the theorems to the large boys while he crowed with delight.

Hood was a graduate of Rugby and Clare College, Cambridge, a bachelor who had come to Australia for health reasons. He was not universally popular, as Willie's recollections confirm. Others remembered him as a good teacher but with sadist tendencies, and as a stern disciplinarian who could be ruthless

[56]W L Bragg, Autobiographical notes, pp. 12–13.

and sometimes inebriated. He left no records, either of himself or the school over the next thirty years: no collected magazines or class lists or examination results or newspaper cuttings or sporting records survive.[57] Yet there are snippets that paint a different picture. A *Queen's School Prospectus* from early in Hood's tenure shows that the academic programme was significant: 'Queen's school follows as closely as possible the lines of an English Public school. Boys receive a thorough grounding in all elementary subjects, special stress being laid upon English, Mathematics and Modern Languages. Latin, Greek, Chemistry and Physics are also included in the school Course...Gymnasium classes are held once a week'.[58] Hood was headmaster until 1926, its longest serving owner, and the record of old scholars from the school was outstanding: six Rhodes Scholarships between 1906 and 1927, for example. One old scholar recalled that Hood 'was an exacting but thorough and painstaking tutor...he was extremely fair, and the number of pupils who went through Queen's and later gained eminent and distinguished positions while he was head bears witness to his...training'.[59]

Willie attended in 1899 and 1900, aged nine and ten, and he clearly benefited from the academic programme. He was already showing his precocious talent for mathematics. I have also found one sporting record, showing that in 1900, at the Queen's School Sports on the Adelaide Oval, W L Bragg won the under 11, 100 yards flat race from a handicap of 4 yards. Provided he could compete with boys his own age, Willie showed that, like his father, he had an aptitude for sport.[60] The school struggled on as Queen's College through the Depression of the 1930s, but it was forced to close in 1949, the last of the personally owned schools in South Australia. It failed not as a school but as a business with inadequate capital.

[57] O'Connor, n. 54.
[58] *Prospectus, Queen's School, North Adelaide*, no date [late 1890s], in private hands.
[59] O'Connor, n. 54.
[60] *Register*, 15 September 1900, p. 11.

10
Aftermath

With a new family home rising and his boys placed in school, William could concentrate on other matters. Although his report to the government on education appears not to have survived, the *Register* newspaper carried a lengthy editorial and an even longer summary. Noting that England had been 'outdistanced in educational progress both by Scotland and Wales and her commercial competitors on the Continent', the *Register* praised William's report and its restriction to 'the means provided for carrying the children from the primary schools on to the higher primary and secondary schools'. It strongly emphasized that 'Professor Bragg refutes the notion that the success of Germany in recent years has been due to her schools... teaching things of a purely technical nature. On the contrary, the results secured have been mainly due to the soundness and thoroughness of the general education imparted at the schools'. The newspaper noted that Professor Bragg had stressed the need for greater education of teachers, and that 'If the University can see its way clear to provide... even one year's instruction... it will perform a beneficent and urgently-called-for public service'.[1]

William's report concentrated on three matters: the provision, under State authority, of education higher than primary, improvement in the methods of providing and training teachers, and, in an appendix, a copy of his letter to the Council of the School of Mines on 'Schools of domestic economy for girls'. On the first he noted, 'there now exist, in most Continental countries, complete and continuous systems of education... for those children that go to work at an early age... to keep their minds open to improvement, [and] to prevent them from becoming deadened by the monotonous nature of the work'. He stressed again that the base principle should be 'to educate the faculties, to open the mind, to inculcate habits of careful observation [and] of careful drawing of conclusions'. 'The second most important question', he continued, 'is the training of teachers... It is felt that the training of primary schoolteachers... is too narrow; though the instruction in technique is admirable, yet the mind lacks the broadening and strengthening that comes from contact with cultivated minds and liberal studies'.[2] William's observations on domestic economy are discussed below.

[1] *Register*, 9 May 1899, p. 4.
[2] Ibid., p. 7.

During William's absence, negotiations between the university and the Ministry of Education concerning teacher training had advanced only slowly, but the building of the Conservatorium of Music had made better progress, with the letting of the contract and the laying of its foundation stone by the Governor on 26 September 1898.[3] At the Council meeting in November it was reported that William Henry Bragg Esq. MA had been elected by the Senate of the university to fill one of its six vacancies on the University Council, and in December he was appointed to the Education Committee.[4]

Teacher training and the new Conservatorium attracted major attention during William's first year on the Council. In September the previous year, 'the Chancellor of the University seized the opportunity afforded him . . . when opening the annual Teachers' Conference of delivering trenchant observations upon the need of greater facilities for the education of the pupil teachers of the State Schools. . . . he frequently moved the audience to loud applause'.[5] Within the Education Department the university's proposal met with qualified acceptance, but it was reluctant to relinquish all control over the training of its teachers. A compromise was hammered out during 1899, and in January 1900 a new scheme emerged under the misnomer 'University Training College'. There was no college in a physical sense and the university was not the major player. Recruits were to spend two years at the Pupil Teachers' School in the city, two years in schools as junior teachers, and finally one or two years at the university, where they would undertake lectures on the science and history of education and a selection from Greek, English, psychology, physics, mathematics, and modern history. The scheme was funded by the Elder bequest at no cost to the Education Department, but there were many initial difficulties. It was 1905 before the scheme was fully implemented.[6]

It was a major step forward, however, and Professors Bragg and Mitchell were part of a special Education Board set up to provide administrative co-ordination. Indeed, later scholars have attributed the overall scheme to Mitchell and Bragg, one calling it the Mitchell-Bragg Plan.[7] William renewed his contacts with local teachers. He presided over meetings of the Collegiate Schools' Association and was re-elected President of the Teachers' Guild.[8] He sympathised with them when the activities of their union became bolder. They

[3] Council Minutes, meetings of 28 April and 26 August 1898, UAA, S18, vol. VI, pp. 185 and 208 respectively.

[4] Ibid., meetings of 25 November and 8 December 1898, pp. 227 and 236 respectively. The Senate was composed of all master and doctor graduates of the university and of all other graduates of three years' standing.

[5] *Register*, 20 September 1898, p. 4.

[6] B K Hyams, 'State school teachers in South Australia, 1847–1950: A study of their training, employment and voluntary organizations', unpublished Ph.D. thesis, Flinders University of South Australia, 1972, ch. 5; see also Reports of Library and Special Committees, vol. II, 1898–1899, UAA, S148, pp. 37–42.

[7] See, for example, C Thiele, *Grains of Mustard Seed* (Adelaide: Education Department, 1975), p. 63; V A Edgeloe, 'Mitchell, Sir William (1861–1962)', in B Nairn and G Serle (eds), *Australian Dictionary of Biography* (Melbourne: MUP, 1986), vol. 10, p. 535.

[8] *Register*, 26 April 1902, p. 10 and 25 October 1902, p. 4.

campaigned for raising the compulsory school age to fourteen and for increasing the minimum attendance of pupils, and they were hurt and restive when new regulations greatly reduced promotion opportunities.[9]

The building of the Elder Conservatorium of Music was largely complete by March 1900. Built on steeply sloping ground, the whole of the main floor was occupied by a spacious concert hall (Elder Hall), with an open hammer-beam roof and polished wood ceiling. Entered at ground level from the south front, it contained a large stage with a recess for an organ and dressing rooms each side. The basement on the north side included six teaching rooms, a small concert room, and a large room for the organ engine and other items. Hollow brick walls, double doors, and silicate-cotton ceilings were included for sound-proofing but were singularly ineffective. The newspapers lauded the beauty of the building.[10]

It was inevitable that William would soon become involved with the new building. He was a member of the Board of Musical Studies that negotiated the building and installation of the organ through 1899. The Board asked a small sub-committee to recommend a design for the organ case and suggested that a gas engine be used to blow the organ.[11] When the matter came to the Council in December, however, it was unhappy with the submitted design and heard from Professor Bragg that he believed a large electric motor might be preferable to a gas engine.[12] At the April 1900 meeting of the Council, William submitted his own design for the organ case, which was approved, and he was appointed to a small sub-committee to oversee the project.[13] Thereafter Bragg's case adorned the Elder Hall until the installation of a new organ in 1979.[14]

There was widespread dissatisfaction with the performance of the Professor of Music, Joshua Ives, which may explain why William personally undertook design of the organ case. Ives had done much for music in Adelaide: he established the university degree course and a system of public examinations, he inaugurated an annual series of organ recitals in the Town Hall, and he instituted a comprehensive programme for the new Conservatorium. However, money seems to have been a constant source of concern. Ives was accused of spending too much time at the stock exchange, and he argued with the university constantly about his income and fees and with private music teachers about his Conservatorium policies. His students found his teaching unsatisfactory, his public relations were strained, and he clashed vehemently with the

[9] *Register,* 30 June 1903, p. 8; Thiele, n. 7, p. 72.

[10] S Marsden, P Stark and P Sumerling (eds), *Heritage of the City of Adelaide* (Adelaide: Corporation of the City of Adelaide, 1990), p. 269; T Nurmela, 'Elder Conservatorium of Music: Architectural critique', 1979, UAA, S315.

[11] Board of Musical Studies Minutes, meetings throughout 1899, UAA, S129, vol. II.

[12] Council Minutes, meeting of 19 December 1899, UAA, S18, vol. VI, p. 325.

[13] Ibid., meeting of 4 April 1900, p. 363.

[14] V A Edgeloe, *The Language of Human Feeling: A Brief History of Music in The University of Adelaide* (Adelaide: Friends of the Elder Conservatorium of Music, 1985), pp. 55–6. The old organ, complete with its original case, is now in St Mark's Cathedral, Port Pirie, South Australia.

Chancellor. When he refused to continue as Director of the Conservatorium beyond 1900, as the university required, the Council had had enough. Ive's appointment beyond 1901 was terminated, although not without a fight. John Ennis succeeded him from 1902.[15]

On the matter of the organ blower, William presented a lengthy report to the Board of Musical Studies in May 1900, in which he recommended a turbine driven by an electric motor. Such a scheme was new, however, and there was considerable uncertainty as to when a reliable supply of electricity would be available in Adelaide.[16] By August William's fears had been confirmed; the supply of electricity was still uncertain.[17] It was decided to order a turbine and power it with a gas engine, which necessitated a separate shed for the gas engine, an arrangement that stayed in place until 1902.[18]

More frustrating were the very poor acoustics in the Conservatorium teaching rooms, manifested immediately the building was occupied. Such a problem had come to William's attention ten years before, when he had been asked to assist the colonial government 'to improve the acoustic properties of the House of Assembly'.[19] Bragg had told his acoustics students, from the time of his arrival in Adelaide, that 'Sound is always reflected from a flat surface, especially if hard and smooth. Thus the bare walls of a room reflect sound, but this is stopped by hangings or carpets or furniture or people'.[20] In the House of Assembly, therefore, 'drapings of crimson cloth lined with American cloth were hung across the building from the top of the pilasters supporting the ceiling…and there was a perceptible improvement'.[21] Further experiments by Professor Bragg followed, including the use of strips of red and white calico and of quilted velvet,[22] the outcome of which was the permanent addition of various hangings around the chamber.[23]

At the university, following a request from the Board of Musical Studies, Bragg and Ives undertook some experiments. The problem was that 'Mr Bevan's room was, in particular, spoken of as being useless through echoes'.[24] William reported that 'Messrs Miller Anderson & Co. were instructed to hang an arras of art surge right around the room, to put curtains on the two windows and the door, and to lay two carpets on the floor'. The result was that 'the room

[15] Ibid., pp. 25–30; see also, *Register*, throughout December 1901.

[16] Letter W H Bragg to Board of Musical Studies, 11 May 1900, UAA, S200, docket 514/1900.

[17] Letter W H Bragg to Finance Committee, 29 August 1900, UAA, S200, docket 1222/1900.

[18] Board of Musical Studies Minutes, meetings during 1900 and 1901, UAA, S129, vols II and III; Finance Committee Report 2/1902, 27 February 1902, Reports to Council 1902, UAA, S150, vol. 3.

[19] *Advertiser*, 8 July 1890, p. 4.

[20] W H Bragg, 'Acoustics' (notes for lectures at Adelaide), lecture no. 2, July 1886 ff, RI MS WHB 31A/2.

[21] *Advertiser*, n. 19.

[22] *Advertiser*, 11 July 1890, p. 4.

[23] See, for example, a photo of the interior of the House of Assembly, 1890, in Marsden, Stark and Sumerling, n. 10, p. 250.

[24] Letter W H Bragg and J Ives to Board of Musical Studies, 11 May 1900, UAA, S200, docket 504/1900.

was now so dead that singing in it was painful', and it was therefore necessary to discover what reduced combination of the additions would give the best result.[25] Bragg and Ives then reported that the curtains and two small rugs produced a satisfactory result. They also discovered that 'curtains hung from the ceiling of the concert room in a simple fashion [produced] a very great improvement'.[26]

Eight years later William was asked to tackle the poor acoustics of the Elder Hall itself. He reported to the Council that the reverberation time was four seconds when the hall was empty, 'much too long for good hearing'. This he reduced to two seconds by using wall hangings and sawdust on the floor, and he therefore recommended that permanent draperies be installed and coconut matting placed on the floor and on the stage. For music he suggested that the draperies could be drawn aside and the matting rolled up.[27] These recommendations were implemented, but the acoustics of the hall continued to be a problem down the years. Significant improvements were made in the 1950s, but it was not until the 1970s, at the centenary of the university, that the problem was finally solved by an internal reconstruction of the hall.[28]

The science of sound had attracted William during his student days and the fascination would continue thereafter. During the Great War he would attempt to locate German submarines by listening for the sound of their engines, and his elder son would find the position of enemy guns by recording the sound of their firings.[29] When the war was over William delivered a series of lectures at the Royal Institution on the topic 'The World of Sound', and during the third lecture in the series, entitled 'Sounds of the Town', William discussed reverberation in public buildings and noted: 'Sometimes the more distressing features due to too much reflection can be removed with very little trouble. In a very bare and plain room in the basement of the Conservatorium at Adelaide piano recitals were at first almost impossible, because the room rang with shrill and piercing notes. The evil was practically cured by hanging two or three strips of serge two feet wide from the ceiling across the room'.[30]

On his return from leave William also found sympathy for a review of the regulations and curricula for the BA and BSc degrees. It is not clear how this revision was initiated, but in November the Council had before it drafts of new regulations for both the BA and BSc[31] A comparison of the regulations, as published in the university calendars for 1899 and 1900, shows how substantial the changes were. The long-serving Registrar and recent historian of the

[25] Ibid.; Miller Anderson & Co. was a leading draper and home-wares emporium from 1840 to 1988, for 148 years the oldest department store in South Australia.

[26] Letter W H Bragg and J Ives to Board of Musical Studies, n.d. [June 1900], UAA, S200, docket 596/1900.

[27] Letter W H Bragg to Council, 18 December 1908, UAA, S200, docket 929/1908.

[28] Edgeloe, n. 14, pp. 50–4.

[29] See chapter 17.

[30] Sir William Bragg, *The World of Sound* (London: Bell & Sons, 1920), p. 83.

[31] Faculty of Science Minutes, meeting of 4 August 1899, UAA, S202, vol. II, p. 19; Council Minutes, meeting of 24 November 1899, UAA, S18, vol. VI, p. 311.

university, Vic. Edgeloe, called them a 'root-and-branch revision of the arts and science curricula'.[32] Under the old regulations for the BA there were five subjects at first year, all compulsory (Latin, Greek, Elementary Pure Mathematics, Elementary Physics, English Language and Literature), four compulsory subjects at second year (Latin, Greek, Applied Mathematics, Logic), and three of four subjects were required at the third-year level (Classics and Ancient History, Mathematics, Mental and Moral Science, Modern Languages). Under the new regulations the requirements for the degree were simplified while at the same time providing a wider choice: a student was simply required to complete six of fifteen subjects over the three years. Many of the subjects extended over two years, and detailed suggestions were provided for students specialising in some areas.

Under the old regulations for the BSc, four elementary subjects were compulsory at first year (Pure Mathematics, Physics, Chemistry, and either Biology and Physiology or Applied Mathematics), while specified combinations of these subjects, together with Geology, provided a choice of three subjects at second year and two subjects at the third-year level. Under the new regulations students were required to complete two of the following five subjects over the three years: Pure and Applied Mathematics, Physics, Chemistry, Physiology, and (Geology, Mineralogy and Botany). Each subject, however, extended over two or three years and involved a wide variety of topics. Overall the BSc course became more specialised.

There was also a major reorganisation during 1900 of the public examinations conducted by the university for secondary school students. In the 1890s there had been four such examinations: Preliminary, Junior, Senior (matriculation) and Higher. The Higher had been abolished for 1900 but was now reintroduced, the Preliminary was redrafted and renamed the Primary examination, and regulations for the Junior and Senior examinations were rewritten. William took the leading role in this process and was chair of the Special Professorial Committee concerned, which consulted with teachers and accepted many, but not all, of their suggestions.[33] At the end of 1901 William was again elected President of the South Australian Branch of the Teachers' Guild.[34]

The university, sandwiched between the new Art Gallery and Conservatorium, was also feeling an accommodation pinch. As university teaching began early in 1900 the large increase in student numbers, particularly in the Chemical Laboratory and in the advanced Electrical Engineering course, could not be accommodated in the existing facilities. William reported on the difficulties to the Council, which appointed a Special Building Committee.[35]

[32] *Calendar of the University of Adelaide for the Year 1899* (Adelaide: University, 1899), pp. 55–8, 66–70; ibid., 1900, pp. 58–66, 73–82; V A Edgeloe, personal communication.

[33] Special Professorial Committee Minutes, meetings of June through August 1900, UAA, S22, vol. I, pp. 136–68; Council Minutes, meetings of 29 June, 27 July, and 13 August 1900, UAA, S18, vol. VI, pp. 383, 390, and 397 respectively.

[34] *Register*, 9 November 1901, p. 8.

[35] Council Minutes, meeting of 16 March 1900, UAA, S18, vol. VI, p. 353.

It asked the science and medical professors to prepare a complete scheme of their requirements and, when it reconvened in June, agreed to recommend that the first floor of the university building should be extended over the existing chemical and physical laboratories. The Council agreed,[36] and after prolonged discussions regarding the cost,[37] major additions were approved.[38] The medical facilities would be relocated, the small north-east addition to the original building would be enlarged to form the 'Classics Wing', and a further large northward extension would be built, soon to be called the 'Prince of Wales Building'.[39] There would be new rooms for teaching in the arts and for chemistry, biology, geology, museums, and engineering. Most of all there would be a large and well-appointed science lecture theatre to take more than two hundred people, since first-year classes now exceeded one hundred and public lectures were very popular. Since the total number taking physics and electrical engineering was close to two hundred, the existing physical laboratory would be expanded into the vacated space in the original building.[40]

The inauguration of the Commonwealth of Australia was proclaimed in Melbourne on 1 January 1901 by Queen Victoria's grandson, accompanied by his wife: the Duke and Duchess of Cornwall and York, later King George V and Queen Mary. The ceremony celebrated the federation into one country of the separate British colonies—now called States—that only reluctantly and progressively ceded power to the Commonwealth. Later the same month Queen Victoria died, but the Duke and Duchess continued their planned peregrination of Australia. They spent a week in Adelaide, 9–15 July 1901. Excitement rose to fever pitch as the visit approached, and Gwendoline decided to keep a diary of the week's events, in an unused carbon-copy book left over from their trip to England. The top copy was sent to Uncle William and the carbon copy remained in the book as the family's record. It has been published in full and is only summarised here.[41]

The Bragg family was heavily engaged on most days. On the first day they were guests of Dr Charlie Todd and his wife and were able to watch the royal procession from the balcony of their home in the city centre. In the evening they enjoyed the city illuminations provided by the new electricity. Next day, at a Government House levee for the city's notable organisations, Gwendoline was present as a member of the Mothers' Union and had prepared their presentation

[36] Special Committee Minutes, meetings of 27 March and 26 June 1900, UAA, S22, vol. I, pp. 125–6 and 146–7 respectively; Council Minutes, meetings of 29 June and 31 August 1900, UAA, S18, vol. VI, pp. 381, 400 respectively.

[37] Special Committee Minutes, meeting of 16 October 1900, UAA, S22, vol. I, pp. 173–5.

[38] Ibid., meetings of November 1900 through February 1901, vol. I, pp. 179–88 and vol. II, pp. 1–8.

[39] D J Gilbert, 'The Mitchell Building, University of Adelaide, Conservation Study', July 1987, Buildings Branch, University of Adelaide; the Prince of Wales Building was demolished in 1972.

[40] *Register*, 12 July 1901, p. 6.

[41] J G Jenkin, 'The 1901 royal visit to Adelaide: An account by William and Gwendoline Bragg', *Journal of the Historical Society of South Australia*, 1986, 14:19–34; this article provides background to the visit as well as a full transcript of the diary.

address, while William was part of the university group. Charles Todd presented the civil servants' address and Dr Charlie Todd that of the British Medical Association. In the afternoon the boys and their mother attended the annual football match between St Peter's College and Prince Alfred College on the Adelaide Oval, and later 'two weary, happy small boys crawled into bed'. On the third day William and his family were totally immersed in the university activities, while in the early evening Professor and Mrs Bragg were guests at a Government House reception, hosted by the Governor, now Lord Tennyson, son and heir of the poet laureate. Day four was free except for an impromptu dance at Charlie Todd's home, attended by officers from the royal yacht *Ophir.* Saturday was hectic, for there was a major military review and medal presentation on the Victoria Park racecourse, opposite the Braggs' new home, and they entertained 60 to 70 people. There was a State concert at the Exhibition Building that evening, and on Sunday morning ceremonies and a church service at St Peter's Cathedral. It was an exhausting but exciting week.

The University Council established a special committee to organise the royal visit there, for which William bore the major burden and responsibility.[42] Only William seemed capable of obtaining the co-operation of the students, which he secured by allowing them an active role in the ceremony. Likewise, he had impeccable relationships with the whole education community, for whom this was a special day. Since a visit to the new Art Gallery was to precede the university activities, William's membership of the Board of Governors ensured smooth coordination. At the Gallery the Duke unveiled two recent acquisitions and, after the departure of the royal couple, the official party sprinted through a hole in the fence to greet the royal carriage as it pulled up at the university's front porch next door. Here the Duke was to alight but the Duchess was to be driven a further short distance to the Elder Hall. The Duchess emerged at the front steps, however, and there were anxious glances until William came forward and offered to show her the way on foot. Charlotte, her sister, and the boys were watching from William's office, with an excellent view of everything, including, according to Charlotte, 'The Professor walking down with the Duchess, and I say to my sister and my sister say to me, how well the Professor pair with the Duchess, *much* better than with the Dook—yes, yes much better than with the Dook' (see Figure 10.1).[43]

The Duke and the rest of the party followed, stopping to allow the Duke to lay the foundation stone of the extension to the university and to grant it its name, the Prince of Wales Building. All then assembled in the Elder Hall, which was packed. Professor Bensley read the Latin address of welcome he had composed.[44] Printed on a parchment scroll, it was presented to the Duke in an elegant casket, designed by the Director of the School of Design, H P Gill, and constructed from Australian blue gum, carved with university and

[42] Special Committee Minutes, meetings of 6, 14, and 21 June 1901, UAA, S22, vol. II, pp. 12–18; *Register* 12 July 1901, p. 6 had a full account of the university ceremonies.

[43] Jenkin, n. 41, p. 30.

[44] Ibid., p. 31.

Fig. 10.1 Professor Bragg accompanying the Duchess of Cornwall and York to the official opening of Elder Hall and its pipe-organ, University of Adelaide, July 1901. (Courtesy: Mrs S Timbury.)

Australian motifs, lined with gold-embossed kangaroo skin, bound with bands and secured by hasps of Australian copper, all the work of Arthur Rogers in the Physical Workshop.[45] The Duke was then admitted to the degree of Doctor of Laws *ad eundem gradum*, and the Duchess declared the new organ open. The undergraduate students had entertained the audience with happy songs before the ceremony, and they continued to do so at intervals thereafter, but without giving offence as on previous occasions. In fact, as the Chancellor rose to open the organ by command and in the name of the Duchess, the students broke into another song, 'There is a ladye', which became an instant hit. William summarised the event in the family diary:[46]

> There were very few hitches and I don't think anyone saw them except the anxious reception committee... The Duke and Duchess seemed to thoroughly enjoy the whole thing, and sent word that they liked it much better than Sydney or Melbourne. 'There is a ladye' quite won them, and the Royal party sang it two or three times, together with the other songs, one evening at Government House. Copies of the songs and music were in great demand, and had to be supplied to the *Ophir* band and various members of staff. The organ and its novel blower behaved themselves. We had hardly a word of complaint except for some of the members of the Senate, who said their places were not good enough. It was a glorious

[45] J G Jenkin, *The Bragg Family in Adelaide: A Pictorial Celebration* (Adelaide: Adelaide University Foundation, 1986), pp. 40–1.
[46] Jenkin, n. 41, p. 31.

> day so far as weather went, which helped much to make the ceremony a bright success.
>
> [Gwendoline added:]
>
> I don't think Will's words are half strong enough. After the ceremony everyone kept coming up and saying what a brilliant success it was and how perfectly managed, which of course pleased me no end. It was an awful strain though, and when all was over we heaved great sighs of relief.

'There is a ladye' had three verses, the first of which is representative:[47]

> There is a ladye sweet and kind,
>
> Whose winsome face so pleas'd our mind.
>
> We did but see her passing by,
>
> Yet we shall love her till we die.

On the Saturday, the Bragg home was decorated inside and out with red, white, and blue, and William erected a temporary grandstand on a vacant block of land nearby. Their many guests had a perfect view of the passing of the troops and the royal party. William wrote, 'It was a lovely day and I don't think I have ever seen the place look better: hills so blue, grass and trees so green, just enough clouds to make lovely shadows'. There was afternoon tea for the adults and a tea-party for the children.[48]

William continued to be called on in every facet of university life. From their inception the university's sporting teams had had difficulty in obtaining playing venues, and in 1902 William headed a small Council sub-committee in an attempt to solve the problem. Unable to afford the hire of the Jubilee Oval behind the university buildings, they were also denied a strip of land on the southern bank of the Torrens River because it had been promised to the Agricultural Society. William and his two colleagues waited upon the Premier of South Australia, who suggested that they 'might obtain the consent of the Adelaide City Council to the use of the ground on the north side of the Torrens, between the river and Frome Road, being part of the Park Lands. To be successful this plan must include the throwing of a small suspension bridge over the river'.[49] Such a scheme came to fruition in 1910, with the opening of the first university oval, pavilion and change-rooms, and continues to the present time.[50]

Things also moved ahead at the School of Mines and Industries during William's absence. In June 1898 the School became seriously dissatisfied with its accommodation in the Exhibition Building and its Council approached the government for funds for a new, purpose-built home.[51] Sir Langdon Bonython,

[47] *Register*, n. 42.

[48] Jenkin, n. 41, p. 34.

[49] W H Bragg, 'Recreation grounds sub-committee progress report', no date, Reports to University Council, UAA, S150, vol. 3, no. 2/1902; Council Minutes, meetings of 21 March through 25 July 1902, UAA, S18, vol. VII, pp. 205–50.

[50] M M Finnis, *The Lower Level* (Adelaide: Adelaide University Union, 1975), p. 85.

[51] School Council Minutes, meeting of 13 June 1898, University of South Australia Archives, vol. 3, p. 225.

newspaper editor and company director, had been a member of the board that had recommended the birth of the School, and from 1889 until his death in 1939 he was a member and, for fifty years, President of its Council. There is an anecdote in Adelaide that, when Bonython received little support from the government for a new building, he and his friend George Brookman financed and built it on prime land, without government permission for either land or building.[52] Bonython certainly claimed he had no permission for the location, the site of the eastern annex of the Exhibition Building on the corner of North Terrace and Frome Road.[53] Both claims are false. Certainly Brookman, businessman and politician, gave £10,000 to the project (later increased to £15,000), and his wife laid the foundation stone of the new building on 7 March 1900.[54] However, there was a formal School Building Committee, of which William was a member, and there were regular reports in the minutes of the School Council of interaction between the School and the government concerning the project. In April 1900, for example, the Council was shown reworked plans, agreed to them, and sent them to the government for approval; while later in 1900 the plans were still under discussion with the Superintendent of Public Buildings.

In October the government accepted the tender of Mr F Fricker to build the new School for £25,999.[55] The precarious economy reduced the government's contribution to £16,000 and limited the design possibilities. Three and four storeys high, it was constructed of red brick, dressed with limestone in the Gothic-perpendicular style, and built upon a blue-stone base. All the materials were local to South Australia. Construction coincided with federation and the Boer War and the fervour generated found expression in the splendid stained-glass windows added to the building in 1903. Two of the windows commemorated British scientists and engineers: Watt, Newton, Stephenson, and Bessemer in one, Kelvin, Faraday, Wren, and Dalton in the other.[56] A marble plaque inside the main entrance commemorates the opening of the building by Lieutenant-Governor Sir Samuel Way on 24 February 1903, and carries the names of the councillors, including 'Profr. W H Bragg, M.A.' In more recent times the School has expanded and changed its name, to the South Australian Institute of Technology and now the University of South Australia (see Figure 10.2).

William's letter to the School of Mines reporting his investigations in Britain of domestic economy had significant impact, both in Adelaide and later

[52] W B Pitcher, 'Bonython, Sir John Langdon (1848–1939)', in B Nairn and G Serle (eds), *Australian Dictionary of Biography* (Melbourne: MUP, 1979), vol. 7, pp. 339–41; R M Gibbs, 'Brookman, Sir George (1850–1927), in ibid., pp. 429–30.

[53] A Aeuckens, *The People's University, 1889–1989* (Adelaide: S A Institute of Technology, 1989), p. 37.

[54] *Register*, 8 March 1900, p.4.

[55] School Council Minutes, meetings of July 1898 through October 1900, University of South Australia Archives, vol. 3, p. 229 through vol. 4, p. 167.

[56] Marsdan, Stark and Sumerling, n. 10, pp. 272–3.

Fig. 10.2 Visit of Baron Tennyson, Governor-General of Australia (son and heir of the poet and previously Governor of South Australia) to the new School of Mines and Industries building, Adelaide, William Bragg and Langdon Bonython on the extreme left, February 1903. (Courtesy: University of South Australia.)

in Britain. He wrote:[57]

> There are now springing up in England, in many places, schools for the teaching of domestic economy. These schools appear to me to be admirable institutions, worthy, if it possible, of our earnest imitation. I have spent some time recently in studying...two representative schools, and generally in collecting facts respecting them.
>
> Various subjects of domestic economy have been taught for years in primary schools in England...But schools of domestic economy are a

[57] [W H Bragg], 'Instruction in domestic economy', in *The South Australian School of Mines and Industries and Technological Museum, Annual Report, 1899* (Adelaide: School of Mines..., 1900), pp. 186–9.

> comparatively new idea—schools, that is to say, in which *all* the subjects of domestic economy are taught in proper proportion to each other, and the whole course of teaching is devised with the object of training good housewives...the girls are taught cookery, dress-cutting and dress-making, laundry work, housewifery, plain needlework, patching and darning, with some instruction in the laws of health....the schools are mainly intended for girls who are likely to have in the future the charge of working men's homes, and the instruction is modelled to that end.

A recent article has explored the domestic education for young women in South Australia during this period:[58]

> State education was a powerful tool in the attempted socialisation of working-class women into ordered, punctual, efficient middle-class ways....[and] there was a continuing debate about the education of girls after the passing of the Education Act of 1875. Both influential individuals and women's organisations supported domestic instruction...It was initially seen as a solution to the employment problem for poorer women, to the servant shortage for middle-class women, and as a means of improving the lives of working-class families.
>
> The first dedicated domestic education centre, the Domestic Economy Centre, was opened in 1900...by Inspector Alice Hills. Her 1901 report recounts that the centre...was housed in a corrugated-iron building...The centre ran for four years, after which classes for inner-city schools were transferred to the School of Mines. In 1910 the Education Department took over teaching Domestic Arts in the new high schools, with a view to extending it to primary schools.

William reported to the School of Mines Council in November 1900 that he had seen Mrs Hills regarding the establishment of a Domestic Economy Department and that she was enthusiastic. The following June the School Council resolved to establish a class in domestic economy and appointed a small sub-committee, headed by Professor Bragg, to arrange the details and report progress.[59] The School intended to establish a course in 1902 but inconclusive negotiations with government stalled the plan.[60] Late in 1904, however, the Ministry of Education decided to transfer its Domestic Economy Centre to the School of Mines, and the School successfully asked the government for funds and the transfer of equipment and staff.[61] William's suggestions had born fruit.

Nor was this the last time. By 1940 the humble Adelaide professor had become Sir William Bragg, OM, KBE, Director and Superintendent of the

[58] L. Hammond, 'Class and control: Social reform in the West End of Adelaide in the early twentieth century', *Journal of the Historical Society of South Australia*, 2003, 31:5–17 and references therein, particularly J Matthews, 'Education for femininity', *Labour History*, 1983, vol. 45.

[59] School Council Minutes, meetings of 5 November 1900 and 24 June 1901, University of South Australia Archives, vol. 4, pp. 173 and 234 respectively.

[60] Ibid., meetings of 9 September 1901 through 27 October 1902, vol. 4, p. 252 through vol. 5, p. 77.

[61] Ibid., meeting of 10 October 1904, vol. 5, p. 276.

Royal Institution and President of the Royal Society of London. In the early years of the Second World War he became Chairman of Britain's Scientific Committee on Food Policy, and he arranged a series of lectures entitled 'The Nation's Larder'. Various authorities spoke about 'the problem of nutrition, since a nation, if it is to fight with all its strength, must needs be rightly fed'. The lectures received warm support from the Ministry of Food and were published under the title *The Nation's Larder and the Housewife's Part therein*. Topics included: health and medicine, national requirements, the food industry and home production, the housewife and the feeding of children, and the preservation of food.[62] There were Adelaide resonances in so much that Professor Bragg did later.

William was conscious of the welfare of the staff in the institutions he served. At the June 1900 meeting of the School of Mines' Council he gave notice of his intention to move at the next meeting, 'that instructors be appointed to represent the staff of the School on the Education Committee'. At the same meeting the formation of a 'School of Mines Instructors' Association' was approved.[63] At the following meeting it was resolved that certain instructors be nominated by the Council to attend general, but not special, meetings of the Education Committee, and that such nominations would be sought before the next meeting.[64] It was less than William had hoped for, but it was far-sighted and it was representation on an important committee, something for which staff at Australian tertiary institutions would have to fight for the remainder of the twentieth century. In a similar vein the School's instructor in mathematics and physics, Mr J Dalby, received commendation and a significant salary increase, no doubt on the suggestion of Professor Bragg.[65]

The School of Mines had now become well established and prominent in South Australia and a serious competitor of the university for students, funds, and visibility. Having depended very heavily on the university in its founding years, the School was now ready to stand on its own feet. Friction between the two institutions was inevitable. There had been a minor skirmish in 1897, but it was defused when William initiated discussion between the parties:[66] a meeting of representatives clarified subjects at the two institutions.[67] Now, in 1900, William was becoming the meat in the sandwich and a crucial player in maintaining some respect and communication between the two organisations. In May the University Council appointed a small committee to confer with similar representatives from the School.[68] The committee met in June under William's chairmanship, and it became clear that the basis of concern was

[62] *The Nation's Larder and the Housewife's Part therein* (London, Bell, 1940), quotation from Bragg's Preface, p. vii.

[63] School Council Minutes, meeting of 4 June 1900, University of South Australia Archives, vol. 4, pp. 116–17.

[64] Ibid., meeting of 26 June 1900, vol. 4, p. 121.

[65] Ibid., meeting of 1 April 1901, p. 210.

[66] Ibid., meeting of 13 December 1897, vol. 3, pp. 181–2.

[67] Minutes of Joint Meeting, 8 July 1898, separate sheet in ibid., vol. 3, between pp. 238 and 239.

[68] University Council Minutes, meeting of 25 May 1900, UAA, S18, vol. VI, p. 375.

his belief that all engineering should be taught at the university, whereas the School was extending its operations in this area.[69]

In November university students complained about the standard of tuition at the School.[70] In February 1901 the University Council agreed to establish a joint committee of the two institutions to deal with common issues, and it appointed the Chancellor and Professors Bragg and Rennie as the university's representatives.[71] Next month Bragg persuaded the School Council to do the same,[72] but when the six representatives met they simply agreed to appoint a smaller sub-committee.[73] In October Rennie and Bragg wrote jointly to the University Council expressing their frustration following unproductive meetings of the sub-committee. The Council referred the matter to yet a further sub-committee![74] With the issue hopelessly bogged down, larger events sparked action. In the midst of the December graduation season, the *Register* editorialised that 'definite steps should be taken with the object of drawing closer the relations between the School of Mines, the School of Design, and the University by affiliating the former institutions to the latter'.[75] The School of Mines had no intention of allowing this to happen and, on the suggestion of its President, the School Council unanimously agreed, 'That it is desirable to establish a School of Electrical Engineering in connection with the School of Mines and Industries'. William was present. Did he abstain or vote in favour? Was he keeping his powder dry for another day or did he see this concession as worthwhile to secure the rest of engineering for the university?[76]

Letters from the University Registrar to his School counterpart urged further consultation,[77] the School raised a diversion regarding Council membership,[78] and the University refuted it.[79] The School Council agreed to meet the University Council, which meanwhile appointed George Henderson as the new Professor of Modern History and English Language and Literature, following the cloudy resignation of Robert Douglas, and awarded Coleridge Farr the university's first DSc[80] An Instructor in Electrical Engineering was

[69] Special Committee Minutes, meeting of 15 June 1900, UAA, S22, vol. I, p. 140; Council Minutes, meeting of 26 October 1900, UAA, S18, vol. VII, p.11.

[70] Council Minutes, meeting of 30 November 1900, ibid., p. 34.

[71] Ibid., meeting of 22 February 1901, p. 66.

[72] School Council Minutes, meeting of 15 March 1901, University of South Australia Archives, vol. 4, p. 202.

[73] Special Committee Minutes, meeting of 10 August 1901, UAA, S22, vol. II, p. 21.

[74] Letter E H Rennie and W H Bragg to Council, 25 October 1901, UAA S200, docket 1177/1901; Council Minutes, meeting of 25 October 1901, UAA, S18, vol. VII, p. 142.

[75] *Register*, 19 December 1901, p. 6.

[76] School Council Minutes, meeting of 23 December 1901, University of South Australia Archives, vol. 4, p. 291.

[77] Letters University Registrar to School of Mines Registrar, 15 and 24 February 1902, copies in Council Minutes, meeting of 28 February 1902, UAA, S18, vol. VII, between pp. 195 and 196.

[78] Letter School Registrar to University Registrar, 11 March 1902, UAA, S200, docket 284/1902.

[79] Letter University Registrar to School Registrar, 27 March 1902, UAA, S1, Letterbooks, vol. XXII, p. 42.

[80] School Council Minutes, meeting of 7 April 1902, University of South Australia Archives, vol. 5, p. 25; Letter School Registrar to University Registrar, 26 May 1902, UAA, S200, docket 568/1902; University Council Minutes, meetings of April 1902, UAA, S18, vol. VII, pp. 210–30.

appointed at the School and a major donation was received from Messrs Noyes Brothers of Melbourne to equip a new electrical laboratory.[81] Although an outline of a scheme for united action by the two institutions was ready later in 1902, the end of the academic and calendar year blocked further progress.

Early in 1903 frustrated teachers from both institutions met under William's chairmanship and hammered out an agreement, which then bypassed various committees and went direct to the two Councils, where it was approved.[82] The agreement had the following elements, leading to a joint University Diploma in Applied Science and Fellowship of the School of Mines and Industries: four diploma courses in mining and engineering, joint use of the electrical laboratories of the School and university, establishment of a Faculty of Applied Science with equal representation, and formation of a Joint Board of the two institutions to exercise oversight.[83] Inevitably William was appointed a member of both latter entities. It is unnecessary to enumerate the many wrinkles that remained to be ironed out, except to note and applaud William's central role in facilitating the agreement and its subsequent development. He believed that the university was the appropriate body for the higher levels of tertiary study and tensions continued to smoulder, but the agreement was another of William's vital legacies.

It is also noteworthy that, although he was instrumental in urging the January/February meetings so that the agreement could be implemented for the impending academic year, William was also irritated that they interrupted his annual holidays with the family. When he wrote to the university Registrar from the seaside town of Normanville, his tone was courteous but his annoyance was very plain.[84]

Preparations for research

One other matter arising from William's year in England remains to be discussed, and it was to have profound implications for the future: his renewed desire to begin serious research. He did not act immediately on his return; he became embroiled in a range of other issues as outlined above. His intentions became clear, however, when he wrote two long letters to the university's Education Committee in November 1899. They are a controlled but passionate request for more financial and moral support for teaching and research in

[81] School Council Minutes, meetings of 10 March and 12 May 1902, University of South Australia Archives, vol. 5, pp. 16 and 33 respectively.

[82] Council Minutes, meeting of 4 February 1903 with copy of Bragg's undated Report of the Joint Committee of the University and School of Mines, UAA, S18, vol. VII, pp. 305–306; School Council Minutes, meeting of 16 March 1903 with copy of final Agreement dated 3 March 1903, University of South Australia Archives, vol. 5, p. 118.

[83] Ibid.

[84] Letter W H Bragg to Registrar, 23 January 1902, UAA, S200, docket 78/1903.

physics and mathematics. The first said:[85]

> During the past year I have had the assistance of Mr Allen in conducting the classes in Mathematics and Physics: but Mr Allen's appointment terminates with the year, and it will be necessary to make a fresh arrangement for 1900.
>
> It will, in my opinion, be sufficient to appoint someone an Assistant Lecturer and Demonstrator at a salary of £250 per annum... But if the work of the department increases as it has done in the past... it will be necessary to pay a higher salary than this...
>
> The present staff consists of Mr Chapman, an Assistant-lecturer or Demonstrator, and myself. Between us we give the lectures in mathematics and physics to all three years of the Arts and Science courses, as well as... Honours. We also duplicate some of the physics lectures for the sake of evening students. We give the lectures in mining, applied mechanics, surveying, levelling etc., also on acoustics for the Mus.Bac. course, and two sets of lectures on Electrical Engineering. In all we give 32 lectures a week, five or six of these being in the evening; and we have, in addition, the Laboratory work of all the years, some of which is heavy as the classes are large. During the past year there have been more than 100 students in the laboratory, and the mere handling of the apparatus for their experimental work and for the lecture illustrations is sufficient work for one person...
>
> There is another way of adding to the efficiency of the laboratory teaching to which I ask leave to call your attention; viz: the employment of our own students, who have just taken their degrees and are suitable as junior demonstrators at a nominal salary. This plan is commonly adopted in laboratories and has many advantages. To begin with, such an assistant is of great use in the conduct of the practical classes. But this... is far from being the only one. The position would be of value to the aspiring student, and as a reward for good work: it would give opportunities for further study, and for doing original work... and that would be a very desirable thing. We should have much more chance than we have at present of finding students fit to fulfil the conditions of the 1851 Exhibition scholarships and bursaries. Moreover, a young graduate after filling such a position... would stand a better chance of employment in the schools.
>
> Again, it would add greatly to the tone of the laboratory if there were always one or more students doing original, or post-graduate work in it. We ought to do all we can to gather about us students asking for the highest instruction... In fact, I think the advantages of employing our own students in this way are so great that I recommend that the experiment be tried as soon as a suitable opportunity arises.

The last two paragraphs mark a major turning point in William's career. It certainly has a modern tone, for it is this system of academic apprenticeship

[85] Letter W H Bragg to Education Committee, 21 November 1899, UAA, S200, docket 936/1899.

that is common in university scientific laboratories around the world. It has been emphasized above that William undertook some research much earlier and that he had been encouraged by a range of events to take it up more seriously thereafter. Other demands on his time, however, and his own uncertainty and lack of confidence of how to begin, and on what subject, had combined to thwart his ambition. Now his doubts—practical, professional, and personal—had been largely cleared away. He had extensive laboratory experience, he was a scientific leader in Adelaide and around Australasia, and his British peers had recognised and welcomed him. Personally he had matured into a family man and public identity. The period of leave had given him the last ingredients he needed to begin a serious research programme.

His second letter concerned mathematics in the schools and harked back both to his leave and to his own experiences at King William's College. Perhaps he also had his own sons in mind: 'I believe it is time that some mathematical examination was open to the schools of a higher grade than the Senior Public... clever boys in English schools reach this level some years before they leave... I believe there are many boys here who could go further than they do in their mathematical studies at school, and would do so if they had encouragement... When boys with a bent towards mathematics are under the impression that, for them, it is time to leave school for the university as soon as they have passed the senior, then it is impossible for us to do them the good we might do. If in any way we can induce them to stay longer at school, it will be an advantage to them and to us'.[86] As mentioned earlier, the Higher Public Examination was reinstated in 1901 for precisely this and related purposes. Again William had contributed to a significant change.

In response to William's concerns the University Council resolved to renew the appointment of Bernard Allen for 1899 with increased responsibilities.[87] When, in November, the Council considered William's long letter in detail it also agreed to appoint an Assistant Lecturer and Demonstrator at £250 per annum, to pay the present lad (Bromley) fifteen shillings per week, and that 'the appointment of students who have just graduated as Junior Demonstrators in the Laboratories be tried, and that one such assistant be appointed to the Chemical and one to the Physical Laboratory at the first suitable opportunity at an honorarium of 10/- [shillings] per term'.[88]

In January 1900 Bernard Allen was appointed for another two years,[89] but this did not adequately solve the teaching problem. In March William wrote to the Council pleading for further assistance in view of the 'large and unexpected' increase in the number of physics students, due to the introduction of the teacher training scheme and increases in the number of students from the School of Mines, in the science and medical courses, and in electrical

[86] Letter W H Bragg to Education Committee, 22 November 1899, UAA, S200, docket 937/1899.

[87] Council Minutes, meeting of 24 March 1899, UAA, S18, vol. VI, p. 253.

[88] Ibid., meeting of 24 November 1899, p. 311.

[89] Ibid., meeting of 26 January 1900, p. 329.

engineering.[90] Following a further letter in July emphasizing the excessive teaching loads of Chapman and Allen, approval was given for the appointment of Laurence Birks, previous winner of both the Angas Engineering Exhibition and Scholarship and recently returned to Adelaide because of the death of his father.[91] Chapman was now devoting the major part of his time to engineering students. Yet a further request from William in August was also approved: for payment of third-year student Geoffrey Duffield as Junior Demonstrator, for glass tubing and electrical apparatus, and for the appointment of a boy to replace Bromley and an apprentice to assist Rogers in the Physical Workshop.[92]

In December the *Register* reviewed the university year favourably. It noted, 'The growing preponderance of the science school is the most notable fact which can be gleaned from an inspection of the University degree list this year'.[93] The university had centres in Perth in the far-west and at Broken Hill in the north. Thus, when the Perth Observatory was established early in 1896 by the appointment of William Ernest Cooke, following a glowing reference from Charles Todd for his able protégé, Cooke also agreed to act as Secretary of the 'West Australian Centre of the University of Adelaide'. The centre was designed to provide some of the services expected of a university before the foundation of Western Australia's own in 1911: public examinations for secondary-school students, extension lectures to satisfy community demand, and some tertiary courses. Cooke had earlier completed a BA degree in Adelaide, with a mathematics major under Lamb, and then an MA by coursework under Bragg. Likewise, after the Perth Technical School was established in 1900, Bernard Allen was appointed on William's strong recommendation to teach mathematics and physics, and he was active in encouraging his best students to complete Adelaide undergraduate examinations in Perth and then to go on to Adelaide to complete the final year of their degrees.[94]

These appointments are testimony to the generosity and breadth of vision of both Bragg and Todd. Struggling to establish viable departments of their own, they willingly encouraged their junior staff to move to even more needy institutions elsewhere, to enhance their young careers, and to expand the disciplines in Australia. William's comment the he would 'feel the pinch when he [Allen] goes' was a substantial understatement.[95] To their adopted homeland

[90] Letter W H Bragg to Council, 29 March 1900, UAA, S200, docket 1279/1900.

[91] Letters W H Bragg to Education Committee and Council, 13 July 1900, UAA, S200, dockets 680/1900 and 698/1900 respectively; Council Minutes, meeting of 27 July 1900, UAA, S18, vol. VI, p. 393.

[92] Letter W H Bragg to Finance Committee, 29 August 1900, UAA, S200, docket 827/1900; Council Minutes, meeting of 25 January 1901, UAA, S18, vol. VII, p. 56.

[93] *Register*, 13 December 1900, p. 4.

[94] J G Jenkin, 'Frederick Soddy's 1904 visit to Australia and the subsequent Soddy-Bragg correspondence: Isolation from without and within', *Historical Records of Australian Science*, 1985, 6:153–69, particularly pp. 153–5; D Hutchison, 'William Ernest Cooke, astronomer 1863–1947', ibid., 1981, 5:58–77.

[95] Letter W H Bragg to C Jackson, 21 March 1901, J.B. Allen personal file, W.A. Education Department, no. 239/1901, un-numbered folio, Battye Library, Perth.

both he and his father-in-law gave unstintingly of their time, effort, expertise, and graduates and staff.

On 31 December 1900, in summarising 'The nineteenth century', the *Register* suggested, 'We live in an age of science miracles. Almost every day produces its new wonder, and so we work at high pressure'.[96] The new century certainly brought no slackening in pace. Discussing 'The problem of radio-activity', for example, the *Register* observed that 'the discovery of the Röntgen rays, and the subsequent proof by the French physicist Becquerel of the existence of somewhat similar rays which need no electric battery for their production, brought about something like a revolution in scientific thought. Old theories required to be, to a large extent, revised'.[97] Later in the year, in reporting the presidential address of Arthur Rücker to the British Association for the Advancement of Science, the *Register* noted that 'atoms are no mere figments of the mathematical imagination...they are physical realities;...like matter itself, electricity is also atomic in its nature'.[98]

Fortunately, Allen's departure was not as debilitating as William had anticipated. He wrote urgently to his colleagues in Melbourne and Sydney and was immediately able to obtain the services of an outstanding replacement, J.P.V. Madsen. John Madsen was born in country New South Wales in 1879 and was educated at the University of Sydney, where he studied science and engineering concurrently. He graduated BSc with first-class honours in physics and mathematics and the University Medal in mathematics in 1900, and a year later added a BE degree, again with first-class honours and the University Medal.[99] In April 1901 William offered Madsen the combined position of Assistant Lecturer in Mathematics and Assistant Lecturer and Demonstrator in Physics, which he accepted. At the same time William was given permission to appoint one of his previous students, Isaac Boas, as a Junior Demonstrator in Physics.[100]

In August William wrote to the university Finance Committee to plead for funds 'for roofing in a portion of the yard at the back of the main university building in order to provide a new workshop'.[101] Unlike his earlier neat handwritten correspondence, this letter is hasty and contained a number of corrections. William was in a hurry! He had Madsen's advice that the allocated space was inadequate for the new apparatus arriving for the electrical engineering course. Space currently occupied by the mechanical workshop would solve this

[96] *Register*, 31 December 1900, p. 4.

[97] *Register*, 18 January 1901, p. 4.

[98] *Register*, 19 October 1901, p. 6.

[99] Sir Frederick White, 'John Percival Vissing Madsen', *Records of the Australian Academy of Science*, 1970, 2:51–65; D M Myers, 'Obituary: John Percival Vissing Madsen', *The Australian Physicist*, 1970, 7:14.

[100] Letter W H Bragg to Finance Committee, 25 April 1901, UAA, S200, docket 445/1901; N Rosenthal, 'Boas, Isaac Herbert (1878–1955)', in B Nairn and G Serle (eds), *Australian Dictionary of Biography* (Melbourne: MUP, 1979), vol. 7, pp. 332–3; Boas was subsequently Chief of the Division of Forest Products of the Australian Council of Scientific and Industrial Research (CSIR), 1928–44.

[101] Letter W H Bragg to Finance Committee, 29 August 1901, UAA, S200, docket 930/1901.

problem, and a new larger workshop, obtained by roofing a portion of the yard, could then be built at a cost of only £230. 'Such a quantity of work... is now done in the workshop', William wrote, 'that it seems very desirable to provide a comfortable and convenient room'.[102] The request was approved and William now had resources that were adequate for what he had in mind, although they still compared unfavourably with those he had seen in Sydney fifteen years earlier.[103]

The teaching of engineering was now assuming greater significance, and the next year (1902) William gave Madsen overall responsibility for the four-year electrical engineering programme, which would soon be recognised by the Institution of Electrical Engineers in London as sufficient qualification for its Associateship.[104] In August Madsen's application for leave-of-absence during the southern summer, in order to visit the UK and USA at his own expense to better prepare himself for teaching these classes, was approved.[105] In September his further request to be designated 'Lecturer in Electrical Engineering' and to be granted an augmented salary was also agreed.[106] On his return Madsen reported to the university on his findings, suggesting that the course 'will enable men to enter a profession which has... a scope hitherto unequalled', and that it therefore needed to be adequately supported with internal resources and external experience for its students.[107] William had now gathered the resources he needed, both for a greatly extended teaching programme and for a serious tilt at research.

[102] Ibid.

[103] Council Minutes, meetings of 27 September and 4 October 1901, UAA, S18, vol. VII, pp. 122 and 130 respectively.

[104] Council Minutes, meeting of 27 June 1902, UAA, S18, vol. VII, p. 242.

[105] Ibid., meeting of 29 August 1902, p. 256.

[106] Ibid., meeting of 26 September 1902, p. 262.

[107] J P V Madsen, 'Report on a recent tour through England and America', 24 April 1903, UAA, S200, docket 422/1903; Madsen also spent some time as Assistant Engineer on the Adelaide staff of the Melbourne-based Electric Lighting and Traction Company (letters of recommendation from the Adelaide Chancellor, Professor Bragg, and Manager of the company regarding Madsen's 1904 application for a position at the University of Sydney, in possession of Madsen family).

11
Front-rank research: alpha particles

At the turn of the century South Australia was a new state, with great hopes for the future but a legacy of past difficulties. The severe depressions of the 1880s and 1890s could not be overcome quickly, particularly in a state subject to drought and dependent on primary production. Unemployment and its consequences had been serious. In the new century, however, secondary industries became more firmly established, and there were better seasons and higher prices for rural products. In December 1902, for example, 'monsoonal' rain, the best for ten years, broke a prolonged drought.[1] Adelaide became something of a magnet, now offering a good standard of living: efficient water supply, extensive drainage and sewerage, increasing gas and electricity reticulation, post, telegraph, and telephone facilities, a growing train and horse-tram system, and a lively arts and entertainment culture. Theatre and other amusements thrived, as did sports of all kinds. On the other hand there were continuing social problems: poverty and unemployment as well as growing affluence, street gangs as well as many churches, crowded slums as well as pleasant suburbs. Overall, however, the first years of the new century were ones of steady development—until 1914, when drought and especially a world war interrupted progress and brought back distress.[2] Little of this seems to have affected the Bragg family directly; certainly there is nothing of it in their reminiscences. Willie, for example, remembered a happy home and uninhibited holidays:[3]

> While I was still at Queen's we moved into our new home. I started there the love of gardening which has always stayed with me. Bob and I were given two small plots at the back of the house and I remember vividly the thrill of seeing a green tip appear from a daffodil bulb I had planted, and eventually the formation of the flower. I had a prolific peach tree in my garden, but its pride and glory was an immense vine. Someone had given me a foot-long twig of vine which I planted, and it grew until it covered the trellis along the whole back of the house, with gigantic bunches of black grapes, though they were rather tasteless... I experienced a curious coincidence in connection with this vine. Once, when waiting at London

[1] *Register*, 19 December 1902, p. 9.
[2] R M Gibbs, *A History of South Australia* (Adelaide: Southern Heritage, 1984), chs XI and XII.
[3] W L Bragg, Autobiographical notes, pp. 13–17.

Airport for the departure of our plane to America, I started chatting to my neighbour, found he was an Australian from Adelaide...and asked him if he had done all the missions on his programme. 'All except one', he said, 'I have a friend who lives in a house on East Terrace which has an enormous vine growing over the back. There is a story current that the vine grew from a cutting of the famous vine at Hampton Court. My friend knows that a man called Bragg once lived in that house and asked me to track him down in England and find out if the story is true, but I have not been able to do this'. I, of course, was able to say, 'I planted that vine'. I saw it during our stay in Adelaide in 1960, and in sixty years it had grown into a noble vine, with a trunk as thick as a man's leg...

At Catherwood House [in Adelaide] there was a shed of galvanised iron in the back-yard, which was allotted to us as a workshop. Dad arranged for one of the assistants in the laboratory to show us how to carpenter. We made endless gadgets, particularly electrical ones. I made a simple form of motor...power was supplied by a bichromate battery...I remember well how astonished I was when it worked (I had only read about it in a book)...Then Bob and I rigged up an electric bell in the workshop, with a push-button in the nursery so that Charlotte could summon us when tea was ready...I made a telephone of the original Bell type, a clock and, inspired by a new instrument just then installed at the Observatory, a beam seismograph. This last did not record earthquakes but was very sensitive to small tilts produced by our walking about the room.

We had wonderful holidays, sometimes in the hills of the Mount Lofty range but more often at the seaside in St Vincent's Gulf. Adelaide was a very hot place in summer...It was a great relief to escape to the relative coolness of the hills or the sea around Christmas time. Places we stayed at...were Port Noarlunga, Aldinga and Yankalilla. Port Noarlunga at that time had only one farm, the Pocock's, where we stayed, and a fisherman's hut. The river Onkaparinga ran out to sea at Port Noarlunga and was tidal where the bridge crossed it. We had a boat on the river and had happy days fishing and exploring the reefs at low tide. It was so deserted that Bob and I often wore no clothes except our hats with the essential fly-veils around them. The flies were a curse, but they could be kept at bay by the veil, which was a very open string net with tassels hanging on it; a jerk of the head whisked the flies away.

Port Noarlunga is about twenty-five miles [40 km] from Adelaide...In those times it took a good part of a day to get there. We started out from Hill's Coaching Yard in a coach and five. Bob and I, together with our fox-terrier Tommy, generally managed to secure a place on the driver's seat...The driver had a pile of mail for the houses along the road beneath his legs, and we were impressed by the way he could flip a packet of letters so that it sailed over the fence and fell on the mat before the front door. We changed horses after twelve miles or so, and when we came to a steep hill the passengers got out and walked. At Willunga [sic, Noarlunga] we were met by Mr Pocock in his buggy and were driven to the farm...

One summer my mother was expecting a baby (my sister Gwendy) and my father hired a buggy with two horses and a driver from the livery

> stables so that my mother could go for an excursion. Before breakfast Bob and I rode the ponies bare-back into the sea until they swam and we were towed by their manes, and afterwards galloped on the sands till we were dry... [One] summer was also memorable because two girls of about our age, Hilda Fisher and Frances Hawker, were invited to stay with us on the Pocock farm. It was the first time Bob and I really met girls, other than at formal occasions like dancing classes, and what a time we had showing off all the features of the place to them.

The man asking about the Black Hamburgh vine was, in fact, the well-known Adelaide surgeon and medico-politician, Sir Henry Newland. The suggestion that the vine had royal parentage was a myth.[4] The Bragg family had other pets in addition to Tommy, including a cat, a parrot, a canary, and a sparrow called Tim.[5] Most notable in this extract, however, is Willie's growing self-confidence and his parents' encouragement to engage in adventurous activities. In a short paragraph on his father's radio experiments, Willie also recorded that 'Bob and I took a great interest in these experiments, especially because it meant a picnic on Sunday afternoons, when my grandfather and father drove to Henley Beach with us in the official Post Office wagonette to see the signals coming in'.[6] Science and adventure enjoyed together.

William continued to attend meetings of the AAAS, even if he was not presenting a paper (Melbourne 1900, Hobart 1902). The final trigger for him to take up the research that he had long contemplated was his agreement to give the presidential address for Section A of the January 1904 Dunedin (New Zealand) meeting.[7] He began preparation during the long summer vacation of 1902–03, when he opened a new notebook: a literature review in his characteristically neat and economical handwriting, that he continued throughout 1903.[8] He scanned the major English journals taken by the university library, particularly *Nature, Philosophical Magazine,* the *Proceedings* and the *Philosophical Transactions of the Royal Society of London*, and *The Electrician*. The library also took major European journals such as *Comptes Rendus de l' Académie des Sciences, Physikalische Zeitschrift and Annalen der Physik* but, since William had no formal training in German and only occasionally translated from French, he relied largely on the regular abstracts, summaries, and translations in *The Electrician* for the important European publications.

[4] The story is contained in *Advertiser*, 19 August 1950, p. 2 and in correspondence between I J Ball and W L Bragg, July 1958, RI MS WLB 57A; for Newland see N Hicks and E Leopold, 'Newland, Sir Henry Simpson (1873–1969)', in G Serle (ed.), *Australian Dictionary of Biography* (Melbourne: MUP, 1988) vol. 11, pp. 8–9.

[5] J G Jenkin, *The Bragg Family in Adelaide: A Pictorial Celebration* (Adelaide: University of Adelaide Foundation, 1986), p. 35.

[6] W L Bragg, Autobiographical notes, pp. 17–18.

[7] W H Bragg, 'On some recent advances in the theory of the ionization of gases', in G M Thomson (ed.), *Report of the Tenth Meeting of the Australasian Association for the Advancement of Science, held at Dunedin, 1904* (Wellington: AAAS, 1905), pp. 47–77; the central importance of this address was acknowledged by Bragg himself in W H Bragg, *Studies in Radioactivity* (London: Macmillan, 1912), p. 1.

[8] W H Bragg, Notebook, untitled, RI MS WHB 12/1.

The notebook contains about 200 separate references to contemporary articles, almost all of which are from these eight journals. From the earliest pages it is clear that William had already determined the general area of his interest: the new physics of X-rays, radioactivity, the electron, positive ions, and the ionization of gases. Radioactivity was a phenomenon in which, according to the very recent work of Rutherford and Soddy, unstable atoms decayed to lighter, stable species by the emission of one or more types of rays. Since these rays were initially unidentified, scientists had labelled them by the first three letters of the Greek alphabet: alpha (α), beta (β), and gamma (γ). The betas were increasingly thought to be identical to J.J. Thomson corpuscles, and to cathode rays and electrons, particles of mass only one-thousandth that of the lightest atom (hydrogen), negatively charged, a component of all atoms, and an indicator of an atomic inner structure. The natures of the alpha- and gamma-rays were less clear, the former perhaps heavy, particle-like, and of uncertain charge, the latter thought by some to be similar to X-rays. All this was very new, very exciting, and very perplexing.[9] William was diving in at the deep end; his year in England in 1898 had been decisive.

As became his practice, William summarised the content of the papers he read on the right-hand pages of the notebook and then entered his own observations on the left. The first article that caught his close attention was a 1903 paper by Rutherford from McGill University in Canada, where the young student who had visited William in Adelaide was now a professor.[10] Rutherford, at times in company with the English chemist Frederick Soddy, had become a leader in the new field of radioactivity.[11] William paid particular attention to articles by Rutherford and John Townsend, but the range of authors and topics was large. An early, brief note headed 'Curies' report in Vol. III of Paris Congress' was later given a more extensive elaboration in the notebook, under the heading 'Rapports au Congrès'.[12] The notes concerned the paper of Pierre and Marie Curie that appeared in the reports of the 1900 Paris international congress on physics,[13] and predated Marie's doctoral thesis of 1903.[14]

From his earlier reading, William was aware that the alpha rays behaved differently and were far less studied than the beta. He entered his own

[9] There are numerous books and articles on this revolution in physics; a recent one is H. Kragh, *Quantum Generation: A History of Physics in the Twentieth Century* (Princeton: Princeton University Press, 1999).

[10] E Rutherford, 'Excited radioactivity and the method of its transmission', *Philosophical Magazine*, 1903, 5:95–117.

[11] See, for example, T J Trenn, *The Self-Splitting Atom: The History of the Rutherford-Soddy Collaboration* (London: Taylor and Francis, 1977).

[12] W H Bragg, n. 8, pp. 67, 127–31.

[13] P Curie et Mme. Curie, 'Les nouvelles substances radioactives et les rayons qu'elles emettent', in C E Guillaume and L Poincare (eds), *Rapports Présentes au Congrès International de Physique, Paris, 1900*, (Paris: Gauthier-Villars, 1900), 4 vols, vol. 3, pp. 79–114; the section 'Pouvoir pénétrant des rayons non déviables', due to Marie alone, is especially relevant, pp. 101–3.

[14] Mme. Curie, *Recherches sur les Substances Radioactives* (Paris: Gauthier-Villars, 1903), which appeared in English translation as 'Radio-active substances', *Chemical News*, 1903, vol. 88, nos 2282–91, and as *Radioactive Substances* (New York: Philosophical Library, 1961) and (Westport, Conn.: Greenwood, 1971).

observations far more expansively than before: 'The α-rays of polonium show marked diminution at a certain distance [from the radioactive source], and the absorption by a plate is greater the further it is away from the source... [Remarks on α-ray behaviour, including their exponential (very rapid) decrease in intensity with increasing distance from their source, due to scattering from the atomic electrons of the material they were traversing]...The α-rays will have more complicated effects. Active electronic collision will occur 1000 times as often as for the β-ray, but we could hardly expect actual deflexion to occur in such a case, so perhaps the α-ray will go on. If so α-rays would rather tend to stop when their energy ran out, and so stop rather suddenly, as in Curie's experiments'.[15] The nature of the alpha particle was an even greater mystery than its behaviour,[16] and what light might these questions shed on the structure of the atoms from which the rays originated? Here was a field to which a new researcher might contribute.

The first fruits of this review were three extension lectures William gave during the winter of 1903 on 'The Electron and the Radio-activity of Radium, Thorium and other substances'.[17] The first lecture covered the role of the electron in electrical phenomena and the colourful effects of an electrical discharge in a glass tube that was progressively evacuated. 'There was not a single hitch in the arrangement of the experiments... [during] his lucid and convincing lesson'.[18] The second lecture discussed the known properties of the electron, and then turned to demonstrations of the effects of X-rays, including the ionization of air, so that, when the air pressure was rapidly reduced, water droplets formed on the charged particles produced. This last is direct evidence of the influence of William's visit to the Cavendish Laboratory; namely, the recent work of C T R Wilson and his cloud chamber. The Adelaide audience was amazed at the sophistication and success of the experiments.[19] William's final lecture was reported at length, during which he discussed the three radiations emanating from thorium. He explained that alpha rays only travelled a short distance and were largely uninfluenced by a magnet, beta rays were penetrating and were strongly deviated by a magnet, and gamma rays were even more penetrating but were poorly understood. The *Register* reporter thought the lecture series 'afforded a treat never before presented to a South Australian audience... greatly augmented by a number of remarkable experiments'.[20]

At the forthcoming meeting of AAAS in New Zealand, South Australia would be represented by Bragg and Rennie, the joint honorary secretaries of the South Australian committee, and about forty others, chiefly lady teachers.[21] Professor Bragg, accompanied by Mrs Bragg and Miss Lorna Todd, departed

[15] W.H. Bragg, n. 8, pp. 127–8.

[16] Later in this notebook (p. 105, entry date unknown), Bragg maintained the view that the 'α-particle contains 1000 or so electrons'.

[17] Extension Lecture Circulars and Syllabuses, UAA, S102, vol. 1, 1902–1925.

[18] *Register*, 18 June 1903, p. 7.

[19] *Register*, 24 June 1903, p. 5.

[20] *Register*, 1 July 1903, p. 6.

[21] *Register*, 8 December 1903, p. 4.

for Melbourne by train on 28 December, and left there for Dunedin by steamer on the thirtieth.[22] The boys were left in Adelaide in the care of Charlotte and the Todd family. They spent time at the seaside suburb of Semaphore, enjoyed picnics and games on the beach, caught whiting and yellow-tail while fishing, visited a German mail steamer with Uncle Hedley Todd, saw the Lendon family at their Mount Lofty home, and then returned to school after the long summer vacation.[23]

William Bragg's Dunedin address drew on all aspects of his literature review and was entitled 'On some recent advances in the theory of the ionization of gases' for, as he pointed out, 'the phenomenon itself furnishes one of the principal methods by which the strange new properties of radio-active substances are made manifest and studied'.[24] There followed, in thirty dense pages of the conference report, discussion of ionization, the nature of positive and negative ions, and the stopping behaviour of fast electrons, with particular emphasis on Becquerel's experiments with radium beta rays. Not until the last four pages do we meet: 'it is very interesting to compare the penetrating and ionizing powers of the α and γ rays with those of the β ray'.[25] A summary of the known properties of the alpha particle followed, including the observation that the 'α ray penetrates more than a thousand times... and yet moves in a straight line;... at the high speed which it possesses when it leaves the parent body it breaks down the defence of any molecule it encounters and passes through'.[26] That the gradual stopping of the alpha particles through expenditure of energy was a more important cause than scattering was confirmed by two curious results due to the Curies, William explained.[27] This material was elaborated in Marie's doctoral thesis.[28]

The basis of the apparatus was common to many experiments at this time. The two plates of a condenser, acting as an ionization chamber, were arranged so that the lower plate admitted rays from a radioactive source, whose distance from the chamber could be varied incrementally and so that any charge accumulated by ionization (and hence the number of particles reaching the chamber) could be measured (using an electrometer). In addition to the air, thin sheets of various absorbing materials could also be placed between the radioactive source and the ionization chamber. Marie Curie showed that, while the radiations from most radioactive substances were very complex, polonium, which she had isolated, emitted only one group of alpha particles. 'I found the absorbability of the rays to increase with increase of thickness of the matter traversed... contrary to that known for other kinds of radiation', she recorded.[29] The alpha particles all travelled a similar distance 'by rectilinear propagation'

[22] *Register*, 29 December 1903, p. 5, and 8 December 1903, p. 4 respectively.
[23] Letters W L Bragg to parents, January and February 1904, RI MS WLB 95B.
[24] W H Bragg, n. 7, p. 47.
[25] Ibid., p. 74.
[26] Ibid.
[27] Ibid., p. 75.
[28] Mme. Curie, *Recherches*, n. 14.
[29] Mme. Curie, *Radioactive Substances*, n. 14, p. 50.

and, as Rutherford had suggested, appeared positively charged. Much else had 'not been studied in detail'.[30]

William had now come to the tentative conclusions that 'the α particle should pursue a perfectly rectilinear course, passing without deflection through all the atoms it [meets]', that 'the number of α particles penetrating a given distance does not alter much... until a certain critical value is passed, after which there is a rapid fall', and that 'the energy of the α particles... gradually decreases... and dies out at the same critical value'.[31] These were matters he had decided to test in a new research programme when he returned to Adelaide and, in preparation, he had written to W G Pye & Co. in Cambridge before departing for the meeting: 'My dear Mr Pye, I wrote to you last week about a Dolezalek electrometer. I hope the letter reached you safely. My next want is some radioactive material. I had a little present the other day, and I think I can spare £20, which should be enough to procure an amount sufficient for many forms of experiment. I have no idea... as to where and how radium compounds are obtained. But I gather that it is on the market... I am sure Professor Thomson or Mr Searle would be good enough to advise... I will ask our Registrar to forward a request to the Agent General that he will advance about £25 on your request: £20 for the radium material and £5 for the electrometer, at a guess'.[32]

The 'little present' had been supplied through the generosity of a constant friend of the University of Adelaide, Robert Barr Smith,[33] and may have been part of Barr Smith's July 1903 donation of £500 towards the purchase of instruments.[34] 'Three tubes of 5mgms each Radium bromide pur. cryst.' were shipped from Cambridge by registered post on 12 May 1904,[35] and arrived in Adelaide on 14 June,[36] just in time for William to incorporate them in his extension lectures on 'The electron and the atom'.[37] These three lectures were a continuation of his 1903 series and were similarly popular. The first dealt with the new understanding regarding the electron composition of atoms.[38] In the second lecture William explained the removal of electrons from atoms of gas (ionization) to leave positive ions, the consequent ability of the gas to pass an electric current, and the production of X-rays when the released electrons were suddenly stopped. Radium gave out three different radiations: one like electrons, one like the heavier ions, and one like X-rays. The first two were

[30] Ibid., pp. 51, 55, 57.

[31] W H Bragg, *Studies in Radioactivity* (London: Macmillan, 1912), p. 4.

[32] Letter Agent General for South Australia to Registrar, 13 May 1904, enclosing account for radium and copies of correspondence between Professor Bragg and Mr Pye, UAA, S200, docket AG9/1904.

[33] W H Bragg, n. 31, p. 5.

[34] Finance Committee Report to Council, no. 7/1903, 30 July 1903, UAA, S150, vol. 4.

[35] Letter W G Pye to W H Bragg, 12 May 1904, n. 32.

[36] In a public lecture that evening, Bragg announced that the radium had arrived earlier the same day: *Register*, 15 June 1904, p. 6.

[37] Extension Lectures Circulars and Syllabuses, UAA, S102, vol. 1, 1902–1925; *Register*, 15 June 1904, p. 6.

[38] *Register*, 15 June 1904, p. 9.

material but could pass through matter largely unhindered, and it seemed therefore that atoms and molecules were mostly empty, perhaps like miniature solar systems.[39] William's last lecture, again to a crowded theatre including the Governor and other community leaders, focused on his new radium sample, including the large energy release. The lecture concluded with an outline of J J Thomson's new model for the atom—elsewhere called the 'plum-pudding' model—with electrons embedded in a sphere of equal positive charge. Indeed, most of William's material for these lectures was based on Thomson's Silliman Lectures at Yale University in May 1903.[40]

On Saturday, 30 July 1904, William opened his first research notebook and began his first experiment on the alpha particles from radium.[41] In his first international paper on this work, William pointed out that radium was the only radioactive source available to him, although he would have preferred polonium for reasons of simplicity.[42] In fact, the complexity of the decay of radium opened up, luckily and at once, a more interesting and fruitful research project. From the start William was assisted by his first research student, Richard Kleeman. Kleeman, the eldest of nine children of a German farming family, had left school at age thirteen and entered the coopering trade in the local wine industry. He had also read mathematics and physics privately with the help of his Lutheran pastor. In 1897 he began sending short papers to Professor Bragg, who was impressed and arranged for his special admission to the university. Kleeman graduated BSc with first-class honours in physics in 1905, studying while also lecturing and demonstrating in the subject and assisting the professor with his research.[43]

The very next day, Sunday 31 July, another chance encounter took place— akin to that involving Bragg and Rutherford nine years before—and it, too, was to have profound implications for the future of William's research. Frederick Soddy visited Adelaide for just a few hours and, he later recalled, 'was met by Professor Bragg and taken off to dinner at his home with Mrs. Bragg and some of the Adelaide University folk'.[44] Soddy had completed his pivotal research in radioactivity with Rutherford in Canada as well as his important studies with William Ramsay at University College London, and he had not yet taken up his new lectureship at Glasgow. He had willingly accepted an invitation to visit Western Australia to

[39] *Register*, 22 June 1904, p. 7.

[40] *Register*, 29 June 1904, p. 6; for Thomson's work see J Z Buchwald and A Warwick (eds), *Histories of the Electron: The Birth of Microphysics* (Cambridge, Mass.: MIT Press, 2001), J J Thomson, *Electricity and Matter* [The Silliman Lectures, 1903] (Westminster: Constable, 1904), and chapter 13 below for further discussion of the 'plumb-pudding' model.

[41] W H Bragg, Adelaide University Laboratory [research] Notebook, vol. 1, RI MS WHB 38/1; this first page is shown in J G Jenkin, *The Bragg Family in Adelaide* (Adelaide: University of Adelaide Foundation, 1986), p. 44.

[42] W H Bragg, 'On the absorption of α rays, and on the classification of the α rays from radium', *Philosophical Magazine*, 1904, 8:719–25.

[43] J G Jenkin and R W Home, 'Kleeman, Richard Daniel (1875–1932), physicist', in C Cunneen (ed.), *Australian Dictionary of Biography: Supplement 1580–1980* (Melbourne: MUP, 2005), pp. 218–19.

[44] M Howorth, *Pioneer Research on the Atom: The Life Story of Frederick Soddy* (London: New World, 1958), p. 137.

Fig. 11.1 Frederick Soddy, during the year he lectured in Western Australia and visited William Bragg in Adelaide, 1904. (From L Badash, *Rutherford and Boltwood*, Yale University Press, 1969, plate 3. Courtesy: *British Journal of Radiology*.)

give several series of lectures on 'electricity' (interpreting the topic broadly), in Perth and at a number of country centres.[45] After this strenuous task was complete, Soddy returned to Britain via the Pacific and the Atlantic, stopping at ports along the way, including Adelaide. There he learnt of William's plans, whose success he was to assist substantially in the next few years (see Figure 11.1).

William subsequently wrote a book-length account of his Adelaide research and there is a recent long analysis of it;[46] only the major elements of his scientific progress are needed here. His first experiments were undertaken

[45] J G Jenkin, 'Frederick Soddy's 1904 visit to Australia and the subsequent Soddy-Bragg correspondence: Isolation from without and within', *Historical Records of Australian Science*, 1985, 6:153–69.

[46] W H Bragg, n. 31, chs I–X; J G Jenkin, 'William Henry Bragg in Adelaide: beginning research at a colonial locality', *Isis*, 2004, 95:58–90.

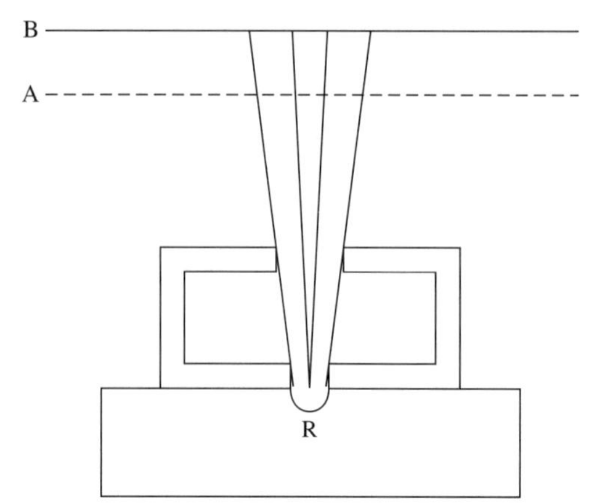

Fig. 2. –*A B* is the ionisation chamber; the upper wall of metal, the lower of metal gauze. The α rays stream up from the radium at R.

Fig. 11.2 An early form of William's alpha-particle apparatus, Adelaide, 1904. (From W. H. Bragg, *Studies in Radioactivity*, Macmillan & Co., 1912, p. 5. Courtesy: Palgrave Macmillan.)

with simple apparatus (see Figure 11.2). It was, however, an improvement on Curie's equipment, in that it defined a narrow pencil of alpha radiation that would wholly enter the ionization chamber and be detected (provided it was not stopped earlier), regardless of the chamber's distance from the radiation source. Curves of the variation in ionization in the chamber with changing distance from the source were plotted and then analysed in terms that took into account the fact that many of the alpha particles initially lost energy in escaping from the thick radium source he used. A second set of experiments was conducted with a spherical ionization chamber and a much thinner (and weaker) source, and the meaning of the results then became clearer. The most important of these were: that when radium decayed it passed through several changes, four of which appeared to expel an alpha particle, with all the particles of any one change having the same energy, and with the first change producing alpha particles of the lowest energy (thereby confirming a suggestion by Rutherford and Soddy[47]); that the alpha particles passed through the matter they traversed without appreciable deviation and lost their extraordinary energy by ionization along a straight trajectory; and a new and important finding, that alpha particles were much more efficient in ionizing the material through which they passed near the end of their course, when their energy was much reduced.

[47] See papers in Sir James Chadwick (ed.), *The Collected Papers of Lord Rutherford of Nelson O M, F R S*, 3 vols. (London: Allen and Unwin, 1962), vol. 1, pp. 565–657.

The first of these results was won only after the sort of fright that experienced researchers know only too well: 'Then I got a hint from Professor Soddy, who was passing through Adelaide, that I should dissolve the preparation in water, which would wash away three of the active substances but leave radium itself, the parent of them all. So I did, but, horror of horrors, as I brought my measuring apparatus up towards the radium in the way I had learnt to do, there was *no radiation at all* when I was well within the old range. However, with a very downcast spirit, I pushed the apparatus closer still, and closer; and suddenly a tremendous effect flashed out. The radium itself sent out the particles of the *shortest* of the four ranges, not the longest as I had thought; and, free from overlying impurities [its radioactive daughters], was shooting with great effectiveness. My assistant, Dr Kleeman, and I were excited enough!' .[48]

Progress had been very rapid and William first presented the results to a regular meeting of the Royal Society of South Australia, which promptly published an abstract of his conclusions.[49] Two papers describing the work left Adelaide two days later in a letter to Frederick Soddy, who responded at once, the first items of an ongoing correspondence between the two men, only the Soddy to Bragg half of which survives: 'I got your long & interesting letter this morning & write at once to congratulate you on the extremely successful turn your work has taken. I need hardly say that I shall study the papers... with great interest; indeed, I await their appearance that I may be able to quote them to my own (theoretical) ends. To this end I have written the Phil. Mag. people asking them to get them out quickly if possible. I was chagrined to read that lack of space prevented Nature publishing your full letter & was on the point of writing the Editor to that effect, but now... I think on the whole it is best that it was not published in the unsatisfactory form of a short letter'.[50] From the beginning Soddy provided encouragement and crucial support in speeding William's papers through the publishing process; not least because, when the editor of the *Philosophical Magazine* replied to Soddy, agreeing to get William's papers out quickly, he also stated that he would send proofs and any technical queries to Soddy, thus avoiding the very long round trip by sea to consult the author directly.[51] The two papers appeared consecutively in the December 1904 issue of the journal.[52]

Just two days after his first letter, when Soddy was in London to conclude his association with University College, he wrote again to William with two pieces of news that must have made his heart jump for joy: 'Excuse this scrawl, but I just want to catch the mail with a small tube of a milligram or so of my

[48] From a popular account of this early work see W H Bragg, 'In the days of my youth', *T P's & Cassell's Weekly*, 3 April 1926, p. 834 (copy at RI MS WHB 39/2).

[49] W H Bragg, 'On the absorption of α rays, and on the classification of the α rays of radium', *Transactions of the Royal Society of South Australia*, 1904, 28:298–9.

[50] Letter F Soddy to W H Bragg, 18 October 1904, RI MS WHB 6A/23; regarding Nature's inability to publish Bragg's letter, see *Nature*, 1904, 70:445.

[51] Letter F Soddy to W H Bragg, 2 November 1904, RI MS WHB 6A/25.

[52] W H Bragg, 'On the absorption of α rays, and on the classification of the α rays from radium', *Philosophical Magazine*, 1904, 8:719–25; W H Bragg and R Kleeman, 'On the ionization curve for radium', ibid., pp. 726–38.

pure Radium Bromide. I had a little over at Uni. Coll. & you mention in your letter that you have had difficulty in getting effects powerful enough for your electrometer & I thought this small qty. might prove useful. I wrote Stark, the Editor of the Jahrbuch der Radioaktivitat, that he should get you to write an article on your work & I understand he has done so'.[53] The radium sample was superior in both strength and purity to what William had used thus far and would be important in his next experiments.

The invitation from Johannes Stark would be central in alerting German scientists to William's recent and future research work, and Stark first wrote on 22 October 1904: 'I am informed by Prof. Soddy that you are conducting important investigations of α-rays. New results on these insufficiently-studied rays seems highly desirable, and I wish to ask you to write a brief report of your work for the Radioactivity Yearbook. In view of the great distance between us on the one hand, and the desirability of an early publication of your results on the other, I should be glad to receive your reply by return mail'.[54] William prepared an account of his research and, translated by Kleeman, it appeared in the 1905 Yearbook.[55] Soddy wrote again in January 1905, assuring William that the radium he had sent could be used 'in any way you think will help you at all', telling him that he had had to make only a very minor correction to his papers and providing a long report of his own views on the nature of radioactive change.[56] I have given the name 'advocate' to the pivotal role Soddy had now assumed on William's behalf.[57]

Nor was Soddy the only influential scientist to receive an account of the new results. As early as 10 August 1904 William wrote excitedly to J J Thomson, discussing the contrasting behaviours of beta and alpha rays and his two forthcoming papers, adding 'I am sending home a good mathematician this year, and have ventured to give him a letter to you. His name is Wilton'.[58] William also sent accounts of his research to an increasing number of his Australian and overseas friends and colleagues.[59] There is a hand-written list from a little later, indicative of his circle of correspondents: seven Adelaide University colleagues, nine ex-students, six Sydney and Melbourne university staff, William Sutherland in Melbourne,[60] Uncle William and William's surviving brother Jimmy, four at the Cavendish Laboratory, and twenty-three elsewhere (including four in

[53] Letter F Soddy to W H Bragg, 20 October 1904, RI MS WHB 6A/24.

[54] Letter J Stark to W H Bragg, 22 October 1904, RI MS WHB 6B/8.

[55] Letter W H Bragg to J Stark, 1 December 1904, Staatsbibliothek, Berlin; W H Bragg, 'Die α-Strahlen des Radiums', *Jahrbuch der Radioaktivität und Elektronik*, 1905, 2:4–18.

[56] Letter F Soddy to W H Bragg, 12 January 1905, RI MS WHB 6A/26.

[57] Jenkin, n 46.

[58] Letter W H Bragg to J J Thomson, 10 August 1904, Thomson papers, Cambridge University Library, Add MS 7654, B70.

[59] R W Home, 'The problem of intellectual isolation in scientific life: W H Bragg and the Australian scientific community, 1886–1909', *Historical Records of Australian Science*, 1984, 6:19–30.

[60] T J Trenn, 'Sutherland, William', in C C Gillispie (ed.), *Dictionary of Scientific Biography* (New York: Scribner's Sons, 1976), vol. XIII, pp. 155–6; R W Home, 'Sutherland, William (1859–1911)', in J Ritchie (ed.), *Australian Dictionary of Biography* (Melbourne: MUP, 1990), vol. 12, pp. 141–2; R W Home, 'William Sutherland and the "Sutherland-Einstein" diffusion relation', *Australian Physics*, 2005, 42(2):53–60, and *Historia Scientiarum*, 2005, 15(2):125–38; W A Osborne, *William Sutherland* (Melbourne: Lothian, 1920).

Canada, four in England, seven in Germany and Austria, two in Paris, four in America, and two unidentified). Many of the names are well known in the history of physics: Thomson, Campbell, Richardson, Laby, Rutherford, Hahn, Callendar, Lamb, Stark, Schmidt, Hess, Becquerel, Langevin, and Boltwood.

From the beginning and along with Soddy, however, William's second vital correspondent was Ernest Rutherford. In another August letter William presented his new results and emphasized his belief that the stopping of alpha particles was not exponential. 'I know of course that this is rather contrary to your theories', he wrote, 'and yet I think you will be pleased with what I want to tell you because my results are so beautifully explained on your theory of radioactive change'. Finally, 'I have read of you so often since you went through Adelaide many years ago that I seem to have never lost touch with you entirely. But let me take this opportunity to congratulate you on your magnificent work'.[61] These congratulations were not gratuitous, since William was nearly ten years older than Rutherford; but the assertive Rutherford would remain the senior partner during most of their long and friendly relationship. He became William's second influential 'advocate'.

Rutherford responded generously, supporting William's conclusions, encouraging publication, and reporting his recent work. Furthermore, he conceded that, 'I am all the more interested [in your work] in that I was independently attacking the question...but had not your data'; and then, 'I shall of course keep away from the subject until you are through'.[62] The nature of the alpha particle remained a mystery, however, for its ionization characteristics might be explained, Rutherford suggested, by the hypothesis that 'the α particle is initially uncharged & does not begin to ionize much until an electron is liberated;...I feel convinced it is helium but the measurements are very difficult'.[63] In November of the same year Rutherford wrote again, saying he was preparing a second edition of his *Radio-Activity* text and inviting William to forward the results of his work, 'as I would like to include it in the new edition'.[64] William responded, thanking Rutherford for a copy of his Bakerian lecture and regretting that Rutherford was visiting New Zealand but could not get to Adelaide.[65] It is hard to imagine that anyone had such a breathtaking beginning to a new research program as William enjoyed during the last six months of 1904; his long years of preparation were bearing fruit.

On 19 January 1905 William again wrote to Rutherford, attaching a 24-page, hand-written copy of his AAAS address, together with ten pages of additional notes and concluding, 'I hope this letter is in time [to catch the overseas mail]. I am rather afraid that yours was delayed; it was marked

[61] Letter W H Bragg to E Rutherford, 31 August 1904, CUL RC B354.
[62] Letter E Rutherford to W H Bragg, 23 October 1904, RI MS WHB 26A/1.
[63] Ibid.
[64] Letter E Rutherford to W H Bragg, 14 November 1904, RI MS WHB 26A/2; E Rutherford, *Radio-Activity*, 2nd edn. (Cambridge: CUP, 1905).
[65] Letter W H Bragg to E Rutherford, 18 December 1904, CUL RC B355.

"Too Late"'.[66] This is an example of William's ever-present worry: that his distance—from Europe in general and England in particular—would delay his letters and manuscripts so that he was out of touch with the latest scientific developments and that his priority would be jeopardised. In mid-year he again wrote to Rutherford, now in New Zealand, pleading with him to visit Adelaide: 'I should be delighted if you could spend a little time with me; I have house room. It would be a keen pleasure to me to have your society for as long as you can give me; and the chance of hearing your opinion on some of the radioactive problems'.[67] Personal contacts were perhaps the greatest loss felt by the distant, colonial scientist.

In February Rutherford had written to the journal *Nature* to report experiments that 'undoubtedly show that the α particles do carry a positive charge';[68] but this had been followed a week later by a letter from Soddy that quoted William in suggesting that the alpha particles were emitted uncharged but immediately became positive on losing an electron.[69] It would be 1908 before Rutherford and his collaborators could settle the argument: the alpha particle was a helium ion carrying two positive charges.[70]

In June 1905 William reported to the Royal Society of South Australia on the next phase of his research, and in summarising his previous work he explicitly uses the term 'range' for the first time: 'each [alpha] particle possesses a definite range in a given medium, the length of which depends on the initial velocity of the particle and the nature of the medium... all the particles of any one group have the same initial velocity and the same range'.[71] He gave precise results for the ranges in air of the four groups of alpha particles from radium, and concluded that their loss of range was nearly proportional to the square root of the atomic weight of the atoms they traversed or, in the case of molecules, to the sum of the square roots of the atomic weights of the constituent atoms.

The papers that followed in the *Philosophical Magazine* continued to be channelled through Frederick Soddy, who was pleased to 'see them through the Phil. Mag. as you request'.[72] The next paper gave details of the major improvements that had been made to the apparatus, thanks to the expertise of Arthur Rogers and the use of Soddy's radium bromide sample. The results were therefore 'much

[66] Letter W H Bragg to E Rutherford, 19 January 1905, CUL RC B356.

[67] Letter W H Bragg to E Rutherford, 7 June 1905, CUL RC B357.

[68] E Rutherford, 'Charge carried by the α rays from radium', *Nature*, 1905, 71:413–14.

[69] F Soddy, 'Charge on the α particles of polonium and radium', *Nature*, 1905, 71:438–9.

[70] E Rutherford, 'The nature and charge of the α particles from radio-active substances', *Nature*, 1908, 79:12–15 and references therein.

[71] W H Bragg and R Kleeman, 'On the alpha particles of radium, and their loss of range in passing through various atoms and molecules', *Transactions of the Royal Society of South Australia*, 1905, 29:132–3. Spectacular visual verification of these conclusions regarding range came in 1912 with Wilson's cloud-chamber pictures: see C Chaloner, 'The most wonderful experiment in the world: A history of the cloud chamber', *British Journal for the History of Science*, 1997, 30:357–74, 372–3.

[72] Letter F Soddy to W H Bragg, 30 July 1905, RI MS WHB 6A/29.

more accurate, and supply much more information'.[73] Bragg and Kleeman were able to argue convincingly that, for the absorption of alpha particles, 'there is no... absorption coefficient, nor any approach to an exponential law'.[74] Having studied the passage of alpha particles through thin foils of six different metals (from aluminum to gold) and through seven different gases (from hydrogen to ether), they were able to conclude that a material's ability to reduce the range of alpha particles was described by the atomic-weight relationship given above.

William apparently arrived at this result empirically, describing it as 'remarkable';[75] more recent work has shown it to be fortuitous and restricted to a low energy range.[76] His conclusions that 'the α particle makes the same number of ions during its course no matter what the gas which it traverses', and that 'the energy required to make a pair of ions is always the same', are now known to be only approximately true.[77] Soddy wrote to William twice in June 1905, urging him to 'push on with the absorption experiments... I think you are probably on the eve of something pretty fundamental... With your theoretical insight I shall expect a great elucidation'.[78] He also volunteered to assist in obtaining the best possible metal foils to replace the poor-quality samples William had been using, and he sent two pieces of his own platinum foil as an interim measure.[79] As always, his letters were full of praise and encouragement.

Although not published specifically until later, these papers embodied what became known as the 'Bragg ionization curve' for alpha particles. Figure 11.3 depicts the two forms of the curve presented by William in his later account of the Adelaide work: (a) William's own method of presentation, and (b) the conventional illustration due to Hans Geiger (see Figure 11.3).[80] The name lives on in a wider context, as in the modern medical procedure called 'Bragg peak therapy'.[81] In the treatment of a tumour, for example, the energy of a beam of charged particles is adjusted so that the Bragg peak, and therefore the greatest damage, occurs at a specified distance (range) below the surface of the skin.

Rutherford wrote from New Zealand in July, regretting that there was no hope of a side trip to Adelaide and saying, 'I will keep my men as clear as possible of the line of work you wish to take if you let me know in time—as I know it is a drawback for publication living in Australia'.[82] The letter did admit, however, that one of his staff had repeated the Adelaide work at Montreal and found it to be 'completely corroborated'. There was also notification of a significant

[73] W H Bragg and R Kleeman, 'On the α particles of radium, and their loss of range in passing through various atoms and molecules', *Philosophical Magazine*, 1905, 10:318–40, 325.

[74] Ibid., p. 328.

[75] Ibid., p. 332.

[76] R D Evans, *The Atomic Nucleus* (New York: McGraw-Hill, 1955), pp. 652–3.

[77] Bragg and Kleeman, n. 73, p. 339.

[78] Letter F Soddy to W H Bragg, 1 June 1905, RI MS WHB 6A/27.

[79] Letter F Soddy to W H Bragg, 14 June 1905, RI MS WHB 6A/28.

[80] W H Bragg, n. 31, pp. 29–30;

[81] See, for example, M Nitschke, 'The discrete charm of exotic nuclei', *New Scientist*, 1989, 121:58–9.

[82] Letter E Rutherford to W H Bragg, 3 July 1905, RI MS WHB 26A/3.

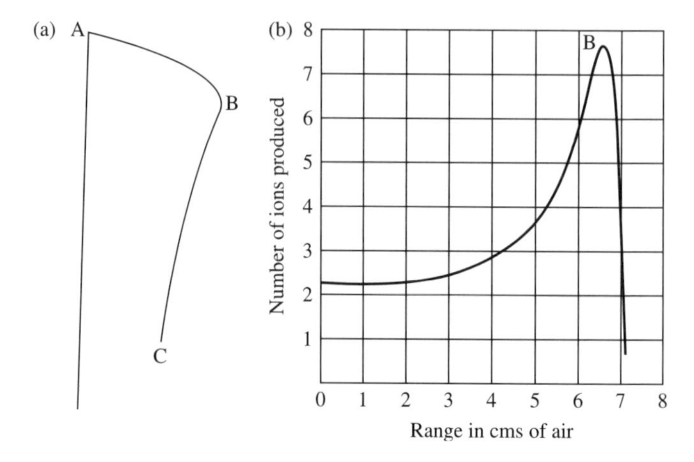

Fig 11.3 Two forms of the 'Bragg ionization curve': (a) Bragg's own style, with range vertically and ionization horizontally. (b) The form that soon superseded (a), with the axes reversed. (From W H Bragg, *Studies in Radioactivity*, Macmillan & Co., 1912, pp. 29, 30 respectively. Courtesy: Palgrave Macmillan.)

problem that was best stated in a letter from Soddy the same month: 'Going on to the last part of [your] paper I am, of course, stopped dead by the great difference between your views on the energy remaining to the α particle after it loses its power of ionising and those in Rutherford's last paper. You assume that when it ceases to have this power the energy has fallen to an amount which is very small compared to its initial energy. R deduces that at this point the velocity is still 60% & the energy 40% of the initial'.[83] William responded with a letter to the editor of the *Philosophical Magazine*, pointing out what changes needed to be made to his recent paper in view of these findings.[84] Soddy queried the supposition,[85] and although William appeared to accept Rutherford's criticism, later events were to show that he was not convinced.

The years 1904 and 1905 witnessed William's continuing commitment to his previous activities as well as extensive time devoted to his new research, although difficulties continued. The university's financial position, never strong, deteriorated further due to expenditure on new buildings and programmes. The Council and its Finance Committee deliberated hard and long, but the inevitable result was that recurrent expenditure had to be cut at just the time when the university was in an expansionary phase. The State government was not helpful. In William's wide field of responsibility existing staff were reassigned to cover emergency gaps, and a request from Madsen for a salary rise was refused.[86]

[83] Letter Soddy to Bragg, n. 72 (he is referring to E Rutherford, 'Some properties of the α rays from radium', *Philosophical Magazine*, 1905, 10:163–76).

[84] W H Bragg, 'On the α particles of radium', *Philosophical Magazine*, 1905, 10:600–2.

[85] Letter F Soddy to W H Bragg, 9 August 1905, RI MS WIIB 6A/30.

[86] Letter J P V Madsen to Registrar, 25 March 1904, UAA, S200, docket 251/1904; Council Minutes, meeting of 16 May 1904, UAA, S18, vol. VIII, pp. 30–1.

William's own request in March 1904 for relief from evening physics lectures was granted by the appointment of Kleeman to take the class,[87] but when he pleaded at length and in exasperation for increased expenditure on the electrical engineering programme, and otherwise threatened dire consequences, the Finance Committee responded in kind: 'To base upon the incertitudes of conjectures a financial calculation supposed to be reliable is not feasible; and belief in such calculations, far from being counted to the Council as sound financing, will, in the degeneracy of modern financial life, lead the believers to diverge from the rugged path to Paradise into the broad and smooth highway of further borrowings, which ends at the too-hospitable door of the Insolvency Court. It is quite easy, of course, to build up on hypotheses a system of mathematics, but a scheme of finance needs to rest on a surer foundation'.[88] William had never received such a rebuff, but his future requests for modest financial assistance were treated with respect and usually obtained agreement, as they had in the past.[89]

On the wider university scene there were now other outstanding professors who shared the broader load. In addition, Douglas Mawson was appointed by invitation as Lecturer in Mineralogy and Petrology.[90] Befriended by William, Mawson was destined to become Australia's greatest Antarctic explorer and long-term Professor of Geology and Mineralogy at Adelaide.[91] Tensions in the relationship with the School of Mines continued. The School generously donated a selection of tools to the physical workshop at the university, but then announced its intention to appoint its own teacher of physics and remove its students from William's first-year physics class.[92] In December 1905 students again disrupted the annual commemoration ceremony, this time with blasphemous and affronting behaviour, and it was decided to exclude them from future ceremonies.[93]

William's three extension lectures in 1905 addressed the topic of 'Radium' and continued the popular account of his burgeoning research. As the university's advertising leaflet announced, 'A preliminary explanation will be given of the general phenomenon of radio-activity... [and] recent work and discoveries will then be described. In particular, an account will be given of some phenomena which have been under investigation at the University of Adelaide during the past year...Models and lantern slides and a few experiments will be employed to illustrate the phenomena described'.[94] The *Register* noted

[87] Letter W H Bragg to Council, 11 March 1904, UAA, S200, docket 218/1904; Council Minutes, meeting of 25 March 1904, UAA, S18, vol. VIII, p. 16.

[88] Finance Committee Report to Council, 5/1904 (meeting of 6 May 1904), UAA, S150, vol. V, pp. 85–91.

[89] See, for example, letter W H Bragg to Finance Committee, 27 April 1905, UAA, S200, docket 394/1905; Council Minutes, meeting of 28 April 1905, UAA, S18, vol. VIII, p. 123.

[90] Council Minutes, meeting of 30 September 1904, UAA, S18, vol. VIII, p. 65.

[91] P Ayres, *Mawson: A Life* (Melbourne: MUP, 1999).

[92] Council Minutes, meetings of 29 April and 16 May 1904, UAA, S18, vol. VIII, pp. 29 and 31 respectively.

[93] Council Minutes, meeting of 15 December 1905, UAA, S18, vol. VIII, pp. 189–90; whether this referred to graduating as well as non-graduating students is unclear.

[94] Extension Lecture Circulars and Syllabuses, UAA, S102, vol. 1, 1902–1925.

continued large audiences and reported the lectures at length.[95] The newspaper also included reports of radioactivity research overseas; the new physics was attracting a significant public audience.

During this period the *Register* also reported the death of Archdeacon Farr, the retirement of Sir Charles Todd at the age of seventy-eight and amid much respect and affection, a conference of astronomers that discussed whether the separate State observatories should be handed over to the new Commonwealth, and a major university deputation to the South Australian government seeking increased financial support.[96] Professor and Mrs Bragg continued to exhibit at the annual exhibition of the Society of Arts and became involved in the new kindergarten movement through the Kindergarten Union of South Australia, Mrs Bragg on the Executive Committee and Professor Bragg on the large General Council.[97] Gwendoline's artistic talents were evident at a Government House Garden Fête for charity.[98]

Todd's retirement raised the question of his success: to what was it due? William answered in 1911 following his father-in-law's death the previous year: 'He had no commanding personality; at a first glance it might have been difficult to discover the source of his power... [Many] soon recognised his sense of proportion, his strong grasp of essentials, his acute understanding, and untiring energy. Yet... it was his conviction that all those who served under him or with him were as enthusiastic as he himself for the success of the work to which they were pledged... [and] he rarely failed to find what he thought to see. The whole of his great department was infected with his sense of duty and loyalty, his kindly courtesy and good humour'.[99] Similar words were used when William himself died many years later. Charles Todd had become, after his father and Uncle William, the third influential male figure in William's life.

William was now a widely recognised worker in the field of radioactivity. His research turned to study the ionization process itself and he had ambitious plans. 'My general idea is to attack the question of molecular and atomic structure by examining the absorption effects of the various atoms', he wrote to Rutherford.[100] As had now become his pattern, in September/October 1905 William presented his latest results to the Royal Society of South Australia,[101] and then wrote detailed letters to some of his colleagues and his advocates,

[95] *Register*, 15, 16, 21, and 28 June 1905, pp. 4, 6, 6, and 7 respectively.

[96] *Register*, 8 February 1904, p. 6, January 1905, *passim*, 11 May 1905, p. 4, and 9 November 1905, p. 6 respectively.

[97] For the new kindergarten movement in Adelaide see C Dowd, The *Adelaide Kindergarten Teachers' College: A History 1907–1974* (Adelaide: South Australian College of Advanced Education, 1983), ch. 1; for the Bragg's initial involvement see *Register*, 2 November 1905, p. 3.

[98] *Register*, 13 October 1905, p. 6.

[99] W H Bragg, 'Sir Charles Todd, K C M G 1826–1910', *Proceedings of the Royal Society of London*, 1911, 85:xiii-xvii; also G W Symes, 'Todd, Sir Charles (1826–1910)', in G Serle and R Ward (eds), *Australian Dictionary of Biography* (Melbourne: MUP, 1976), vol. 6, pp. 280–2.

[100] Letter W H Bragg to E Rutherford, 16 July 1905, CUL RC B358.

[101] W H Bragg and R D Kleeman, 'On the recombination of ions in air and other gases', *Transactions of the Royal Society of South Australia*, 1905, 29:187–206.

Soddy and Rutherford,[102] including a draft of his paper for *Philosophical Magazine*.[103] The passage of alpha particles through a gas produced ions, and if the voltage was not sufficient to sweep all the ions to one or other electrode, then some of the ions recombined in the chamber at a rate proportional to the product of the number of positive and negative ions per unit volume. Noting that theory and experiment did not always agree for this 'general recombination', Bragg and Kleeman suggested that there was also another, faster process involved, in which some of the electrons very quickly recombined with their parent ion, a process they christened 'initial recombination'. By reducing the pressure in the chamber general recombination was negligible and initial recombination became obvious, thus providing evidence to support their hypothesis.

William's most able Australian colleague was William Sutherland, a freelance theoretical physicist and physical chemist who lived quietly in Melbourne but whose prolific publications were highly regarded internationally.[104] He responded to William's draft manuscripts, particularly the result that the stopping power of matter was proportional to the square root of its atomic weight, which he felt should throw light on the inner dynamics of the atom.[105] Sutherland proposed explanations that involved temperature and pressure dependency and suggested measurements to test them, but William subsequently found stopping power to be independent of both.[106] In the same letter Sutherland alerted William to J J Thomson's suggestion that an alpha particle ceased to ionize when its charge was neutralised by combining with an electron. This 'neutral pair' concept would re-emerge in William's next research project.

It was at this point that William invited Madsen to participate in his widening research programme by verifying the result of one of Rutherford's experiments.[107] Initially Madsen was unable to do so, and William postulated that the positive-negative pair from the same atom or molecule might remain in semi-detached suspension for some time before initial recombination took place.[108] In further work, however, Madsen was unable to confirm his disagreement with Rutherford and William's suggestion therefore lapsed.[109] Madsen's work was nevertheless thought worthy of a DSc degree.[110]

[102] For example, letter W H Bragg to E Rutherford, 8 September 1905, CUL RC B359.

[103] W H Bragg and R D Kleeman, 'On the recombination of ions in air and other gases', *Philosophical Magazine*, 1906, 11:466–84.

[104] See n. 60.

[105] Letters W Sutherland to W H Bragg, 1905–1908, RI MS WHB 6B/18–23.

[106] Letter W Sutherland to W H Bragg, 5 February 1906, RI MS WHB 6B/19; W H Bragg and Kleeman, n. 103, pp. 482–4; the Thomson reference is *Nature*, 1905, 73:191.

[107] R W Home, 'W H Bragg and J P V Madsen: Collaboration and correspondence, 1905–1911', *Historical Records of Australian Science*, 1981, 5:1–29, 2.

[108] W H Bragg 'On the ionization of various gases by the α particles of radium', *Philosophical Magazine*, 1906, 11:617–32.

[109] See chapter 13.

[110] J P V Madsen, 'The ionization remaining in gases after removal from the influence of the ionizing agent', *Transactions of the Royal Society of South Australia*, 1908, 32:12–34; Home, n. 107, pp. 2–3; the thesis examiners were Lyle and Pollock, professors of physics at Melbourne

Focusing on a wide range of gases and metals and on the two independent parameters, *R* and *I*, the range of a particular alpha-particle group and the ionization it produced in his apparatus, William demonstrated that *R* could be accurately measured, but that, 'It is much more difficult to measure *I* accurately', and 'the total ionization of a gas is not simply dependent on the weights of the atoms of which it is composed. Molecular structure counts for something'.[111] How much future science was contained in these last five words! This was a time when even the structure of the atom was unknown, and it would be decades after the solution of that problem before the outer electronic structure of molecules and solids would shed light on the complexity of the problem.

The paper finished with two postscripts. The first was a strong rebuttal of a paper in which Becquerel had claimed to show that one of William's earlier suggestions was false. William analysed Becquerel's experiment in detail and convincingly demonstrated that Becquerel had misinterpreted it.[112] His theory of alpha-particle energy loss remained, and he conveyed the same message to the journal in which Becquerel's original paper had appeared.[113] The second postscript concerned the charge on the alpha particle. William thought the alpha particle was emitted uncharged and quickly became positive by losing an electron, whereas Rutherford was now convinced that the alpha particle was intrinsically positive.[114] As happened so often, Rutherford later proved to be correct, although at this time the evidence was inconclusive.

Having provided invaluable support for William in the earliest stages of his research, Richard Kleeman now left for Cambridge on an 1851 Science Research Scholarship. Thereafter, with the aid of this and other scholarships, he enjoyed an extraordinarily productive period of research at the Cavendish Laboratory. He began with experiments closely related to his work with William and then went on to publish more papers than any other Cavendish researcher in the years around 1910.[115] When, in 1914, his German heritage seemed problematic, he moved to the United States and an appointment at Union College, Schenectady, New York. He died in New York in 1932, at the age of only fifty-seven.[116] William now began to acknowledge two new research associates—but not experimental collaborators—John Madsen and Herbert

and Sydney respectively, and their recommendation was approved by the Adelaide University Council on 29 November 1907, Council Minutes, UAA, S18, vol. VIII, p. 368.

[111] W H Bragg, n. 108, p. 622.

[112] Ibid., 627–31; the Becquerel paper was 'Uber einige Eigenschaften der α Strahlen des Radiums', *Physikalische Zeitschrift*, 1906, 6:666–9.

[113] W H Bragg, 'Die α-Strahlen des Radiums', *Physikalische Zeitschrift*, 1906, 7:143–6.

[114] W H Bragg, n. 108, p. 632; however, Bragg's view then quickly moved to: 'insofar as it can be observed, α-particles are positively charged from the very starting point of their flight path'—see W H Bragg, 'Uber die α-Strahlen des Radiums', *Phisikalische Zeitschrift*, 1906, 7:452–3, 452.

[115] R D Kleeman, 'On the ionization of various gases by α-, β-, and γ-rays', *Proceedings of the Royal Society of London*, 1907, 79A:220–33; id., 'On the recombination of ions made by α, β, γ, and X rays', *Philosophical Magazine*, 1906, 12:273–297; the assessment is from D-W. Kim, *Leadership and Creativity: A History of the Cavendish Laboratory, 1871–1919* (Dordrecht: Kluwer, 2002), p. 164.

[116] See n. 43.

Priest.[117] An Adelaide science graduate in 1902, Priest was a demonstrator and lecturer in physics while assisting William's research. William's earlier pleas for staff and the introduction of research students had at last been heard and heard clearly.

In his September letter to Rutherford, William doubted Rutherford's claim to have used a very thin film of radium in one of his experiments. 'If I may venture to say so', William wrote, 'it is in fine crystals of fairly uniform thickness... [of] many thousands of atoms'.[118] After receiving Rutherford's reply of 4 November,[119] William responded at once, thanking him for the copy of the second edition of *Radio-Activity* and saying, 'I must thank you also for all your kindly references to my own work'.[120] He wrote again near Christmas, from the balcony of his family's seaside holiday residence: 'It is very jolly to have your comments on my paper; most reassuring and welcome. You see, I am a little out of the world here, and do not hear very much; and so I sometimes wonder whether those who understand the subject are approving of what I have done. You and Soddy have been most kind and have quite kept me going'.[121]

Soddy wrote to William twice towards the end of 1905. The first letter responded to the recent papers of Bragg and Rutherford and was warm in its praise of William's achievements.[122] The second began, 'I am going to ask you if you can give me any help in an important point which arises in connection with my own work', which he then outlined. This was followed by an acknowledgement of copies of William's recent papers and the response, 'the proof corrections you make so much of are nothing at all, as I got them most carefully revised and only one or two queries to attend to'.[123] As is also apparent from the Rutherford correspondence, William had now clearly grown in confidence and stature, his advice was sought, and his work was regarded as authoritative by leaders in the field. Soddy had been including William's work in his *Annual Progress Reports on Radioactivity to the Chemical Society* since their inception for 1904,[124] and other authors in addition to Rutherford were including William's findings in their books on radioactivity.[125]

Two aspects of William's research program now called for further attention: use of his method to examine the alpha particles of radioactive

[117] Bragg's Adelaide research notebooks indicate that he undertook nearly all aspects of the experiments himself, and his papers were now largely single-author.

[118] Bragg to Rutherford, n. 102.

[119] Letter E Rutherford to W H Bragg, 4 November 1905, RI MS WHB 26A/4.

[120] Letter W H Bragg to E Rutherford, 22 November 1905, CUL RC B360.

[121] Letter W H Bragg to E Rutherford, 21 December 1905, CUL RC B361.

[122] Letter F Soddy to W H Bragg, 10 October 1905, RI MS WHB 6A/31.

[123] Letter F Soddy to W H Bragg, 6 November 1905, RI MS WHB 6A/32; two subsequent letters also sought Bragg's advice on Soddy's research, namely 3 and 10 March 1906, RI MS WHB 6A/33 and 34 respectively.

[124] See T J Trenn (ed.), *Radioactivity and Atomic Theory: Annual Progress Reports on Radioactivity 1904–1920 to the Chemical Society by Frederick Soddy F R S* (London: Taylor and Francis, 1975).

[125] See, for example, Walter Makower, *The Radioactive Substances* (London: Paul, Trench, Trübner, 1908), pp. 93–108, and Norman R Campbell, *Modern Electrical Theory* (Cambridge: CUP, 1907), pp. 186–91.

species other than radium, and further exploration of the complex process of ionization. Regarding the first, Soddy had asked William to determine the range of alpha particles from uranium and thorium,[126] to which request William had willingly acceded: 'I am so much in his debt that I was glad to try to do what he asked'.[127] Because the best uranium and thorium sources he was able to obtain were much weaker than his radium source, William had to amend his experimental procedure. No longer able to use only those alpha particles travelling in one direction, he had to accept a wide angular range, redesign the apparatus, and formulate a new mathematical model to interpret the results.[128] The approximations he had to make caused less accuracy and, having satisfied Soddy's request, he did not return to these elements again. Soddy received William's paper, sent it on to the editor of the journal, and added, 'the sad news has just been received of the death of [Pierre] Curie, run over & killed in the streets of Paris'.[129] This was valuable time-saving editorial work for William and news that kept him in touch with events in Europe.

William now began to grapple further with the complexities of the ionization process.[130] The relative amounts of ionization produced by the alpha particles of radium in some eighteen different gases were determined, from which William was able to deduce an empirical equation summarising his results. In a long letter to William in February 1906, Rutherford agreed, 'we really know mighty little about ionization when we come down to it'. He reported that, 'Hahn is working out the Bragg curves for radiothorium', and he noted, 'by the way, Soddy seems to have rather broken loose from his scientific moorings to judge from his recent letters to Nature'.[131] It was the beginning of the decline in Soddy's prestige in scientific circles, but he remained a generous assistant and supporter for William. William's letter to Rutherford in April reported the recent work of Kleeman and Madsen and informed Rutherford that a radioactive mineral deposit had been discovered in outback South Australia.[132]

Soddy's letter to William in July 1906 observed that William's method of studying alpha particles was now widespread and that the Bragg style of ionization curve 'is filling the Phil. Mag. just now'.[133] Soddy also reported

[126] Soddy to Bragg, n. 123.

[127] Letter W H Bragg to E Rutherford, 5 April 1906, CUL RC B363.

[128] W H Bragg, 'The α particles of uranium and thorium', *Transactions of the Royal Society of South Australia,* 1906, 30:16–32, *and Philosophical Magazine*, 1906, 11:754–68.

[129] Letter F Soddy to W H Bragg, 20 April 1906, RI MS WHB 6A/35.

[130] W H Bragg, 'On the ionization of various gases by the α particles of radium—no. 2' (c.f. n. 108), *Transactions of the Royal Society of South Australia*, 1906, 30:166–87, and *Philosophical Magazine*, 1907, 13:333–57; the manuscript for the latter was communicated to the editor by the Physical Society of London, where the paper was read on 22 February (*Proc. Phys. Soc.*, 1905–7, 20:523–50).

[131] Letter E Rutherford to W H Bragg, 24 February 1906, RI MS WHB 26A/5.

[132] Bragg to Rutherford, n. 127; R D Kleeman, 'On the recombination of ions made by α, β, γ, and X-rays', *Philosophical Magazine*, 1906, 12:273–97; R D Kleeman, 'On the ionization of various gases by α-, β-, and γ-rays', *Proceedings of the Royal Society of London*, 1907, 79A:220–33.

[133] Letter F Soddy to W H Bragg, 9 July 1906, RI MS WHB 6A/36; for a European example see B Kucera and B Masek, 'Uber die Strahlung des Radiotellurs', *Phsikalische Zeitschrift*, 1906, 7:630–40.

that he was 'pushing some experiments on the magnetic deviation of the α particle', and that, 'so far... the α particle is always deviated', implying that it 'is initially charged on expulsion'. A few days later, however, William received an international cable from Soddy proclaiming triumphantly 'Proved alpha uncharged',[134] and this message was augmented in a long letter to *Nature*.[135] By March 1907, however, Soddy was forced 'to withdraw what I have already published'.[136] Subsequent letters from Soddy buoyantly reported his engagement to Miss Winifred Beilby and were loud in criticism of J J Thomson's atom model and other 'Cavendish Crudities'. He arranged the purchase and shipment of a liquid air plant for William's Adelaide laboratory (again donated by Robert Barr Smith and to be used for experiments where gas purity was vital), he continued to shepherd William's papers through the publication process, and he reported that Rutherford had been appointed to Manchester: 'it will make a great difference having him over here'.[137]

In examining the influence of the velocity of the alpha particle upon the stopping power of metals, William was hampered by the fact that his determination was only a relative one: the ratio of the stopping power of a metal to that of air. The variation of the former with speed was therefore clouded by changes in the latter. Nevertheless, he was able to conclude that the relative stopping power was a changing function of the particle speed, proportional to the square root of the atomic weight of the metal. So the relative stopping power of aluminum was almost independent of speed while that of gold increased markedly with increasing speed.[138] If this were so, then the order in which an alpha particle traversed a pair of dissimilar metals should be significant, and this William confirmed, despite the contrary findings of other workers.[139] Similarly, the relative stopping power should have decreased with increasing speed for a substance with an atomic weight less than that of air. This William also verified with the assistance of a colleague from the Adelaide chemistry department, W Ternent Cooke.[140] Cooke had graduated in science in 1900, won an 1851 bursary and then scholarship, worked with William Ramsay at University College London, and in 1905 obtained Adelaide's first D.Sc. degree in chemistry. In 1906 he was appointed the university's first lecturer in chemistry.[141]

[134] Cable F Soddy to W H Bragg, 26 July 1906, RI MS WHB 6A/37.

[135] F Soddy, 'The positive charge carried by the α particle', *Nature*, 1906, 74:316–17.

[136] F Soddy, 'The positive charge carried by the α particle', *Nature*, 1907, 75:438.

[137] Letters F Soddy to W H Bragg, 4 September, 27 November, 9 December, 10 December 1906, 10 January, 16 January 1907, RI MS WHB 6A/38–43, quotation from letter of 10 January 1907; Robert Barr Smith generously provided £500 for the liquid air plant (letter W H Bragg to University Council, 30 November 1906, UAA, S200, docket 916/1906).

[138] W H Bragg,' The influence of the velocity of the α particle upon the stopping power of the substance through which it passes', *Philosophical Magazine*, 1907, 13:507–16.

[139] Ibid.; see also, L Badash, *Radioactivity in America: Growth and Decay of a Science* (Baltimore: John Hopkins UP, 1979), ch. 15.

[140] W H Bragg and W T Cooke, 'The ionization curve of methane', *Philosophical Magazine*, 1907, 14:425–7.

[141] R J Best, *Discoveries by Chemists* (Adelaide: University of Adelaide Foundation, 1987), pp. 28–30.

In a short letter later in 1907 to the journal *Physikalische Zeitschrift*, William suggested that the alpha scattering explanation recently offered by Lise Meitner for a result obtained earlier by Marie Curie was incorrect, and that the correct interpretation was the one he had given recently in the *Philosophical Magazine*.[142] He now felt able to debate with anyone! However, while William had told Rutherford that there was much more to be discovered about ionization, he also knew that it was 'mysterious',[143] and he correctly concluded that his current studies had now exhausted their productivity. He began to consider an alternative project and this would take him into another realm. William's own later summary of this exciting time, when he first ventured into front-rank research, is more transparent than its predecessors:[144]

> I had to begin my study of electricity on the boat out [to Australia in 1886], and to learn by degrees the use of the apparatus in the well-found laboratory which my predecessor had left me. There was little time at first for anything except routine teaching...
>
> At that time [1904] the scientific world was deeply interested in the new knowledge of the electron and of the wonderful facts of radioactivity...So I thought that I would gather together the facts that had been proved in the hope that my New Zealand audience would be interested...It seemed to me that certain experiments made by Mme. Curie were only capable of explanation if it were supposed that one atom could and did sometimes pass through another, without either of the two being permanently the worse...Some months later I succeeded in obtaining a little [radium] from England, just a few grains of dust as it appeared to be; and the time for the test had come...The experiment behaved as it should do, and then went on to show other effects that I had not expected. I found that as I pushed the chamber nearer and nearer there was not only one point at which the firing of alpha particles became effective; *there were four*, the other three being met in succession as the distance diminished. It was as if the army approaching the castle came first within the range of the big guns, then small guns, then rifles, then pistols.

Several interesting observations arise from this short but intense and highly successful period of research. First, the facilities available for William's work were limited and followed closely the pattern of Victorian experimental science suggested by Graeme Gooday.[145] William was allocated a small room in the basement of the original university building for his research, happily close to Rogers' workshop and with two slate work benches supported on solid brick piers, but otherwise devoid of sophisticated facilities (see Figure 11.4).[146]

[142] W H Bragg, 'Uber die Zerstreuung der α-Strahlen', *Physikalische Zeitschrift*, 1907, 14:425–7.

[143] Bragg to Rutherford, n. 127.

[144] Sir William Bragg, 'A page from my life: My adventures among atoms', *The Graphic*, 1927, March 12, p. 408.

[145] G J N Gooday, 'Instrumentation and interpretation: managing and representing the working environment of Victorian experimental science', in B Lightman (ed.),*Victorian Science in Context* (Chicago: University of Chicago Press, 1997), pp. 409–37 and references therein.

[146] J G Jenkin and R W Home, 'Horace Lamb and early physics teaching in Australia', *Historical Records of Australian Science*, 1995, 10:349–80, 364–7.

Fig. 11.4 William's research laboratory in the basement of the University of Adelaide building, (L to R) Priest, Bragg, and Madsen, William facing his apparatus for investigating the nature of X- and γ-rays (see chapter 13 and figure 13.1), with a mechanical vacuum pump on the box next to Madsen and several pieces of equipment used for gas handling and the alpha-particle research in the foreground, July 1906. (From *Supplement to The Critic.*)

Second, the important paper on ionization discussed above is notable for the presentation for the first time of precise details of the apparatus William had been using (see Figure 11.5).[147] It had evolved continuously from the earliest days of his program, and its description was presumably delayed until it had reached a level of development that satisfied William's demanding standards. A comparison of Figures 11.2 and 11.5 portrays the very substantial development that had occurred. Problems—of producing and maintaining an adequate vacuum, of sustaining the pressure of an introduced gas, of leakage, of temperature control, of movements in such an environment, and more—were ever present, and William's papers testify to the very high level of experimental expertise he had acquired to overcome such problems. His earlier years in Adelaide had been essential to the success he now achieved. On the matter of William's experimental and presentation skills, two later remarks by his elder son and scientific partner are revealing:[148]

[147] W H Bragg, n. 130.

[148] The first paragraph is from letter W L Bragg to Sir Cyril Hinshelwood, 26 January 1965, and the second from letter W L Bragg to Sir Mark Oliphant, October 1966, RI MS WLB 53A and 54A respectively.

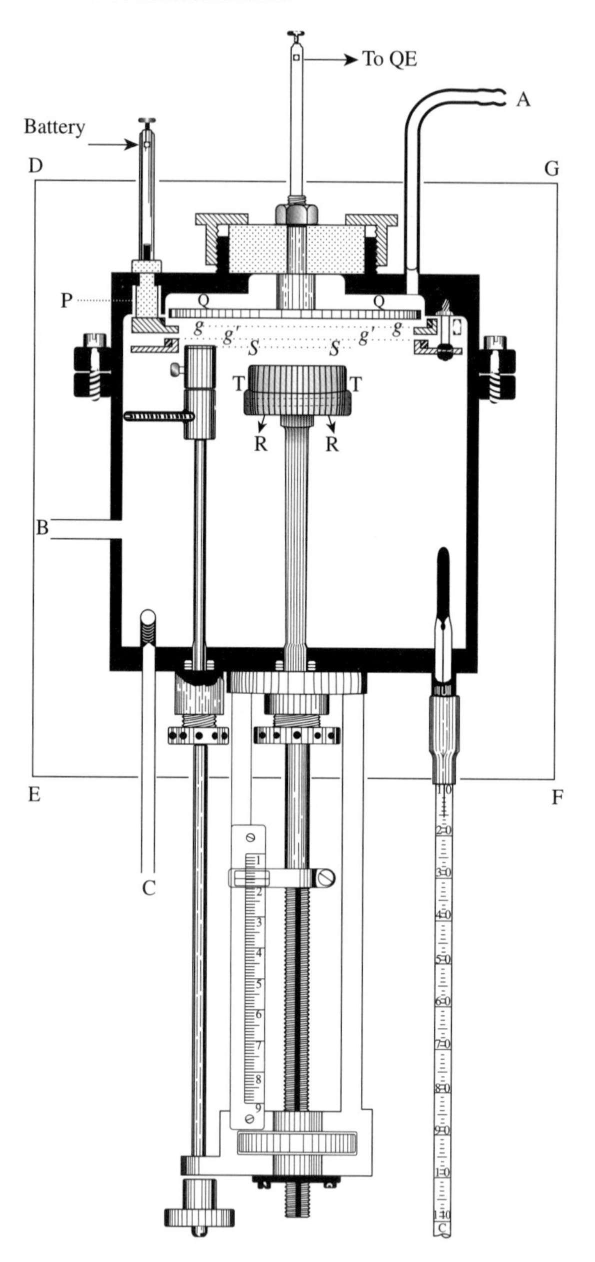

Fig. 11.5 The final form of William's alpha-particle apparatus, Adelaide, 1907. A cylindrical chamber, in which RR = radium source, ggQQ = ionization chamber for detecting alpha-particles, the distance between it and RR being variable, SS = metal sheet to interpose between them, QE = connection to quadrant electrometer, B & C = gas/vacuum connections, A = connection to gas-pressure manometer, DEFG = oven to vary temperature, measured by the thermometer. (From W H Bragg, *Studies in Radioactivity*, Macmillan & Co., 1912, p. 14. Courtesy: Palgrave Macmillan.)

> [1] Arriving in Adelaide...he arranged to get instruction from a firm of instrument makers, learned to use the lathe and made the apparatus for his classes. I always think this early training was reflected in the beautifully-designed apparatus which he used for his various researches.
>
> [2] He was unusual in two ways. He came to Physics as a mathematician and always looked at Physics from that point of view, with an almost Greek way of expressing his ideas in logical language. And in sharp contrast to this theoretical approach, he entirely escaped the 'string and sealing wax' tradition. He had high ideals about good tools, good design, and good workmanship....Much of Rutherford's early apparatus was markedly in the tin-can and sealing-wax tradition—my father's never....His note-books, and the curves he plotted, were always my envy and admiration. Points just fell on a smooth curve as if by magic when he read [the data] and plotted.

Regarding William's apparatus and the nature of his experimental difficulties, the following note from someone who joined the Adelaide physics staff a few years after William's departure is indicative (although towards the end of his radioactivity research William did acquire 'a high-tension battery' as part of another Barr Smith donation,[149] and the university acquired an electricity supply):[150]

> When I first started work as demonstrator in Physics, I was given a small room in the basement. The shelves supported some thousands of medicine bottles which had been decapitated to form small jars, each jar contained two small metal plates. I discovered from Rogers that...Bragg needed a source of very high e.m.f. He asked his students to collect all the medicine bottles they could lay their hands on. Rogers cut the tops off. Adding two dissimilar metals and acid to each jar and joining them up, Bragg had his source of high e.m.f....The wooden shelving was stained dark brown from the effects of acid spills.

William was also very conscious of the debt he owed to his expert mechanic, Arthur Rogers, and a generous acknowledgement was included in the same paper that included the final form of the apparatus.[151] Indeed, many years later William wrote to Rogers from University College London saying, 'I still have the alpha apparatus that you made me, and the students use it in their radioactive experiments. I think that when I retire I shall have it silvered and keep it in a prominent position in my house'.[152] An extensive search has failed to locate it, however. Perhaps its chances of survival were doomed by radioactive contamination.

During the three years (1904–07) that he had been engaged in experiments on alpha particles and the ionization they induced, William's star had risen

[149] Letter W H Bragg to E Rutherford, 21 August 1907, CUL RC B368.

[150] W H Schneider, autobiographical notes, private communication from Mr M P Schneider.

[151] W H Bragg, n. 130, in which Bragg says, 'I owe my thanks to my assistant, Mr A L Rogers, for the great care and skill with which he has made the apparatus used in this work, and drawn the plate illustrating this paper'.

[152] Letter W H Bragg to A L Rogers, 2 June 1920, Bragg (Adrian) papers.

to world-wide prominence, illustrated by two outcomes of major significance: his election as a Fellow of the Royal Society of London, and offers of senior appointments elsewhere. Horace Lamb had followed William's career with interest, and as early as May 1906 he wrote: 'Some of your scientific friends on this side the equator are of opinion that the time has come when you ought to be put in nomination for the Royal Society. If you do not object, will you kindly let me have a list of your papers, and a few details about your career. I will draw up the necessary certificate, and get some weighty signatures. The candidature will be warmly supported by J J Thomson, who will help in getting signatures in Cambridge and elsewhere'.[153] Three certificates were lodged before Christmas 1906: the first had been signed by R J Strutt, C T R Wilson, Oliver Lodge, John Townsend, and Arthur Schuster 'from general knowledge', and by Horace Lamb, J J Thomson, A R Forsyth, Hugh Callendar, A E H Love, H H Turner, T R Fraser, R Threlfall, and W H Preece 'from personal knowledge'; the second, from Australia, was signed by R L J Ellery, Orme Masson, E C Stirtling, W Baldwin Spencer, A Liversidge, and W A Haswell 'from personal knowledge'; and the third bore just one signature, 'E Rutherford'.[154] It was an extraordinary list of supporters.

William's election was announced on 1 March 1907, and Lamb sent a telegram to 'University Adelaide' the same day saying, 'Bragg selected Royal'.[155] Rod Home, who has written extensively on the scientific network and patronage system embodied in the Royal Society of London, has pointed out that, by the end of the nineteenth century, the pressure on places had increased inexorably and that election was not assured even after several years of candidacy.[156] 'New candidates tended to find themselves joining something of a queue of fellow specialists, and only the most outstanding (for example, Bragg in 1907) managed to leap-frog those ahead of them'.[157] William was certainly surprised when he was elected on the first occasion.[158] Local newspapers carried the story.[159] There were letters from academic colleagues around the country and from his former students in Australia and overseas, octogenarian Robert Barr Smith sent him a note in very shaky hand-writing, and his fellow Adelaide professor, George

[153] Letter H Lamb to W H Bragg, 5 May 1906, RI MS WHB 4A/2.

[154] Royal Society of London, Certificate of a Candidate for Election, W H Bragg, nos 240–2.

[155] Cable, H Lamb to University Adelaide, 1 March 1907, Bragg (Adrian) papers.

[156] R W Home, 'A world-wide scientific network and patronage system: Australian and other "colonial" Fellows of the Royal Society of London', in R W Home and S G Kohlstedt (eds), *International Science and National Scientific Identity* (Dordrecht: Kluwer, 1991), pp. 151–79; R W Home, 'The Royal Society and the empire: The colonial and commonwealth fellowship, Part 1: 1731–1848' and 'Part 2: after 1847', *Notes and Records of the Royal Society of London*, 2002, 56:307–32 and ibid., 2003, 57:47–84 respectively.

[157] Ibid., p. 166.

[158] See, for example, letter W H Bragg to G Duffield, 4 April 1907, Duffield papers, Basser Library, Canberra, 4/7: 'I scarcely like to think about it until I get letters. I have seen a list of the candidates, and am lost in amazement at my good fortune'; also Royal Society of London, Bragg's letter of acceptance, 6 June 1907.

[159] For example, *Register*, 4 March 1907, p. 6.

Henderson, left a message on William's desk acclaiming, 'Hurrah, hurrah for Bragg and his University. Splendid. G C H'.[160] It was a great achievement.

Rutherford's appointment to Manchester became known early in 1907. He would leave McGill in mid-year and Howard Barnes would be promoted to the Macdonald chair;[161] but Rutherford thought an additional person was needed 'to give the Graduate School a good start and fill a gap in the mathematical teaching', and that Bragg was 'exactly the man that is wanted at McGill'.[162] Accordingly, Rutherford wrote to William in July: 'Now in regard to an important matter I want to talk to you about... they are anxious, between ourselves, to get somebody to carry on my particular lines of work with vigour... and there is considerable need of a man to take the work in applied mathematics... I have been asked to sound you out on the question... Montreal is not so isolated as Adelaide & you can run down as often as you like to attend the meetings of the Amer. Phys. Soc. who are glad to welcome all good Canadians. Let me know what you think of it'.[163]

Soon thereafter William wrote to Lamb: 'It is a very tempting offer, and I feel strongly inclined to accept it, when actually made... I seem to have got on to a certain line of research work which is worth pursuing. It has been possible to get on with it here during the last two or three years because I have made rather a big effort which I could hardly keep up; I have done it, so to speak, out of hours.... In any case I am rather cut off from others in the same line, intercourse with whom would help me very much, and might save me from many mistakes.... [details of the likely Montreal offer]... I believe Montreal is a healthy place, and it is only six days from England, where I hope that my boys will soon be at college.... I am writing to Rutherford this mail to say that I will consider an offer with every intention of accepting. It would be a terrible wrench to move from here; there never was a kinder lot of people nor a nicer little city, but perhaps it is good to be stirred up'.[164] Lamb responded: 'The conditions of the post seem very attractive... I think you are quite right to consider it favourably'.[165]

William's next letter to Rutherford stated 'the reasons for and against a move from Adelaide to Montreal... That which attracts me most is the chance which seems offered me to do better work. I should be glad to be free from the ties of elementary classes and the burden of administration work which is unavoidable in a small but rather ambitious and enthusiastic university.... Here I am somewhat isolated, for though I have many good friends, there is none except William Sutherland of Melbourne whose work is on at all the same lines as mine. Moreover,... the number of advanced students is almost

[160] Bragg (Adrian) papers.

[161] Letter E Rutherford to W H Bragg, 12 March 1907, RI MS WHB 26A/8.

[162] L Pyenson, 'The incomplete transition of a European image: Physics at Greater Buenos Aires and Montreal, 1890–192', *Proceedings of the American Philosophical Society*, 1978, 122:92–114, 108.

[163] Letter E Rutherford to W H Bragg, 5 July 1907, RI MS WHB 26A/10.

[164] Draft letter, W H Bragg to H Lamb, n.d., RI MS WHB 10A/6.

[165] Letter H Lamb to W H Bragg, 1 October 1907, RI MS WHB 4A/5.

negligible[166]... The most serious difficulty is that of moving my household. We have built our own house and altogether have struck our roots very deep'.[167] In the end, however, other events determined the outcome. University budgetary constraints, caused by fires that destroyed the engineering building and the medical school at McGill, led to a much less attractive offer that William had little difficulty in declining.[168]

The Bragg-Soddy correspondence continued during 1907: they exchanged new-year greetings and portrait photographs, Soddy continued to guide Bragg's papers through the publication process, Bragg sent Soddy a well-received wedding present, and Soddy sent Bragg congratulations on the birth of a daughter, commenting, 'I remember your two boys quite well. They must be now quite grown up'.[169] The surviving correspondence ends abruptly here except for one later letter,[170] and it seems likely that it ceased altogether when William left Australia. Soddy was not sympathetic to William's forthcoming particle concept of radiation, William's physics and Soddy's chemistry drifted apart, and when William returned to England the principal raison d'être of the correspondence ceased.[171] In the following years both men reached the high point of their careers, but there seems no evidence that they reactivated the closeness of earlier years. Indeed, towards the end of his life, sad and sometimes bitter, Soddy struck out at his old friend: 'One often hears that nowadays physics and chemistry are one subject. God forbid! But even were it so, it does not give physicists any right to steal the work of chemists, or for that matter the Fullerian Professorship in Chemistry at the Royal Institution either!' He was referring to William Henry Bragg. His biographer, Mrs Howorth, then added: 'Since the foregoing was written, another distinguished physicist has been appointed to the Fullerian Professorship of Chemistry at the Royal Institution'. She was referring to William Lawrence Bragg![172]

[166] In fact, Australian universities did not introduce the PhD degree until after the Second World War: see I D Rae, 'False start for the PhD in Australia', *Historical Records of Australian Science*, 2002, 14:129–141.

[167] Letter W H Bragg to E Rutherford, 21 August 1907, CUL RC B368.

[168] L Pyenson, n. 162, p. 108; letters W H Bragg to E Rutherford, 17 December 1907 and 22 January 1908, CUL RC B369 and 370 respectively; letter E Rutherford to W H Bragg, 24 October 1907, RI MS WHB 26A/12.

[169] Letters F Soddy to W H Bragg, 5 February, 18, 21 and 23 April, 23 July 1907, RI MS WHB 6A/44–48 respectively, quotation from letter of 23 July 1907.

[170] Letter F Soddy to W H Bragg, 25 May 1908, Bragg (Adrian) papers.

[171] Jenkin, n. 45, p. 164.

[172] Ibid., p. 165.

12
Willie and Bob's Australian education

Many of the oldest and best-known educational institutions in Australia have a similar early history. The secondary schools, colleges, and universities founded in the nineteenth century were often established by well-meaning and dedicated supporters, but in an environment where financial support and the number of eligible and willing pupils were small. Many of these institutions struggled for several decades before becoming viable, then modestly healthy, and finally secure. Only a very few became prosperous. This was certainly true of the University of Adelaide and of the small number of secondary schools and colleges that provided its matriculants. When William Bragg began to consider where he might enrol his sons for their secondary education there were no government high schools and the choice quickly shrank to the two most prominent Protestant schools for boys, the Anglican St Peter's College and the Methodist Prince Alfred College (PAC). The latter was within walking distance of the family home on East Terrace, just across the East Park Lands, while St Peter's was further way, in the suburb of Hackney, but still within reach.

Several factors decided for St Peter's. It was Anglican and it had always prided itself on producing Christian gentlemen by following the traditions of the great English public schools. It was the school of the best-known Adelaide families, including the Todds, and there were teachers on the staff whom William knew. The choice could not have been easy, however. Prince Alfred College was then led by its outstanding headmaster, Frederic Chapple, who was Warden of the University Senate from 1883 until 1922 and whose school had already achieved better academic results than St Peter's: in the 1880s and 1890s its students often secured two or even all three of the annual scholarships to attend the university, and the Angas Engineering Exhibition was awarded to PAC boys six times in the first seven years of its competition, 1889 to 1895.[1]

St Peter's College had begun its first class in July 1847 with just eleven boys, in the schoolroom attached to Trinity Church in the city. It was given the name 'The Church of England Collegiate School of St Peter's Adelaide' with

[1] R M Gibbs, *A History of Prince Alfred College* (Adelaide: Prince Alfred College, 1984), chs 5–6, p. 90.

its Act of Incorporation in 1849; reduced in 1889 to 'The Church of England Collegiate School of St Peter', or colloquially 'St Peter's College'.[2] The term 'collegiate' was included because the early proprietors envisaged an institution that combined a grammar school and a college (unrealized) for training Anglican clergy. The addition of 'St Peter's' was made by the high churchman and Bishop of Adelaide, Augustus Short, who had been educated at St Peter's College, Westminster, and consecrated Bishop in Westminster Abbey on St Peter's Day 1847.[3] The school motto, Pro Deo et Patria (for God and Country), speaks of the inspiration and ambition of the college, as do its colours, royal blue and white.

The first few years at St Peter's were very difficult, but by 1854 there were sixty-five boys, of whom thirty-one were boarders, and a magnificent property upon which attractive buildings were rising. The school found strong and thoughtful leadership in Rev. George Henry Farr, who was mentioned above in his later role at the university. By the end of Farr's twenty-four years at St Peter's it was a leading Adelaide school, with its own ethos and good prospects for the future: 'Farr's pioneering work cannot be measured solely by the fact that he doubled...the size of the school. He organised religious life, built a chapel, increased the buildings, appointed highly-competent staff, liberalised and extended the curriculum, obtained examination successes and initiated most of the existing school sports'.[4]

The years 1879 to 1893 have been called 'the years of adversity...possibly the most difficult in the history of St Peter's'.[5] The school tradition of appointing its headmasters from clergy of the Church of England was not always successful, it was a period of financial depression in the State, and the popular Second Master, J H Lindon, resigned to found Queen's School. Late in 1893, however, the situation eased in two directions: Robert Barr Smith lent the school £8,000 while it awaited revenue from the munificent Da Costa Estate, and a cable arrived from England saying, 'Excellent man, Girdlestone, Stroke Oxford Eight, Honours, Maths and Science, leaving P O R M S Paramatta'.[6]

Canon Henry Girdlestone, headmaster from 1894 to 1915, was, according to Price, 'great in character, ability, physique and teaching capacity, and he was great in achievement, for he lifted St Peter's from a small, struggling school to a large and flourishing college...Girdlestone was a strong but kindly disciplinarian, and his religious outlook was one of "broad churchmanship" and of muscular Christianity...in appointing new masters he showed an ability to judge men which he maintained up to the end of his career...The Girdlestone régime saw extensive improvements in grounds, buildings and general school

[2] A Grenfell Price, *The Collegiate School of St Peter, 1847–1947* (Adelaide: Council of Governors of St Peter's, 1947), ch. 1.

[3] J Tregenza, *Collegiate School of St Peter Adelaide: The Founding Years, 1847–1878* (Adelaide: Collegiate School of St Peter, 1996), pp. 17, 40–1.

[4] Price, n. 2, pp. 15–16.

[5] Ibid., p. 25.

[6] Ibid., ch. 3; for Girdlestone see S M Owen, 'Girdlestone, Henry (1863–1926)', in B Nairn and G Serle (eds), *Australian Dictionary of Biography* (Melbourne: MUP, 1983), vol. 9, p. 19.

facilities...[including] turf wickets on the school oval [and] the Chemical Laboratory'.[7] Others noted a later deterioration in Girdlestone's influence and his predilection for corporal punishment on the English public-school model,[8] but the college has been notable for its ability to perform well and progress steadily through the terms of a variety of headmasters, thanks to the steadfastness of its Council of Governors and a core of outstanding teachers.

Willie entered St Peter's in 1901; Bob joined four years later. They were also the years of their father's first major research programme. Willie's written recollections were brief:[9]

> At about eleven [in February 1901, just before his eleventh birthday], I was sent to St Peter's College, the premier Church of England school in South Australia. It must have had between three and four hundred boys in my time, with about seventy boarders. It was a good school. The headmaster was Girdlestone, a vast and impressive man with a china-blue eye and small yellow beard. He carried in summer a small baton with a switch of hair at the end of it. Fresh from Queen's, I supposed it to be some instrument of punishment, but actually it was for attacking flies and not small boys, though I am not sure the Head thought them very different. He had a way of saying, 'Boy, you are a humbug', which shrivelled one up. He had a passion for good English and spared no pains in correcting the essays of the boys in the sixth form. I have always felt very grateful for this training.
>
> We did not specialise so much in those days, as boys have to do now. I took English Language, English Literature, French, Latin, Greek, Scripture, Mathematics and Chemistry, all to an equal level. The only subjects on the school curriculum which I did not take were German and Physics, and I have always been sorry I learnt no German [a mirror of his father's experience].
>
> My great friend at St Peter's was Bob Chapman, the son of the Professor of [Engineering] at Adelaide University. The master of our maths class was rather feeble. Bob Chapman and I were rather hot on mathematics and we had the advantage of two fathers ideally equipped to help us with our homework problems. Often the maths master could not himself understand the answer to the problem he had set us, and he got out of his difficulty by making Bob Chapman or myself run over it in class.
>
> I had the same handicap at St Peter's due to my being so immature and bad at games on the one hand, and on the other being precocious in my lessons, which led to my being in a class of much older boys. I got into the sixth form when I was fourteen, and at fifteen my father decided that a further stay at school would not be profitable and I entered Adelaide University. The games at St Peter's were played in the lunch break, and the boys were divided into groups of roughly the same ability who played

[7] Price, n. 2, ch. 4.

[8] See, for example, R S Ellery, *The Cow Jumped Over the Moon* (Melbourne: Cheshire, 1956), ch. II.

[9] W L Bragg, Autobiographical notes, pp. 14–16.

> together. My prowess, or rather the lack of it, indicated my being placed in one of the lowest sets, while my fellow sixth-formers were the prefects and glorious heroes of the school teams. I just could not disgrace myself playing games with the little boys in the lowest forms, and so had a lonely and aimless time in the lunch break. Boys who do not fit into the normal pattern do not have an easy time at school, though it must be said that there was a fine spirit of tolerance at St Peter's. I was regarded as an amusing freak and interesting specimen, and had very little teasing or persecution.
>
> I found my physical outlet in rowing. We rowed on the River Torrens, which was really a dammed-back lake in the old river-bed running through the parks between Adelaide and North Adelaide. The headmaster was an Oxford Blue and had a beautiful style. He used to stroke us in a four, and as the seats were off-set and he weighed as much as the rest of us put together, we used to fill a large can with water and put it well to starboard to balance him. The Head was clever with his hands, and we did much of the repairing of the boats under his direction. The great race of the year, between St Peter's and Geelong Grammar School in Victoria, was rowed on the Port River.

Based on English public school principles, St Peter's sought boarding students from South Australia and interstate. However, about eighty percent of its pupils were 'dayboys', who lived at home and travelled to and from the school each day. Those families who lived too far away from Adelaide to allow their sons to live at home predominated in the boarding house. The English practice of tearing children away from their families when very young was not adopted in Australia, except where it could not be avoided. Willie lived at home during all his school and undergraduate studies. Generally he walked to St Peter's, but on wet days he sometimes 'drove to school with Charley Hay' in Charley's horse and cart.[10]

The same letter also recorded that, 'We bought a football yesterday at John Martin's with the shilling you gave us and some little other money out of the pink money box. We got [it] very cheap, as it was very dirty from the soaking it got at the fire. I got on very well at school yesterday... We are all very well, but Robbie, when he kicked the football yesterday, got something in his eye, which is not out yet'.[11] John Martin's was, for generations, one of Adelaide's largest department stores, and on Easter Sunday, 6 April 1901, it was severely damaged by a large fire.[12] This letter was followed by another, from Bob to his parents, reassuring them that his eye was now better and that grandfather (Charles Todd) had been to dinner.[13]

[10] Letter W L Bragg to his parents, n.d. [April 1901], RI MS WLB 37A/2/13; Willie's wife later recalled, in regard to their own children, that 'W.L.B. had no tradition of boarding preparatory schools, or for that matter public schools, and thought the whole business unnatural' (A G J Bragg, Autobiographical notes, p. 195).

[11] Ibid.

[12] C Thiele, *The Adelaide Story* (Adelaide: Peacock, 1982), p. 102.

[13] Letter R C Bragg to parents, n.d., RI MS WLB 37A/2/17.

In his first year at St Peter's Willie was placed in a lower fifth form, well ahead of his age group. He was eighth in his French set, fourth in Greek, and won the form prize for Latin.[14] Later in the year he passed the Primary Public Examination, which involved English (grammar, composition, and dictation), and arithmetic (both compulsory), and a minimum of two from geography, English history, Greek, Latin, French, German, algebra, and geometry. Willie passed in six subjects: English, arithmetic, history, Latin, algebra, and geometry.[15]

Of the several public examinations set by the university,[16] the 'Primary' was a test of basic competence and the 'Junior' a modest advance thereon. These examinations were intended for students leaving school in their mid-teens to take up employment. The 'Senior' was the matriculation examination for the university, while the 'Higher' was intended for students wishing to prepare themselves more deeply for university study. It was not intended that students would take all four examinations during their school days, but this was what happened. Examination results had become the criteria by which the success of students and their schools were measured, and schools wishing to enhance their reputation therefore encouraged their students to do as many examinations as possible. Chapple, with his London University and practical Methodist background, adopted this approach very successfully at PAC. He thought both students and schools benefited from such testing. Girdlestone, an Oxford graduate, strongly disagreed. He spoke passionately in his annual St Peter's College Speech Day addresses between 1901 and 1905 of 'the number of boys who pass examinations, as if that was the only thing a school had to do', of 'some misguided person who, like the man who introduced rabbits and snails into Australia, conceived the idea of schools' examinations', and of 'a school's work is threefold: it should develop the intellect, the body, and the character, and the first two are but means towards the third'.[17] A member of the University Council, Girdlestone also spoke there of his concerns regarding the public examinations.[18]

Willie's recollections are reminiscent of his father's: overly modest and somewhat misleading. He did not participate in team sports such as football and cricket, presumably for the reason he gave; namely that he was disconnected from his age group in class and felt unable to join them in team games. His slightly short and crooked left arm may have been a problem too. Nevertheless, he *was* a good athlete. At the School Sports held on the Adelaide Oval on 18 September 1901, in the 120-yards flat race for under-thirteen boys,

[14] School Lists, 1901, St Peter's College Archives, pp. 20–7.

[15] *Calendar of The University of Adelaide, 1902* (Adelaide: University of Adelaide, 1902), p. 408.

[16] Regulations for the public examinations and syllabi for the various subjects are given each year in the relevant *Calendar of The University of Adelaide*.

[17] School Lists, 1902, St Peter's College Archives, p. 41; ibid., 1904, p. 36; ibid., 1905, p. 27 respectively.

[18] For example, regarding the difficulty of the examinations, Council Minutes, meeting of 30 June 1905, UAA, S18, vol. VIII, p. 134.

Fig. 12.1 The Bragg family in front of their new Adelaide home, (L to R) Willie, Gwendoline, Bob, and William, *circa* 1902. (Courtesy: Dr S L Bragg.)

'an excellent race ended in a win for Bragg, with Muirhead second and Bagot third'.[19] Further successes followed later.

In 1902 Willie was in the upper fifth form, finishing eleventh of the twenty-five boys in the final order. He again did well in languages, Greek, French, and Latin. In the Junior Public Examination only five subjects were required, although good students took more. That year Willie passed in eight subjects: English literature and algebra, both with credit (the only form of recognition for high achievement), and English history, Latin, French, geometry, arithmetic, and inorganic chemistry. The best St Peter's results were obtained by two boys who passed nine subjects with four credits and by a third boy who passed eight subjects with six credits.[20] Willie's results were not beyond those of his fellow students, except for the fact that, at twelve, he was three or more years younger than his peers. Bob was sixteen when he took the Junior Examination in 1908.

In 1903 Willie passed into the lower sixth form, where he was now sufficiently mature to realize the ability that was to characterize the rest of his life. He was second in the final form order, and also in French and Latin. He won the scripture prize for his form and was *proxime accessit* for an Open Farrell Scholarship. Two such scholarships, arising from the will of Dean Farrell, a foundation governor and benefactor of the college, were awarded each year.[21]

[19] *The St Peter's School Magazine*, no. 46, September 1901, pp. 318–19.
[20] School Lists, n. 14, 1902, *passim*.
[21] Ibid., 1903, *passim*.

The *School Magazine* recorded that, 'A greater interest is now taken in the welfare of the Museum by the present scholars, as will be seen by the number of specimens given by boys at school... W L Bragg has given a piece of stalactite and several shells'.[22] As a result of the many family holidays spent at the seaside, Willie had developed a love of shell collecting, a passion that was to have a significant outcome a little later, when he was a university student.

The year 1904 was to prove pivotal. Only thirteen years old, the letters he wrote to his parents in New Zealand and his rise to the middle sixth form testify to an intelligent and observant youth. He and Bob were invited to spend one weekend at the Lendon's home at Mount Lofty in the Adelaide hills. Willie's letter to his parents reported:[23]

> It was beautiful up there. They have a fine house on the side of a hill with an enormous stable... They have a dear little pup, a fox terrier, only 8 weeks old, who plays and sleeps all day. He attacks the cat and hauls at its tail, till the cat boxes his ears... Guy has a beautiful model windmill made of wood, which comes to pieces and packs in a little box. It really goes round and works two little hammers...
>
> The little creek always runs. It is quite bushy round the house... Bob saw two real snakes and about half-a-dozen ordinary ones. He could not go to sleep because he dreamt snakes were on the drawing-room floor, and he was there too.
>
> The Rhymills drove up in their motor and gave us rides. They have a new oil motor like this [drawing]... The machinery is all in a box in front and the oil in a tank behind, and a battery at the side, and a coil to make exploding sparks, and a snake in front to cool the water, which cools the machinery. The machinery gets so hot that, when water is put on it, it fizzes. It has two gears, and when it changes gear it stops almost quite still and rattles. Mr Rhymill changes gear every 10 seconds and so we go like a porpoise, with jumps. It is all right fun going down hill with the high gear. It will climb any hill, even where a horse has to walk.

At the end of the year Willie was first in the final form order, won the Christchurch Scholarship for classics and divinity, and form prizes for mathematics, chemistry, and French. In the important Senior Public Examination, Willie passed nine subjects, although only five were required: arithmetic and algebra, geometry, trigonometry, and inorganic chemistry, all with credit, and English literature, history, Greek, Latin, and French. He was eighth in the state overall.[24] Late in the year the headmaster took the opportunity at Speech Day to comment: 'In the list published this morning I see that Bragg takes the eighth place in the honours lists. Glancing at them I am more than ever struck with the large number of candidates who are over age. I think it is becoming more apparent every year that some of these University examinations are less

[22] *The St Peter's School Magazine*, no. 53, August 1903, p. 544.
[23] Letter W L Bragg to parents, February 1904, RI MS WLB 95B/6.
[24] Public Examinations Board Manual for 1904, UAA, S14.

suitable for schools than they used to be. If a boy is naturally gifted above his fellows he can take a good place without any undue pressure; but the normal boy must either be restricted to one or two subjects or driven over his strength to obtain a distinction. What we want is an examination for the ordinary schoolboy, if we must have one at all'.[25] Charles Todd was also delighted at his grandsons' academic successes. He wrote from Melbourne, where he was attending a conference:[26]

> My dearest Willie
>
> I am overflowing with love & pleasure on your great success in your Exams, both University Senior and Collegiate. Your position in the Senior is most gratifying, whilst your winning the Scholarship of £10 p.a. for 2 years and four prizes show how diligently you have been in your studies.
>
> Keep up your credit by renewed & energetic efforts for future success & you will be the pride of your family & the delight of your beloved parents. I can realize their feelings as regard both you & dear Robbie, who also has done so well. I am proud of you both, & I hope you will have a nice restful & healthy holiday...I must write a few lines to dear Robbie, so thanking you for your nice kind letter, I must say good-bye. With fondest love & best wishes for your future,
>
> Your loving & proud Grandfather, C Todd.

Willie remembered the headmaster's insistence on good English expression, and it left a legacy that was valuable to him throughout his professional life. Max Perutz, who was very close to him in later years, envied Willie's ability: 'Bragg possessed a remarkable degree of physical insight into natural phenomena...He would illustrate his conclusions in a series of neatly drawn sketches, and then write the accompanying paper in a lucid and vivid prose. Some scientists produce such prose as a result of prolonged redrafting and polishing, but Bragg would do it in one evening, all ready to be typed the next day'.[27]

At the school sports held on the Adelaide Oval on 24 September 1904, W L Bragg won the open 220-yards handicap flat race from the out-mark of 30 yards and was second in both the under-fifteen 100-yards and 150-yards flat races. As a result he was runner-up in the Junior Championship, another example of his difficult place in the school, stranded between the sixth form for lessons and the junior athletics programme.[28] Although not selected in the two crews for inter-school competition, Willie was a capable oarsman, given his youth and modest stature (he weighed less than nine stone or 57 kg). He participated eagerly in intra-school competitions and in the school regatta held in December 1904. Five masters accepted the boys' invitation to participate and Willie rowed in the three seat of a four stroked by the headmaster. They

[25] School Lists, n. 14, 1904, p. 36.

[26] Letter C Todd to W L Bragg, 18 December 1904, RI MS WLB 95C/1.

[27] M F Perutz, 'Bragg, protein crystallography and the Cavendish Laboratory', *Acta Crystallographica*, 1970, A26:183–85, 184.

[28] *The St Peter's School Magazine*, no. 57, December 1904, p. 6.

Fig. 12.2 Charles Todd with some of his grandchildren, *circa* 1904 (refer family tree, Figure 0.2). Back row (L to R): Willie and Bob Bragg, 'Tom' Todd (son of Hedley and Jessie); front row: Alice and Mary Masters (daughters of Maude and Rev. Masters). (Courtesy: Mr S Gillam-Smith and State Library of South Australia, SLSA: B 69996/13.)

won their first race but were defeated in the final 'after a very even race, by a quarter of a length'.[29]

Willie's post-matriculation year at school (1905) was, as it had been for his father, rather stationary academically; but, unlike his father, it was a year of substantial personal maturation and blossoming. In a small upper sixth form of nine boys Willie was second overall, doing well in Latin and Greek and winning the prizes for chemistry and applied mathematics. In the Higher Public Examination late in the year a maximum of five subjects were allowed from the twelve offered. Willie was sixth in the state overall.[30] Mathematics and

[29] Ibid., p. 21.
[30] Public Examinations Board Manual for 1905, UAA, S14.

chemistry had emerged as his major disciplines, although he continued to read, in their original languages and with much pleasure, the ancient classics and selected French literature throughout the remainder of his life.[31]

In his annual report for 1905 the headmaster noted: 'Turning to the intellectual side, in the higher public examination, where boys are limited to five subjects, in history we obtain the third place... In Greek Barton won first place... In Latin Barton obtained honours and the second place. In pure mathematics two of our boys—Bragg and Thomson—are bracketed fourth. In applied mathematics Bragg obtained the second place. In chemistry Thomson and Bragg obtained the fifth and sixth places'.[32] The reluctant headmaster was now accepting the inevitable focus on examinations and devoting more time in his reports to the results of St Peter's College boys.

Willie recalled that the mathematics teaching he received at St Peter's was poor, but the senior chemistry teacher was outstanding. Edward Wainwright (BSc, London) was appointed to St Peter's College in 1883. His students obtained excellent results in the public examinations throughout his time at the school, and in 1900 the headmaster described him as 'a skilled teacher in his one subject [chemistry] with long experience. Discipline complete'.[33] Chemistry class lists were not recorded in the School Lists, but other evidence confirms that Wainwright taught Willie nearly all his school chemistry.[34] On Wainwright's death in 1919 the school published 'An Appreciation' that is both glowing and heartfelt:[35]

> Of all the masters of his day, few perhaps have left with us such a vivid set of recollections. Mr Wainwright was gifted highly, with an inborn capacity for teaching... Probably no master relied more for success on his power of teaching instead of on a boy's capacity for learning; no master set so little homework, and few had better teaching success judged either by examinations in the School or University or by the boys' appreciation of the knowledge gained. His quiet, orderly and systematic methods of class management conduced to good discipline, which was enhanced by the natural seriousness of the teacher. Yet... there was always a strong, freakish undercurrent of merriment and fun. Mr Wainwright... knew exactly when to divert this current from below to the surface and how to submerge it again; knew just how and when to relax or tense the minds of his class and so keep them alert and eager for fresh ideas. So his memory stays with us, not so much as a reminder of what we learnt but with what pleasure we learnt it.

Willie, in many later recollections of his journey to X-ray crystallography and the Nobel Prize, regularly credited his St Peter's chemistry teacher with

[31] Patience Thomson (née Bragg), personal communication.

[32] School Lists, n. 14, 1905, p. 28.

[33] Report of headmaster to Council of Governors, 1900, St Peter's College Archives.

[34] Wainwright was the only chemistry teacher listed in the school during 1902–5, and the School Lists confirm that he taught Willie's upper sixth form in 1905, information kindly supplied by the school historian, Ms Katharine Thornton.

[35] *The St Peter's School Magazine*, no. 101, September 1919, p. 7.

inspiring his interest in science. Thus he noted:[36]

> My interest in science started when I was at school, and I think the main reason was that my chemistry master taught in an interesting way...Our chemistry master at school was very formal and precise. I remember at our first practical class he said: 'Boys, take up your mortars—now take up your pestles and see how much noise you can make banging your mortars'. We did so. He then said: 'Now you have found out how much noise you can make, let me never hear that noise again'.
>
> During the lunch break he used to let me into the laboratory, and I set out all the experiments for the afternoon while he had a little nap, lying on a form with two fat books under his head. He fascinated me when talking about the properties of atoms, and he did not mind my asking questions.

After 1905 James Thomson was in charge of the chemistry teaching and, like his predecessor, 'there has been no subject in the public examinations in which St Peter's boys have met with more honours and successes'.[37] Some years later another important Australian scientist, Howard Florey, also attended St Peter's College and, like Willie Bragg, was inspired to a life of science by the chemistry teacher, this time 'Sneaker' Thomson. Given that Bragg went on to participate in the invention of 'the analysis of crystal structure by means of X-rays', and of molecular biology and the biological revolution still in progress, and that Florey went on to participate in 'the discovery of penicillin and its curative value in various infectious diseases', and thus the antibiotic revolution in medicine, it is hard to believe that there have ever been two school science teachers at one school who exerted a more profound influence on human welfare around the world. And in 2005, yet another St Peter's old scholar won the prize: Robin Warren (with Barry Marshall), the Nobel Prize for Physiology or Medicine for the discovery that bacteria cause stomach ulcers, not stress. He, too, remembered 'my chemistry teacher who was German and also happened to be a superb fencer'.[38]

St Peter's College followed Willie's progress with interest, and by 1915 it was aware of his outstanding success. James Thomson wrote to him, but Willie was in France, embroiled in the First World War, and the letter took some time to be delivered. It was not until 5 March 1916 that Willie wrote to his mother saying, 'Will you please send a photograph of mine to Mr Thomson, the science master at St Peter's in Adelaide; he wanted one and asked me for it. I think I showed you the letter, but forget whether I asked you to send the photo. He wants to stick it up in the class room'.[39] The photo was sent and was later joined by one of Florey, both being remembered by later students. With three Nobel Prizes awarded to former students, St Peter's College now belongs to a very

[36] Sir Lawrence Bragg, *The Start of X-ray Crystallography* (London: Longmans, 1967), p. 1.

[37] *The St Peter's College Magazine*, no. 150, December 1935, p. 5.

[38] The quotations are the Nobel Prize citations; for Warren see 'Bronx v. St Peter's: two Nobel causes', *Sunday Mail* (newspaper), Adelaide, 4 December 2005.

[39] Letter W L Bragg to G Bragg, 5 March 1916, RI MS WLB 37B/1.

tiny and select group of schools that can boast more than one laureate among its old scholars. Indeed, St Peter's is second in the world to the Bronx High School of Science in New York, which has produced seven, all in physics.[40]

Rain fell for the greater part of the afternoon during the school sports on Saturday 23 September 1905. Willie entered several events contributing to the award of the prestigious College Cup, then run on a handicap basis. He won the 100-yards flat from the front mark of nine yards but was unplaced in the 150-yards hurdles and 440-yards flat handicaps. Since not one of the event winners was placed in another event, the cup was decided on a repeat of the 100-yards handicap, which 'Bragg won easily... [he] ran very well indeed, and should develop into a very good runner, although at present he starts slowly and loses a lot of ground in the first 15 yards'. The system was rather unsatisfactory and was soon changed to a greater number of races run from scratch, but Willie Bragg's name remains as the winner of the College Cup in 1905.[41] In rowing he stroked the school's maiden four with fixed seats at the State regatta.[42]

This was also the year in which Willie found his social connection to the school. He was prominent in the Cadet Corps and in the Literary and Debating Society. The senior cadet corps paraded publicly in November, and two intra-school shooting matches were held, in which Willie shot well but his team was beaten comfortably.[43] At a meeting of the sixth form held on 19 June 1905 a Literary and Debating Society was formed. Willie was present at a further meeting on 26 June, when rules drawn up by a provisional committee were confirmed. The first formal meeting of the Society was held on Saturday 1 July, with a debate on the motion, 'That this House views with apprehension the rapid advance of the Japanese nation'. Willie led the side opposing the motion and his 'points were excellent'. The debate finished dramatically when a small, yellow-coloured dog entered the room and lay under the chairman's table, at which the mover of the motion concluded his reply by observing, 'The yellow peril has come to stay!' Willie Bragg's side narrowly won the Society's first debate, however. He took part in three further debates: on adult female suffrage, on the reality of supernatural appearances (ghosts), and on the motion 'That Adelaide is behind the times'. In a similar vein he found new delight in play readings: as Nerissa in a reading from *The Merchant of Venice*, and as Captain Absolute in a reading from *The Rivals*.[44]

Having done well at Canterbury School,[45] Robert Bragg entered St Peter's College in Willie's last year (1905) and thereafter travelled a much more conventional path. Beginning in the lower fourth form, he was placed ninth of twenty-two in the final class order, and this ranking level continued for the

[40] *Sunday Mail*, n. 38.
[41] *The St Peter's School Magazine*, no. 60, December 1905, pp. 30–2.
[42] Ibid., pp. 34–5.
[43] Ibid., pp. 6–7.
[44] *St Peter's School Collegiate Literary and Debating Society: Minute Book*, St Peter's College Archives, *passim*.
[45] For example, Robert won a special prize for languages in 1903: *Register*, 17 December 1903, p. 7; the school closed permanently in 1906 (see *Register*, 16 May 1906, p. 5).

next three years.[46] He did well in the university's public examinations. In the 1907 Primary Examination (minimum four subjects) he passed seven, while in 1908, for the Junior Examination (five subjects or more), he again passed seven: geometry and physics, both with credit, and French, arithmetic, algebra, chemistry, and drawing.[47]

In view of the family's earlier decisions to bypass school physics, Bob's result is interesting, especially as he was placed first in the State on the 'Special Honours List for Physics (under 16)',[48] a separation of candidates by age that responded to Girdlestone's concerns regarding the dominance of students of mature years. Indeed, St Peter's was so pleased by the science results of its students in the various public examinations in 1908 that it arranged for a photograph to be taken of 'St Peter's Science Credits, 1908', when only the prefects and its sporting teams were normally granted this privilege. Thirteen students, with handsome three-piece suits, watch chains, and celluloid collars, are shown with the chemistry master, James Thomson. Two boys are noted as being absent, including 'R C Bragg, 1st Place Junior Physics under age'.[49] Perhaps the picture was not taken until early in 1909, by which time Bob and his family had left Australia and were on their way to England.

Outside the classroom Bob followed his elder brother into the cadet corps, onto the athletic track, and into school rowing shells. He participated in the 1905 shooting competition for junior cadets,[50] and although light in weight at less than nine stone (57 kg), he was quickly into the bow seat of the college's second four in 1907, and he therefore occupied the same seat for the school eight. His four had an eight-length win in the city's Summer Regatta late in the year.[51] The following year Bob stroked an intra-school crew and occupied the bow seat of the College's first four, although the *School Magazine* gave no details of their performance in inter-club competition.[52]

It was in athletics, however, that Bob showed the most promise of matching his father's sporting prowess. The *School Magazine*, and several St Peter's College athletic medals preserved amongst the Bragg Archive at the Royal Institution in London, show that at various school sports Bragg ii won a number of events in 1905 and 1906 and was second in the Junior Championship in 1907.[53] The special attention accorded the various sporting competitions between St Peter's College and Prince Alfred College—known down the years as Intercollegiate or Intercoll. contests—had become established, and at the tenth Intercollegiate Athletics meeting Bob 'showed immensely improved form' and was the outstanding competitor in the Junior (under-fifteen) races.

[46] School Lists, n. 14, 1905–1908, *passim*.
[47] Public Examinations Board Manuals for 1907 and 1908, UAA, S14.
[48] Ibid.
[49] Photograph in the St Peter's College Archives.
[50] *The St Peter's School Magazine*, no. 60, December 1905, p. 8.
[51] *The St Peter's School Magazine*, no. 66, December 1907, pp. 6–8; ibid., no. 67, May 1908, pp. 43–45.
[52] Ibid., no. 69, December 1908, pp. 2–3.
[53] *The St Peter's School Magazine*, *passim*.

He was equal third in the high jump, equal first in the 100-yards flat race, and outright first in the 120-yards hurdles,[54] for which an Intercollegiate Athletic Sports medal can be found in the Bragg Archive.

When Bob left the college the headmaster recorded that 'Robert Charles Bragg . . . leaves us with a good record. He has always been a good worker and bears an excellent character to which I can testify without any reservation'.[55] The two boys retained affection for their old school long after they had left Australia, and the *School Magazine* regularly carried accounts of their successes. Willie visited the school in 1960, during his only later visit to his birthplace, and he chaired the old scholars' dinner in London as late as 1964.[56] Regarding the 1960 visit, a student of that time, who attended Willie's talk to the senior students, provided an interesting insight:[57]

> I attended St Peter's all my school days [and] I *do* have a very clear recollection of the end of his talk . . . I didn't have the sporting student's ready acceptance and endured a couple of years of misery. Bullying seemed to be accepted as part of the school ethos . . . Bragg spoke of his time at St Peter's [and] described how he was then very keen on the sciences . . . But Bragg also told us that, because of his helping the teacher, he was ridiculed and picked on by the other students. I remember his description well. He told us that, at that time, 'stag' was the derisory term used to describe boys on the outer . . . He told us that he was known as 'Willie Bragg, the little Stag' . . . It was his attempt to use his visit and his honoured status to ask us not to bully those who might seem the odd ones out . . . It was also memorable because he started to say something like 'Don't pick on the boys who seem to be different because they . . .', but about there he faltered, trying to say something like 'they might turn out to be successes', but he had trapped himself into sounding as though he was boasting . . . I recall that, as he faltered, he blushed with embarrassment, and I can still remember vividly this very red face under the very white hair . . . He recovered and was applauded . . . I was much taken by him . . . I felt that [he] was on my side.

The similarities between the childhood and schooling of Willie and his father now strike one with particular force. Both were blessed with a substantial gift for mathematics and achieved outstanding results in all its facets. They also enjoyed and achieved excellent results in English and ancient and modern languages. Like his father, who did not study either German or physics at King William's College, Willie also bypassed these subjects at St Peter's. Both would later find a career, high achievement, and fame in physics, and both would come to regret their lack of German. Both were promoted well in

[54] *The St Peter's School Magazine*, no. 66, December 1907, pp. 8–12.

[55] St Peter's College departure card with personal annotations by Henry Girdlestone, MA Oxon., Head Master,1 December 1908, RI MS RCB 35.

[56] *The St Peter's School Magazine*, no. 86, August 1914, p. 74; ibid., no. 91, December 1915, pp. 30–31; *The St Peter's College Magazine*, no. 197, December 1960, *passim*; ibid., no. 201, December 1964, p. 41; ibid., no. 208, December 1971, pp. 14–15.

[57] Dr Rob. Morrison, personal communication.

advance of their age group, and both then suffered isolation and disconnection from their social peers. Both participated in debates and play-readings at school, and both were capable sportsmen, whose achievements were reduced by their social separation from their sporting fellows. Both were first-born sons and both exhibited characteristics common to such children: they were high-achieving, serious, and self-contained within the boundaries defined by their guardians or parents. They were also subject to more pressure and higher expectations than their siblings. William was shy and possessed a deep humility; Willie was timid and self-conscious. Both were happy in their own company. William had, of course, suffered the early loss of his mother, whereas Willie grew up in the embrace of a loving family. That security allowed him a level of independence; although conventional in pose, in all the photos of the family group during this period it is notable that Willie stood a little aside, while Bob was in close physical contact with his mother, even when he was no longer a child. On the other hand, Willie's independence should not cloud the very substantial control that his father exercised over his academic and extra-curricular activities. Willie would later attempt to assert similar control over his own children's education, with the result that he resisted for some time the requests of his elder daughter and her teachers to abandon mathematics in favour of arts subjects, he was angry when she chose to major in history, and he was furious when she went to Oxford instead of Cambridge![58]

There were differences, however. Willie's family and his colleagues have insisted that in later years his personality and his identity were completely different from those of his father, and later writers support this opinion. What happened during the life experiences of William Lawrence Bragg to produce an adult of dramatically different personality?

Willie at the University of Adelaide

At the end of 1905 the family faced a dilemma regarding Willie's education. He had already done one post-matriculation year and academically the school could offer little more. In addition, William's memories of his own final school year were painful and he saw no reason for his son to risk a repetition. Willie, however, was only fifteen years old, and chapter IX of the statutes of the University of Adelaide required a person to be 'not... less than sixteen years of age' when signing the university Roll Book and becoming a matriculated student. He would be sixteen on 31 March 1906, soon after the start of classes earlier in the month, and this offered a solution. Willie recalled that 'my father managed to wangle me in while I was still only fifteen',[59] and presumably his father argued that his son had satisfied all the requirements to a very high level, was ready for university work, and was within a month of meeting the age

[58] Margaret Heath (née Bragg), personal communication.
[59] W L Bragg, Autobiographical notes, p. 18.

requirement. Indeed, there had been previous cases of the same kind,[60] and Willie satisfied the letter of the statute by signing the Roll Book on 2 May 1906, after his birthday.[61]

Willie's autobiographical notes are generally reliable, but sometimes not on points of detail. His extraordinarily brief recollections of his studies at the University of Adelaide are typical: 'My course was Honours Mathematics, in which I got a 1st class in 1908. I had my first introduction to Physics at the University, Physics and Chemistry being the required subsidiary subjects for an Honours Mathematics degree. My father was Professor of both Mathematics and Physics, so most of my instruction came from him'.[62] Although Willie enjoyed languages and could take mathematics and some science subjects as part of an arts degree, in 1906 he enrolled for the BSc degree.[63] This offered the advantage that 'Students who at the Higher Public Examination have shown special excellence in Mathematics may be exempted from... Compulsory [first year] Mathematics'.[64] Willie therefore enrolled in second-year pure mathematics as well as first-year physics and chemistry.[65] At the end of the year he obtained first-class passes and was placed first in all three subjects, in classes averaging eighteen students.[66] A careful search has failed to reveal the lecturers and examiners of Willie's subjects, except that the relevant Calendars all list Bragg, Chapman, and Madsen together for mathematics and physics.[67] There is Willie's own recollection quoted above, however, and he also recorded that, 'I was fascinated by these [Bragg–Threlfall] letters because I attended my father's Electricity lectures as a student in Adelaide, and I can trace in them the origin of his wonderful lucid way of putting the main points to us'.[68] William had also determined the content of all the mathematics and physics courses. His influence on Willie's education continued to be overwhelming (see Figure 12.3).

The Regulations for the BSc degree were further clarified in 1907, and they required a major study of two of the five subjects available: mathematics, physics, chemistry, physiology, and geology and mineralogy. Willie nevertheless

[60] Two such cases occurred, in 1886 and 1891: Nicholas Hopkins (see his letter to the Chancellor and the Chancellor's consenting direction to the Registrar, UAA, S200, dockets 84 and 85/1886), and Alfred Chapple (see his letter to W H Bragg, 7 June 1931, RI MS WHB 10C/19, in which he recalls his under-age admission to the University of Adelaide 40 years earlier).

[61] *Student Roll, 1876–1963*, UAA, S425, vol. 4, 1902–1908 (no. 1891), p. 383.

[62] W L Bragg, Autobiographical notes, p. 18.

[63] List of Students, 1906, *Calendar of the University of Adelaide, 1907* (Adelaide: University of Adelaide, 1907), p. 57.

[64] Regulation III for the B.Sc. degree, *Calendar of the University of Adelaide, 1906* (Adelaide: University of Adelaide, 1906), p. 103.

[65] Academic record of William Lawrence Bragg on student record cards and in *Register of B.Sc. Students, 1898–1908*, UAA, S246, p. 53; the subject labels varied between courses but were brought under a common numbering scheme in 1907.

[66] Pass Lists, 1906, *Calendar*, n. 63, pp. 354–7.

[67] Examiners, 1908, University of Adelaide Examination Papers, UAA, S15; in addition Chapman and Madsen now had major responsibility for the engineering courses.

[68] Sir Lawrence Bragg and Mrs G M Caroe, 'Sir William Bragg, F.R.S., 1862–1942', *Notes and Records of the Royal Society of London*, 1962, 17:169–82, 174.

Fig. 12.3 Willie Bragg during his first year at the University of Adelaide, 1906. (Courtesy: The Royal Institution, London.)

decided to continue with three subjects, and it would appear that he also failed to notice the BSc requirement to complete elementary biology in addition to first-year mathematics, physics, and chemistry. While his 1906 enrolment was relatively light, his 1907 programme was very heavy: third-year pure mathematics, applied mathematics, second-year physics, theoretical chemistry part II, and practical chemistry part I.[69] Was he trying to complete an honours degree in three years when normally it took four? His results were again outstanding: first place and a division I pass in pure mathematics (only two students), in applied mathematics (nine) and in physics (nine), second and a

[69] See n. 65; also n. 63, pp. 96–7, 158–63; a major study of chemistry required practical work at both first- and second-year levels, and, in addition to elementary biology, Willie also may have overlooked the opportunity to undertake practical chemistry part I the previous year.

division I pass in chemistry (seven), and equal third and a division II pass in practical chemistry (six students).[70]

Examination papers give a reliable indication of the content of courses.[71] Willie's physics papers were similar in style to my own in the mid-twentieth century, but very different from current examples. They were descriptive rather than mathematical, with emphasis on practical instrumentation. The two second-year papers contained, in addition, more difficult questions on optics, heat and thermodynamics, much electromagnetism, and a little sound, but they were less challenging mathematically than today. The complex units of measurement reminded me of that earlier agony, before the adoption of the metric system in Australia.[72] The mathematics papers now seem straightforward and old-fashioned. Geometry (now largely banished) predominated, with the calculus in second-year pure mathematics typical of mid-twentieth-century papers. Similar remarks apply to the two third-year pure mathematics papers, except that high manipulative skills were required and the calculus was 'quite tricky' in places.[73] John Scott agreed: 'although demanding, the questions are pretty straightforward applications of work that must have been presented in lectures and tutorials. They are testing and fair, but should have been readily attempted by an assiduous student'.[74] The applied mathematics papers of 1907 were 'more physics than mathematics' and covered statics, friction, Newtonian laws of motion, and hydrostatics and dynamics. All these papers showed a strong affinity to the Cambridge Mathematical Tripos examinations that William Bragg sat in the 1880s and leave no doubt as to their source. Mathematics teaching is now very different, including, for example, analysis, abstract algebra, topology, probability and statistics, quantum mechanics, and numerical analysis.[75]

Willie's degree course now took a strange turn, and there appear to be no records to explain what happened; I can only guess that his father played a pivotal role. William himself had completed the equivalent of an honours degree in a little over three years and was seriously considering an overseas appointment. His son might have only one more year in Adelaide and it was hoped he would complete an honours degree before leaving. Honours mathematics was the focus, but Willie lacked two subjects for the award of an honours science degree—elementary biology and either third-year physics or practical chemistry part II. Such a combined load might have been too great even for Willie. Could he possibly transfer to an arts degree and graduate with an honours BA degree as his father had done? Not according to the regulations for the BA degree, which required the study of at least four introductory subjects:

[70] Pass Lists, 1907, *Calendar of The University of Adelaide, 1908* (Adelaide: University of Adelaide, 1908), pp. 363–6.

[71] The examination papers Willie sat are in University of Adelaide Examination Papers, UAA, S15.

[72] I am grateful to Em. Prof. Robert Leckey for comments on the physics papers.

[73] I am grateful to Dr Peter Stacey for advice on the mathematics papers.

[74] Rev. Dr John Scott, personal communication.

[75] See n. 73.

from English language and literature, Greek, Latin, French, German, history, economics, and mental and moral science.[76]

Willie had passed five of these at the Senior Public Examination and two at the Higher, but he had studied none of them at the university. Despite this apparently insuperable hurdle, in 1908 Willie was allowed to enrol for the Honours Degree of Bachelor of Arts in the Department of Mathematics.[77] Perhaps his school results were given weight, and the enrolment did satisfy the requirement that 'Candidates in [Honours] Mathematics shall pass in Pure Mathematics, Applied Mathematics, and Physics as prescribed for the Ordinary Degree'. Third-year physics was not required for the BA honours degree whereas it was for an honours BSc.[78] Only a brief outline of the honours mathematics syllabus was published: 'Subjects for the final examination—analytical geometry, infinitesimal calculus, analytical statics, dynamics of a particle, hydrodynamics, elementary rigid dynamics'.[79] Willie was the only student and the course was devised, taught, and largely examined by his father.

There were five three-hour examination papers at the end of the year, one each on solid geometry, analytical statics, dynamics, differential equations and hydrodynamics, and 'general problems'.[80] John Scott commented on these papers: 'advanced theory for the time, but straightforward... not too hard for anyone who had studied the requisite theory'. The university formally listed Kerr Grant of Melbourne and Professor Chapman as examiners for honours mathematics in 1908, but they were the external examiners.[81] Willie obtained excellent results and graduated with an Honours BA degree with first-class honours in mathematics.[82] He was only eighteen years old.

Willie later recalled that he was 'a misfit' at Queen's School and that he had 'the same handicap at St Peter's', being precocious in lessons beyond his age. The same difficulties confronted him at the university and his recollections of that period are even more painful, to the extent that they have been removed from some versions of his autobiographical notes:[83]

> Although I was fifteen when I entered Adelaide University, I think my emotional age was about twelve or less, and my fellow students were mature young men and women. Such a disparity has a cumulative effect. Anyone handicapped in this way is debarred from taking part in those normal activities of his age group, and the very fact that he cannot enter

[76] See n. 70, pp. 89–90, 97–8.

[77] Student record cards and *Register of B.A. Students, 1898–1908*, UAA, S247, p. 195.

[78] *Calendar of The University of Adelaide, 1908* (Adelaide: University of Adelaide, 1908), pp. 89–90, 97–8.

[79] Ibid., p. 157.

[80] Examination Papers, n. 71.

[81] 'Examiners 1908', ibid; in 1908 Kerr Grant was a Lecturer in Natural Philosophy at the University of Melbourne.

[82] Pass Lists, 1908, *Calendar of The University of Adelaide, 1909* (Adelaide: University of Adelaide, 1909), p. 385.

[83] W L Bragg, Autobiographical notes, extract generously provided by Stephen Bragg, personal communication.

> into their plans, schemes, differences of opinion, exercise of authority and so forth means that he loses the earlier experience which would teach him how to take his place later in life in the world of affairs. He loses touch with what is going on round about him, and he thinks of the people who guide the course of events as 'they', not as 'we'. He develops a defence mechanism to hide his inexperience from those he meets, and this again makes him shy of asking the questions, the answers to which would keep him in touch. He is like a hermit crab, with a formidable array of whiskers and claws in front, but with a soft white tail which it has to conceal in a protecting shell.
>
> How is such a one to face, in later life, the demands made upon him when his position is one of authority? He is greatly helped and fortunate if he can find a colleague who will act as a 'blind man's dog', who provides a pair of eyes to guide him through the jostling contact with other people and across the roaring traffic of current events... His disability has a compensation in that his very lack of appreciation of what is going on around him gives him a power of intense concentration and enables him to excel in the subjects of his interest... I can never be sufficiently grateful to those colleagues who have so often helped and guided me.

At this stage in his life Willie was even more isolated from his fellow students than his father had been in Cambridge, an isolation compounded by studying in his father's office at the front of the university building, a 'strange' practice instituted by William, his daughter thought.[84] William did not have the ability to empathize with his student son's relationship to his peers, a legacy of his own childhood. Felix Barton, another St Peter's student mentioned publicly by the headmaster for his outstanding results in the classics, also described himself as 'lonely' and 'non-gregarious' and claimed that he and Billy Bragg 'saw a great deal of each other... all through my university days'.[85] We can only guess, however, what other students said about the professor's son who was first in nearly every subject, many taught by his father, and who remained largely aloof from them. Willie came to recognize that these experiences were at least partly responsible for his inability and unwillingness, when later in positions of authority, to become immersed in the administration, politics, and personalities of academic life, especially when fascinating problems awaited him in the research laboratory. His elder son remembered that, 'he disliked committees,... he had little interest in politics or in the theories of management and hated any form of deviousness or deceit'.[86]

For now Willie was the holder of an honours degree and an outstanding academic record, but much personal growth and an acquaintance with the

[84] G M Caroe, *William Henry Bragg, 1862–1942: Man and Scientist* (Cambridge: CUP, 1987), pp. 77–8.

[85] Copy of typescript 'Extracts from reminiscences of Felix Kingston Barton, B A Hons Classics', in possession of the author.

[86] S L Bragg, 'A personal view by his elder son', in J M Thomas and Sir David Phillips (eds), *Selections and Reflections: The Legacy of Sir Lawrence Bragg* (London: The Royal Institution, 1990), p. 142.

hurly-burly of adult life lay ahead. These he would find, along with great academic success, in his father's old college, laboratory, and university: Trinity College and the Cavendish Laboratory, Cambridge. In Adelaide he found other outlets for his energy and enthusiasm, and they were largely solitary pursuits, including shell collecting, seismology, and assisting his father on the golf course:[87]

> My main hobby at that time was collecting shells, and I built up quite an interesting collection of about 500 species. The shells were not as large and magnificent as those in tropical waters but there were many interesting and striking ones. I found a new species of cuttlefish, which was named *Sepia braggi* by Dr Verco in Adelaide. He had the finest collection of shells in South Australia and was very kind to me as a young enthusiast. I remember when I took some specimens of the 'bones' of these little cuttlefish to him, and he had verified that there was no previous description of the species, he said, 'I will call it *Sepia gondola* because of its shape'. Then, seeing my face fall, he rapidly altered the name to *Sepia braggi*.
>
> I also found new species of ishnochiton, and several species which were new records for South Australia. St Vincent's Gulf was protected from the ocean rollers, so the shells were washed up on the beach in good condition . . . I gave my collection to the Manchester Museum.

Joseph Verco reported the find and its naming to a meeting of the Royal Society of South Australia on 6 August 1907, and details appeared in the society's *Transactions* later that year. The type cuttlebone was described thus: 'It is 60 mm long by 11 mm broad at its widest part, with a maximum thickness of 4.75 mm. The dorsal surface is very slightly convex in its anterior two-thirds but markedly curved in the posterior third . . . The type was found at Glenelg by Master Bragg, and we have pleasure in naming it after him, and at the same time complimenting his father, Prof. Bragg, one of our most honoured Fellows, who has just been distinguished by the Fellowship of the Royal Society of London'.[88] Later descriptions were less technical: 'the extremely narrow, delicate, elongated [and curved] cuttle-bone of Bragg's Cuttle . . . occurs fairly commonly on beaches in South Australia, Victoria and Western Australia'.[89] The bones I have collected in recent years have been even smaller and more delicate than the type specimen, and they have largely disappeared from beaches close to Adelaide.

The pride inherent in the reference to Willie's father in Verco's description is understandable, but the young man might have wished that his first piece of scientific research would be recognized for himself alone. Indeed, many years

[87] W L Bragg, Autobiographical notes, pp. 18–19.

[88] J C Verco, 'Notes on South Australian marine mollusca, with descriptions of new species, part VI', *Transactions and Proceedings and Report of the Royal Society of South Australia*, 1907, XXXI: 213–30, 213–14.

[89] J. Allan, *Australian Shells* (Melbourne: Georgian House, 1959), p. 447; the species has recently been re-examined in detail by A L Reid, 'Australian cuttlefishes (Cephalopoda: Sepiidae): the "doratosepion" species complex', *Invertebrate Taxonomy*, 2000, 14:1–76.

later, in responding to a suggestion that 'such a hobby may be a useful indication of aptitude for research', Willie wrote, 'It is even possible that biology rather than physics might have been my trend had there not been such a strong family tradition', although 'I think that perhaps in many cases, as it was in my own, the instinct is rather that of the collector than the scientist'.[90] The type specimen remains in Adelaide, although it was assumed for many years that it had gone to Manchester. Willie's pleasure in the study of nature continued throughout his life and included astronomy and physical geography, shells and shell-collecting, plants and gardening, birds and butterflies, and sketching and painting outdoors, all interests he communicated to his children.[91]

William's interest in golf as a healthy form of relaxation continued throughout the period of the boys' schooling. The long-term prospects of the North-east Park Lands for the Adelaide Gold Club were poor, and late in 1895 it was agreed to amalgamate with the Glenelg Golf Club. Even at Glenelg, however, conditions were far from ideal. William's handicap was reduced to scratch early in 1896.[92] On their return from leave William again took up golf and played regularly in the weekly competitions, with a handicap between five and ten. An un-attributed newspaper cutting in the club's 1896–1902 Minute Book describes a mixed foursome on the Glenelg links in the following light-hearted terms: 'Had a cynic stood by the last flag he would have found ample material for his cynicisms—a well-known city merchant toiling over the hill... whose courage failed him when he found his ball half-buried in the sand; doctors hacking their balls... with as much complacency as if the innocent rotundities were so many patients; Lawyers fairly thick... who found a broken club was apt to make them forget their surroundings—but why prolong the list? Golf makes different creatures of us all—some worse, some better, but all enthusiastic!'[93]

Dissatisfaction with the arrangement came to a head in September 1904, when it was decided to buy land in a seaside area on which to build a new course. The committee found a suitable site at Seaton; the sandy location was ideal and it was served by a good train service from the city. The Adelaide Golf Club had finally reached its permanent home, the train line through the middle of the course being a well-known feature to the present day. Early in 1905 William wrote to the committee, and at its next meeting it was recorded that his 'offer be accepted with thanks and Professor Bragg be informed that the details of carrying out his scheme be left to the incoming committee... Alterations to the course as shown on the plan were approved'.[94] William's letter has not survived, but we can infer its content with confidence, for Willie remembered that '[My father] was a fine golfer. I used to caddy for him as a boy, and I remember

[90] Letter W I B Beveridge to Sir Lawrence Bragg, 23 September 1949, and reply W L Bragg to Beveridge, 28 September 1949, RI MS WLB 50A/414 and 415 respectively.

[91] P Thomson, 'William Lawrence Bragg: the education of a scientist', *Adelaidean: News From The University of Adelaide*, 1999, 8(8):2; id., personal communication.

[92] J G Jenkin, 'William Bragg in Adelaide: and finally golf', *The Australian Physicist*, 1986, 23:138–40.

[93] Quoted ibid., p. 139.

[94] Ibid.

going round with him when he was planning a new course at Seaton, near Adelaide, which later became a well-known course'.[95]

Up to £500 was borrowed for a new clubhouse, and a trestle platform was erected alongside the railway line after the Railway Commissioner gave approval for trains to stop at the new clubhouse. The first competition was held in August 1905 for a trophy donated by Professor Bragg, who continued to nominate a succession of family members, university staff, and other acquaintances for membership. Early in 1906 a Sydney professional was engaged to suggest improvements to the course, and he was scathing in his criticisms of what we may assume was William's initial design. Against the wish of most of the members his plan to raise the course to championship standard was adopted, although finally a layout suggested by the club captain and secretary, and much closer to William's original design, was implemented. The result was a course of about 6,000 yards (5,500m), bogey 80, with the ninth 'crater' hole and a short fourth receiving particular praise. The number of members rose beyond 100 and a professional coach was engaged.

William won the club's Senior Medal for the 1906–07 season and participated in club and state championships when they were played at Seaton. In the final of the 1907 club championship, for example, he was beaten 'four up and two to play' by Professor Henderson, after they had been 'all square' after the morning round. The club became the Royal Adelaide Golf Club by the grace of King George V in 1923. It is surely unique in having been laid out initially by two future Nobel laureates, the elder noting the terrain and stepping out distances as he played his round and the younger carrying the bag of clubs and making notes as his father called them out.[96]

Among Charles Todd's multitude of responsibilities was the South Australian Observatory, which itself embraced a number of important functions, including astronomical observation, weather recording and forecasting, and the local time service. Yet another responsibility—seismological recording—fell to the state observatories from the first AAAS meeting in 1888, with the establishment of a committee on 'Seismological Phenomena in Australasia'. Todd and other state astronomers were members of the committee for many years. Both Sydney and Melbourne observatories acquired early-model seismographs, and in 1890s the British Association launched a campaign to establish an imperial network of recording stations equipped with standardized horizontal seismometers designed by the leading British expert, John Milne.[97]

After the disastrous San Francisco earthquake in April 1906, the South Australian government could no longer resist the call for a local seismograph,[98]

[95] Bragg and Caroe, n. 68, p. 171.

[96] Jenkin, n. 92.

[97] R W Home, 'Defining the boundaries of the field: early stages in the physics discipline in Australia', in R MacLeod and R Jarrell (eds), 'Dominions Apart: Reflections on the culture of science and technology in Canada and Australia, 1850–1945', *Scientia Canadensis*, 1994, 17:53–70, 62–3.

[98] *Register*, April 1906, *passim*, 2 May 1906, p. 4, and 28 June 1906, p. 6.

and a Milne instrument was purchased and installed in the Adelaide observatory by Charles Todd and his son-in-law in 1908. Todd suggested to his grandson that he might like to take an active interest in the new facility. Willie investigated the matter and was invited to present his findings and the new instrument to one of the regular meetings of the Astronomical Society of South Australia. The society had followed the radio experiments of Todd and his son-in-law with interest, had heard the professor discuss 'The atom as a planetary system' in April 1906, and would now hear from Todd's grandson.[99]

The instrument was housed in a small building erected for the purpose, and in June 1908 the society expected to hear from Willie on the topic 'Seismology and the Seismograph'. The general meeting was postponed, however, 'by reason of Mr Bragg being unable, on account of illness, to be present... [and] the attendance being very small'.[100] In August twelve members and sixteen visitors were present, and Willie dealt 'with a difficult subject in an able and masterly manner... illustrating it with a model of the Milne instrument and comprehensive diagrams. The paper was afterwards discussed by members present, Mr Bragg answering many questions submitted to him and was accorded a hearty vote of thanks'.[101]

Willie had worked hard in preparing his first public lecture: he had studied the topic thoroughly and had prepared a model of the instrument and several diagrams, including a record of the San Francisco quake, to illustrate his talk.[102] Like his father, Willie too would later become renowned for his ability to give interesting and informative lectures to non-specialist audiences, particularly young people. His insightful writings on the topic have been collected together in a work that compares his skills favourably with those of the British pioneer in the area, Michael Faraday.[103] Like his father, Willie's Adelaide experiences appeared again and again throughout his life. In December 1966, for example, in addressing a meeting of the American Association for the Advancement of Science in Washington, DC, he discussed the art of lecturing and referred specifically to 'the movement of a recording instrument, for instance, a seismograph'.[104]

The Observatory complex in Adelaide, including the Todd family who presided there, was a major influence on Willie, as it was on his father. With its bubbly family life, scientific facilities, open spaces, Aboriginal visitors, piles of 'junk', and most of all its freedom, it symbolized all that was best in his childhood. Willie's sense of humour mirrored that of his grandfather: childlike, it included puns and spoonerisms, recognized comic situations, and was

[99] B Waters, *A Reference History of the Astronomical Society of South Australia Inc.* (Adelaide: the author, 1980–1982), in four volumes, vol. 1 (1891–1901) and vol. 2 (1902–1911).

[100] Ibid., vol. 2, p. 239.

[101] Ibid., p. 241.

[102] *Register*, 13 August 1908, p. 3.

[103] Sir George Porter and J. Friday, *Advice to Lecturers: An Anthology Taken from the Writings of Michael Faraday & Lawrence Bragg* (London, The Royal Institution, 1974).

[104] W L Bragg, 'The art of talking about science', *Science*, 1966, 154:1613–16, 1615, an article adapted from the address.

supported by a roar of laughter that was sometimes inappropriate.[105] He loved his Aunt Lorna, who was later a regular visitor to England. One of Willie's grandchildren has recorded that, 'Only once did my grandfather, by then a Nobel Prize winner in physics, reveal his roots. When a rural dean asked him if he was descended from convicts, he hit him'![106] Always determined to keep his emotions under firm control, only rarely was Lawrence provoked to respond, even to such ignorance.

[105] Stephen Bragg and Alice Bragg, personal communications.
[106] Alice Thomson, *The Singing Line* (London: Chatto and Windus, 1999), p. 13.

13
Further research: X-rays and γ-rays

A heat wave swept South Australia during the Christmas/New Year period of 1905–06, with long spells of extraordinarily hot weather.[1] The Bragg and Todd families followed their tradition of finding summer relief in a holiday at the seaside; a time of relaxation and fun with his family which William treasured. He reacted testily, for example, to letters from the Registrar that interrupted this precious period, and he was usually absent from University Council meetings in January.[2] Adelaide was now more than 'the city of churches', with the growth of the education precinct along North Terrace testifying to its maturity. Extensions had been made, were in progress, or were planned for all the major institutions on its north side: to the Institute Building (housing the Public and Circulating Libraries, the Royal Society of South Australia, the South Australian Society of Arts, the local branch of the Royal Geographical Society, and the Astronomical Society), and to the Museum, the Art Gallery, the University, and the School of Mines and Industries.[3]

While William was contributing somewhat less to many of these institutions than in previous years and was devoting more time to his research and family, he must have viewed their progress with pleasure, for he had contributed significantly to all of them. This was especially true of education outside the university, where his influence had been substantial but was now in the hands of other university staff, a new generation of government officers, and the broader society, which was taking a closer interest in the education of its young people. While problems were widely recognised, it was the Teachers' Union that became the catalyst for change, through its annual conferences. William had spoken regularly at these gatherings, which had become foci for criticism and debate, for new ideas and practices, and for social interaction among teachers.[4]

Major reforms were seen as necessary. The Headmaster of the Norwood Public School and President of the Union, Alfred Williams, electrified his

[1] *Register*, 23 January 1906, p. 4.

[2] See, for example, letter W H Bragg to Registrar, 8 January 1902, UAA, S200, docket 27/1902; Council Minutes, UAA, S18.

[3] See, for example, new extensions to the Institute Building described in *Register*, 11 June 1907, p. 4, 12 June 1907, p. 7, and 13 June 1907, p. 6.

[4] C Thiele, *Grains of Mustard Seed* (Adelaide: Education Department of S.A., 1975), chs 5 and 6.

audience at the 1904 congress with a passionate call for change, and the floodgates were prised open. Early in 1906 the State's first Labor Premier, Tom Price, took over the education portfolio and revolutionised its administration. He created a new office of Director of Education and, when Professor George Henderson declined the position, he appointed Alfred Williams. The spirit of the department lifted and major policy developments followed, particularly regarding secondary and technical education.[5] Throughout 1906 and 1907 the local newspapers were constantly discussing educational reform,[6] and when William spoke to the 1906 conference about 'Recent discoveries in radio-activity' and dealt particularly with the implications of the radium present in the earth, he must have felt the change in the air.[7] Early in 1908 a state high-school system was launched, to be followed by improvements to teacher training through greater university access as well as practice, increased salaries for teachers, renovation of school buildings, and the provision of improved school facilities generally.[8]

Inside the university William was as busy as ever. Indeed, whenever the university wanted a wise, sympathetic, and experienced representative it turned first to Professor Bragg. He was regularly reappointed to the University Council, the selection committee for the Rhodes Scholarship, and the committee to negotiate with undergraduate students regarding their participation (or not) in the annual commemoration ceremony. He was frequently appointed to other committees: for example, to approve exemptions to the new compulsory Sports Association/Union fee, to consider further improvements to the acoustics of the Elder Hall, and to address problems arising from the public examinations.[9] William was asked twice by the Council to write to Richard Kleeman to express its congratulations on the extension of his 1851 Exhibition scholarship for a second and then a third year,[10] and he also reported to the Council that Robert Barr Smith had promised £500 for the purchase of apparatus,[11] and that he had attended a conference in Melbourne regarding the transfer of meteorological services from the States to the Commonwealth.[12]

William further proposed that a chair of engineering should be established, with Robert Chapman as the first professor, noting the good status of the Engineering School, the healthy student numbers, and Chapman's excellent work and service to the university for twenty years. The Council agreed with the proposal unanimously, but declined a salary increase.[13] Its finances were in a poor state, and it had been negotiating with the government for a number of

[5] Ibid.
[6] See *Register* throughout 1906 and 1907.
[7] *Register*, 4 July 1906, p. 8.
[8] Thiele, n. 4.
[9] Council Minutes, meetings throughout 1906 and 1907, UAA, S18, vol. VIII.
[10] Ibid., meetings of 31 August 1906 and 30 August 1907, vol. VIII, pp. 264 and 346 respectively.
[11] Ibid., meeting of 30 November 1906, vol. VIII, p. 287.
[12] Ibid., meeting of 31 May 1907, vol. VIII, p. 327.
[13] Ibid., meeting of 22 March 1907, vol. VIII, p. 310.

years, requesting it to buy the country lands held by the university to alleviate its financial position. Like the statue to the university's first benefactor, Sir Walter Watson Hughes, unveiled on 27 November 1906 and still standing in front of the first university building today,[14] a few things in Australian universities never change!

William's academic load was lightened when it was agreed that the university professors should be relieved of their responsibility of marking all the public examination papers in their discipline,[15] but it increased when he became Dean of the Faculty of Science for 1907.[16] William and the other Council members were stern-faced when a photographer from *The Critic*—a social, sporting, and literary weekly—arrived on Friday 29 June 1906 to photograph 'The Council in Session'. Other photographs had been taken that week, all for a large pictorial supplement on the 'University of Adelaide'.[17] The Council members are shown seated on simple wooden chairs around a long trestle table; the academic staff number 34, the administrative staff 20, and the rooms and facilities look adequate if not generous for the 234 students in their photograph. Teaching and research were provided in many branches of arts, science, engineering, medicine, law, and commerce, and there were additional staff and facilities for music.

At the School of Mines and Industries William continued as a Council member, where he was warmly congratulated on his election to the Fellowship of the Royal Society and appointed to the School's education and museum committees.[18] He was present when the Governor, now Sir George Le Hunte, opened the new facility for chemistry and metallurgy and named it the Bonython Building in honour of the School's president.[19] At the annual Visitors' Day, when 3,000 members of the public attended the School, 'the imitation thunder and lightning from the electric spark and the delicate beauty of the lights in the vacuum tube vied with the Röntgen rays and many other attractions in compelling attention'.[20]

William's research on radioactivity suddenly acquired a practical reality and notoriety when, in May 1906, it was reported that a radioactive ore had been discovered at Olary, 260 miles (420 km) north-east of Adelaide, close to the New South Wales border. *The Register* reported with some excitement:[21]

> On Thursday morning the Commissioner for Crown Lands handed the following report from the Government Geologist (Mr H Y L Brown) to the press:

[14] *Register*, 28 November 1906, p. 6.
[15] Council Minutes, meeting of 27 April 1906, UAA, S18, vol. VIII, p. 231.
[16] Ibid., meeting of 17 December 1906, vol. VIII, p. 297.
[17] Ibid., meeting of 29 June 1906, vol. VIII, p. 245; 'University of Adelaide', a *Supplement to The Critic*, 11 July 1906.
[18] School Council Minutes, meetings of 11 March and 8 April 1907, University of South Australia Archives, vol. 6, pp. 15–16 and 29 respectively.
[19] *Register*, 3 December 1907, p. 10.
[20] *Register*, 14 December 1907, p. 8.
[21] *Register*, 4 May 1906, p.5.

> 'I have just received some specimens of a mineral which has been analysed at the School of Mines and determined as carnotite... Carnotite contains Uranium oxide, 63 to 65 percent... The importance of this discovery centres in the fact that pitchblende (oxide of uranium) is the mineral from which the rare substance radium is obtained... [P.S.] Since handing in the above, a sample of the mineral was forwarded to Professor Bragg of the University of Adelaide, who on experiment declares it to be strongly radio-active.'...
>
> – Chat with Professor Bragg –
>
> Professor Bragg was busy lecturing on Thursday morning, but he had a telephonic conversation with a reporter concerning the discovery. He said the find was exceedingly interesting, but he preferred to wait for a complete analysis before committing himself to a definite opinion... Professor Bragg was seen again at the Observatory in the evening, and he said, 'The matter is really one of simple arithmetic. Radium is found in uranium minerals, and the proportion of radium to uranium in almost all the substances hitherto investigated is always the same. The reason is probably that radium is a product of uranium, and that, where the mineral has been undisturbed for countless years, the uranium has had time to form slowly the radium that goes with it... Radium is worth something like £5 to the milligram, or roughly £100,000 an ounce [28 gm], but it is, of course, a costly process to extract it'.

The following day a further report was published, including information about the appearance of the specimens, the pegging of new claims, and a long interview with Douglas Mawson:[22]

> Mr Mawson was working in Sydney two years ago with Mr T H Laby, and they then discovered the first radium in Australia. 'It has not previously been mentioned', Mr Mawson remarked, 'that a most important discovery has been made at Moonta in the old copper workings [north-west of Adelaide]. It was made by Mr S Radcliffe [assistant chemist of the local mining company] about three months ago... Several samples have been sent from time to time to the University for further examination, and Professor Bragg has applied electrical tests with positive results... The present discoveries should not cause undue excitement because, so far, they do not warrant immediate commercial notice'.

Mawson's work at Sydney University with Thomas Laby referred to preliminary studies undertaken while they were both young demonstrators in chemistry.[23] Mawson now took up the new matter and, at a meeting of the Royal Society of South Australia on 4 September, both he and William (on behalf of Radcliffe) read papers, Mawson on the Olary carnotite find and William on radium in the

[22] *Register*, 5 May 1906, p. 7; for Douglas Mawson see P Ayres, *Mawson: A Life* (Melbourne: MUP, 1999).

[23] D Mawson and T H Laby, 'Preliminary observations on radio-activity and the occurrence of radium in Australian minerals', *Journal and Proceedings of the Royal Society of New South Wales*, 1904, 38:382–9.

Moonta copper mines.[24] Mawson reported on the visual geological observations he had made at Olary and set them in a global context. He was supported by a preliminary analysis of two of his mineral specimens, undertaken in the university's chemistry department. He named the particular locality of the find: 'As this spot has so far remained unnamed, "Radium Hill" seems appropriate'.[25]

Sydney Radcliffe had been examining the ore deposits at and around Moonta for a year and thought it now 'desirable to give some preliminary account of them'.[26] He had discovered that some were active, and he had analysed them chemically in an attempt to isolate the active constituents. One of his resulting samples 'was forwarded to Professor Bragg in March last, and he made its activity to be about nine times that of U_3O_8...I should like to take this opportunity of expressing my thanks to Professor Bragg for the interest he has taken in this research throughout, and for the time and trouble he has expended in making measurements on active products...[a sample] was forwarded to Professor Bragg for examination, and his measurements...indicated that the major part of the activity was due to radium'.[27] However, 'In considering this question of concentration of activity, the extremely small amount of radium present in the ore—roughly one part in twenty million—must be taken into account'.[28]

Earlier, in August, William had responded to an invitation to discuss the new discoveries at Moonta itself, speaking in the Moonta Institute.[29] However, the Moonta site proved unproductive and there was no immediate success at Radium Hill, despite a flurry of activity on the Adelaide Stock Exchange.[30] There was one report in 1909 of the successful extraction of radium salts from Radium Hill uranium ore—by Sydney Radcliffe, now at the Bairnsdale School of Mines in Victoria—and of its sale overseas,[31] but there is no other evidence of significant success. The onset of The Great War in August 1914 terminated both government and private-enterprise activity. It resumed in the 1920s, and subsequently there were productive activities at Radium Hill and elsewhere.[32]

[24] *Register*, 5 September 1906, p. 9.

[25] D. Mawson, 'On certain new mineral species associated with carnotite in the radio-active ore body near Olary', *Transactions of the Royal Society of South Australia*, 1906, 30:188–93; E.H. Rennie and W.T. Cooke, 'Preliminary analytical notes on the minerals described in the preceding paper', ibid, p. 193.

[26] S. Radcliffe, 'Radium at Moonta mines, South Australia', ibid., pp. 199–204.

[27] Ibid., pp. 201–2.

[28] Ibid., p. 204.

[29] *The Yorke Peninsula Advertiser*, 20 July 1906, p. 2, and 24 August 1906, p. 2; Bragg knew Moonta well, he had been an examiner in 'Magnetism and Electricity' at its School of Mines in 1895 and 1896 (see *Moonta School of Mines, South Australia, Annual Report, 1895* (Adelaide: Moonta School of Mines, 1896), p. 4; ibid., 1896 (1897), p. 4).

[30] B O'Neil, *In Search of Mineral Wealth: The South Australian Geological Survey and Department of Mines to 1944* (Adelaide: S A Department of Mines and Energy, 1982), pp. 152–4, 166–7.

[31] L. Prendergast (ed.), *Scatter the Light: A Century of Technical Education in Bairnsdale, 1890–1990* (Bairnsdale: the editor, 1990), pp. 1–2.

[32] See n. 30, pp. 274–7, 310; also J Cooper (ed.), *Records and Reminiscences* (Adelaide: University of Adelaide Department of Geology and Geophysics, 2000), ch. 3, and G M Mud, 'The legacy of early uranium efforts in Australia, 1906–1945: from Radium Hill to the atomic bomb and today', *Historical Records of Australian Science*, 2005, 16:169–98.

Although not involved directly in these later activities, William maintained contact with Douglas Mawson for many years. Mawson wrote to him (in Leeds) from London in May 1911, bemoaning the lack of financial support for his forthcoming Australasian Antarctic Expedition, and he also quoted William's strong support in an unsuccessful appeal to Thomas Barr Smith for £10,000 to £15,000 sterling for the same purpose.[33] On 1 November 1913 Mawson wrote to William from 'Winter Quarters, Commonwealth Bay, Adelie Land' in Antarctica to say:[34]

> I cannot help writing you just a line to let you know that even this rigorous and unparalleled climate does not leave us unmindful of those who have helped either myself or this expedition in the past...
>
> Our misfortune...was in having such a rigid land as Adelie Land for our main base to work upon; this has allowed of the work of this base being accomplished only by the greatest effort. The accident to Ninnis and its consequences are the fortunes of war; a thing which is always liable to happen; a risk which is part of the game and inseparable from pioneering in ice-covered land. Of course I greatly regret this extra year's detention...
>
> You will be glad to hear that the Adelaide men *all* did well:...Madigan, Kennedy and Moyes from the University, and Hodgeman from the Public Works Dept. It will be a very busy time for me on return until a popular account of the expedition has been put through the press...
>
> Please give my most kind regards to Mrs Bragg and accept the same for yourself.

It would appear that, after the war, Mawson also sent William a copy of his popular account of the 1911–14 Antarctic expedition, *The Home of the Blizzard*, for William sent Mawson a very warm letter of thanks in which he said: 'I never thought you would be so kind as to give me that magnificent book of yours. I can only say that it will be treasured always. It is such a wonderful story: I can hardly believe you were able to get through those experiences...The book is greatly admired by everyone who comes. Thanks ever so much. I am just delighted to have it...Give my love, will you, to the old gang at the University [of Adelaide]'.[35] Early in 1933, when William was at the Royal Institution in London, he arranged for Mawson to give a Friday Evening Discourse on 'The new polar province [Antarctica]'.[36]

[33] Letter, D Mawson to W H Bragg, undated but probably May 1911, Mawson Collection, South Australian Museum, Adelaide, AAE11; draft cable, D Mawson to T Barr Smith, undated, ibid.

[34] Letter D. Mawson to W.H. Bragg, 1 November 1913, ibid.

[35] D Mawson, *The Home of the Blizzard* (London: Heinemann, 1915); letter W H Bragg to D Mawson, 3 May 1919, Mawson Collection, South Australian Museum, Adelaide.

[36] Letter W H Bragg to D Mawson, 21 February 1933, ibid.; for the lecture (on 3 March 1933) see S K Runcorn (ed.), *The Royal Institution Library of Science: Earth Sciences* (London: Applied Science, 1971), vol. 3, pp. 261–76; for Mawson's 1932–33 visit to England see Ayres, n. 22, pp. 204–6.

Despite all these other activities, William's major focus was now very firmly on his physics research. The results seemed to require radical changes to late nineteenth-century classical physics and its complacency, and William continued to speak publicly about his own research and its wider implications. In April 1906 he spoke to the Astronomical Society on the topic 'The atom, a miniature planetary system', in June he presented two extension lectures in Adelaide on radioactivity, and in August 1907 he addressed the university's Scientific Society on the subject 'Recent advances in the science of radio-activity'. The first lecture predates by about five years Ernest Rutherford's definitive publication on this subject,[37] and it is therefore of particular interest:[38]

> In the course of an able and much appreciated address the lecturer pointed out many analogies and parallelisms between the vast systems of the universe and those systems on the extremely small scale which are called atoms. The subject was reviewed from the standpoint of the new science of radio-activity. The electrons which compose atomic systems are in size and distance proportionate with members of the planetary systems. As the planets revolve around their primary in one plane, so also it is most likely that the members of atomic systems move in a similar manner. Again, it would be possible for one planetary system to pass through another without necessarily involving any collision or cataclysm, and from experiments made at the Adelaide University it has been shown that atomic systems can pass through one another; e.g. the helium atom goes through everything, and though some disturbance is caused by its passage, it is not usually so violent as to do any damage.

There is an extensive literature on the emergence of the nuclear model of the atom, in which William was a noted player, but this lecture has not been noticed previously. It is an early public indication of William's belief that his research offered the prospect of shedding light on the inner structure of the atom.[39] The source of William's atomic model was not Nagaoka's 1903–04 planetary atom—with electrons circulating around a positive centre—for when in 1911 William alerted Rutherford to its existence, it had just been brought to William's attention by his Leeds colleague, Norman Campbell.[40] Indeed, as Heilbron has noted, at this time both Bragg and Rutherford subscribed to J J Thomson's view

[37] E. Rutherford, 'The scattering of α and β particles by matter and the structure of the atom', *Philosophical Magazine*, 1911, 21:669–88.

[38] *Register*, 13 April 1906, p. 6.

[39] He had indicated as much privately, however, in the notebook recording his initial literature search (RI MS WHB 12/1), and in letters to Thomson on 10 August 1904 (CUL Add MS 7654 B70) and to Rutherford on 16 July 1905 (CUL RC B358).

[40] J L Heilbron, 'The scattering of α and β particles and Rutherford's atom', *Archive for History of Exact Science*, 1967–68, 4:247–307, 257; see also J L Heilbron and T S Kuhn, 'The genesis of the Bohr atom', *Historical Studies in the Physical Science*, 1969, 1:211–90, 241–2; these and other references (for example, L Badash, 'Nagaoka to Rutherford, 22 February 1911', *Physics Today*, 1967, 20:55–60; P M Heimann, "Rutherford, Nagaoka and the nuclear atom", *Annals of Science*, 1967, 23:299–303) make it clear that the Nagaoka model, which was not widely known, was rejected because it was mechanically unstable and that the Rutherford atom appeared to suffer from the same defect until Bohr introduced the quantum hypothesis.

that the atom was a vast 'hive' of electrons, with the huge numbers required to make up the atomic mass, circulating freely through a diaphanous 'jelly' of positive charge sufficient to make the atom neutral overall. Under the attractive and repulsive forces involved, Thomson had shown, the electrons arrange themselves in concentric rings or spheres.[41] In August 1907 William again spoke about the constitution of the atom, this time to another meeting of the Scientific Society, of which he was President.[42]

William's two extension lectures in June 1906 were a hurried response to the discovery of radioactive ores in South Australia for, alone of all his extension lectures in Adelaide, there are no advertising leaflets for them in the university archives. Characteristically, William responded generously to the need for public information, and the newspapers covered his lectures in detail. He chose to focus on advances in the science of radioactivity rather than on practical matters such as discovery, testing, and extraction. The first lecture began by noting the worldwide interest in radioactivity, which arose because 'the new science revealed wonders hitherto unsurpassed and also because it dealt with a series of phenomena not previously touched by scientific discovery... The new science was distinguished from the old in that it dealt with the processes occurring within the atom itself'.[43] The second lecture was devoted to a discussion of 'the method by which the life of radio-active substances was measured'.[44] Radium passed through a series of changes until it decayed to lead but, since radium had a life of only a few thousand years whereas uranium's was millions of years, it seemed likely that 'uranium was the original source from which radium was derived'.[45] Recent studies of local radioactive ores confirmed this suggestion. William commented on the radical implication of radioactive processes for atomic transmutation, when it had been believed for two millennia that atoms were indivisible and unchangeable,[46] and he concluded by referring to recent work by Richard Kleeman and Norman Campbell at the Cavendish Laboratory.

The January 1907 AAAS meeting was held in Adelaide. William had been Honorary Secretary and Treasurer for the South Australian local committee in the 1880s, and from 1893–94 until 1905–06 he and Edward Rennie were joint local Secretary/Treasurers.[47] For the 1907 Adelaide congress, however, it was necessary to spread the load more widely: Walter Howchin, an Adelaide University geologist, and John Madsen were appointed joint Honorary

[41] Ibid., p. 300; see also J L Heilbron, 'J J Thomson and the Bohr atom', *Physics Today*, 1977, 30:23–30.

[42] *Register*, 13 August 1907, p. 8.

[43] *Registe*, 19 June 1906, p. 7.

[44] *Registe*, 27 June 1906, p. 7.

[45] Ibid.

[46] Ibid.

[47] *Australian and New Zealand Association for the Advancement of Science, S A Inc.*, State Library of South Australia. Adelaide, SRG 48, minute books, congress papers, correspondence, balance sheets, press cuttings, etc.

General Secretaries and William Bragg Honorary Local Treasurer.[48] Bragg and Kleeman each presented papers to Section A—Astronomy, Mathematics and Physics, although whether Kleeman was in Adelaide on holiday from Cambridge and presented his paper in person or his mentor presented it on his behalf is unknown. The congress *Report* contains only abstracts of the two papers.[49] William referred to a paper from Prague,[50] and reported that he had undertaken new experiments that disagreed with it.[51] Kleeman's paper referred to his earliest work in Cambridge and concerned the recombination of electrons with their parent ions.[52]

Professor Bragg was also a Vice-President of Section A and a member of the publication, recommendation, and reception committees for the Adelaide meeting. Of most social interest is a photograph with the hand-written annotation: 'Australian Association for the Advancement of Science, Adelaide Meeting 1907', which shows a very large group of well-dressed men and women: many of the men in frock-coats and top hats, the ladies in full-length summer dresses and large, elaborate hats. The garden background suggests that this photograph was taken in the grounds of Government House.[53] William sits relaxed and smiling on the lawn, clasping a long cane, but his wife is not nearby and cannot be identified elsewhere in the photograph. Gwendoline almost certainly excused herself, as she had done for the January 1890 Melbourne meeting, and for the same reason. William Lawrence Bragg was born on 31 March 1890, and on this occasion Gwendoline was even further advanced in her third pregnancy. On 26 February 1907 she and William welcomed the safe arrival of a daughter, whom they named Gwendolen Mary.

'Gwendy' would be a beloved sister to her brothers, her mother's helpmate, and later her father's companion, Royal Institution hostess, and affectionate biographer. When William's fellowship of the Royal Society was announced early in March 1907, the Chief Inspector of Schools wrote to him with con gratulations on both his FRS and the birth of his daughter, saying 'though you have done a great deal for science, you have done much more for men and women'.[54] This is generous contemporary evidence of the contribution William had made to South Australian society and to education in particular. In a leading editorial, the *Register* surveyed the 1907 meeting of the British Association for the Advancement of Science and noted: 'To the specialization which has become such a pronounced feature of modern scientific research

[48] W Howchin (ed.), *Report of the Eleventh Meeting of the Australasian Association for the Advancement of Science, held at Adelaide*, 1907 (Adelaide: AAAS, 1908), pp. ix–x.

[49] Ibid., pp. 318 and 319 respectively.

[50] B Kucera and B Masek, 'Uber die Strahlung des Radiotellurs II', *Physikalische Zeitschrift*, 1906, 7:630–40.

[51] See chapter 11; other results of Kucera and Masek are in general agreement with those of Bragg and Kleeman.

[52] See chapter 11; also n. 48, p. 319.

[53] Special Collections, Barr Smith Library, University of Adelaide.

[54] Letter M M Maughan to W H Bragg, n.d. [March 1907], Bragg (Adrian) papers; for Maughan see Thiele, n. 4, *passim*.

may be ascribed most of the wonderful discoveries upon which our present conceptions of the Cosmos are based'.[55] One hundred years ago and lacking nearly all the modern modes of communication, a reader of the *Register* could nevertheless be well informed about the latest developments in science, in Britain and around the world.

X-rays and γ-rays

Towards the conclusion of his major research project on radioactivity, William must have been pleased to read the acknowledgement of his work and the appreciation of its value in the local press, as exemplified by this extract from *The Australasian* and the *Register*: 'Professor Bragg has long been known in Australian scientific circles as a highly distinguished teacher and an able worker in the domain of electrical theory; his investigations—carried out during the past five years in connection with the most obscure problems of the new science of radio-activity—have now brought him a world-wide reputation, and earned for him the highest scientific distinction that a British subject can attain [FRS]... All Australian workers will rejoice in the honour which has come to a colleague whose ability in research only equals the modesty and self-repression with which the results are put forward'.[56] William correctly foresaw, however, that these studies had become less productive and he turned to other possible topics. His thoughts went back to his 1904 Dunedin review of the literature on the ionization of gases.[57] It had become reasonably clear since then, in Britain if less so in Germany, that cathode rays in gas discharge tubes, beta particles from radioactive decay, J J Thomson's corpuscles, the so-called δ-rays, and electrons were all synonyms for the tiny, negatively-charged particles that were the basic constituents of all atoms; and that canal rays were the positive ions, of varying sizes, left after an atom lost one or more electrons, with alpha particles probably positively-charged ions of a light-weight atom such as helium.[58]

Visible, infrared, and ultraviolet light were different. They were thought to be waves, of differing frequencies and wavelengths, which propagated through the ether (aether) that filled space. The fact that these lights could be reflected and refracted and suffer interference, diffraction, and polarisation implied that

[55] *Register*, 21 September 1907, p. 8.

[56] *Register*, 4 May 1907, p. 9.

[57] W H Bragg, 'Some recent advances in the theory of the ionization of gases', in G M Thomson (ed.), *Report of the Tenth Meeting of the Australasian Association for the Advancement of Science, held in Dunedin, 1904*, (Werllington: AAAS, 1905), pp. 47–77.

[58] Three very useful accounts of the topics and times covered in this section are: B Wheaton, *The Tiger and the Shark: Empirical Roots of Wave-Particle Dualism* (Cambridge: CUP, 1983); R H Stuewer, 'William H Bragg's corpuscular theory of X-rays and γ-rays', *British Journal for the History of Science*, 1971, 5:258–81; M C Malley, 'From hyperphosphorescence to nuclear decay: A history of the early years of radioactivity', Ph.D. thesis, University of California, Berkley, 1976 (University Microfilms, Ann Arbor, USA, 1984), particularly chapter V.

they were periodic, transverse vibrations in the ether, quite unlike the particles. The nature of X-rays and the recently recognised gamma (γ) rays was yet a different matter apparently. Initial interest focused on X-rays, and there were numerous suggestions as to their nature. Many scientists assumed that they were high-frequency electromagnetic (light) waves, but other suggestions included material particles and more. William remembered Threfall's early unsuccessful attempt in Sydney to solve this problem.[59] By the early twentieth century most researchers had adopted a form of extreme-frequency ether waves; that is, X-rays were either transverse vibrations of the ether (like light), longitudinal ether waves, or ether impulses. That they were waves and not particles was supported by the facts that they were undeflected by electric and magnetic fields. A majority favoured the third option and thought that X-rays were pulse-like; that is, impulses that propagated spherically out from their source, not periodically but limited in space and time like the skin of a stationary but expanding balloon. Only the huge number of pulses gave the appearance of continuity to an X-ray beam. There were several suggestions as to the shape of this pulse but the details are not important here; they can be found in Wheaton's account.[60]

There were significant problems with all the proposed models, however, including the impulse hypothesis. X-rays suffered no measurable reflection, refraction, interference, or polarisation; it was the fact that they produced photographs that persuaded most to believe that a new form of light had been discovered. Then, in 1904 the Englishman Charles Barkla showed that X-rays could be polarised, confirming their transverse wave property, and in 1905 their velocity was found to be close to that of light. For the few disbelievers, X-ray pulses also had some characteristics akin to particles: localised in time, they collided with atoms like particles. 'Part of the early appeal of the impulse hypothesis', Wheaton noted, 'was this ability, chameleon-like, to express characteristics of both particles and waves'.[61]

Once the alpha and beta particles emerging from radioactive decay had been separated—for example by their opposite deflections in a magnetic field—it became clear that there was a third, highly-penetrating, and undeviated component, the so-called gamma-rays. Curie thought they were high-energy X-rays, produced in the radioactive material itself by the escaping beta particles, and a little later Rutherford agreed, since the two always seemed

[59] R Threlfall and J A Pollock, 'On some experiments with Röntgen's radiation', *Philosophical Magazine*, 1896, 42:453–63.

[60] Wheaton, n. 58, chs 1–3. A transverse vibration is one in which the ether vibrates at right angles to the direction of the wave (as when a wave travels along a rope); refraction is the bending of light as it passes through the interface between two transparent mediums (so a stick standing in water appears bent); interference occurs when two light beams meet, so that the crests and troughs of one add to or cancel out the crests and troughs of the other (used by some radio stations to beam their signal in particular directions); diffraction is the process by which light bends (a little) around the edge of any opaque object (important in camera design because of the aperture); and polarisation occurs when the vibrations of a light wave are confined to a single plane (the brightness of sunlight is reduced by polarising sunglasses).

[61] Ibid., quotation p. 16.

to accompany each other and in proportional amounts. By 1905 a series of supportive results had confirmed the view that gamma rays were transverse electromagnetic impulses in the ether; that is, high-energy X-rays.[62]

Just when some order seemed to be emerging in this exciting but troubled field, however, two problems emerged that quickly acquired major significance; Wheaton called them 'the paradox of quantity' and 'the paradox of quality'.[63] When X-rays or γ-rays interacted with an atom, electrons and secondary X-rays were emitted, and this provided new uncertainties. First, the rays ionised only a small fraction of the gas molecules over which they passed, when expanding pulses should have affected the whole 'quantity' of molecules equally. Second, it was assumed that spreading waves would impart only a small fraction of their energy ('quality') to the electrons they ionised, whereas such electrons received *all* the energy of the pulses as if the pulses were neat packages and not spreading waves. Indeed, the energy of the primary electrons that had initially produced the X-rays, and which had been transferred to the spreading waves, now suddenly reappeared undiminished in the secondary electrons. One suggested solution to both these paradoxes became widely accepted, that 'photoelectrons' ejected from atoms in this way already possessed a large kinetic energy in the atoms before they were released; the X- or γ-rays only 'triggered' their escape. Furthermore, only a very small number of atoms had electrons of this energy, explaining the small number of photoelectrons.[64]

William had touched on this problem in his 1904 Dunedin address, when he had noted that waves and particles are absorbed differently, particles in proportion to the number of electrons they encountered but waves less readily because they interacted with atoms as a whole and not with individual electrons. Studying the existing literature, he came to the conclusion that gamma rays acted on individual electrons in atoms and must therefore be localised in space. Since they suffered no electric or magnetic deflection and had great penetrability, he was reluctant to call them particles, but he did say, 'if they are waves, the only possible supposition would seem to be that they are waves so small as to be unable to act [on] a whole molecule or atom at once'.[65] Now, three years later, his research had been recognised worldwide and he possessed new confidence, both personally and professionally. Isolated from the persuasion of his European peers, William felt able to trust his own judgement and go his own way. Furthermore, he now had three 'lads' to assist in Rogers' workshop, where there was 'a great deal of work to be done now: some of it due to the natural increase of the University classes, some due to the fact that several problems of research are being worked out in the laboratory'.[66] He would courageously question the impulse model and its triggering hypothesis and propose that γ-rays and hard (high-energy) X-rays were particles.

[62] Ibid.
[63] Ibid., ch. 4.
[64] Ibid.
[65] W H Bragg, n. 57, p. 77; see also Wheaton, n. 58, p. 83.
[66] Letter W H Bragg to Finance Committee, 15 February 1907, UAA, S200, docket 118/1907.

Willie, a senior secondary-school student and then a university undergraduate able to understand his father's α-particle and γ-ray research, later recalled this time clearly: '[My father] tried out his ideas and his papers on me. I remember his telling me of a great new idea which had just come to him, that the γ-rays were neutral particles, as we were mounting into the horse-tram to take us to the Observatory. His experiments were carried out in the basement of the main University block. His instrument maker Rogers was a real genius, and the α-ray apparatus was a gem. My father aimed at a high standard of perfection in design and construction... He even tamed that intractable instrument, the quadrant electrometer, which he used to measure ionization'.[67]

The reader may wonder why Albert Einstein's 1905 suggestion that light sometimes behaved like a particle—his so-called 'light quantum hypothesis'—has not been mentioned. It was designed, in part, to explain how radiation could liberate photoelectrons from solids, and it would appear ideally suited to explain the paradoxes of quantity and quality. In fact, Einstein's paper referred to ultraviolet light, and the intimate connection between light and the new rays was not at all clear. Furthermore, Einstein was an unknown patent clerk at the time, and nearly all the physicists who read his suggestion regarded it as untenable. William, unable to read German, explained later that, 'it is easy to miss a single reference when one is in a very isolated laboratory'.[68] Indeed, even had William read it, it is doubtful he would have found it relevant because 'I thought of the X-ray and γ-ray problems as distinct from that of light'.[69]

Nevertheless, in his first paper on the subject, William was unknowingly close to Einstein's spirit. As had now become his custom, William first presented his new thoughts to meetings of the Royal Society of South Australia; also characteristically, they were in a carefully considered form that contained all the major ingredients of the hypothesis that was to capture the attention of the scientific world. His model was described to successive monthly meetings of the Society, on 7 May and 4 June 1907, and subsequently published in a combined form in the *Philosophical Magazine* in Britain and in the annual report of the Smithsonian Institution in the USA.[70] William described his 'neutral-pair hypothesis' in the following terms:[71]

> Radioactive substances emit both positive and negative particles. It does not seem at all out of place to consider the possibility of the emission of neutral particles, such as, for example, a pair consisting of one α or positive particle and one β or negative particle... The α particle loses speed as it penetrates atoms [due to] the field which is about it... but if a β particle

[67] W L Bragg, Autobiographical notes, pp. 19–20.

[68] W H Bragg, 'Photo-electricity', *Proceedings of the Royal Institution*, 1928, 25:338–48, 341.

[69] Ibid., p. 343; R H Stuewer, n. 58, p. 258–9.

[70] W H Bragg, 'A comparison of some forms of electric radiation', *Transactions of the Royal Society of South Australia*, 1907, 31:79–93; id., 'The nature of Röntgen rays', ibid., 1907, 31:94–8; W H Bragg, 'On the properties and natures of various electric radiations', *Philosophical Magazine*, 1907, 14:429–49, and *Smithsonian Institution, Annual Report*, 1907, pp. 195–214.

[71] W H Bragg, ibid. (*Phil.* Mag.), pp. 440–8; the doubly-charged nature of the α-particle was not clear at this time.

> is associated with the α particle the chief cause of the stopping of the α particle has been removed. The penetrating power of a pair might be very great indeed, and its ionizing power correspondingly reduced... Such a pair would be incapable of deflection by magnetic or electric fields, and would show no refraction. It is conceivable that it might show a one-sided or polarization effect, for if it were ejected from a rotating atom it would itself possess an axis of rotation.
>
> When X-rays were first investigated, and again when γ rays were discovered, it was often suggested, in each case, that the radiation might consist of material particles... But it was always felt that the difficulty of accounting for the great penetration of these radiations was insuperable. It seems now that this difficulty was quite exaggerated, and even imaginary.
>
> Assuming, then, that the neutral pair has [these characteristics] it does so far conform to the properties of the γ ray. And further,... it may at last suffer some violent encounter which will resolve it into a positive and a negative, an α and a β particle. Of these the β particle would be the one possessed of much the greater velocity, and would appear as a secondary ray... If the gradual disappearance of a stream of γ radiation were caused by collision in this way, the number disappearing in any unit of length of the course would be proportional to the total number in the stream, so that an exponential law would result. It appears, therefore, that all the known properties of the γ ray are satisfied on the hypothesis that they consist of neutral pairs...
>
> If the X-ray is an aether pulse, it is difficult to understand why the spreading pulse affects so few of the atoms passed over... why the high-speed secondary cathode rays are ejected with a velocity which is independent of the intensity of the pulse, and why it should be able to exercise ionizing powers when its energy is distributed over so wide a surface as that of a sphere of say 10 or 20 feet radius...
>
> To sum up, it is clear that a stream of X rays contains some aether pulses, but it is not easy to explain all the properties of X rays on the aether-pulse theory. The explanations are easier if the rays are supposed to consist mainly of neutral pairs; and the existence of such pairs is not improbable a *priori*.

The power and plausibility of William's case was arresting to his contemporaries, although they still saw the issue in terms of either the pulse theory or the neutral-pair hypothesis. Even today, William's desire to retain both models in the one entity seems strange, given that the models appear so contradictory. Yet in 1907 the evidence already spoke clearly in favour of both models, and William responded accordingly. He saw, before his contemporaries, that some combination of the two views might be necessary.

William searched for an experimental method to demonstrate the superiority of his neutral-pair hypothesis, and he decided that a study of the secondary radiation induced in various materials by primary rays from a radioactive source offered the best opportunity. He therefore invited Madsen to collaborate with him and initially encouraged him to investigate the experimental technique to be used. It was not straightforward, for previous use of small ionization chambers to measure the energy of the primary and secondary beams in such experiments had been flawed, first because many of the β particles escaped the small measuring chamber, and second because the results were heavily dependent on the material giving rise to the secondary rays as well as the nature of the primary and secondary radiations themselves. Bragg and Madsen reported their findings to the Royal Society of South Australia and to the *Philosophical Magazine*.[72]

Gripped by the excitement of a new research challenge they pushed on vigorously, and two reports of their initial findings were written in December and received at the offices of *Nature* and of the Royal Society of South Australia in January the following year (1908). They had been further sparked into action by the appearance of a letter in *Nature* from Charles Barkla, a lecturer in physics at the University of Liverpool and an expert on X-rays since his scholarship days in the Cavendish Laboratory.[73] Barkla was convinced by the ether-pulse theory and, in seeking to refute William's particle suggestion, thought he had found a crucial experiment in the form of the angular variation of the scattered X-radiation with respect to the direction of the primary beam. J J Thomson had developed a formula for X-ray scattering on the assumption that the incident beam caused the atomic electrons to vibrate and to re-radiate, and Barkla used this ether-pulse model to predict that the ratio of the resulting 'Thomson scattered' radiation in the direction of the beam and at right angles to it would be two to one. The neutral-pair model gave a different distribution, with a ratio of infinity to one. Barkla then made such a measurement under experimental conditions that would have yielded ratios of 1.9 to 1 and 8 to 1 under the two models. His result of 1.6 to 1 was, he claimed, 'conclusive evidence in favour of the ether pulse theory'.[74]

William penned a vigorous reply. Barkla, he said, 'makes the assumption that the probable direction of motion of a neutral pair on emergence from an atom with which it has been entangled is independent of its original direction of motion... [and] there is no justification for this assumption. It does not even appear to be probable... With the assistance of Mr J P V Madsen, of this University, I have been comparing the secondary radiations issuing from the two sides of a plate through which γ rays are passing. On the ether pulse theory there should be complete symmetry... [whereas] if the γ rays are material, it is quite possible, though not necessary, that the secondary radiations on the two sides of the plate should be different... As a matter of fact, there is the most

[72] W H Bragg and J P V Madsen, 'The quality of the secondary ionization due to β rays', *Transactions of the Royal Society of South Australia*, 1907, 31:300–4; id., *Philosophical Magazine*, 1908, 16:692–7.

[73] P Forman, 'Barkla, Charles Glover', in C C Gillispie (ed.), *Dictionary of Scientific Biography* (New York: Scribner's Sons, 1970), vol. I, pp. 456–9; H S Allen, 'Charles Glover Barkla, 1877–1944', *Obituary Notices of Fellows of the Royal Society of London*, 1947, 5:341–66; R J Stephenson, 'The scientific career of Charles Glover Barkla', *American Journal of Physics*, 1967, 35:140–52; papers by Barkla and Sadler in *Philosophical Magazine*, 1907–11.

[74] C G Barkla, 'The nature of X-rays', *Nature*, 1907, 76:661–2.

remarkable want of symmetry, and this is fatal to the ether pulse theory of the γ rays... the experimental proof of the material nature of the γ rays carries with it, almost surely, a corresponding proof as regards the X-rays. The points of similarity are too numerous for it to be otherwise'.[75]

Working over the precious Christmas/New Year period to elaborate this letter for the local Royal Society and the *Philosophical Magazine*, William wrote: 'The object of this paper is to give a preliminary account of an investigation which appears to us to show that [γ rays] are material in nature. Secondary radiation, which is excited in an atom by a passing wave or pulse, must be distributed symmetrically... This is a well-recognized principle... Supposing, therefore, a pencil of γ rays to pass normally through a plate so thin that its absorption may be neglected, the secondary radiation should be exactly the same on the two sides of the plate—in amount, in quality and in distribution'.[76]

The apparatus was described and pictured in diagrams drawn and labelled in William's neat and clear hand (see Figure 13.1). A deep ionization chamber was closed, top and bottom, by two plates, with A and A' alike and B and B' alike but of different material, and a pencil of gamma rays passed vertically down the axis of the chamber, the resulting ionization being measured by an electrometer as before. They reported a change in ionization when A and B were inverted, as also when A' and B' were interchanged, and a long discussion followed. The use of four plates now seems unnecessarily complex, when a single plate, alternatively top and bottom of the ionization chamber (the other side covered with a wire mesh or thin foil, for example), should have been adequate to compare the 'emergence' radiation, emerging from the under side of the plate, with the 'incidence' radiation emerging from its top side. The end result was unmistakable, however: 'On the aether-pulse theory we ought to find perfect symmetry in the secondary radiations from the two sides of the plate; but experiment shows nothing of the kind'.[77] Finally, William recognised that the mass of the positive component of his neutral pair posed a problem for his model, since the radiations were thought to have a very high velocity. 'Probably it becomes necessary to consider it as small compared to the mass of the negative', he concluded in the paper printed by the Royal Society of South Australia, but the sentence was omitted in later versions.[78]

Barkla responded at once to William's January letter and focussed on a weakness in William's case; namely, his explanation of the polarisation of X-rays on the neutral-pair model, which required very specific assumptions regarding the rotation of the pair. Barkla again stressed the agreement between the ether-pulse theory and the measured angular variation of the secondary

[75] W H Bragg, 'The nature of γ and X-rays', *Nature*, 1908, 77:270–1.

[76] W H Bragg and J P V Madsen, 'An experimental investigation of the nature of the γ rays—No. 1', *Transactions of the Royal Society of South Australia*, 1908, 32:1–10; id., 'An experimental investigation of the nature of the γ rays', *Philosophical Magazine*, 1908, 15:663–75; also published in *Proceedings of the Physical Society* of London, 1907–09, 21:261–75, and in *Chemical News*, 1908, XCVII:162–5.

[77] Ibid. (*Phil. Mag.*), p. 670.

[78] Bragg and Madsen, n. 76 (TRSSA), p. 8.

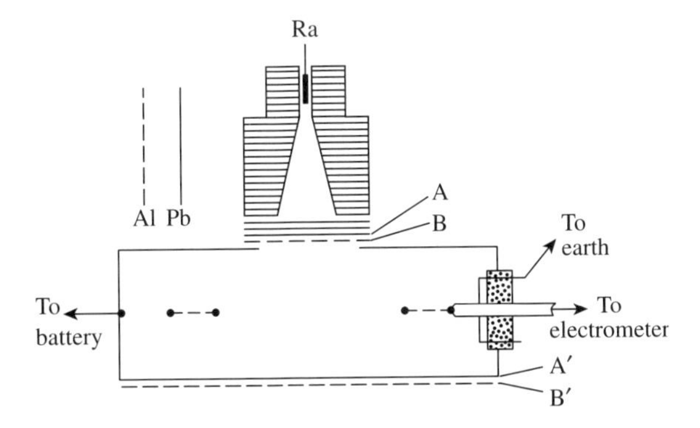

Fig. 13.1 The cylindrical chamber used by Bragg and Madsen for their studies of the nature of X- and γ-rays; also see Figure 11.4. (Redrawn from *Philosophical Magazine*, 1908, 15:665 and 16:921. Courtesy: Taylor and Francis Ltd., www.informaworld.com.)

radiation resulting from the interaction of soft (low-energy) X-rays with matter, and he referred to his fuller exploration of the topic as reported in a recent paper in the *Philosophical Magazine*. Finally, Barkla made an important observation: 'My argument has not been concerned with γ rays, but with the type of radiation with which I am experimentally more familiar—X-rays of ordinary penetrating power'.[79]

William then replied, and with uncharacteristic spice: 'I have to thank him for the admission that his experiments are not so contrary to the neutral pair theory as he had at first supposed. Mr Barkla still concludes, however, in favour of the ether pulse theory...[but] it is no compliment to the ether pulse theory to describe such incomplete successes as "absolutely conclusive evidence"...He wishes to avoid arguments founded on an experimental study of the γ rays. Evidence of this sort cannot be avoided by resolutely facing the other way...If I admit the existence of ether pulses, I do not thereby weaken my contention that the most important and effective part of γ and X-ray radiation is material. We know that ether pulses exist; it does not follow that they do everything...there is a danger that the *post hoc* has been confused with the *propter hoc*'.[80]

Barkla again replied quickly, and it is clear that the dispute had now become sour, both between the two physical models and between the two proponents. Barkla wrote, 'Prof. Bragg in a recent letter (NATURE, April 16) credits me with the admission that the experiments I made on the intensity of secondary (scattered) X-rays are not so contrary to the neutral pair theory as I at first supposed. Will you permit me to correct this by saying that all the evidence I have obtained has verified the ether pulse theory in a more striking way than I

[79] C G Barkla, 'The nature of Röntgen rays', *Nature*, 1908, 77:319–20; id., 'Note on X-rays and scattered X-rays', *Philosophical Magazine*, 1908, 15:288–96.

[80] W H Bragg, 'The nature of γ and X-rays', *Nature*, 1908, 77:560.

ever anticipated, and I cannot think of a single experimental result obtained in researches on secondary X-rays which gives any support to his theory?... As Prof. Bragg is apparently not convinced, I venture to recommend the consideration of the following evidence obtained in investigating secondary X-rays'.[81] Barkla then listed nine items and the evidence was persuasive, beginning with the polarisation of X-rays and the experimental angular variation of secondary X-rays, and concluding with new insights into the nature of secondary X radiation. This 'homogeneous radiation [was] characteristic of the element emitting it, and independent of the penetrating power of the primary radiation producing it', he said, a foretaste of his discovery of X-ray spectral lines characteristic of each element and the consequent award of the 1917 Nobel Prize in Physics.[82] He concluded the letter with a conciliatory and perceptive paragraph: 'Regarding the nature of γ rays, or even of very penetrating X-rays, the direct evidence is much less conclusive'.[83]

In the meantime an American physicist, Charlton Cooksey, had repeated William's experiment with pairs of metal plates bordering an ionization chamber but using X-rays rather than γ-rays, and 'found that in every case the ionization due to the 'emergence' secondary kathode rays was greater than that due to the 'incidence' rays', in agreement with William's results.[84] However, Cooksey found that, 'I cannot agree with Prof. Bragg that the evidence is conclusive that X-rays and γ rays must consist of some type of radiation other than electromagnetic pulses'; for, like the neutral pair, 'an electromagnetic pulse possesses momentum also in the direction of propagation [and] it is not unreasonable to suppose that an ether pulse could contribute some of its momentum to the secondary kathode particles, causing them to go more in the direction of propagation of the primary than in any other'.[85] This is a very important point, missed by many of Cooksey's contemporaries for two decades thereafter.[86]

William's newfound confidence found less space for Frederick Soddy's role as advocate. In one of his last letters to William, Soddy wrote: 'I read your two interesting communications on the X-rays etc. with zest & am glad you are going into the question... I don't agree altogether with the view of the X-rays but it is, as you say, very instructive to put the other side'.[87] In addition, William now had a direct route to rapid publication through the pages of *Nature*, which he again used to announce his latest experimental results. He wrote to its editor and read the related paper to the Royal Society of South Australia on the same

[81] C G Barkla, 'The nature of X-rays', *Nature*, 1908, 78:7.
[82] See n. 73.
[83] Barkla, n. 81.
[84] C D Cooksey, 'The nature of γ and X-rays', *Nature*, 1908, 77:509–10.
[85] Ibid., p. 510.
[86] J G Jenkin et al., 'The development of angle-resolved photoelectron spectroscopy, 1900–1960', *Journal of Electron Spectroscopy and Related Phenomena*, 1981, 23:187–273, especially Appendix 3, p. 266; Kleeman, however, noted the effect of the momentum of ultraviolet light on the direction of motion of ejected electrons (R D Kleeman, 'On the direction of motion of an electron ejected from an atom by ultra-violet light', *Proceedings of the Royal Society of London*, 1910–11, 84:92–9).
[87] Letter F Soddy to W H Bragg, 23 July 1907, RI MS WHB 6A/48.

day, 5 May 1908.[88] Australian colleagues commented on his draft papers, including William Sutherland, who wrote, 'Your grossly material theory of the gamma rays seems to me to be prospering';[89] and in a subsequent letter Sutherland concluded that, 'you have abundant encouragement for your heterodoxy so far, although you have not given orthodoxy its coup de grace yet'.[90] Norman Campbell's first book, *Modern Electrical Theory*, had appeared in 1907,[91] and he wrote thanking William for his 'very kind letter concerning my book', and emphasizing that 'It is only too clear from its irrelevance [to the rest of the book] that the last chapter [on radiation] is the one which has the greatest interest for me.... May I say how glad everyone here is at the prospect of your being restored to England. I trust that it will not be long before we meet'.[92]

As had been the case with his alpha-particle work, William now improved the apparatus and focused more tightly on the aim of the experiment and its implied results.[93] He had earlier confirmed to Rutherford that, 'it is wonderful how apparatus improves when you are always handling it'.[94] The radiation from the radioactive source and impinging on the ionization chamber was now cleansed by the use of a strong magnetic field to deflect the beta particles, and a strong radium sample was employed. Just one plate and a thin aluminium foil were used to enclose the ionization chamber, in alternating positions in order to determine the incidence and emergence radiations (largely β particles) resulting from the impact of both soft and hard gamma rays. For these radiations, and for plates of eight different substances from carbon to lead, the results were unambiguous: 'The figures here given show the very large want of symmetry between the radiations on the two sides of a plate'. William had summarised the findings in his letter to *Nature*, listing the properties of the secondary radiation and showing that 'these properties are readily explained if the γ rays are supposed to be material, but are not easily to be reconciled with the aether-pulse hypothesis'.[95]

Bragg and Madsen wrote at length about the likely inaccuracies in their results, but it is now unproductive to go into further detail, for their experiment contained numerous significant effects of which they and their contemporaries had no knowledge. The asymmetry in their data was real, but its causes and its implications were hidden. As I have pointed out elsewhere and at length,

[88] W H Bragg and J P V Madsen, 'An experimental investigation of the nature of γ rays—No. 2', *Transactions of the Royal Society of South Australia*, 1908, 32:35–54 (the date given on this paper may be incorrect, see Royal Society of South Australia, Minutes of Ordinary and Annual General Meetings, 1853–, meeting of 5 May 1908, State Library of South Australia, Adelaide, SRG 10/1); W H Bragg, 'The nature of the γ and X-rays', *Nature*, 1908, 78:271 (letter dated 5 May 1908).

[89] Letter W Sutherland to W H Bragg, 11 January 1908, RI MS WHB 6B/23.

[90] Letter W Sutherland to W H. Bragg, 22 January 1908, RI MS WHB 28B/1.

[91] N R Campbell, *Modern Electrical Theory* (Cambridge: CUP, 1907).

[92] Letter N R Campbell to W H Bragg, 23 July [1908], RI MS WHB 2C/1.

[93] W H Bragg and J P V Madsen, 'An experimental investigation of the nature of γ rays—No. 2', *Philosophical Magazine*, 1908, 16:918–39.

[94] Letter W H Bragg to E Rutherford, 1 June 1907, CUL RC B367.

[95] W H Bragg, n. 88 (*Nature*); the quotations are from Bragg and Madsen, n. 93, pp. 922, 918–19.

analysis of the experiment in modern terms is impossible. For example, the incident γ-ray beam was complex, with at least half-a-dozen strong lines in the energy range 600 to 2200 keV, and three separate energy-dependent absorption mechanisms are significant for all the materials used: the photoelectric effect, Compton scattering, and pair production.[96]

Following his 23 July letter to *Nature*, but before its supporting paper appeared, William again wrote to the editor to rebut Barkla's most recent letter.[97] His criticisms were not uniformly successful, since a number of the effects listed by Barkla were not easily explained on either hypothesis. Barkla soon pointed out the inadequacies of several of William's criticisms and, in the same issue of the journal, William noted that his letter predated access to Barkla's latest results.[98] 'The tyranny of distance' was ever-present.[99]

Some time during 1908 William made a decision to expand his research team. Confident in the abilities displayed by Madsen, he gave him a project of his own and invited one of his new young graduates, Joseph Glasson, to join him in another aspect of the work. Having researched as thoroughly as he could the angular distribution of secondary beta particles produced by γ rays irradiating a variety of materials, William sought to reinforce his conclusions by a comparison of the forward and backward yield of secondary γ rays produced in the same way, and also of secondary X-rays resulting from X-rays impinging on different materials. The first project he gave to Madsen, the second to Glasson under his close supervision. Glasson had been born in country South Australia and was just completing a science degree with a major study of mathematics and honours in physics, his results being uniformly 'first class'.[100]

Madsen reported the conclusions of his experiments in a paper read to the Royal Society of South Australia and in a letter to *Nature*,[101] and their lengthy justification then appeared successively in the *Transactions* of the Society and the *Philosophical Magazine*.[102] References were made to the recent work of Richard Kleeman.[103] The results were unambiguous, he claimed, and he summarised his results as follows: '. . . 3. Secondary γ radiation appears on both sides of a plate which is penetrated by a stream of γ rays. There exists a marked

[96] Jenkin et al., n. 86 and references therein; readers unfamiliar with these technical terms can happily ignore this sentence.

[97] W H Bragg, 'The nature of γ and X-rays', *Nature*, 1908, 78:293–4; the Barkla letter is n. 81.

[98] C G Barkla and W H Bragg, 'The nature of X-rays', *Nature*, 1908, 78:665, two letters under the same heading and on the same page.

[99] This phrase is the title of a very well known book on Australian history: G. Blainey, *The Tyranny of Distance* (Melbourne: Sun Books, 1966).

[100] Register of Bachelor of Science students, 1898–1908, UAA, S246, p. 53; see also 'Glasson, Joseph Leslie', in R W Home, *Physics in Australia to 1945: Bibliography and Biographical Register* (Melbourne: University of Melbourne and Monash University, 1990), p. 70; Royal Commission for the Exhibition of 1851, Imperial College Archives, London, Glasson file, no. 315.

[101] J P V Madsen, 'The nature of γ rays', *Nature*, 1908, 79:67–8.

[102] J P V Madsen, 'Secondary γ rays', *Transactions of the Royal Society of South Australia*, 1908, 32:163–92, and *Philosophical Magazine*, 1909, 17:423–48.

[103] R D Kleeman, 'On the secondary cathode rays emitted by substances when exposed to γ rays', *Philosophical Magazine*, 1907, 14:618–44; id., 'On the different kinds of γ rays of radium, and the secondary γ rays which they produce', ibid., 1908, 15:638–63.

lack of symmetry between the amount of secondary radiation which proceeds from the two sides; 4. A lack of symmetry exists in the case of some substances between the quality of the radiation on the two sides; [and] 5. The last results seem very difficult to reconcile with a pulse theory'.[104] Most of this would be confirmed in the years ahead, but the interpretation would not be nearly so straightforward.

The results of the other investigation concerning secondary X-rays were reported to the Royal Society of South Australia and the Physical Society of London, and thereafter published following William's now well-established custom. A carefully-constructed experiment by Bragg and Glasson, in which X-rays irradiated a variety of 'radiators' to produce the secondary X-rays, convinced them that a 'want of symmetry does exist, that it is sometimes very pronounced, and that it is in keeping with expectations based on Madsen's study of the secondary γ rays. Hard γ rays show a very large difference between the quantities of emergent and incidence radiations; for soft γ rays the difference is smaller. Since X-rays are to be looked on as a very soft form ofγ rays, the difference should be smaller still; and this is what we have found to be the case'. They concluded, 'On a material theory of X- and γ rays the effect is easily explained... But if the X- and γ rays consist of energy bundles of very small volume, as suggested by J J Thomson, then these bundles must be capable of deflexion in going through atoms... just as neutral pairs would be in virtue of their electric fields. It seems hard to understand the distinction between such bundles and entities generally classed as material'.[105]

As McCormmach has shown, Thomson had been wrestling with the 'structure of light' intermittently since 1893, and he now had a discrete construction based on Faraday's lines of force and Maxwell's theory of the electromagnetic field.[106] William did not favour the Thomson model, but it was the source of another link with Norman Campbell, for Campbell was then conducting experiments in Thomson's Cavendish Laboratory on the nature of light. He was fully aware of William's work and was looking forward to discussing it with him.[107]

At the end of the paper printed in the *Proceedings of the Physical Society*, the discussion that took place when the paper was presented orally in London on 23 April 1909 was reproduced.[108] 'Prof. C H Lees said that Prof. Bragg had given a lucid account of his theories of γ and X rays. His researches would make physicists more careful in accepting the aether-pulse theory, [and] Mr

[104] Madsen, n. 102 (*Phil. Mag.*), pp. 447–8.

[105] W H Bragg and J L Glasson, 'On a want of symmetry shown by secondary X-rays', *Transactions of the Royal Society of South Australia*, 1908, 32:300–1, *Philosophical Magazine*, 1909, 17:855–64, and *Proceedings of the Physical Society of London*, 1907–09, 21:735–45.

[106] R McCormmach, 'J J Thomson and the structure of light', *British Journal for the History of Science*, 1967, 3:362–87.

[107] Ibid., p. 378.

[108] Bragg and Glasson, n. 105 (*Proc. Phys. Soc.*), p. 745; William's own account of the meeting is contained in letter W H Bragg to J P V Madsen, 29 April 1909, Basser Library, Australian Academy of Science, Canberra.

C A Sadler pointed out that whatever lack of symmetry might exist in the emergence and incidence secondary X radiations from a plate of a substance which was a source of scattered primary radiation, Professor Bragg's own results conclusively proved that such lack of symmetry did *not* exist when the plate was a source of homogeneous radiation'. William commented briefly on Lees' remarks but chose not to respond to Sadler, Barkla's collaborator. It was the first evidence in the scientific literature that William was now in England and attending meetings in London.

Despite their declining personal correspondence, Soddy continued to give generous exposure in his annual progress reports on radioactivity to William's research after it moved into the new realm. In his report for 1907 Soddy simply noted that 'Bragg advocates the view that the γ- and X-rays are...neutral particle[s] composed of a pair of positive and negative electrons';[109] while for 1908–09 he gave extended coverage to 'the theory that they also, like the β-rays, consist in the emission of discrete particles'.[110] He also noted that, 'A tendency from this point of view is to regard the β- and γ-rays as mutually convertible in their passage through matter'.[111]

The Bragg-Rutherford correspondence continued unabated during the same period, despite the distraction of Rutherford's move from Montreal to Manchester. William first communicated his hypothesis to Rutherford in June, with the hope that 'you will not think I have quite taken leave of my senses! The ether pulse theory of the γ and X rays is so universally accepted that it seems a bold thing, or perhaps I ought to say a rash thing, to question it'.[112] Rutherford, however, was sufficiently impressed to give the model a paragraph in the review of recent advances in radioactivity that he was preparing for the Royal Institution.[113] For the rest of 1907 their letters concerned the deteriorating conditions at McGill University in Canada and the possibility of an English offer.[114] In December William penned further long letters on his particle model,[115] to which Rutherford now responded.[116] 'I was very glad to get your budget of news and to hear you were progressing satisfactorily', he began. 'Your results re the γ rays are very exciting and seem certainly to support your view'; but he then shrewdly observed, 'The decision depending on secondary

[109] F Soddy, *Annual Progress Report on Radioactivity to the Chemical Society for 1907*, 1908, 4:311–43, 321.

[110] F Soddy, *Annual Progress Report on Radioactivity to the Chemical Society for 1908/09*, 1910, 6:232–67, 242.

[111] Ibid., p, 243.

[112] Letter W H Bragg to E Rutherford, 1 June 1907, CUL RC B367.

[113] E Rutherford, 'Recent advances in radio-activity', *Nature*, 1908, 77:422–6, a discourse delivered to the Royal Institution on Friday 31 January 1908.

[114] Letters W H Bragg to E Rutherford, 21 August 1907 and 22 January 1908, CUL RC B368 and B371 respectively; letters E Rutherford to W H Bragg, 5 July, 23 September, 24 October, 10 November and 3 December 1907, RI MS WHB 26A/10–14 respectively.

[115] Letters W H Bragg to E Rutherford, 17 and 26 December 1907, CUL RC B369 and B370 respectively.

[116] Letter E Rutherford to W H Bragg, 26 January 1908, RI MS WHB 26A/15.

radiation effects has certain drawbacks, especially as the laws of secondary radiation appear so complicated'.

The overall message was now clear. The wave theorists, focused on the nature of soft X-rays, and the particle proponents, centred on γ-rays, were locked into their positions. Both sides had seen a need to accommodate the other model but no one could see how this could be done. There was a stalemate, and some in the scientific community had grown weary of the argument. At the end of Madsen's letter in the 19 November 1908 issue of *Nature* the journal's editor appended the following note: 'As there are few opportunities in Australia for an investigator to place his views quickly before a scientific public, we print the above letter, but with it the correspondence must cease. The subject is more suitable for discussion in special journals devoted to physics than in our columns'.[117]

[117] Madsen, n. 101, p. 68.

14
Goodbye Australia!

Early in 1907, at about the same time that the possibility of a position for William in Canada first emerged, similar events were unfolding in England. On 24 January Frederick Soddy wrote to Arthur Smithells, Professor of Chemistry, Dean of the Faculty of Science, and Pro-Vice-Chancellor of the University of Leeds, regarding the possible retirement of its Cavendish Professor of Physics, William Stroud: 'When recently in Manchester I was told that the Chair of Physics in the University of Leeds is likely soon to become vacant. Will you excuse me taking the liberty of writing you about Prof. Bragg of Adelaide University whose claims in the case of any vacancy of this kind might otherwise be overlooked? I am a great admirer of his work in radioactivity... With Rutherford coming to Manchester... no doubt an effort will be made in Leeds to get a strong man also. Bragg... is I believe a thoroughly all-round physicist & mathematician... The developments of this work have been surprising and his methods have been followed by many during the past year... While there I was much struck with the spirit he has created around him... Of course I write unbeknown to Bragg... & I should be glad if you will consider my letter confidential... He is married & has a family'.[1] Following a reply from Smithells, Soddy wrote again: 'I am glad to get your letter... and I am delighted to hear that you have already had him in mind in consequence of his work in radioactivity. Perhaps I gave you a wrong impression of his age... The memory I carry away of him is of a man in the full flower of his energy and activities & the last thing I should have thought of would have been that he is too old'.[2]

Stroud had intimated that he planned to take early retirement to enter into a business partnership, and Soddy soon wrote to William to alert him to recent events: 'I got a letter from you today sending me greetings for the New Year which I warmly reciprocate. How different it must be in Australia now to what we are having here. We can hardly keep the homes warm & skating is in full swing & also death-dealing smoke smog smelling of SO_2 & the dreariest colour... You will have heard of the vacancies of Manchester & Dundee in the physics chairs. I have heard since that the physics chair at Leeds is shortly to be vacated by Stroud. I took the liberty, off my own bat, of writing to Smithells...

[1] Copy of letter F Soddy to A. Smithells, 24 January 1907, RI MS WHB 10A/2.
[2] Ibid., 28 January 1907, 10A/3.

I had a very cordial reply & he gave me to understand that the suggestion was not altogether new as he had himself been wondering about you . . . you will perhaps pardon my interference'.[3]

At the October 1907 meeting of the Leeds University Council, a small but senior committee was appointed 'to consider what steps should be taken to fill the vacancy which will be made . . . on the resignation of Dr Stroud'.[4] The committee decided to appoint by invitation and not advertisement, at a salary of between £800 and £900 per annum, and it agreed on a short-list of two: Professor William Bragg of the University of Adelaide and Professor Harold Wilson of King's College, London.[5] After favourable testimonials were received from J J Thomson, Joseph Larmor, Arthur Schuster, and Rutherford, the position was offered to William at a salary of £850 a year and the university retiring allowance. The letters of negotiation between William and the university have apparently not survived but, in response to his concern regarding research facilities and equipment, it was agreed to make available up to £2,000 for physical apparatus. The fact that William would not be free to take up the appointment until early in 1909 ceased to be a problem when Stroud was willing to continue in the position until that time.[6] Lamb wrote to William in November saying the Leeds Vice-Chancellor had reassured him that, regarding laboratory arrangements, 'they are building a large temporary building in addn to the existing accommodation', and that 'they would not burden you with too much teaching'.[7]

As William wrestled with the prospect of moving to the northern hemisphere he confided most to Rutherford. Regarding Leeds, he told Rutherford in January 1908, he had been 'extremely perplexed'. He was primarily concerned about the possible state of the research facilities, even after the £2,000 had been spent. He offered to give up his claim to a retirement allowance if Leeds would add its likely maximum value to his research allowance,[8] but an assurance of additional funds in the future made this sacrifice unnecessary.

On the other hand, William noted, 'if I did get a good lab. at Leeds I think I should do well to go there. I would be glad to go to England for many reasons . . . one of these is to be near to people like yourself. Again, I want to send my boys to Cambridge, and my wife and I want to be near them. I believe the country north of Leeds is fine'. He may also have wanted to test himself on a larger stage, and to go before it was too late. Then again, Adelaide had

[3] Letter F Soddy to W H Bragg, 5 February 1907, RI MS WHB 6A/44; William Stoud and Archibald Barr entered into partnership, and over time they developed the successful precision optical instruments company Barr & Stroud Ltd (W Stroud, 'Early Reminiscences of Barr & Stroud Rangefinders', in *Newsletter—Special Issue*, Institute of Physics (UK) History of Physics group, October 2005, pp. 34–45).

[4] Document entitled simply 'The University, Leeds, 19th May 1908' regarding Bragg's appointment and contained in his personal file, Central Filing Office, University of Leeds, CFO 3/0780.

[5] Committee on Cavendish Professor of Physics, University of Leeds Archives, Leeds University Minute Book No. 9: Committees (1907–10), pp. 99, 117, 118, 168.

[6] See ns 4 and 5.

[7] Letter H Lamb to W H Bragg, 10 November 1907, RI MS WHB 4A/6.

[8] Letter W H Bragg to E Rutherford, 22 January 1908, CUL RC B371.

its advantages, including: 'South Australia is really a very generous country, giving with both hands in the way of gardens and good houses, fruit and flowers; and though I could forego these myself for the sake of research opportunities, I do hesitate to ask my wife to give up all her friends and her life in this sunny place'.[9] When he wrote in April to tell Rutherford that he had decided on Leeds rather than Montreal, he admitted: 'I find many things drawing me to England'.[10] William wrote to Leeds in April 1908 expressing his willingness to accept the appointment.[11] At the same time he also wrote to the Council of the University of Adelaide:[12]

> I have the honour to inform you that I have received from the Vice-Chancellor of the University of Leeds a request that I should accept a nomination to the Cavendish Chair of Physics in that University. After very much anxious consideration I have decided to ask you to relieve me of my duty to you at the end of this year in order that I may be able to accept the offer made to me.
>
> You will readily understand with what unwillingness I take a step which puts an end to my long engagement here. My life and my work in this country have been so singularly happy...The reasons which prompt me to seek a change relate only to the nature of my work. At Leeds I shall be asked to profess only the subject of physics: not physics and mathematics together. The amount of elementary teaching will be less, and the number of advanced students will be greater. Moreover I am offered facilities for research which are more than this University could afford if all the other work of the chair which I now occupy is to be carried out as it should. And again I shall have constant opportunities in England of meeting men who are engaged on the same lines of investigation as myself. These are very real advantages, and I believe that I ought not to set them aside.
>
> I am, however, so conscious of what I risk in leaving Adelaide that I am loath to make the step irrevocable until I am assured that it is right...I may perhaps be allowed to ask for leave for ten months...If I then returned to Adelaide I should be the better for my year's experience in England and be more capable of fulfilling my duties to the University. In this matter I can only place myself in your hands.

The Adelaide Council held a 'Special meeting' the same day, 'To consider an urgent and important communication from the Professor of Physics'.[13] William's letter was read and 'The Council also conferred with the Professor on the subject matter of his letter and the question of carrying on his work. In granting the Professor's request the Council expressed their very high appreciation of the services rendered by Professor Bragg during the past 22 years, especially in regard to the development of the School of Physics. The Council

[9] Ibid.
[10] Letter W H Bragg to E Rutherford, 5 April 1908, CUL RC B373
[11] See n. 4.
[12] Letter W H Bragg to Council, 11 April 1908, UAA, S200, docket 341/1908.
[13] Council Minutes, meeting of 11 April 1908, UAA, S18, vol. VIII, pp. 399–400.

also expressed their very great regret at the severance of Professor Bragg's connection with the University; and they heartily congratulated him on so distinguished an appointment where he could devote the greater part of his time to research…The question of arrangements to carry on the work of the Mathematics & Physics chair was deferred for further consideration. Professor Bragg was asked to draft a scheme'.[14]

As to William's request to defer his final resignation from the university, the Chancellor told the *Register* that Professor Bragg's removal to England would be 'a very great loss to the University, but there is the counterbalancing advantage in the knowledge that a professor from Adelaide is being promoted to an English chair'.[15] Sir Samuel Way, like most Australians at this time, could not conceive that William would want to return. He was going 'home', and there was apparently no reason to take seriously his request for leave-of-absence; it was just another example of his modesty and carefulness.

It is unclear whether William was aware of the history of the University of Leeds when he agreed to become its next professor of physics. If he was, he must have been struck by the similarities between it and that of Adelaide. Although begun for different reasons, the foundation years of the University of Leeds and the University of Adelaide were identical, 1874.[16] The Yorkshire College of Science, as Leeds was known initially, was founded in direct response to British fears of the loss of technical supremacy, as prompted by its disappointing performance at the Paris International Exhibition of 1867. The Yorkshire brothers George and Arthur Nussey visited the exhibition to report on woollen textiles and, disturbed by what they saw, began a movement to establish a local college 'to embrace all the trades in the district and to be a centre of manufacturing instruction for the North of England…there is no reason why the present School of Arts should not be improved, and a School of Weaving and Designing established at once'.[17] A Committee for the Establishment of a Yorkshire College of Science was established with a strong technical, practical focus and, following the examples set by University College London and Owens College, Manchester, it was non-sectarian and looked no further than the county for its students when it was founded a few years later. Lord Frederick Cavendish was the first President of its Council.[18]

The initial appeal for funds was disappointing, so that only three professors could be appointed and classes had to be held in rented premises in the centre of the city, mirroring the Adelaide experience. Like Adelaide, too, the college and then the university were troubled for decades by inadequate funding. Leeds did obtain high-quality men for its foundation chairs: A W Rücker

[14] Ibid.

[15] *Register*, 13 April 1908, p. 4.

[16] Most of the following information is taken from the centenary history of the University of Leeds: P H J H Gosden and A J Taylor, *Studies in the History of a University, 1874–1974: To Commemorate the Centenary of the University of Leeds* (Leeds: Arnold & Son, 1975).

[17] M Sanderson (ed.), *The Universities in the Nineteenth Century* (London: Routledge & Kegan Paul, 1975), pp. 156–7.

[18] Gosden and Taylor, n. 16, ch. 1.

for physics and mathematics, T E Thorpe for chemistry, and A H Green for geology and mining. They were scientists rather than technologists, and each went on to a prestigious position in English science. Early technical instruction depended largely on external funding: for textile manufacture and dyeing from the Clothworkers' Company, and for mining from the Drapers' Company, reflecting the central importance of the textile industry in this part of the country. The main purpose of the pure science departments, soon including biology, was to provide elementary service courses.[19]

The next appeal for building funds brought a generous response from the Clothworkers' Company, which enabled the purchase of a large parcel of land close to the city-centre. The erection of a range of buildings followed, along Beech Grove Terrace, later named College Road and then University Road and still the centre of the modern university. Alfred Waterhouse, England's most celebrated architect of public buildings, was commissioned to provide structures of economy, utility, and good taste, and he produced a design of 'fourteenth-century Gothic stone buildings transmuted into red-brick'.[20] The scheme was changed and built progressively, the first building being opened in 1880. The student body was small and heterogeneous; many were aged between fourteen and seventeen and undertook secondary and matriculation studies, and there was a high failure rates at the end of first-year tertiary courses, which trained students for London or Durham University degrees. There was one area of homogeneity, however: nearly ninety percent of students came from local (Yorkshire) homes, again mirroring the Adelaide situation. In 1884 the existing School of Medicine in Leeds amalgamated with the College, adding to its status and lifting its confidence, and in 1885 it sought affiliation with Victoria University (previously Owens College), only to be rejected, finally joining in 1887.

The next group of buildings included a Great Hall and Library, completed in 1894, the extravagance being criticised but also providing the college/university with a focus that Adelaide lacked for another four decades. Teacher training, of both new and existing teachers, was a priority and, again like Adelaide, electrical engineering was closely associated with the teaching of physics. Thus the physics department at Leeds acquired improved equipment in 1890 for 'students desiring to be trained as electrical engineers'.[21]

In 1903 the Manchester and Liverpool components of Victoria University were granted separate university status and, although their Yorkshire cousin was reluctant, it had no option but to follow in 1904. Again the College recognised that its resources were problematic; the library holdings, for example, were inadequate for university teaching let alone research, and three-quarters of the laboratory space used by the physics department was made up of artificially lighted underground rooms and corridors. Horace Lamb, now at Manchester, wrote to the Adelaide Chancellor regularly if infrequently, and

[19] Ibid., ch. 6.
[20] Ibid., ch.4, quotation pp. 146–8.
[21] Ibid., ch. 6, p. 257.

twice commented on the Leeds institution. In 1900 he mentioned Liverpool's wish to become independent saying, 'I don't myself object, provided we could get rid of Leeds, which is rather a feeble institution'. Then in 1904 he noted, 'The Leeds University has been constituted and Manchester now stands alone... I don't think we shall suffer much from the change, but the position of Leeds is unhappy, as they have to keep up appearances, with not much money and no great local enthusiasm at their backs'.[22]

Even the College's desire to be titled 'The University of Yorkshire' was denied by Sheffield's ambition, so that it was left to adopt the name The University of Leeds. As previously, this reluctant change of status was followed by a period of renewed enthusiasm and increased numbers and quality of students. Between 1894 and 1897 there was only one student who completed an honours degree in physics; by 1903 some forty-five percent of day students were reading for degrees.[23] In 1904 the university introduced a new degree structure in arts and science, and honours programmes became available in a wide range of disciplines, both pure and applied. The number of students completing such degrees rose only slowly, however: BSc numbers were less than twenty and honours physics graduates less than five per year before the First World War.[24] Adelaide was similar.

At Leeds, further work was commissioned from Paul Waterhouse, Alfred's son, who had taken over his father's business. The Charter group of buildings included an eastern extension of the earlier Baines frontage and, at the extreme east end, not a clock tower as originally envisioned but temporary accommodation, largely for physics. The 'physics sheds', as they became known, consisted of little more than walls and roofs of glass and corrugated iron on a timber framework. When completed in 1908 it was hoped they would soon be replaced by a permanent structure, but the intention was not fulfilled until after two world wars. Here William would have to carry on the laboratory teaching and his research as best he could (see Figure 14.1).[25]

By 1909 the new university was able to cancel its matriculation classes, special grants had been made to the library, and there were new buildings, new degree structures, and a new desire to foster research. Professor William Bragg was appointed to lead the academic transition to a new-generation, twentieth-century university. He had done it before, but not even he could have imagined what the next few years would bring!

When all the decisions had been made, Rutherford wrote to William to welcome the news: 'I shall be delighted to have you nearby so that we can foregather occasionally'.[26] Others also sent their congratulations. Lamb now welcomed the decision: 'There are many very strong reasons in favour of the

[22] Letters H Lamb to Sir Samuel Way, 7 November 1900 and 26 June 1904, in Papers of Sir Samuel Way, State Library of South Australia, Adelaide, PRG 30/3 (letters received 1870–1914).
[23] Gosden and Taylor, n. 16, pp. 261, 262.
[24] Ibid., pp. 266, 267.
[25] Ibid., ch. 4.
[26] Letter E Rutherford to W H Bragg, 24 July 1908, RI MS WHB 26A/16.

Fig. 14.1 Part of the University of Leeds, (L to R) the Baines block, the 'physics sheds', and the Brotherton Library, mid-1930s. (Courtesy: University of Leeds Archives.)

course you have taken, and I cannot doubt that it is on the whole the right one. I am glad that we shall have yourself and Mrs Bragg as not very distant neighbours... Your old pupil Duffield takes his D.Sc. degree "for research" here next month. He will, I suppose, be shortly going out to try and get interest and money for solar physics in Australia'.[27] William Sheppard and Walter Workman, William's fellow Cambridge mathematics students, wrote to offer 'heartiest congratulations'. Colleagues from around Australia and the Governor of South Australia applauded William's contributions and wished him well.[28]

William Sutherland wrote to William to say: 'Let me congratulate you on the substantial recognition of your work conveyed by this morning's news of your invitation to Leeds. Of course I read the news with mixed feelings... However, I have always strongly maintained that it is an excellent thing for some Australian professors to win promotion nearer to the scientific centres; I mean an excellent thing for the Australian universities which still suffer too much from an impression amongst their staffs that they have finished their careers in a blind alley... It will also be a delight to you to meet so many kindred spirits full of ideas on your work. Sometimes you will regret sunny, free and easy Australia being left behind, but on the whole I hope you will be decidedly happier in the large mental field of influence opening out to you'.[29]

Coleridge Farr best summed up the feelings of William's friends and colleagues when he wrote to Rutherford late in 1908: 'You will soon be welcoming to the sister University of Leeds a very old friend of mine, Bragg. I do not know if you have met him; he is one of the finest fellows I know. He is as unassuming as he is brilliant and would be an acquisition to any university. We are sorry to lose him from this hemisphere, for lately he more than any other man has helped to shift the centre of gravity of scientific research a little to the south'.[30]

[27] Letter H Lamb to W H Bragg, 13 June 1908, Bragg (Adrian) papers.
[28] Bragg (Adrian) papers.
[29] Letter W Sutherland to W H Bragg, 13 April 1908, Bragg (Adrian) papers.
[30] Letter C C Farr to E. Rutherford, incomplete, no date [late 1908], CUL RC F15.

Before William could contemplate his new life in detail, however, there were many jobs to be done in Adelaide. At Council meetings, Madsen's DSc was approved and Priest was promoted to Assistant Lecturer and was granted permission to take part of William's lecture load during 1908, as William was 'very anxious to continue some research work'. Kleeman was awarded his DSc.[31] The university continued to push the government to purchase the university's country lands and thereby alleviate its financial difficulties, and late in 1908 the government relented and agreed. William, along with the members of the Council and the rest of the university, breathed a sigh of relief.[32] Perhaps the state's generosity was prompted by recent good years: 'this State has been passing through fat years', the *Register* noted early in 1908, but it also wondered 'Will Prosperity Continue?'.[33] Later in the year it recorded that, in the current prosperity, the South Australian population had grown steadily towards four hundred thousand, still not large.[34]

The thirtieth annual conference of the South Australian Public School Teachers' Union was held mid-year, with a large and enthusiastic attendance. There was a new mood of optimism.[35] Professor Bragg addressed the conference on the second morning, but before his lecture, and at its conclusion, several remarks were made regarding his contribution to education in the State:[36]

> In view of his forthcoming departure they wished to thank him most earnestly for the assistance he had given to the teachers here. Whatever result he might achieve in the old land, South Australians would never forget his great work in their own State, his splendid personal character, and the kindness he had shown to all who had come in touch with him. (Applause)...Those who had come under his tuition at the University felt something more than respect for his learning and exalted position. It was sentiment of real affection for himself. Professor Bragg had the sympathetic disposition of the true teacher. (Applause)...they wished him godspeed.
>
> Professor Bragg in reply said he did not know how to thank them...It had been a happy thing for the University to have the teachers associated with it. Most of those in the University who had had anything to do with them realized...that their presence had been a great help—not merely that they helped to fill the lecture rooms, but that they had been such appreciative listeners. When these students went into the country to take charge of schools...he had friends in practically every town, and they helped to spread the influence of the University and give it a greater hold on the

[31] Council Minutes, meetings of 29 November 1907, 13 December 1907, and 27 March 1908, UAA, S18, vol. VIII, pp.368, 373, 390, and 396 respectively; quotation from letter W H Bragg to Council, 13 March 1908, UAA, S200, docket 231/1908.

[32] Council Minutes, meetings throughout 1907 and 1908 and especially 30 October 1908, UAA, S18, vol. IX, p. 54.

[33] *Register*, 2 January 1908, p. 4.

[34] *Register*, 25 August 1908, p. 9.

[35] *Register*, 30 June 1908, p. 6.

[36] *Register*, 1 July 1908, p. 5.

> State... Wherever he was he would always look back on this State and its teachers with feelings of the warmest affection. (Applause)

The content of William's address to the teachers was similar to that of the two extension lectures he had given a month earlier to a general audience,[37] and concerned: 'A problem which was at the present time greatly discussed throughout the scientific world—the question of the nature and existence of the ether of space... The wave theory of light... had finally triumphed over its rival—the corpuscular theory which Newton advocated... [and] they were led to think of a great ocean of ether filling space... [However] If such a thing were true, they ought to be able to find a change in the velocity of light depending on whether it was going in the direction of the earth's motion or the opposite way... No such effect could be found, although most carefully sought for... There was one other strange line of attack upon the ether problem. The famous X-rays and the gamma rays of radium were supposed to be irregular waves... some experiments made in the University of Adelaide had shown that the electrons which were ejected by the rays always moved at first in the direct line in which the rays had come, suggesting, in fact, that the rays were not pulses or waves at all, but were material things... (Applause)'.[38] William was here placing his research in a broader context than he had done before, again illustrating his awareness of its larger significance as well as its specialised importance.

During 1908 William was often absent from meetings of the Council of the School of Mines and Industries, but one matter arose that claimed his atten tion and reminded him of past battles. The School had taken over the teaching of first- and second-year electrical engineering and Madsen wrote asking to give part of his third-year course during the day rather than entirely in the evening. The School Council deferred a decision until enrolment was complete and resolved that it was desirable for the course to be given at the School by a School instructor.[39] The university responded that it would offer no opposition to the proposal but suggested that other re-arrangements be made to compensate it for the loss of fee income.[40] The School was uncooperative.[41] The university suggested that specified courses be exchanged between the two institutions, but the School Council rejected the proposal.[42] William attended only one further meeting of the School Council, in October, and at the December meeting the President, Sir Langdon Bonython, spoke of William's extended service to the School:[43]

> As one of our members will shortly be retiring... it is only proper that... we should record in our minutes an expression of our sincere appreciation

[37] *Register*, 10 June and 17 June 1908, p. 6.
[38] Register, 1 July 1908, p. 7.
[39] School Council Minutes, meeting of 9 March 1908, University of South Australia Archives, vol. 6, pp. 103–4.
[40] Ibid., meeting of 25 May 1908, pp. 124–5.
[41] Ibid., meeting of 20 July 1908, pp. 138–9.
[42] Ibid., meeting of 31 August 1908, p. 150.
[43] Ibid., meeting of 21 December 1908, pp. 171–2

> and complete realization of service rendered and loss sustained by his departure. I refer of course to Professor Bragg... The Professor has been a member of this Council since 1890, so that he is one of our oldest members. We all feel that he has been most generous in the way he has devoted time and attention to the affairs of this School and we shall miss him very much indeed. As President, I have on many occasions found his advice of the utmost value. But, apart from service, his association with the School has been of advantage by reason of his eminence in the scientific world... Professor Bragg's last act in Australia will be to occupy the chair as President at the meeting of the Australasian Association for the Advancement of Science to be held next month in Brisbane. There is a singular fitness in this arrangement of things. The presidency will be a credential of the best kind... to the people of England as to his scientific status in Australia. Professor Bragg will carry with him the hearty good wishes of the members of this Council.

At the university William was refused respite by three important matters: the future of astronomy in the State, the future of his disciplines and their staff and students, and preparation of his Brisbane address. In addition, the acoustics of the Elder Hall were still a problem, and at home there were many tasks associated with relocating his family. Fortunately there were pleasant distractions too.[44] Professor and Mrs Bragg attended the University Sports Association Ball in 1908,[45] and Gwendoline continued to exhibit at the annual Society of Arts exhibition.[46]

Final tasks

When the Australian states federated in 1901 to form the Commonwealth of Australia they became subject to a new Australian Constitution, and this made provision for various powers to be ceded to the Commonwealth: some immediately (customs), some on dates to be determined (defence, post, and telecommunications), and some that required the enactment of special legislation (for example, astronomy and meteorology). As in South Australia under Charles Todd, there had always been a close association between these last two services in the Australian states, but the tie was severed when the federal government passed its Meteorological Bill (1906), which provided for the establishment of a meteorological department (from 1908 the Commonwealth Meteorological Bureau) but made no mention of astronomy.[47] State officers had met in May 1905 and opposed the move, some recognising correctly that

[44] *Observer*, 25 August 1906, p. 28, which has a photograph of 'The stewards, judges and officials' for the 1906 University Athletic Sports, including Professor Bragg.

[45] *Register*, 30 May 1908, p. 6

[46] *Register*, 11 April 1908, p. 10.

[47] R W Home and K T Livingston, 'Science and technology in the story of Australian federation: the case of meteorology, 1876–1908', *Historical Records of Australian Science*, 1994, 10:109–27.

the loss of meteorological functions would lead to a decrease in support for astronomical work, but the federal parliament was adamant. The astronomical observatories in Sydney, Melbourne, Adelaide, and Perth were under threat.

Consideration of these implications arose first in private discussions between Professor Bragg and the Chancellor, and on 20 January 1908 William wrote to the Chancellor to answer the three questions Way had put to him. First, he reported that quite a number of universities in Europe and America had chairs of astronomy and taught the subject.[48] As regards finance, he thought they could manage with a minimum of one professor of astronomy and mathematics, one assistant, and a cadet, and that the University should be glad to contribute a little money since the new arrangement would materially reduce the difficulty of crowding all the mathematics and physics into one chair. 'Now as to the general point', he continued, 'the value of astronomical teaching is almost unique. It deals with such wonders and grandeurs that it is a valuable mental discipline, most healthy for a man's soul [and] the interest of the public would be greatly increased...No subject could be of more value than astronomy to school teachers in training as a degree subject: it would fill their minds with ideas with which they could fascinate their children, very much to the advantage of both teachers and taught'. At a University Council meeting a few days later the Chancellor stated that he had been in contact with Sir Charles Todd, Professor Bragg, and the State Government regarding the possible appointment of a professor of astronomy and the use of the Observatory by the university. After William's letter was read the Council approved these actions and authorised continuance of the negotiations.[49] Charles Todd supported the plan.[50]

A deputation from the university, including Professor Bragg, then waited on the State Treasurer on 3 February 1908 and requested the transfer of the Observatory to the university, with funding for the Observatory and a new professor and use of the ample land for a students' sports ground. The Treasurer quickly rejected the third request and noted that, while the British nation had always supported astronomy and the State was willing to continue the astronomical work, funding for a new professorship was unlikely. Other sciences should be considered, the professorship would lose some of its rationale when the Commonwealth took over, and other sources of funds should be explored. William objected: 'it should be understood that that the University supplied a great amount of the expenditure out of its own earnings, and that in this country the State contributed less to it than in almost any community in the world'.[51]

[48] Letter (draft and typed forms) W H Bragg to Chancellor, 20 January 1908, UAA, S200, one of a number of documents in docket 75/1908; the letter was subsequently published in the *Register*, 29 January 1908, p. 5. (There was considerable public interest in the letter and three days later the *Register* published the results of a long interview with Professor Bragg in an article entitled 'Original Research, Adelaide University Professors, What They Have Done', *Register*, 1 February 1908, p. 9.)

[49] Council Minutes, meeting of 24 January 1908, vol. VIII, p. 382.

[50] Letter C Todd to Chancellor, 27 January 1908, UAA, S200, docket 75/1908.

[51] *Register*, 4 February 1908, p. 7.

In the days that followed, William canvassed the possibility of raising money by public subscription in honour of Sir Charles Todd, Todd and Ernest Cooke of the Perth Observatory wrote to the South Australian Treasurer to urge the maintenance of the South Australian Observatory, and Melbourne and Sydney professors wrote to William in support of the attachment of the State observatories to their respective universities.[52] But the battle was lost. The university did not appoint such a professor, although the State Observatory did survive under its Director and Government Astronomer, George Dodwell, until the early 1950s. At that time it was demolished and buildings for Adelaide Boys' High School constructed on the site. It was hoped to recreate the facility at the university but the observatory was dead, overtaken by Federal Government facilities elsewhere.[53]

In 1951, in reply to an appeal from Dodwell for support, Willie recalled his childhood and youth at the old Observatory, where he and Dodwell had met fifty years before, Dodwell having graduated from the university in 1905 and been an assistant at the Observatory since 1899: 'I was very grieved to hear from you that the Adelaide Observatory has ceased to function. So many of my most vivid early memories are bound up with the old buildings and my grandfather's work there, and I regret deeply that not only has the old building [the house] been swept away, but that nothing has been created to take its place and carry on its magnificent early pioneering work...I would like to add how pleased I was to hear from you again. It is very sad to think that the old building has gone; I can picture vividly every detail of it now, both of the house and the Observatory buildings. I remember so well accompanying my grandfather on his regular rounds at the weekends to read the instruments. I had always been hoping that when I do come to Australia again, which I hope will not be long now, I would be able to revisit these old haunts'.[54]

William's readiness to embrace these possibilities for local astronomy was prompted by letters he had received earlier from one of his former students, Geoffrey Duffield. Duffield had completed a science degree at the university in 1900 with an 'undistinguished' record and third-class honours in physics and mathematics. So small was the student population, however, that he won the university's Angas Engineering Scholarship and with its assistance completed the Cambridge Mechanical Sciences Tripos in 1903, followed by a year at the Engineering Laboratories of the National Physical Laboratory on a student assistantship.[55] Now attracted to academic life, Duffield joined Arthur Schuster's Physical Laboratories at Owens College on a scholarship and began

[52] Letters W H Bragg to Chancellor, 4 and 7 February 1908; letter C Todd to Chancellor, 8 February 1908, with copy of letter to Treasurer attached; copy of letter W E Cooke to Treasurer, n.d.; copy letter J A Pollock to W H Bragg, 8 February 1908; copy letter T R Lyle to W H Bragg, 9 February 1908: all in UAA, S200, docket 75/1908.

[53] Letter G F Dodwell to W L Bragg, 30 March 1951, RI MS WLB, 56A/481.

[54] Letter W L Bragg to G F Dodwell, 18 April 1951, RI MS WLB 56A/482.

[55] R Love, 'Science and government in Australia, 1905–14: Geoffrey Duffield and the foundation of the Commonwealth Solar Observatory', *Historical Records of Australian Science*, 1985, 6:171–88.

research in the new field of atomic and molecular spectroscopy. He was particularly impressed by the work of the Central Bureau of the International Union for Co-operation in Solar Research housed in the laboratory building. Inspired, Duffield determined to establish a facility for solar research in Australia.

In 1906 William had reported that Duffield's research 'is of special interest in that it helps elucidate the constitution of the atom and the spectra it emits'.[56] In March 1907 Duffield wrote again; he was about to attend a meeting of the Solar Union and included a copy of the draft resolution he intended to present there, on the importance of a solar research facility in Australia to complete the international chain of observatories. William promptly handed the resolution to the local press and it was published on 3 April.[57] At the Observatory, Dodwell welcomed the suggestion and offered Adelaide as a suitable location;[58] William suggested that Duffield write to the Premier of South Australia on the matter.[59] At the conference a forthright, redrafted resolution was carried unanimously under the name of Sir Norman Lockyer, President of the Solar Union.[60]

With this resolution in hand, Duffield now approached all the major solar observatories around the world and received warm support and the offer of instruments. The Royal Society strongly supported the proposal, and at Duffield's prompting it said, 'The Royal Society suggests that if this new…solar observation could be affiliated to the Observatory and the University of Adelaide the case would be met'.[61] Duffield and Bragg were now clearly working in concert, but their efforts would be in vain. Henry Hunt, the first Commonwealth Meteorologist, saw the proposal as a threat to the resources for his new Bureau, and the relevant Minister then rejected the scheme on the grounds that 'it is not desirable at the present juncture to incur the heavy expenditure involved'. The South Australian Government saw it as a Commonwealth responsibility and William communicated the stalemate to Duffield in July 1908.[62] A final approach to the State Governor by the Chancellor brought the same negative response.[63]

In October Duffield returned to Australia with a Manchester MSc and DSc, and with some assistance from Bragg he worked tirelessly throughout 1909 and the first half of 1910 to lay the groundwork for future government acceptance of his plan. From mid-1910 he was Professor of Physics at University College, Reading, where he continued to push the scheme until 1923, when the Commonwealth Solar Observatory was finally established on Mt Stromlo

[56] Quoted ibid., p. 175.

[57] Ibid., p. 177; *Advertiser*, 3 April 1907.

[58] *Advertiser*, 4 April 1907.

[59] Letter W H Bragg to W G Duffield, 4 April 1907, Duffield papers, Basser Library, Australian Academy of Science, Canberra.

[60] Love, n. 55, p. 177.

[61] Quoted ibid., p. 178.

[62] See respective quotations ibid., pp. 178–179; letter H M Hunt to W H Bragg, 8 May 1908, and letter W H Bragg to W G Duffield, 2 July 1908, Duffield papers, n. 59.

[63] Draft paragraph W H Bragg to Chancellor, n.d., and 'Return', Chancellor to Governor, 4 August 1908, both in UAA, S200, docket 597/1908; for the Governor's position see letter G R Le Hunt to the Earl of Crewe, 4 August 1908, Duffield papers, n. 59.

near Canberra and Duffield became its first Director. In more recent times the Mt Stromlo and Siding Spring Observatories have become one of the premier astronomical organisations in the world.[64] Duffield's persistence finally paid off handsomely.

At the university Madsen received an offer of appointment as Lecturer in Electrical Engineering at the University of Sydney. He had applied unsuccessfully for a position there in 1904, when he had returned to Sydney for his marriage;[65] but prompted by William's forthcoming departure he had now been successful and had submitted his likely resignation.[66] Madsen's request to borrow the apparatus he had been using in Adelaide in order to continue his experiments in Sydney was approved and he continued his research for two years, principally a study of the scattering of radium β-rays.[67]

William was tardy in informing the Education Committee of his suggestions for the future of his disciplines, partly because he was seeking advice from colleagues interstate.[68] On 9 October 1908 he wrote to the Education Committee: 'I have the honour to make the following recommendations in regard to the work of the mathematical, physical, and engineering schools during 1909... I suggest (1) that Mr Kerr Grant of Melbourne be asked to take seven lectures a week in physics and the practical work in physics, and two in honours mathematics, at an honorarium of £400, and that he be offered the title of acting professor; (2) that Mr H. J. Priest be asked to take nine lectures a week in mathematics, including one in honours mathematics, at an honorarium of £375, and that he be offered the title of acting professor. Mr Priest is at present taking junior work at a salary of £170 per annum: if the new scheme of distribution of the teaching work comes into effect, his services will not be required after 1909...'.[69] These recommendations involved an annual saving of about £100 and were accepted unchanged by the University Council. A possible rearrangement of professorial chairs and lectureships was deferred to a later meeting.[70] Held on 6 November, the Council then had one report regarding an increase in professorial salaries and another by Professors Bragg and Mitchell suggesting that there should be a Professor of Pure and Applied Mathematics (who should also have charge of Mining Engineering), a Professor of Physics (who should also have charge of

[64] R W Home, *Physics in Australia to 1945: Bibliography and Biographical Register* (Melbourne: Melbourne and Monash Universities, 1990), pp. 56–58; Love, n. 55.

[65] Madsen was a Sydney graduate and his 1904 application is inferred from three letters of recommendation from W H Bragg, S J Way (Chancellor), and F W Clements (Electric Lighting & Traction Co.), dated July/August 1904, copies made available to the author by the Madsen family.

[66] Letter J P V Madsen, to Registrar, 24 September 1908, UAA, S200, docket 693/1908.

[67] Letters J P V Madsen to Registrar, 18 December 1908 and 26 January 1909, UAA, S200, dockets 926/1908 and 43/1909 respectively; Council Minutes, meeting of 18 December 1908, UAA, S18, vol. IX, p. 76; R W Home, 'W H Bragg and J P V Madsen: collaboration and correspondence, 1905–1911', *Historical Records of Australian Science*, 1981, 5:1–29.

[68] Education Committee Minutes, meetings of 14 August and 11 September 1908, UAA, S23, vol. VII, pp. 109 and 111 respectively.

[69] Letter W H Bragg to Education Committee, 9 October 1908, UAA, S200, docket 750/1908.

[70] Council Minutes, meeting of 30 October 1908, UAA, S18, vol. IX, p. 52.

Electrical Engineering), and a competent lecturer for each.[71] The Council, ever conscious of the financial situation, resolved to accept the reports but added that it could not commit itself to their complete carrying out, being 'guided therein, from time to time, by due consideration of the finances of the University'.[72]

The future of William's staff was now assured for 1909, and that would allow time for the situation to be reassessed, but Arthur Rogers was disconsolate at news of the family's departure. He had become much more than head of the Physical Workshop. In addition to constructing apparatus superbly suited to lecture demonstrations and research projects, he had assisted at public lectures, built 'a little camera for Bobby Bragg' and a range of miniature items for baby Gwendolen's dolls' house,[73] helped medical practitioners with their X-ray photography and by sharpening their scalpels,[74] clashed with Madsen,[75] and trained a series of apprentices. He could be temperamental and difficult and his health was often poor. William was sympathetic and understanding and Rogers responded with warmth, loyalty, and superb craftsmanship. At the December 1908 Council meeting William applied for leave-of-absence for Rogers, who was unwell and had been advised by his doctor to take a break, and it granted him two months' leave during the long vacation.[76] Later the same month William wrote to Rogers from his holiday-house at Semaphore: 'Just a line to wish you and your family a really happy Christmas and New Year...I shall miss you dreadfully in my new laboratory. You have been such a wonderful help to me in all my work...I hope you will do very many more years of good work, old fellow...My wife has a couple of little books for the children'.[77] Rogers disliked William's successor and recorded satisfyingly in his 1909 diary, 'Kerr Grant blown up!'.[78]

Kerr Grant had graduated MSc from the University of Melbourne with first-class honours in mathematics and physics and had already developed a reputation for his enthusiasm and excellent teaching. Following William's departure he remained acting professor for two years and was then appointed Elder Professor of Physics, a position he held until his retirement in 1948. While not a strong researcher, he did publish irregularly and became well known as a teacher and prominent public figure.[79] Herbert Priest's career is not

[71] Council Minutes, meeting of 6 November 1908, UAA, S18, vol. IX, p. 57; Education Committee Minutes, meeting of 9 October 1908, UAA, S23, vol. VII, p. 116.

[72] Ibid.

[73] Camera reference in entry for 31 August 1899, A L Rogers annual personal diaries, 1896–1910, in possession of Rogers family; dolls-house items in possession of Lady Adrian, Cambridge, UK.

[74] Ibid. (Rogers' diaries), *passim*.

[75] Ibid., entries for 17 July 1901 ('Madsen presumptuous and generally objectionable'), and 15, 18, and 19 November 1907 ('Madsen makes rude remarks [re self and boys in workshop]... unpleasantness with Madsen...I speak to Madsen about his remarks').

[76] Council Minutes, meeting of 16 December 1908, UAA, S18, vol. IX, p. 71.

[77] Letter W H Bragg to A L Rogers, 23 December 1908, Bragg (Adrian) papers.

[78] Entry for 19 May 1909, Rogers' diaries, n. 73.

[79] S G Tomlin, 'Grant, Sir Kerr (1878–1967)', in B Nairn and G Serle (eds), *Australian Dictionary of Biography* (Melbourne: MUP, 1983), vol. 9, pp. 77–9; Home, n. 64, pp. 71–3.

well documented. He graduated BSc in 1902 and BA with honours in mathematics in 1904, and from 1903 he held a number of casual positions at the university as demonstrator in physics and assistant lecturer in mathematics. William used the modest funds available to employ recent graduates in these roles, especially but not exclusively those who assisted him with his research. Priest's appointment as assistant professor was for only one year, and thereafter he studied and travelled overseas and became a lecturer in mathematics at the University of Queensland. Priest's health was never strong, however, and from 1927 he spent periods of sick leave in England and in Adelaide, where he died in 1930.[80]

As an economy measure, Robert Chapman was appointed Elder Professor of Mathematics and Mechanics in 1910, a position he held until he returned to the engineering chair at the end of 1919. His teaching load was extraordinarily heavy and, while he was a superb solver of applied problems, he had little interest in pure mathematics. He was succeeded in mathematics by Raymond Wilton, who had graduated from Adelaide in 1903 with first-class honours in mathematics and physics. On William's advice Wilton had gone to Cambridge, where he graduated as fifth Wrangler in the Mathematical Tripos of 1907 and in the following year was awarded first-class honours in physics in the second part of the Natural Sciences Tripos. After research in the Cavendish Laboratory, Wilton taught mathematics at the University of Sheffield, where he published prolifically (DSc, Adelaide, 1914). During WWI he was a pacifist, suffered a breakdown, and later joined the Society of Friends, to whose philosophy he became dedicated. In 1919 he was appointed to the Elder chair of mathematics at Adelaide, a brave selection so soon after the war. His first task was to revise and modernise the secondary-school and university courses. His research productivity continued, in analysis and number theory, and he was awarded a Cambridge ScD and the Lyle Medal in Australia in the 1930s. G H Hardy regarded his work as that of 'a fine mathematician, with admirable taste and a natural inclination towards deep and difficult problems'. A stroke in 1941 so impaired Wilton that he had to relearn his mathematics, even the multiplication tables, and just as he was returning to teaching in 1944 he suffered a second, fatal stroke.[81]

Student numbers were small throughout William's time in Adelaide, although his teaching to medical, engineering, and music students meant his influence was wider than it might have been. Similarly, the number of students continuing into honours courses and then postgraduate work was tiny. Nevertheless, there were a number of such graduates who went on to distinguished careers and who

[80] Unattributed obituary, *The Advertiser* newspaper, Adelaide, 4 December 1930, p. 12; R B Potts, 'Mathematics at the University of Adelaide, 1874–1944', copy of typescript in possession of the author.

[81] R B Potts, 'Wilton, John Raymond (1884–1944)', in J Ritchie (ed.), *Australian Dictionary of Biography* (Melbourne: MUP, 1990), vol. 12, pp. 533–4; W N Oates (ed.), *These Three: Love Faith Hope: The Collected Addresses of J. Raymond Wilton, Sc.D.* (Adelaide: The Advertiser, 1945) also contains a useful biography.

retained fond and appreciative memories of Professor Bragg's years in Adelaide. The careers of Richard Kleeman, Ernest Cooke, Bernard Allen, Coleridge Farr, John Madsen, Herbert Priest, and Geoffrey Duffield have been discussed earlier, and Eric Jauncey's experiences will become important shortly when his unsung but crucial role in unravelling the wave-particle dispute is revealed. Others won 1851 Science Research Scholarships[82]—J L Glasson[83] and J A Gray (1909), and G E M Jauncey (1911); three won Rhodes Scholarships—N W Jolly (1904)[84], W R Reynell (1906),[85]and H H L A Bröse (1913)[86]; and A Chapple (1895) won the Angas engineering scholarship.[87] Many others took subjects from Professor Bragg but did not study mathematics and physics as major subjects, including I H Boas[88] and T Brailsford Robertson.[89]

To William's chagrin, in 1908 the acoustic properties of the Elder Hall were still a source of complaints from lecturers and musicians. He undertook some further experiments that were unproductive and was then authorised to expend up to £50 for further study.[90] When he reported to the Council in December William was authorised to continue the investigations still further and to prepare a scheme for any suggested improvements.[91] He replied from his holiday house at the seaside suburb of Semaphore: 'In respect to the acoustics of the Elder Hall I have asked Liberty & Co. to submit plans and an estimate for permanent hangings similar to the temporary arrangements which we all thought successful'.[92] It was an improved but still a makeshift solution.

The impending departure of the Bragg family was now widely known, and new arrangements and appropriate farewells were organised. The University of Cambridge wrote inviting Adelaide to be represented at their 1909 commemorations of the centenary of Charles Darwin's birth and the fiftieth anniversary of

[82] *Record of the Science Research Scholars of The Royal Commission for the Exhibition of 1851, 1891–1960* (London: Commissioners, 1961).

[83] Prince Alfred College Archives, Adelaide; *Record*, n. 82, p. 43; Home, n. 64, p. 70; University of Tasmania Archives; for the neutron experiment see L Badash, 'Nuclear physics in Rutherford's laboratory before the discovery of the neutron', *American Journal of Physics*, 1983, 51:884–9, 886, and N Feather, 'A history of neutrons and nuclei', *Contemporary Physics*, 1959–60, 1:191–203, 257–66, 262.

[84] N B Lewis, 'Jolly, Norman William (1882–1954)', in B Nairn and G Serle (eds), *Australian Dictionary of Biography* (Melbourne: MUP, 1983), vol. 9, p. 504.

[85] V A Edgeloe, 'The first twelve South Australian Rhodes scholars', *Journal of the Historical Society of South Australia*, 1989, 17:134–51, 138–9.

[86] J G Jenkin, 'Henry Herman Leopold Adolph Brose: vagaries of an extraordinary Australian scientist', *Historical Records of Australian Science*, 1999, 12:287–312.

[87] H T Burgess (ed.), *The Cyclopedia of South Australia* (Adelaide: Cyclopedia Co., 1909), vol. II, p. 31; R M Gibbs, *A History of Prince Alfred College* (Adelaide: P A C, 1984), *passim*.

[88] N Rosenthal, 'Boas, Isaac Herbert (1878–1955)', in B Nairn and G Serle (eds), *Australian Dictionary of Biography* (Melbourne: MUP, 1979), vol. 7, pp. 332–3; C B Schedvin, *Shaping Science and Industry: A History of Australia's Council for Scientific and Industrial Research, 1926–1949* (Sydney: Allen & Unwin, 1987), pp. 103–7, 225–6.

[89] G E Rogers, 'Robertson, Thorburn Brailsford (1884–1930)', in G Serle (ed.), *Australian Dictionary of Biography* (Melbourne: MUP, 1988), vol. 11, pp. 420–1; T B Robertson, *The Spirit of Research* (Adelaide: Preece & Sons, 1931); Schedvin, ibid., *passim*.

[90] Education Committee Minutes, meeting of 14 August 1908, UAA, S23, vol. VII, p. 110.

[91] Council Minutes, meeting of 18 December 1908, UAA, S18, vol. IX, p. 76.

[92] Letter W H Bragg to Council, 30 December 1908, UAA, S200, docket 7/1909.

the publication of *The Origin of Species*, and William was appointed Adelaide's representative.[93] It was the first of many occasions on which he would represent and assist the University of Adelaide in the United Kingdom. In June 1909, in a long letter to the Adelaide University Registrar, William reported that he had attended the Darwin celebrations but had been prompted by the non-arrival of the illuminated address from Adelaide to compose an address himself and send it 'up to London to be written and illuminated in all sorts of lovely colours'. He also described two examples of the elaborate academic dress at the celebrations, including 'a violet velvet hat like this [sketch] with violet strings all round, carrying violet blobs at the end that must have tickled him a lot, but I have no doubt it kept the flies off successfully'; a very Australian remark.[94]

At the December 1908 Council meeting William also made two personal requests: 'Will you kindly allow me to take away some special pieces of apparatus which I have used in my research work? I propose to continue the work in England, and it will be a great advantage to me if I may avail myself of the apparatus I have used before'; and 'A small quantity of radium was bought by the University four years ago for £21. There is a "corner" in radium at present... I may have the utmost difficulty in procuring any in England. Would you consent to sell to me a portion of that which you have?... I would propose to purchase the unopened tube for a price to be settled by the Finance Committee'.[95] The Council resolved to make a present of the apparatus and allow the purchase of the radium as requested.[96] In response to an invitation from Johannes Stark to review his work on the nature of radiation for the *Jahrbuch der Radioaktivität*, William explained his difficulty, arising from his forthcoming departure from Adelaide, but did promise to 'write a short account of the γ ray work... during the voyage home'.[97]

Early in November William's closest colleagues at the university joined him for a picnic in the Adelaide hills, arranged by Edward Rennie, the long-serving professor of chemistry. Rogers recorded in his diary, 'Picnic to Prof. Bragg leaves University about 1 o'clock. A great success. Given by Dr R; most enjoyable',[98] and a photograph recorded the occasion (see Figure 14.2). The students said good-bye to Bragg and Madsen during a farewell dinner at Ware's Exchange Hotel early in December. Professor Bragg received a case of razors and Dr Madsen a tobacco jar from the students, and in response William said that 'for nearly a quarter of a century he had been associated with the scientific school of the University, and the growth of the school during that time was

[93] *Register*, 31 August 1908, p. 4.

[94] Letter W H Bragg to Registrar, 29 June 1909, UAA, S200, docket 542/1909.

[95] Letter W H Bragg to Council, 18 December 1908, UAA, S200, docket 929/1908; the radium was returned in 1911, after several delays (letter W H Bragg to Registrar, 13 April 1911, UAA, S200, docket 338/1911, including news of Australian visitors and the beauty of the countryside around their country cottage).

[96] Council Minutes, n. 91.

[97] Letter W H Bragg to J. Stark, 30 December 1908, Staatsbibliothek, Berlin; see next chapter for outcome.

[98] Entry for 3 November 1908, Rogers' diaries, n. 73.

Fig. 14.2 William's closest University of Adelaide colleagues at a picnic to farewell him, Adelaide hills, November 1908. Back row (L to R): Priest, Madsen, Fuller, Benson, Hodge, Cooke; centre row: Rogers, (hidden), Naylor, Henderson, Mitchell, Bragg, Stirling, Adams; front row: Eardley, Higgins, Brown, Chapman, Rennie. (Courtesy: University of Adelaide Archives.)

sufficient reward for his colleagues and himself. He was pleased to have had a hand in the making of the school'.[99]

One of William's first Adelaide students, David Hollidge, had become a leading South Australian educator. In 1902 he had opened a private secondary school, Kyle College, and had persuaded William to become chairman of its council. At the college's seventh annual speech day William presided, although 'he was the victim of an unfortunate throat affliction and had to ask the principal of the school...to read the Chairman's address', in which 'he believed that the difference between boys of apparently diverse mental capacity was often...one of mental discipline'.[100] So fondly was William remembered by the teaching profession after his departure that the Teachers' Union prepared an address 'to be sent to Professor Bragg at the Leeds University...which is handsomely got up [and] contains an album of views of educational institutions in South Australia, as well as beauty spots in the State'.[101]

William's inability to read his speech is not surprising when his programme for that week is considered. On Tuesday afternoon the family attended the St Peter's College speech day, and in the evening they were tendered a farewell

[99] *Register*, 4 December 1908, p. 6; invitation/programme for farewell dinner in Records of Henry Brose, Barr Smith Library, University of Adelaide, file 2.2.

[100] *Register*, 18 December 1908, pp. 5 and 7; in 1919 Kyle College became Scotch College, which is now a prominent Adelaide private school.

[101] *Register*, 14 June 1909, p.4; I have not located the address.

social by the members of St John's Church. On Wednesday the annual university commemoration was held in the afternoon, and William and Willie were guests of the Chancellor at the Adelaide Club in the evening. At St Peter's Bob won a prize for physics, while at St John's a letter outlining the contributions of both Professor and Mrs Bragg to the church was read and gifts were presented. In his reply William hoped that 'Some day... they would see him again, as he had not the least intention of staying away for good (applause) even if he only paid a visit to Australia (Hear, hear). Mrs Bragg also expressed thanks for the presentation which had been made to her'.[102] In fact, neither William nor Gwendoline ever returned to Australia.

The university commemoration was also a notable occasion for all the Bragg family. The Chancellor spoke for the first time in many years, he said, because two forthcoming departures could not be overlooked; namely, those of the Governor and of Professor Bragg. His Excellency and Lady Hunt had been great supporters of the university, while Professor Bragg 'had filled a large place in the University, in its lecture room, its laboratories, its sports and social gatherings, in the movement for extension lectures, in the alliances with the School of Mines and the Education Department, and in the domain of original research... When he took up his work at the University he found two students in the physical laboratory, but now, and for many years past, there had been over 100... No words he might say would be complete without reference to Professor Bragg's relations with the students. (Cheers) He had been as much interested in their social activities as in their progress in the classrooms, and had been always ready to encourage students in their high ideals. (Hear, hear)... he could assure the professor that the people of Adelaide would always regard "Bragg of Adelaide" with pride and affection. (Cheers)'.[103]

Of greater pride and importance for the family, however, was the graduation at the ceremony, 'For the honours degree of Bachelor of Arts... Mathematics—William Lawrence Bragg'.[104] Willie's education had reached an appropriate point for his anticipated entry to Cambridge University, his father's alma mater. Everything had worked out splendidly. That evening Sir Samuel Way hosted a celebratory table at the Adelaide Club. The club had been founded in 1863, modelled on London clubs, had premises in a prime location on North Terrace opposite Government House, and was dominated by Anglicans and pastoralists.[105] The twenty people at the Chancellor's table in addition to Way and Bragg included the Director of Education, four Adelaide professors (Chapman, Henderson, Stirling, and Watson), six Council members (Chapple, Girdlestone, Hamilton, Murray, Poulton, and Talbot Smith), the University Registrar, other academics (Duffield, Fowler, Howchin, and Madsen), and Willie Bragg.[106]

[102] *Register*, 16 December 1908, p. 8.
[103] *Register*, 17 December 1908, p. 8.
[104] Ibid.
[105] D van Dissel, 'Adelaide Club', in W Prest (ed.), *The Wakefield Companion to South Australian History* (Adelaide: Wakefield, 2001), pp. 25–6.
[106] Diary entry for 16 December 1908 and loose 'Plan of Table', in Papers of Sir Samuel Way, State Library of South Australia, Adelaide, PRG 30/1 (diaries of S J Way).

One task only remained after the Christmas and New Year festivities: the twelfth AAAS congress, to be held in Brisbane in January 1909 and for which William had been elected President. The Association had been very important for William's development as a physicist and research worker, and now it gave him its greatest honour. He had attended the early meetings as a novice and now he was acknowledged as their leader. His address would be his farewell contribution to his adopted home; he wanted it to be worthy and he spent time on its preparation. Willie and Bob also went to the congress, Willie being listed amongst the members as H O Bragg because his youthful signature could be misread that way.[107] The boys were now old enough not to feel out of place.

William's address was entitled 'The lessons of radio-activity' and was a very thorough survey of recent developments in the field, highlighting his own particular interests.[108] 'There are two sides of this study to which I would particularly draw your attention', he began. 'We examine the properties of radiation in order to discover on the one hand the nature of radiation itself, on the other the nature and constitution of the atoms or molecules which emit them'. In summarising the early developments of the new science William stressed that:[109]

> we are dealing with the most fundamental characteristics of the atoms, with the building material, and not with the structure; with the inner nature of the atom, and not its outside show; and it is this which differentiates radio-activity from the older sciences. You will remember how Jules Verne in one of his bold flights of imagination drives the submarine boat far down into the depths of the sea. The unrest of the surface, its winds and waves, are soon left behind; the boat passes through the teeming life below, down into regions where only a few strange and lonely creatures can stand the enormous pressure, and, diving still, reaches at last black depths where there is a vast and awful simplicity. Here, where no man 'hath come since the making of the world', the silent crew gazes on the huge cliffs which are the foundations and buttresses of the continent above. It is with the same feeling of awe that we examine the fundamental facts and lessons of the new science.

Has it ever been said better than this? William had now fully developed the talent that enabled him to make science come alive for his listeners and that was to be such a characteristic of his forthcoming years in Britain.

He next discussed the absorption effects of the rays and highlighted the implication that 'the atoms must be very empty things; something like a solar system in miniature, a few significant points or parts, and in between a relatively large amount of almost unmeaning space... now the interior of the atom is no longer a forbidden country, the new radiations pass through the atoms

[107] 'List of Members', in J Shirley (ed.), *Report of the Twelfth Meeting of the Australasian Association for the Advancement of Science, held at Brisbane, 1909* (Brisbane: Government Printer, 1910).
[108] Shirley, ibid., pp. 1–30.
[109] Ibid., pp. 5–6.

with ease...We may look on such transits as journeys of exploration and hope to learn something of the nature of the interior of the atom'.[110] This was followed by a lengthy summary of his own research on the fate of alpha particles once they emerge from radioactive atoms and on the nature of Röntgen and gamma rays. He gave attention to the apparent interchangeable relationships between X-rays and cathode rays, and between gamma and beta rays, topics that he would explore further in Leeds.

Finally, William offered some thoughts on research, as a result of his Australasian experience:[111]

> The discussion of any pure scientific research before an Australasian audience like this naturally brings forward the question as to how far Australasians are themselves justified in spending their time and money on such work...And again, if there is work which should be done, who is to do it?...
>
> First, then, as regards pure science, the one all-important thing to remember is that pure science lies at the root of all applied science. The former throws up and nourishes the stems which bear the latter as their fruit...if we are content in this country to import always the flowers of European or American thought, and to use them in the establishment of our industries and to grow nothing of our own, then we must be continually replenishing our ideas from abroad...That is neither an honourable nor an economic arrangement...
>
> Now it is true that there are branches of scientific research which have a more or less obvious relation to Australasian progress. But we may also aspire to do work which does not appear to advantage our own country more than the world at large. Indeed, if we wish to take our place amongst the progressive peoples of the world, to gain the strength and inspiration which come from sharing in a common advance, and to shun the soul starvation which would follow on a selfish concentration on our own immediate advantage, we must play our part in this sense also, and play it enthusiastically and well. Pure scientific research is necessary not only *to* Australasia but *in* Australasia; to bring in the spirit of the patient and reverent search for truth, to illustrate the searcher's methods, to open up new fields, and to answer the questions that arise and will arise to an ever-increasing degree if the progress of the country is to be sound.

William then went on to discuss the impact of relevant research in mining and agriculture and to encourage the entry of more young people into universities and into research, not least in Brisbane, where the University of Queensland was in labour pending its birth in 1910.

In addition to the local newspapers, William's address was reported internationally. The journal *Chemical News* carried extensive abstracts;[112] Thomas

[110] Ibid., p. 11.
[111] Ibid., pp. 24–25.
[112] W H Bragg, 'The lessons of radio-activity', *Chemical News*, 1910, 101: 101–3, 111–13, 134–7, 148–9.

Laby wrote from Emmanuel College, Cambridge, passing on a request from the editors of *Scientific Progress* for an article and suggesting that it be based on William's [then forthcoming] Brisbane address;[113] and Alfred Deakin, the Australian Prime Minister, commended the address to his British audience. Before he became Prime Minister, Deakin had accepted an offer from the *Morning Post* in London to become its Australian correspondent, writing a weekly letter and sending occasional cablegrams for a payment of £500 per year. Extraordinarily, this continued throughout Deakin's parliamentary career and, immediately after William's address, Deakin composed a letter on 'Australasian science', which was published in the *Morning Post* on 10 March 1909.[114] In a brief but excellent summary of the rationale for the conference and of William's address, Deakin noted that, 'Over 500 of the leading Australasian representatives of science, education and philosophy have gathered in Brisbane…This year's President, Professor Bragg, pointed out in an altogether admirable address that there remains for solution a large number of scientific problems, many of them peculiar to Australia, intimately associated with our chief sources and processes of wealth production'. And he went on to outline William's mining and agricultural examples and to stress that 'All these difficulties can only be met by the patient and determined application of scientific methods'.

On 16 December 1908 the *Register* reported, 'A vessel which during the past few months has been the subject of much discussion in shipping circles is the steamer Waratah, the latest addition to the fleet of the well-known Lund's Blue Anchor Line.…and her arrival at Pt Adelaide…was awaited with a good deal of interest…There were 780 people on board, mostly immigrants…The Waratah complies with the highest requirements of the Board of Trade for passengers, and is classed 100 A1 by Lloyd's'.[115] There followed an extensive description of the ship and her spacious and attractive accommodation for passengers. It was her maiden voyage, and she would soon return to Adelaide from the eastern states on her return journey to Britain. On 22 January 1909 the *Register* reported, 'Professor Bragg…left the outer Harbour on Thursday by the Blue Anchor liner, Waratah. He was accompanied by Mrs Bragg and their three children [and Charlotte, who had become a family institution]. A large number of friends assembled at the wharf to say goodbye'.[116] One person must have been especially sad to see them go. Sir Charles Todd's wife was dead and much of his family was dispersing, far away from Adelaide. He was old, and he knew it was unlikely that he would ever see his daughter Gwendoline, his scientist son-in-law, and his Bragg grandchildren again.

[113] Letter T Laby to W H Bragg, 21 August 1908, RI MS WHB 4A/1; the invitation was not taken up.

[114] J A La Nauze, *Federated Australia: selections from letters to the Morning Post, 1900–1910* (Melbourne: MUP, 1968), letter no. 116, pp. 253–4; for Deakin see J A La Nauze, *Alfred Deakin: a biography* (Melbourne: MUP, 1965), particularly vol. 2, ch. 15.

[115] *Register*, 16 December 1908, p. 5.

[116] *Register*, 22 January 1909, p. 5.

15
Hello England!

The *Waratah* was a luxurious, coal-fired vessel, constructed for the emigrant-cargo route between Britain and Australia, its name a reflection of its affinity with the antipodes, the waratah being the floral emblem of the State of New South Wales. The ship was said to be unsinkable because of eight watertight compartments along its hull. However, Captain Josiah Ilbery and the crew were sufficiently concerned about her stability on the outward leg of the maiden voyage for the ship to be inspected and loaded under the captain's personal supervision before the journey home. The Bragg family were passengers on this return leg.[1]

From Albany, on the south coast of Western Australia, William sent a folded 'letter card' to Arthur Rogers saying, 'Thank you for your letter and good wishes. They must be bearing fruit, for we are having a first class voyage so far!'.[2] In testimony before a later inquiry, however, William reported that he had been 'very alarmed' at periods during the voyage across the Indian Ocean to Durban. 'I thought she was unstable for small displacements', he said, 'my impression was that her metacentre was just slightly below her centre of gravity when she was upright, and then as she heeled over on either side she came to a position of equilibrium'. He reported a list of four or five degrees that would last for several days, after which 'she came upright and then went over and stopped down on the other side'. William recalled frequent breakfast requests to the captain 'to make the ship level'; but he also reported that the ship rolled very little and 'was a most remarkably steady and comfortable boat'. He had heard conflicting statements from engineers on board, saying that 'she was as safe as a church' and that 'she was the tenderest ship he had ever been on'.[3]

[1] A Villiers, *Posted Missing: The Story of Ships Lost Without Trace in Recent Years* (New York: Scribner, 1974), chs 9, 10; P Ilbery, 'The loss of the Waratah, 1909', *Journal of the Royal Australian Historical Society*, 1996, 82(1):73–87; 'Report at Home, 1910/11–1', being the Report of the Inquiry into the Wreck of the SS Waratah, U.K. Board of Trade, 22 February 1911, Marine Library of U.K. Department of Transport, p. 7.

[2] Letter card W H Bragg to A L Rogers, n.d., UAA, Oliphant papers.

[3] Testimony of Professor W H Bragg to Inquiry into the Wreck of the S S Waratah, London, 17 December 1910, Martine Library of U.K. Department of Transport, pp. 121–9; *The Times*, 19 December 1910, p. 3; for a floating object (ship), only if its meta-centre is above its centre of gravity will it tend to right itself from a tilted position (list), *The Penguin Dictionary of Physics* (London: Penguin Books, 1991), p. 296.

William's recollections arose because the *Waratah* had retraced the route via South Africa on her second round trip to Australia and return and had disappeared in a hurricane and heavy seas between Durban and Cape Town on the return journey. Despite wide-ranging searches by many ships nothing was found: no lifeboats, no debris of any kind, no bodies of the 211 passengers and crew on board. Numerous possible explanations emerged from the Board of Trade inquiry held from December 1910 through February 1911, of which instability was perhaps the most plausible, but none had adequate supporting evidence. The court concluded that, 'the ship was lost in the gale of 28th of July, 1909, which was of exceptional violence', but that 'the particular chain of circumstances leading up to this is a matter of mere conjecture'.[4] A number of searches have been made over subsequent decades, but nothing has been found. In 1924 one author dismissively concluded that, 'The *Waratah* disappeared as completely and as mysteriously as if she had never existed. The sea seemed to have opened and swallowed her up'.[5] Now, however, this seems the most likely explanation, for oceanographers have discovered the possibility of a 'hole' in the sea off the southeast coast of Africa. The southwards Agulhas current becomes an 'oceanic river' as it passes between the island of Madagascar and the African mainland, and when it meets a gale and heavy swell coming up from the southwest, 'abnormal waves of up to 20 metres in height, preceded by a deep trough, may be encountered'.[6] Several South Australians whom the Braggs knew perished in the disaster.[7] They must have wondered about their own safe arrival in England.

It is curious that William never returned to Australia later, and that Willie returned only once, after his passage was assured by an invitation to lecture in New Zealand, for references to Australia occurred throughout the remainder of their lives. William 'always spoke with the greatest affection of South Australia',[8] and 'had so become part of the Australian way of life that England was to some extent like a foreign country to him';[9] while Willie's children hold an attachment to the country that is a direct reflection of their father's feelings. An example of a later Australian resonance is the story told by Bill Coates, Willie's lecture assistant at the Royal Institution.[10] Asked by the Director to see how far he could throw a wooden dart down the long 'red' corridor of the Institution, Coates managed about five metres. Using the woomera principle of the Australian Aborigines, Willie attached a length of string to the dart and

[4] Report at Home, n. 1, pp. 1, 21.

[5] J G Lockhart, *Mysteries of the Sea*, 2nd edn (London: Philip Allen, 1925), p. 222.

[6] R Kennedy, 'The ship that fell down a hole', *The Times*, 16 July 1999, p. 11, although this report of the discovery of the *Waratah* was mistaken.

[7] A Laube, *A Lady at Sea: The Adventures of Agnes Grant Hay* (Adelaide: author, 2001); *Register*, articles throughout August 1909; letter C Todd to W L Bragg, 19 August 1909, RI MS WLB 95C/2.

[8] E N da C Andrade, 'William Henry Bragg, 1862–1942', *Obituary Notices of Fellows of the Royal Society of London*, 1942–44, 4:276–300, 282.

[9] Sir Lawrence Bragg and Mrs G M Caroe, 'Sir William Bragg, F.R.S., 1862–1942', *Notes and Records of the Royal Society of London*, 1962, 17:169–82, 175.

[10] W A Coates, 'Sir William [Lawrence] Bragg and his lecture's assistant', in J M Thomas and Sir David Phillips (eds), *Selections and Reflections: The Legacy of Sir Lawrence Bragg* (London: Royal Institution, 1990), pp. 147–9.

hurled it the length of the corridor, only narrowly missing a large oil painting on the end wall. 'Simple leverage', he said, while Coates wondered how he could have explained any consequent damage 'while playing darts with the Director'!

The family disembarked at Plymouth in early March 1909. William and Gwen left Charlotte and the children at lodgings while they went to Leeds for a few days, to find temporary accommodation and make a preliminary survey of the university. William reported to Arthur Rogers:[11]

> I have just heard from Hodge that you have been unwell... My dear old fellow, I am so sorry, and I hope you are better...
>
> The Leeds laboratories are very curious; but I think they can be made quite workable. There is one gigantic room [in the 'physics sheds'], which would take 200 students at a time, & really I think I shall put all the students in that I can, and leave all the rest for private work. I am to have a good bit of money to spend on apparatus and I think I ought to make a good thing of it. It does not look much of a workshop after yours; I wish I might have you for a while.
>
> I had a day with Rutherford in Manchester, and you can imagine it was very interesting! He showed me 250 mmg of Ra in solution in a pump. They draw off the emanation as wanted... He does not like Pye's cells... One of his research students, Geiger, says he knows where to get good cells in Germany, and is writing both for Rutherford and me.
>
> The Leeds people are really very nice: all those I met anyway. The place itself is grimy, even the suburbs; but you can get out into beautiful country to the north... It was awfully cold for some weeks after we got here... but I am glad to say the children, baby & all, stood it well.

Notable is William's immediate visit to see Rutherford in Manchester and to benefit at once from being close to European facilities. He also noted that the beauty of the Yorkshire countryside north of Leeds offered relief from the city's evident grime, and a country cottage was acquired a year later, near Bolton Abbey (Priory) in Wharfedale, about 20 miles (30 km) north of Leeds. Amongst a group of farm buildings, at the end of a stone-walled lane, William rented a small, two-storey stone farmhouse called 'Deerstones', with Kex Beck flowing by and Beamsley Beacon above. It was a happy release from the pressures of Leeds and the family loved it. Gwendy later remembered that 'I loved Deerstones passionately, and I saw the grown-ups happy there too. WHB enjoyed the simplicity and the quiet; it had a special peace filled with the rustle of beech trees and the sound of the beck; and GB [her mother] baked her own bread and painted in watercolours. My brothers went sketching too, and friends came out from Leeds... GB painted purple heather and vivid green trees... she found the emerald green [useless in Australia] most usable; she could not get over the green of England, delighting in it'.[12]

[11] Letter W H Bragg to A L Rogers, 8 April 1909, Bragg (Adrian) papers.

[12] G M Caroe, *William Henry Bragg 1862–1942: Man and Scientist* (Cambridge: CUP, 1978), pp. 65–6; letter W H Bragg to C R Hodge (Adelaide Registrar), 20 July 1910, UAA, S200, docket 563/1910; Bolton Abbey never was an abbey, it was an Augustinian Priory.

Fig. 15.1 Bob Bragg at Deerstones, Yorkshire, *circa* 1913. (Courtesy: Lady Adrian.)

On arrival in Leeds in early April the family lived for some months in rented accommodation.[13] Gwendoline was appalled 'by the dirt, by the smokey dark, by the rows of little poor back-to-back houses... She was horrified by the pasty-faced babies carried folded into their mothers' shawls, and the rickety children... my white toddler's clothes were changed twice a day, the boys had nothing to do, and WHB was stricken by GB's despair'.[14] In the early summer, however, things began to improve. William rented 'Rosehurst', a solid, two-storey house with a billiard room and large garden, in fashionable Grosvenor Road, Headingley, within walking distance of the university and the famous cricket ground.[15] He described it glowingly to Rogers.[16] The garden fascinated Gwendy, two maids entertained her, and her mother began to make friends with the pleasant wives of wealthy industrialists. Gwendoline was an immediate success when elected to the Art Club, and she threw herself into social work as her mother had done in Adelaide. She lost her initial horror of the city.[17]

In contrast, Gwendy paints a gloomy picture of her father during the early years in Leeds. 'For the first three years he was too miserable about his work, the richness in Leeds made him uncomfortable, the poverty saddened him. He had golfing acquaintances and a tremendous supporter in Professor Arthur Smithells, but socially he sheltered behind GB, a habit which continued till

[13] *Kelly's Directory of Leeds, 1909*, and family correspondence.
[14] Caroe, n. 12, p. 53.
[15] *Robinson's Leeds Directory, 1910–11*; Rosehurst is now used as a Leeds University guesthouse.
[16] Letter W H Bragg to A L Rogers, 17 November 1909, Bragg (Adrian) papers.
[17] Caroe, n. 12, pp. 64–5.

the end of her life. Some Leeds friendships made by GB have lasted into the second and third generation'.[18] William was disappointed with what he found, no doubt: poor university accommodation and facilities, an inadequate workshop, a curriculum not of his making, very few honours and no postgraduate students, and few staff—just one assistant lecturer (Allen) and two assistant demonstrators (Shorter and Edmonds). As his predecessor noted, 'the old Yorkshire College in my time suffered from acute penury'.[19] It was early Adelaide all over again. But perhaps Gwendy's picture is too pessimistic. She was only two years old when the family arrived in England and her view is likely to have been formed later at the Royal Institution, when she was her father's companion. By then William had another reason to remember Leeds with abiding sadness. For the present he had a major task before him, nicely captured in a rhyme of the time:[20]

> Here's to Professor Bragg
> Who sailed in from down under
> To make this College wag
> Its physics tail in wonder.

The person who recalled this verse also added, 'Mrs Bragg was an Australian, and though extremely nice, was rather the dominant character. She gave wonderful children's parties and I used to come home laden with presents, given either for winning at games or as compensation for losing'; another Adelaide tradition continued.

There is little information on the undergraduate physics teaching during this period; William presumably continued with the existing curriculum, modifying it from time to time but generally following a conventional programme. The laboratory equipment was soon upgraded, but neither William nor the university spoke of radical reform. William's essential focus was research: his own, that of others in physics and, by his own example, those in the rest of the university, particularly in the science departments. By the middle of the year the university's Annual Report for 1908–09 was able to report:[21]

> The Professor of Physics states that... he began as soon as possible to collect the necessary apparatus for the prosecution of research work in radio-activity. So far about one-third of the special sum promised by the Council has been expended, and the research laboratory is already sufficiently furnished to allow about half-a-dozen independent researches to be carried on at the same time. The equipment had to be provided speedily, for several applications were made for permission to attend at Leeds during the long vacation... Dr Beatty and Dr Kleeman came from

[18] Ibid., ch. 4, p. 65.

[19] W Stroud, *Apologia pro Vita Mea: Being a Record of the Troubles and Pleasures of a Cavendish Professor of Leeds 1885–1909* (Leeds: no publisher, no date), p. 3, reprinted in *Newsletter—Special Issue*, Institute of Physics (UK) History of Physics Group, October 2005, pp. 8–31.

[20] Verse from Leeds University song, circa 1910, courtesy of Stephen Bragg.

[21] University of Leeds Archives, *Annual Report, 1908–09*, pp. 48–9.

> Cambridge early in July. The former took in hand a difficult experiment, which...is now giving interesting results. The latter has nearly completed a preliminary survey of a new field of research, and will probably have results to publish within the next few weeks. Mr Vegard came later from Kristiania [Oslo], and has completed an important experiment on the polarisation of Röntgen rays; he is now extending it. Mr Thirkill, of Clare College, Cambridge, came in September, and has just finished the construction of apparatus for some alpha-ray experiments.
>
> Mr Hartley...has been gaining experience in radio-active work by carrying out a research on the gamma rays. Of our own demonstrators, Mr Keene, before he left, fitted up some apparatus for the Professor which has given most interesting results...Mr Edmonds...is now attacking a problem of considerable interest. Mr Shorter, in addition to work of his own design, is fitting up some of the Professor's own apparatus so as to continue the latter's previous work.

Two things are immediately apparent: the rapid arrival of visiting researchers of considerable ability, and William's urge to justify his appointment and its special funding as soon as possible. Beatty was an 1851 Science Research scholar from Queen's College, Belfast, who had a joint attachment to Cambridge and Leeds, Kleeman was William's first Adelaide research student and had published prolifically since arriving in the Cavendish Laboratory, and Thirkill had recently arrived at the Cavendish and was destined to remain there for many years as a laboratory demonstrator and its voluntary accountant.[22] The Norwegian Lars Vegard was, successively, a student at Kristiania, Cambridge, Leeds, and Würzburg during the years 1899 to 1912, before gaining a Dr. phil. at Oslo in 1913. He was soon to play a vital part in our story.[23]

A year later William was able to report yet further additions to the facilities for teaching and research.[24] In particular, 'the new workshop has been completed and furnished with the principal tools required, and the new instrument fitter, C H Jenkinson, lately one of the foremen of the Cambridge Scientific Instrument Co., has been at work for some time. The Research Laboratories are being gradually equipped from the grant made for the purpose. A special grant of £300 was made by the Council for the purpose of replenishing the stock of teaching apparatus, and most of this has been expended with very welcome results'. In addition, 'The Department has been strengthened very greatly by the presence of Mr Norman Campbell, lately Fellow of Trinity College, Cambridge, an able worker and writer on the modern development of

[22] Kleeman told Rutherford that he had received 'a letter from Prof. Bragg...[and] during the vacation at the end of this term I will go to Leeds to lend him a helping hand to fix things up', letter R Kleeman to E Rutherford, January 1909, CUL RC K47; for Beatty see *Records of the Science Research Scholars of the Royal Commission for the Exhibition of 1851* (London: Commissioners, 1961), no. 293, p. 40; for Thirkill see J G Crowther, *The Cavendish Laboratory, 1874–1974* (New York: Science History Publications, 1974), pp. 207, 226.

[23] 'Vegard, Lars', in *J C Proggendorff's Biographisch-Literarisches Handwörterbuch für Mathematik, Astronomie, Physik,...* (Leipzig: Verlag Chemie, 1926), vol. V: 1904–22, p. 1303.

[24] University of Leeds Archives, *Annual Report, 1909–10*, p. 61.

electrical science; he has asked to be allowed to join in an honorary capacity, and proposes to live and work here, and in particular to take charge of the arrangements and apparatus in the Research Laboratory'. About this time, H L Porter, BSc (London), also joined the department as a demonstrator and began to assist William with his research.[25]

These were long-term additions to the staff, not just summer visitors. Jenkinson is seen as a young man in a photograph (circa 1894) of staff of the Cambridge Scientific Instrument Company, and we may assume that he had a detailed training in scientific instrument making and repair before he rose to become one of the company's foremen.[26] He was to prove an admirable replacement for Arthur Rogers and he stayed with William for the remainder of William's career, including his time at University College London and the Royal Institution. Rogers and Jenkinson were indispensable to William's experimental brilliance and success, building apparatus 'elegant in its simplicity and fitness'.[27]

Norman Campbell's years in Leeds are an interesting episode in the life of a fascinating man. Educated at Eton and Trinity College, Cambridge, he obtained a first-class result in the Natural Sciences Tripos of 1902 (BA, MA 1906, DSc 1912) and was a Fellow of Trinity from 1904 until 1910. He worked closely with J J Thomson in the Cavendish Laboratory on Thomson's major project on the ionization of gases, and also on the weak radioactivity of natural potassium and other elements, in which he referred to William's studies of alpha-particle decay.[28] During 1905–07 Campbell was a major contributor to British debates concerning Einstein's theory of relativity,[29] while in 1907 he published the first edition of his *Modern Electrical Theory*, which particularly attracted William's attention.[30] In February 1909 Campbell began fundamental experiments for Thomson on the nature of light, the results of which were published in the winter of 1909–10.[31] Campbell must have had independent financial support, for he now moved to Leeds and held only an honorary position, which the university formalised early in 1912 with the creation of a position entitled 'Honorary Fellow for Research in Physics'.[32] Keen to further his career beyond Cambridge, Campbell's

[25] University of Leeds Archives, *Annual Report, 1910–11*, p. 55.

[26] M J G Cattermole and A F Wolfe, *Horace Darwin's Shop: A History of the Cambridge Scientific Instrument Company, 1878 to 1968* (Bristol: Adam Hilger, 1987), Figure 3.1 on p. 49; the single reference in the index to Jenkinson is to another man, Francis Jenkinson.

[27] Caroe, n. 12, pp. 30–1.

[28] J Nicholas, 'Campbell, Norman Robert', in C C Gillispie (ed.), *Dictionary of Scientific Biography* (New York: Scribner's Sons, 1971), vol. III, pp. 31–5; the Bragg reference is in N R Campbell, 'The radio-activity of metals and their salts', *Proceedings of the Cambridge Philosophical Society*, 1904–06, 13:282–7, 282.

[29] A Warwick, 'Cambridge mathematics and Cavendish physics: Cunningham, Campbell and Einstein's relativity, 1905–1911, Part I', *Studies in History and Philosophy of Science*, 1992, 23:625–56; ibid., Part II, 1993, 24:1–25; S. Goldberg, 'In defense of ether: the British response to Einstein's special theory of relativity', *Historical Studies in the Physical Sciences*, 1970, 2:89–125.

[30] See chapter 13.

[31] R McCormmach, 'J J Thomson and the structure of light', *British Journal for the History of Science*, 1967, 3:362–87, 378–9.

[32] University of Leeds Archives, Council Minutes, meeting of 21 February 1912; ibid., letter N R Campbell to Vice-Chancellor, 23 February 1912, gratefully accepting the distinction.

independence enabled him to join the man (William Bragg) whose work he most admired. It was a fruitful relationship until the Great War intervened.

Nor was research in Australia forgotten. At the end of his time in Adelaide Madsen had turned to a study of the scattering of radium β-rays from aluminium and gold foils of various thicknesses. He compared the radiations produced on the incidence and emergence sides of the foils, found an expected asymmetry analogous to that found for γ- and X-rays, and noted a close parallel between the scattering of all three rays, in support of William's various claims.[33] In addition, Madsen's apparatus enabled him to compare small- and large-angle scattering for various thicknesses of the absorbers, and he found that large-angle scattering was still significant even for extremely thin foils. 'We are concerned with only a single collision of any γ particle', Madsen concluded.[34] When J J Thomson's theory of γ-ray scattering appeared shortly thereafter, based upon a multiple-scattering hypothesis and apparently supported by an experimental study by J A Crowther, Madsen's contrary result became unexpectedly significant.[35]

William was confident of Madsen's work, recommended it to Rutherford's attention, and supported it vigorously against the Cavendish group, saying Thomson's theory 'seems to me to be inapplicable to the actual case... From the very first, large deflexions must be considered... I think it only by accident that [Crowther's] aluminium curve fits [Thomson's] formula'.[36] In a significant paper on the emergence in 1911 of Rutherford's nuclear model of the atom, historian John Heilbron has classified Madsen's result as 'of exceptional importance',[37] seeing it as contributing to the development of the 'Manchester Approach' to the scattering of α- and β-rays, which rested on the hypothesis of single scattering, sometimes through large angles, and which led Rutherford to his new model. William and Rutherford both urged Madsen to complete his further studies of β-ray scattering, but he was unable to take advantage of the opportunity; he not only abandoned the project but also research in general for the remainder of his career. Madsen devoted his energies instead to a professorship of electrical engineering at the University of Sydney and to his increasingly important role in the promotion and administration of Australian physics and engineering.[38]

[33] J P V Madsen, 'The scattering of the β rays of radium', *Transactions of the Royal Society of South Australia*, 1909, 33:1–10, and *Philosophical Magazine*, 1909, 18:909–15 (also read at the Brisbane AAAS meeting, January 1909).

[34] Ibid., p. 913.

[35] J J Thomson, 'The scattering of rapidly moving electrified particles', *Proceedings of the Cambridge Philosophical Society*, 1910, 15:465–71; J A Crowther, 'On the scattering of β rays from uranium by matter', *Proceedings of the Royal Society of London*, 1907–8, A80:186–206.

[36] W H Bragg, 'The consequences of the corpuscular hypothesis of the γ and X rays, and the range of β rays', *Philosophical Magazine*, 1910, 20:385–416, 414; also *Jahrbuch der Radioaktivität*, 1910, 7:348–86.

[37] J L Heilbron, 'The scattering of α and β particles and Rutherford's atom', *Archive for History of Exact Sciences*, 1967–8, 4:247–307, 283; see also W H Bragg, *Studies in Radioactivity* (London: Macmillan, 1912), ch. VIII; R W Home, 'W H Bragg and J P V Madsen: collaboration and correspondence, 1905–1911', *Historical Records of Australian Science*, 1981, 5:1–29.

[38] Home, ibid., p. 9.

There was regular correspondence between Bragg and Rutherford during this period, not least concerning the β-ray scattering controversy.[39] In addition, what had been a warm association by correspondence across the world now became a close professional and personal friendship. Writing to Madsen five months after his arrival in England, William reported excitedly: 'I had Rutherford staying here for two days and it was great fun: he and his wife came. We had quarters in a jolly old farm house overlooking Bolton woods and with the moors at the back: a glorious place altogether...The moors belong to the Duke of Devonshire and he & the Prince of Wales are coming down to shoot next week. The heather is just coming out. Well, Rutherford and I talked hard, culminating on the last evening: when we both got excited and stamped about the room at intervals, to the amusement of our wives. He is very sympathetic to the material theory...At one time he broke out with conviction, "The old pulse theory (pause), the old pulse theory, Bragg, is as dead as mutton!" And he won't believe in J.J.'s energy blobs: no one does, I think...We talked a lot about Barkla's recent work, which of course is awfully good'.[40]

A year later, when they were both on a committee choosing a foundation professor of mathematics and physics for the University of Queensland in Australia, William again invited Rutherford to come to Yorkshire: 'I wonder if you would come over & stay a night. I am alone; my wife is at Cambridge installing the boy [Willie] into his College rooms & it would be great cheers to have you'.[41] At Christmas he wrote, 'It would be awfully jolly if you would come over and see us, all of you. We have the boys home...The atom sounds very fine. My boy wants to know if he may hear about it too, as he has been going to J J's lectures'.[42] And in the new year, having developed a piece of apparatus to demonstrate the passage of an α-particle through a Thomson 'plum-pudding' atom—with a swinging magnet representing the α-particle and static magnets representing the atomic electrons—William lent it to Rutherford to see if it could also demonstrate the passage through an atom with 'a big positive at the centre', as in Rutherford's new model of the atom.[43]

William's new staff at Leeds quickly became productive. The university's *Annual Report* for 1909–10 listed five publications, two by Kleeman, one by Vegard, and two by Bragg himself.[44] William first replied to a paper in the *Physical Review*, the major American physics journal, which had claimed that his model for a particle structure of X- and γ-rays was not justified because it assumed a positive component of the same mass as the negative. William

[39] Heilbron, n. 37.

[40] Letter W H Bragg to J P V Madsen, 1 August 1909, Basser Library, Australian Academy of Science, Canberra, A C T, Australia.

[41] Letter W H Bragg to E Rutherford, 13 October [1910], CUL RC B379.

[42] Letter W H Bragg to E Rutherford, 21 December [1910], CUL RC B380A.

[43] Letters W H Bragg to E Rutherford, 8, 19 February [1911], 11 March 1911. CUL RC B382, 384, 385 respectively; E Rutherford to W H Bragg, 9, 11 February 1911, RI MS WHB 26A/18, 19 respectively; there are two photos of the apparatus in letter W H Bragg to Rogers, 21 November 1912, Bragg (Adrian) papers.

[44] *Annual Report*, n. 24, p. 73.

denied that he had made such an assumption, but his suggestions regarding the nature of the positive contribution were unconvincing. It was a troubling aspect of his hypothesis. He was on safer ground with his criticism of the assertion that the secondary radiation produced by β-rays was dependent on the chemical composition of the scatterer: 'a β particle acts upon an atom in a way which is independent of the neighbourhood or association of other atoms', he claimed.[45] Kleeman's papers from Leeds, communicated to the Royal Society by William, extended his study of the ionization produced in gases by β-rays. The first paper appeared to add weight to the suggestion, supported by Rutherford but later shown to be erroneous, that the energy of γ-rays from a radioactive substance was causally related to that of the γ-rays from the same element,[46] while the second gave strong support to his mentor's belief that 'the energy necessary to ionise an atom is the same for the β-particle as for the cathode rays'.[47]

Vegard focused on the difficulty of the supposed polarization of X-rays for William's particle model and explored several possibilities: that the primary X-rays consisted of two components, one polarized and one not, that the extent of the polarization depended on the energy of the X-rays, and that the two components differed in their ability to produce secondary cathode rays. The apparatus, designed by William, was complex, the experimental work by Vegard was extensive, and the results confirming these possibilities seemed convincing.[48] The effort required to set up the laboratory, equip the workshop, build the apparatus, and then carry out the experiments was enormous; surely William was largely satisfied with his earliest months in Leeds.

William's second paper of 1910 was an extensive review of his neutral-pair model of X- and gamma radiations. This was prompted by Stark's request for a contribution to his *Jahrbuch*, but it was also a useful survey for William in planning his future programme.[49] He focused on the passage of the new rays through matter, considered a very wide range of phenomena, and emphasized the individuality and locality of the various 'entities', as distinct from the spreading pulses of the wave theories. Something of its flavour is given by William's stated objectives at the beginning of the paper and his conclusions at the end:[50]

> I have myself found it convenient to regard the X ray as a negative electron to which has been added a quantity of positive electricity which neutralizes its charge but adds little to its mass. Whatever view may be taken of the nature of the entity, the acceptance of the corpuscle idea

[45] W H Bragg, 'The secondary radiation produced by the beta rays of radium', *Physical Review*, 1910, 30:638–40.

[46] R D Kleeman, 'The ionisation of various gases by the β-rays of actinium', *Proceedings of the Royal Society of London*, 1909–10, 83A:530–3.

[47] R D Kleeman, 'The total ionisation produced in different gases by the cathode rays ejected by X-rays', *Proceedings of the Royal Society of London*, 1910, 84A:16–24, 20.

[48] L Vegard, 'On the polarisation of X-rays compared with their power of exciting high velocity cathode rays', *Proceedings of the Royal Society of London*, 1909–10, 83A:379–93.

[49] W H Bragg, n. 36.

[50] Ibid., pp. 386, 416.

> modifies our view of the phenomena attending the passage of rays through matter, and alters the language which we use in describing experimental results. I think that it leads to a marked gain in simplicity, and my object in writing this paper is to show, if I can, that this is the case...
>
> In the foregoing pages I have tried to follow out the consequences of adopting the 'entity hypothesis' of X and γ rays... We are to think of each entity as possessing initially a certain store of energy which it spends gradually as it goes along, the result being ionization of the material through which it passes;... the deflexions or turnings being the result of intra-atomic collisions; the β rays are very liable to such deflexions, and the cathode rays even more so. Certain conversions of form may take place, γ into β, X into cathode ray, and so on; but in such cases the energy is handed on and, in some cases at least, the momentum. The essence of it all is the recognition of the individuality of each entity, which is to be followed by itself from its origin through all its changes of direction and sometimes its changes of form, until its gradually diminishing energy becomes too small to render it distinguishable.

There are undertones of quantization here and, although William did not use the word, he did refer to the concept briefly: 'J Stark has recently developed the theory, based on the work of Planck, that an X ray is a bundle of energy travelling without alteration of form'.[51] Prompted by information from Campbell,[52] the German work was at last breaking through William's ignorance of the language. William was correct in noting that Stark's work was significant, for Stark had suggested that when electrons collided with a metal in an X-ray tube each one produced an X-ray quantum in a process conserving momentum. The notable German physicist Arnold Sommerfeld had immediately challenged Stark's theory, saying it was unnecessary to introduce quanta and that the process could be completely explained on the pulse theory.[53]

Sommerfeld's support for the pulse theory was potentially a serious problem for William's particle model and he wrote to Sommerfeld early in 1910, outlining his views boldly, although 'I feel a considerable diffidence in stating my views to so great a master of mathematical analysis'.[54] Sommerfeld replied at once, acknowledging that William was 'among those who have most successfully investigated these matters', but emphasizing the successful aspects of the pulse theory. Sommerfeld then undertook a detailed theoretical study of γ-rays, and he sent William a reprint of the resulting article early in 1911. William responded, acknowledging the value of the electromagnetic treatment but again questioning, 'How do you propose to get the energy back

[51] Ibid., p. 386.

[52] McCormmach, n. 31, p. 378.

[53] R H Stuewer, 'William H Bragg's corpuscular theory of X-rays and γ-rays', *British Journal for the History of Science*, 1971, 5:258–81.

[54] This correspondence is discussed in detail by Stuewer, ibid; for another discussion of William's neutral-pair model, and in a wider context, see M C Malley, 'From hyperphosphorescence to nuclear deacay: a history of the early years of radioactivity, 1896–1914', University of California, Berkeley, Ph.D. thesis, 1976, University Microfilms, Michigan, no. 77–4527.

again from this everspreading ring to a single electron... for the production of a β ray by a γ ray?' He then added, 'I am very far from being averse to a reconcilement of a corpuscular and a wave theory: I think that some day it must come. But at present it seems to me that it is right to think of the X or the γ ray as a self contained quantum', and he referred to C T R Wilson's new and extraordinary pictures of these individual atomic processes. The production of *one* X-ray by *one* electron in an X-ray tube, and the subsequent recreation of *one* electron by *one* X-ray in a photoelectric event, all carrying approximately the same energy as the original electron, became a central plank in William's argument in support of his neutral-pair hypothesis.[55]

William's reputation as a lecturer for non-specialist audiences preceded him, and he was soon approached to speak at the Royal Institution in London.[56] Having declined early in 1910, he accepted a year later, described some relevant experiments and reiterated points made in his review paper, and said, 'These few experiments... may serve to illustrate both the justice and the convenience of placing all these rays, α, β, γ, and X, in one class. We are tempted to consider them all as corpuscular radiations of some sort'. He was forced to conclude with an emerging puzzle, however, 'For it appears that ultra-violet light possesses many of the properties of x and γ rays', but particulate ultraviolet light 'seems to throw away at once all the marvellous explanations of interference and diffraction which Young and Fresnel founded on a theory of spreading waves... The whole situation is most remarkable and puzzling'.[57] William immediately received requests to publish the lecture and it was reproduced in full in *Nature*, *Chemical News*, and *Archives of the Röntgen Ray*.[58] The wave theorists were not persuaded, however, and some were hostile. H A Wilson, Rutherford's successor at Montreal, wrote to Rutherford saying, 'physicists who have the position of the English school of physics at heart ought to make a stand against the rotten claptrap theories that are being ventilated nowadays; e.g. Lodge's density of the ether, J J's bundles of energy, & Bragg's corpuscular Röntgen rays'.[59]

William's review papers were written as he grappled with his first experiments at Leeds, assisted by Porter. He focused on the ionization of materials by X-rays and asserted that ionization was an indirect or secondary process; the appearance of electrons in the material was not caused by the X-rays themselves but by the action of the electrons that were released when the X-ray neutral pairs broke up in the material. Similarly, the appearance of secondary

[55] Stuewer, n. 53, pp. 273–4.

[56] Letters W Crookes to W H Bragg, 11 February 1910 and 14 January 1911, RI MS WHB 5B/35 and 36 respectively.

[57] W H Bragg, 'Radioactivity as a kinetic theory of a fourth state of matter', *Proceedings of the Royal Institution*, 1911, 20:1–10; William was very familiar with the wave theory of light, having studied it in detail during Part III of his Mathematical Tripos studies at Cambridge.

[58] Ibid.; *Nature*, 1911, 85:491–4; *Chemical News*, 1911, 104:110–13; *Archives of the Röntgen Ray*, 1911, 15:402–15; letter W D Butcher (Editor, *Archive of the Röntgen Ray*) to W H Bragg, 28 January 1911, RI MS WHB 2A/26.

[59] Letter H A Wilson to E Rutherford, 29 October 1909, CUL RC W49; there is also a friendlier series of letters supporting the pulse theory from O Heaviside to W H Bragg during 1910 at RI MS WHB 3C/35–37.

X-rays, as described by Barkla, was not a problem for the particle model, since these rays were characteristic of the material and not of the primary X-ray and were of lower energy: 'the secondary rays are due to a reconversion of the [released electrons] into X-ray form, after they have lost energy in moving through the gas in which they arise'.[60] Barkla, who had been publishing extensively regarding the characteristic X-radiation that he had discovered, could not let this reference pass, and wrote to point out that both his own detailed ionization results, and William's, were only approximate, and that their agreement with any particular theory was therefore 'quite accidental'.[61] William responded in the same journal.[62]

William spoke at the 1911 Portsmouth meeting of the British Association for the Advancement of Science, and the journal *Nature* reported: 'A discussion was opened by an extremely lucid and persuasive paper by Prof. W H Bragg on corpuscular radiation'. Sir William Ramsay, Dr Lindemann, and others commented adversely, but to the reporter the outcome was clear: 'Pulses of genuine delight ran through the meeting while Prof. Bragg expounded his views, and it was clear that many were impressed by the cogency of his arguments'.[63] William also spoke to the Röntgen Society, and he had a new and dramatic piece of evidence to bolster his case; photographs of the ionization process from C T R Wilson's Cambridge cloud chamber, photos that Wilson had sent to Leeds and which William described as follows:[64]

> C T R Wilson has recently given a brilliant demonstration of the X-ray action. It is based on the fact that when a gas is ionized the positives and negatives pick up water vapour with great ease if it is present, and in this way become the centre of tiny water-droplets. If a space is thoroughly impregnated with water vapour, and if the gas in it is cooled by allowing it to expand suddenly, and if just previous to the expansion it is ionized by some agent such as alpha rays or X-rays, the vapour condenses on the ions, forming streaks of fog along the tracks of the rays. [The figures] show fog deposit, in the case of alpha particles and in the case of X-rays. In the former picture the tracks of the alpha particles appear like fine straight lines of definite length, as they should do, since the alpha particle proceeds through a gas in a perfectly definite straight line...In the other picture,

[60] W H Bragg and H L Porter, 'Energy transformations of X-rays', *Proceedings of the Royal Society of London*, 1911, 85A:349–65; William similarly pointed out that Robert Millikan's experiment suggesting that ionization was due to the direct action of the primary beam could just as easily be explained by his corpuscular theory (W H Bragg, 'The mode of ionization by X-rays', *Philosophical Magazine*, 1911, 22:222–3; also letter R A Millikan to W H Bragg, 2 July 1911, RI MS WHB 4C/40).

[61] C G Barkla and L Simons, 'Ionization in gaseous mixtures by Röntgen radiation', *Philosophical Magazine*, 1912, 23:317–33, 317; many of Barkla's other papers at this time were also in the *Philosophical Magazine*.

[62] W H Bragg, 'On the direct or indirect nature of the ionization by X-rays', *Philosophical Magazine*, 1912, 23:647–50.

[63] *Report of the Eightieth Meeting of the British Association for the Advancement of Science, Portsmouth, 1911* (London: Murray, 1912), pp. 340–1; *Nature*, 1911, 87:501.

[64] Letter C T R Wilson to W H Bragg, 23 April 1911, incomplete, RI MS WHB 7B; W H Bragg, 'The energy of the X-ray', *Journal of the Röntgen Society*, 1912, 8:16–20, 18.

> the short, irregular markings represent the tracks of electrons due to the X-rays going through the gas; they begin, as would be expected on the ideas just described, at points irregularly distributed through the gas; they are irregular and crooked, and on the average a few millimetres long.

Overall, however, the argument was unresolved. The wave and particle theorists each had extensive and growing evidence to support their position, and each could mount persuasive rhetoric to criticise the other. William had communicated aspects of the evolving story to Madsen in a series of letters throughout the period,[65] but it was clear here and elsewhere that a final solution was still out of reach.

In one of his regular letters to William, Rutherford applauded the generous way in which William had 'turned [Ramsay] down with great kindness and firmness' at the Portsmouth meeting, and he then went on to 'a business proposal' in regard to the position of external examiner in physics at Manchester: 'I recall that you asked me to be External Examiner for Leeds; but I trust you will not follow my example and turn down my offer... I think the work will be reasonably light... I think it is an excellent excuse for you for a good holiday for two or three days, and you will be able to enlighten me in the darkest period of the year. I shall consider it a gross dereliction of your duty if you do not accept'.[66] William replied: 'If you want me to be External Examiner, I will of course do the best I can'.[67] The different personalities and contrasting approaches of the two men are vividly illustrated by this correspondence, yet each found the other most useful as a sounding board and both found pleasure in the company of the other. Rutherford supported William's neutral-pair hypothesis—but not to the exclusion of the wave model[68]—and they also agreed about the need 'to form a physical idea of the basis of theory', unlike 'the continental people', who were 'quite contented to explain everything on a certain assumption, and do not worry their heads about the real cause of things'.[69] In the spring of 1912 William accompanied Rutherford and his wife on a three-week motoring holiday through the south of France, saying, 'I can never thank you and your wife enough for that gorgeous trip... I can never forget it, and all the fun we have had'.[70] His daughter recalled, 'Rutherford, as a visiting professor at the RI, was often in the Davy Faraday Laboratory... I remember the roars of laughter... WHB was so gentle and quiet, Rutherford so richly boisterous... Rutherford was continually turning up at our home with an enthusiastic "D'you know, Bragg" '.[71]

It was time to complete what William had been contemplating for some time: a lengthy summary of his Adelaide research work and its more recent

[65] Home, n. 37.

[66] Letter E Rutherford to W H Bragg, 14 October 1911, RI MS WHB 26A/21.

[67] Letter W H Bragg to E Rutherford, 15 October 1911, CUL RC B387.

[68] Draft letter W H Bragg to E Rutherford, 8 December 1911, RI MS WHB 5B/52; letters W H Bragg to E Rutherford, 21 December 1911 and 26 January 1912, CUL RC B388 and 389 respectively; letter E Rutherford to W H Bragg, 23 December 1911, RI MS WHB 26A/23.

[69] Letter E Rutherford to W H Bragg, 20 December 1911, RI MS WHB 26A/22.

[70] Letters W H Bragg to E Rutherford, 19 March and 18 April 1912, CUL RC B390 and B391 respectively.

[71] Caroe, n. 12, pp. 98, 99.

developments. The book appeared midway through 1912 and was entitled *Studies in Radioactivity*.[72] The first half dealt with William's alpha-particle work, the second with the nature of radiation. One reviewer stated that it was 'not a text-book on radio-activity in a general sense... [but] well worth study by those interested in the subject'. Other reviewers noted one of William's final sentences, 'But I should now add that we ought to search for a possible scheme of greater comprehensiveness, under which the light wave and the corpuscular X ray may appear as extreme presentments of some general effect';[73] and they then observed that very recent German work had already provided a major new element in the drama.[74] When Stark wrote seeking another article for his *Jahrbuch*, this time on ionization by X-rays, William simply referred him to 'a little book, which I have just published'.[75]

When again invited to address the British Association in September 1912, William focused on the role of physical models in scientific theories, a typical British approach.[76] He began with reference to the theories of light suggested by Newton and Huygens, and then projected onto a screen C T R Wilson's amazing photographs of the passage of α, β, γ, and X-rays through a cloud chamber. He compared the first favourably with his own earlier notional drawings of γ-particle tracks, noted the relevance of the occasional scattering through a large angle to Rutherford's new model of the atom, and proposed that γ- and X-rays do not ionize but 'merely bring to birth β rays which do'. William yet again emphasized the conversion of an electron into an X-ray and back again to an electron, all of the same energy. He concluded, 'there must be X-ray quanta', and he referred to the work of 'Planck, Einstein and others' with regard to similar observations with light. But 'we ought not to think that in doing so we abandon the wave theory... Rather... it is to our advantage to look at it from every side... and we must work in the hope of finding a new hypothesis of greater compass'.[77]

Norman Campbell also published prolifically from Leeds during this period. Topics included relativity, delta rays, and ionization, and some of the work appeared in German periodicals. As an indication, the *Name Index to the Philosophical Magazine* lists thirteen papers by Campbell in this one journal, between his arrival in Leeds and the end of 1912.[78] In only one of these papers is there a specific acknowledgement to William, but it is a generous

[72] W H Bragg, *Studies in Radioactivity* (London: Macmillan, 1912); a German edition followed promptly, W H Bragg, *Durchgang der α-, β-, γ- und Röntgen-Strahlen durch Materie* (Leipzig: Barth, 1913), translated by Max Ikle.

[73] Ibid., p. 193.

[74] Book reviews in W H Bragg, 'WHB Newspaper Cuttings, 1913–1924', RI MS WHB Cuttings/1, pp. 23–36; see chapter 16 for the new work.

[75] Letter W H Bragg to J Stark, 16 October [1912], Staatsbibliothek, Berlin.

[76] W H Bragg, 'Radiations old and new', in *Report of the Eighty-Second Meeting of the British Association for the Advancement of Science, Dundee, September 1912* (London: Murray, 1913), pp. 750–3; also in *Nature*, 1913, 90:529–32, 557–61, and *Scientific American Supplement*, 1913, LXXV:341–2, 358–9, the latter two with Bragg's figures.

[77] Ibid.

[78] W. Francis (compiler), *Name Index to the Sixth Series (1901–1925) of the... Philosophical Magazine and Journal of Science* (London: Taylor and Francis, 1931), p. 30.

one: 'It is, of course, clear that the brief discussion given of the mechanism of ionization is largely due to the suggestions of Prof. Bragg, to whose inspiration all this work is due'.[79] Campbell greatly valued and enjoyed his time in Leeds. He left in 1916 for wartime work at the National Physical Laboratory, and at the end of the war he wrote to the Vice-Chancellor saying: 'All through the war I hoped that I might return to Leeds when it was over, and it is with the very deepest regret that I find that return is impossible. [He had accepted "a very attractive post" with the new research laboratories of the General Electric Company]...My one hesitation in accepting arose from the necessity it involved of leaving my adopted home in the North, which I love, and living in London, which I loath...I am quite sure that my years in Leeds will always remain the happiest of my life...I shall always remember the kindness of all the university to a mere hanger-on, and I shall always have an affection and loyalty for it in which even my original university, Cambridge, will not share. Pray accept my warmest gratitude for all the kindness which you, personally and as head of the university, have showered upon me'.[80] Amongst many later achievements, Campbell is best remembered for the subsequent editions of, and supplements to, his *Modern Electrical Theory*, and for his books on the philosophy of science: *Physics: The Elements*, republished later as *Foundations of Science*, and *What Is Science?*[81]

The university's annual reports from 1909 to 1912 recorded other activities in which William was engaged. The number of physics students in degree classes was close to two hundred, while the number of his honours students varied from one to four. It was the Adelaide situation yet again, but now William had only five lectures a week, four of them to elementary classes, and no practical sessions.[82] During 1910–11 he joined the University Council as a 'member elected by the faculties', and Michael Sadler, a professor of education from the University of Manchester, was appointed Vice-Chancellor following the death of his predecessor.[83] Outside the university William was not as visible as he had been in Adelaide. In November 1909 he gave a lecture to the Leeds Philosophical and Literary Society on 'Radioactivity, the new science', and in November 1910 he spoke to the Society on 'Radium'.[84] Regarding the latter, *The Yorkshire Post* reported that, 'Despite the unfavourable weather there was a large attendance', and that, 'Professor Bragg, who is in the foremost rank of investigators of radio-activity, dealt with his subject in a popular style and

[79] N Campbell, 'Ionization by alpha rays', *Philosophical Magazine*, 1912, 23:462–83, 483.

[80] Letter N R Campbell to Sir Michael Sadler, 17 October 1919, University of Leeds Archives.

[81] N R Campbell, *Physics: The Elements* (Cambridge: CUP, 1920), republished as *Foundations of Science: The Philosophy of Theory and Experiment* (New York: Dover, 1957); N Campbell, *What Is Science?* (London: Methuen, 1921); for work at GEC see N R Campbell and D Ritchie, *Photoelectric Cells* (London: Pitman, 1929).

[82] Letter W H Bragg to A L Rogers, 18 November 1910, Bragg (Adrian) papers.

[83] University of Leeds Archives, *Annual Report*, 1908–09, 1909–10, 1910–11, 1911–12.

[84] Records of the Leeds Philosophical and Literary Society, University of Leeds, Brotherton Library (Special Collections), Minute Book of General Meetings, 1873–1921.

language, avoiding all technical jargon'.[85] In 1911 he was also appointed to the Council of the Royal Society; 'that will keep me going because... it is a day's journey to London and back', he reported to Rogers.[86] In addition, William continued to assist the University of Adelaide, as its representative at festivities at the University of St Andrews, for example, and on a selection committee for the Adelaide botany chair.[87]

More specifically, however, William was intent upon honouring the academic obligations he felt to Leeds. Returning from a distant colony, Professor William Bragg had quickly established himself as a leading player in British physical science. His inability to solve the wave–particle dilemma was a disappointment perhaps, but he could hardly have blamed himself for that. In addition, he was regularly encouraged by Rutherford, and Rutherford and Soddy—to take just two examples—gave significant publicity to his views.[88] William's commitment was tested when Richard Glazebrook asked if he would be interested in becoming Principal of the new University of British Columbia in Vancouver, Canada, with considerable power and money,[89] but William declined. His decision was strongly supported by Rutherford, who said, 'I think that if I were in the position that I felt tired of physical work and had not an idea left to work on, I should consider it an admirable position to occupy one's declining years'.[90] William's sons had also made a good beginning in their new homeland.

Willie and Bob's further education

Willie remembered the early months in England as 'frittered away'. 'It would have been a grand time to go abroad and learn French or German', he thought, 'or to go to art classes in Leeds... Instead I went to Cambridge for the Long Vacation, when I was in an anomalous category and did little useful work'.[91] This seems a harsh judgement. After strenuous university years in Adelaide, some free time must have been refreshing, and there are sketchbooks from these months that show that Willie was out and about, becoming familiar with his new

[85] *The Yorkshire Post*, 14 November 1910, p. 6.

[86] Letter W H Bragg to A L Rogers, 16 November 1911, Bragg (Adrian) papers.

[87] Letters W H Bragg to Council and Registrar, 1911–1912, UAA, S200, dockets 464/1911 and 202, 246/1912 respectively.

[88] For example, E Rutherford, 'Recent Advances in radioactivity', *Nature*, 1908, 77:422–6; F Soddy, *Annual Progress Report on Radioactivity to the Chemical Society for 1907*, vol. 4, p. 321; ibid., for 1908–09, vol. 6, pp. 242–4; ibid., for 1910, vol. 7, pp. 268–71, all available in T J Trenn, *Radioactivity and Atomic Theory* (London: Taylor and Francis, 1975).

[89] Letter R T Glazebrook to W H Bragg, 20 December [1912?], RI MS WHB 10A/8; Arthur Schuster wrote to William six months later regarding a vacancy at King's College, London, with the same result (letter A Schuster to W H Bragg, 22 July 1913, RI MS WHB 10A/9, and letter R Burrows to W H Bragg, 25 July 1913, RI MS WHB 10A/10).

[90] Letter E Rutherford to W H Bragg, 10 January 1913, RI MS WHB 26A/24; also letters W H Bragg to E Rutherford, 9 and 18 January 1913, CUL RC B393 and 394 respectively.

[91] W L Bragg, Autobiographical notes, p. 21.

homeland.[92] He did feel unexpectedly lonely and disconnected, as did generations of us who made the pilgrimage to Britain after education in a British Empire/ Commonwealth country when, because of our common heritage, we expected to feel immediately at home. But there were compensations. His Adelaide grandfather saw the value of an early arrival in Cambridge: 'I have also to thank you for your letter dated Whewell Court, Trinity, July 8, from which I am glad to hear you have gone to Trinity during the long vacation... as you will have time to make friends with some of the undergraduates before term begins'.[93]

Willie's earliest letters home also paint a generally positive picture. He may have said to his mother, 'I miss you and Dad and Sue [sister Gwendy] most horribly; there is a sort of vacant spot in me somewhere that I feel at times'.[94] However, there are also accounts of many happy times with his Aunt Lizzie and Cambridge cousins—Stevenson ('Stenie'), Alice, and Vaughan Squires, of a tennis tournament and other games with fellow students, of the President of the Dramatic Club 'taking Mr Arthur and myself over the rooms', and of meals with other students who had been asked to call on him. Nevertheless, 'I am quite a home person, I think; I miss you most horribly sometimes'.[95] In another early letter Willie reported light-heartedly, 'I have indulged in one extravagance, a tin bath. It is 6d [pence] every time one has a bath in college, and then you can only get it at unearthly hours... it will last me the whole while I am here... I have baths at the rate of two or three a day, just to feel how many 6ds I am saving'. In addition, 'Yesterday morning I went to breakfast with a man called Pym (they are all men here)... and two others... just clean-looking, English, public-school boys who have had one year here'; and although Willie was then beaten in a three-set tennis match by Pym, who was a tennis Blue, 'I think I stood up to him all right'. He had to fit mudguards to his bicycle after being caught in the rain,[96] and his 'bedder and her help' were a source of great amusement.[97] He was an outsider but he was trying hard to fit in.

Despite his recent Adelaide honours degree and the fact that it 'put me on more or less equal terms with 2nd year men here',[98] Willie continued to follow his father's career path and enrolled in Part I of the Cambridge Mathematical Tripos under the coaching of Robert Herman, Senior Wrangler of 1882. Equally important was Willie's Trinity tutor, Rev. Ernest Barnes, equal Second Wrangler of 1896, winner of the First Smith's Prize in 1898, Fellow and Director of Mathematical Studies at Trinity (with a special interest in mathematical physics), and very soon to become a Fellow of the Royal Society of London.[99]

[92] In possession of Patience Thomson (née Bragg), personal communication.
[93] Letter C Todd to W L Bragg, n. 7.
[94] Letter W L Bragg to mother, n.d. [1909], RI MS WLB 37A/2/24.
[95] Ibid.
[96] Letter W L Bragg to mother, n.d. [1909], RI MS WLB 37A/2/25.
[97] Letter W L Bragg to mother, n.d. [1909], RI MS WLB 37A/2/28.
[98] Letter W L Bragg to mother, n.d. [1909], RI MS WLB 37A/2/24.
[99] J Barnes, *Ahead of his Age: Bishop Barnes of Birmingham* (London: Collins, 1979); E T Whittaker, 'Ernest William Barnes, 1874–1953', *Obituary Notices of Fellows of the Royal Society*, 1954, 9:14–25.

Later Barnes would become Canon of Westminster and then Bishop of Birmingham, an active and controversial Modernist Anglican, the only bishop who was also an FRS.[100] He would be an important influence on the shy Australian as he slowly blossomed in his new environment.[101]

In addition, as a tutor Barnes 'took great pains over his pupils, the details of whose Cambridge careers were all carefully recorded in minuscule handwriting in a special ledger'.[102] It provides an untapped record of Willie's Tripos studies.[103] The nature of the Mathematical Tripos was changing and the examination was now in just two parts: students achieving honours in Part I were placed in three classes but not ranked within them, while those achieving honours in Part II were classified as Wranglers etc., again without rank. Barnes shows that in the Michaelmas term (October–December 1909) Willie attended the lectures of Forsyth on differential equations and Whitehead on mechanics, both of which he 'liked', and Hardy on infinite series, of which he recalled, 'I never really felt sympathetic with Hardy's logical treatment of infinite series and their convergence or otherwise'.[104] In November 1909 Willie enlisted in King Edward's Horse, a Cambridge unit of the Special Reserve, originally called 'The King's Colonial Corps', founded in 1901 and composed of men who had close connections with British colonies. A mounted infantry, concentrating on marksmanship, riding, and the care of horses, Willie trained with them during the year and at summer camps for four years.[105] William reported to Rogers the same month: 'Billie nearly won the Freshers' 100 yards at Cambridge, and the trainer says he is going to do well'.[106]

In the Lent term of 1910 he studied hydrodynamics with Harman, 'Principles of Mathematics (Number and Magnitude)' with Whitehead, and an unspecified course with Hardy.[107] Barnes also recorded that in that term Willie suffered from 'severe bronchitis, Mar[ch] 1–14', and that he was absent for much of the Easter term, 'Ill: went home Apr[il] 21, ret[urned] May 25'. The 'May Term [was] all[owe]d' according to Barnes, but Willie's health was clearly not strong. Indeed, he recalled that, 'In the spring I tried for a scholarship. A short time before we were due to sit for the papers I developed a violent cough, which was diagnosed as bronchitis by Dr Cook and I was told to stay in bed for a few days... [When I reported continuing sickness] He did

[100] P J Bowler, *Reconciling Science and Religion* (Chicago: University of Chicago Press, 2001), particularly pp. 208–14 and ch. 8.

[101] Margaret Heath (née Bragg), personal communication.

[102] Barnes, n. 99, p. 39; Barnes records that Bromwich was Willie's coach in his first (Michaelmas) term and that he did without a coach for the rest of his first year, although all other sources, including Willie himself, grant Herman the role.

[103] Rev. Ernest Barnes, ledger; I am indebted to Sir John Barnes for this information, personal communications, 1988.

[104] W L Bragg, Autobiographical notes, p. 21.

[105] Sir David Phillips, 'William Lawrence Bragg, 31 March 1890—1 July 1971', *Biographical Memoirs of Fellows of the Royal Society of London*, 1979, 25:75–143, 84; see *Register*, 29 November 1901, p. 5.

[106] Letter W H Bragg to A L Rogers, 17 November 1909, Bragg (Adrian) papers.

[107] Barnes' ledger, n. 103; *Cambridge University Reporter*, 9 October 1909, pp. 66–7.

Fig. 15.2 Willie/Lawrence Bragg at Cambridge, *circa* 1913. (Courtesy: Dr S L Bragg.)

not come to see me himself but sent his understudy, who must have thought I was malingering, for he told me to go out in very cold weather for a brisk walk. As a consequence I went down with really bad pleurisy and pneumonia and was very ill. Mother came down from Leeds to nurse me. I still had quite a high temperature when the time for the exams came, but was allowed to take them in bed. I think my brain was stimulated by the temperature. The essays were read by the Master, [Henry Montagu] Butler, and he commented on the brilliant imagination shown in mine! Anyhow, I got a major scholarship in mathematics'.[108] There were also the Tripos examination papers shortly after Willie's return to Cambridge from Leeds in late May. His Adelaide studies clearly stood him in good stead for, despite his sickness for a term and a half, he obtained a Class I pass.[109]

[108] W L Bragg, Autobiographical notes, pp. 21–2; in fact, Willie won a Trinity College Senior First Year Mathematics Scholarship (*Cambridge University Reporter*, 31 March 1910, p. 779), which was worth £100 a year for five years (letter W H Bragg to Adelaide University Registrar, 20 July 1910, UAA, S200, docket 563/1910).

[109] *Cambridge University Reporter*, 14 June 1910, p. 1137.

Barnes noted that 'LV [long vacation] will want rooms', and William reported to Adelaide that 'He is up for the Long now, doing Chemistry and tennis mostly'.[110] Nevertheless, surely Willie spent some of the summer in Yorkshire, recuperating and regaining his strength. We know he spent some time at home because, although 'my tutor expected me to continue in the same line [mathematics], my father strongly urged me to switch to physics', which he did.[111] Part II of the Natural Sciences Tripos was a one-year course, but Willie spread it over two years and packed the time full of activity. It was a major turning point in his life; the maturing of the man and the scientist he was to become.

For the Michaelmas and Lent terms of 1910–11 Barnes recorded that Willie consulted Herman and Whetham, but he does not list Willie's lecturers. William's letter to Rutherford notes that Willie attended J J Thomson's lectures in Michaelmas, and Willie himself remembered Searle's class on heat.[112] Thomson gave two courses in the Cavendish Laboratory that term: 'Properties of Matter' and 'Some recent views as to the nature of electricity and light'.[113] In the Easter term of 1911 Willie attended Searle's lectures on 'Electrical and Magnetic Measurements' and Whetham's on 'Electricity (continued)', implying that he undertook 'Magnetism and Electricity' with Whetham in Lent. He also spent time in the Cavendish Laboratory in C T R Wilson's 'Practical Physics' class. In the summer he was 'LV to co[ach] with Herman'.[114] In his last undergraduate year Willie took: in the Michaelmas term, Thomson's course on 'Some applications of recent researches in physics to chemistry', Jeans' lectures on 'The motion of electrons in the electromagnetic theory of light', and Herman's geometrical optics; while in the Lent term 1912, Thomson's 'Discharge of electricity through matter', Jeans' 'Statistical mechanics in its application to the kinetic theory of gases and theories of radiation', and Larmor's 'General electrodynamic and optical theory'. Barnes is silent on the Easter term except to say 'no LV: wants new rooms'.[115] Willie's own recollections provide both academic and wider insights:[116]

> C T R Wilson ran the Part II practical class and lectured on optics. He taught me most of the physics I learnt. His delivery of his lectures was appalling, but the matter was marvellous...He was also excellent in the practical class. He would not let us rush through experiments; he made each of them into a little research for us. Searle gave deadly dull lectures in Heat...J J gave us stimulating fireworks. I also got very excited over some lectures of Jeans, because they opened up a new world of statistical mechanics. After

[110] Letter W H Bragg to Registrar, n. 108.

[111] W L Bragg, Autobiographical notes, p. 22; for the NST see, for example, R Macleod and R Moseley, 'Breaking the circle of the sciences: the Natural Sciences Tripos and the "Examination Revolution"', in R Macleod (ed.), *Days of Judgement* (Duffield: Nafferton, 1982), ch. 8, pp. 189–212.

[112] Letter, n. 42.

[113] *Cambridge University Reporter*, 8 October 1910, p. 77–8.

[114] Ibid.; Barnes' ledger, n. 103.

[115] Barnes' ledger, n. 103; *Cambridge University Reporter*, 7 October 1911, pp. 67–8.

[116] W L Bragg, Autobiographical notes, pp. 22–7.

> them a strange young man used to draw me aside and explain at enormous length just where Jeans was wrong. This was [Niels] Bohr!...
>
> Two things of great importance happened to me at Cambridge. In the first place, I became a member of a small group of close friends: Townshend was a mathematician, Higham a historian, Tisdall a classicist, and Gossling and I were physicists... We read papers to each other [including one on crystals, see chapter 16], and sat up to the small hours discussing the nature of the universe... This was the first time in my life that I had simple, intimate relationship with a group of kindred spirits, and I revelled in it...
>
> The other formative influence in my life was my friendship with Cecil Hopkinson. He came from a famous family of engineers. His father, John Hopkinson... had been a leader in the development of electric power and lighting. He, with a son and two daughters, had been killed in a tragic Alpine accident. The eldest of the family, Bertie Hopkinson, was at that time Professor of Engineering at Cambridge... Cecil was very much the youngest of the family; he was about my age and studying engineering... it was the attraction of opposites. I had gown up with no experience of physical adventure. There was no tradition of it in the Todd or Bragg families... Cecil, like all the Hopkinsons, loved adventure and hardship spiced with danger ... He introduced me to ski-ing, sailing, shooting and climbing. I well remember how it started. Bob, Cecil and I were walking along Trinity Street together when Cecil... asked if I would join a skiing party... I had my usual hesitation... but dear Bob leapt in and insisted that I should accept...

Here Willie goes on for several pages, reliving the exciting days when a new world opened for him: details of skiing, hunting, boating, sailing and, unfortunately, another bout of pneumonia in Ireland, where some nuns nursed him in an infirmary attached to a workhouse and prayed for his spiritual salvation. He concluded: 'When I returned to England my mother met the boat at Liverpool. This was a tremendous event for her... [she] could not bear ever to be alone... she had to talk about anything with a relative or friend. This went with her being so gregarious, and so very clever at making people enjoy themselves and have a good time... What he [Cecil] gave me was like water in a thirsty land. He dragged me into adventures which... bolstered up the self-confidence in which I was so sadly deficient'. Willie was escaping from the family influence that threatened to become claustrophobic. His own unique personality was beginning to emerge.

When the results of the Natural Sciences Tripos, Part II, 1912 were announced in June, 'Bragg, W L Trin. (Physics)' was placed in Class I and was awarded a Trinity College prize. A classmate, 'James, R W Joh. (Physics)', also obtained a Class I result.[117] The paths of Willie Bragg and Reginald James were to cross many times after this; so that finally Willie became the obvious author for James' Royal Society biographical memoir, a tribute to 'a lovable man'.[118]

[117] *Cambridge University Reporter*, 15 June 1912, p. 1280; *The Cambridge University Calendar for the Year 1912–13*, p. 1129.

[118] W L Bragg, 'Reginald William James, 1891–1964', *Biographical Memoirs of Fellows of the Royal Society*, 1965, 11:115–25, 124; also see later chapters regarding WWI and Manchester.

For the present, feeling relief and satisfaction, Willie enjoyed the summer of 1912. He was quite unaware of the bombshell that awaited him.

While Willie went to Cambridge, Bob was sent to Oundle School in Northamptonshire. It had emerged from the shadows of earlier centuries to be one of England's leading public schools, under its famous and reforming headmaster, Frederick Sanderson.[119] Oundle is a small midlands town, near the centre of a triangle defined by Cambridge, Market Harborough, and Peterborough; William Bragg would have known the area well. The town and its School stand on high ground, where the River Nene makes a large loop. Here, in about 1485, the religious Gild of Our Lady of Oundle was founded, and in 1506 a free school for the sons of members of the Gild was established.[120] The major changes that swept through English education in the middle of the nineteenth century were generally resisted at Oundle. Student numbers declined, results were not commensurate with expense, and the curriculum was not modernised.[121] In July 1892, however, the Oundle Court of the Grocers' Company appointed Frederick Sanderson as the School's new headmaster and everything changed. Sanderson was a mathematician, identified with the development of science, engineering, and applied sciences, and Oundle became a demonstration of his beliefs. His early years were difficult, however; he had not been to a public school, was not athletic, was north country in speech and gesture, and had a violent temper.[122]

Sanderson completely reorganised the school. The junior forms fed into four streams: the Classical Side, the Modern Languages Side, the Science Side, and the Engineering Side. All boys took the certificate examinations of the Oxford and Cambridge Board and the number of certificates and distinctions increased steadily, as did the breadth of extra-curricula activities. Pupil numbers grew and new buildings and facilities appeared. Fundamental to Sanderson was the School's responsibility for every boy: no boy was altogether uneducable, there was always something he could do. He took a particular care to provide laboratories and workshops of all kinds and encouraged the boys to use them. The School became recognised as a pioneer in the teaching of science and engineering.[123] Gathorne-Hardy, critical of much British public-school education, is fulsome in his praise of Sanderson of Oundle.[124] Little wonder that William chose it for his younger son. Robert Bragg arrived in May 1909 for the summer term.

Bob lived in Dryden House, under housemaster Llewellyn Jones, a Jesus College Cambridge rowing and athletics blue, who had joined the School in 1882 and would stay until 1915, founding and running the Boat Club, encouraging athletics, cricket, and rugby, and ruling 'the Lower Fifth on the Classical

[119] H G Wells, *The Story of a Great Schoolmaster* (New York: Macmillan, 1924); [many contributors], *Sanderson of Oundle* (London: Chatto and Windus, 1923).

[120] W G Walker, *A History of the Oundle School* (London: Grocers' Company, 1956).

[121] Ibid., ch. XVII.

[122] Ibid., pp. 478–81.

[123] Ibid., ch. XVIII.

[124] J Gathorne-Hardy, *The Public School Phenomenon* (Harmondsworth: Penguin, 1979), pp. 350–4.

Side with a rod of iron and a heart of gold'.[125] Nicknamed 'Juggie', he was particularly kind to the new boy.[126] Like any new boy Bob was 'doing some things which I have never done before & others which I have done ages ago'. He disliked morning school 'because you have to get up & have an icy bath at half past six';[127] but 'chemistry I like... the best, it is nearly all practical... we are going to extract pure metal from an ore ourselves next time in little muffle furnaces, it will be great fun'.[128] There was practical work in physics too, although here he found the 'practical exams much harder than the theoretical ones because we never did any practical work at St Peter's'.[129]

Bob also enjoyed other activities: 'I am getting to like cricket very much', [and] I think I will be able to join the [cadet] corps very soon'. School food was plentiful, nutritious, and varied.[130] Jones reported to his Aunt Jessie Todd (Hedley Todd's wife) that Bob 'has made an excellent start here... He is doing very well in school, and out of school he... leaves nothing to be desired'.[131] Academically Bob reported to his father, 'I was beaten for 1st place this fortnight by one mark; anyhow I came 1st in maths and 2nd in French'.[132] In a few days he wrote again: 'Thank you so much for that 10/- [shillings]; it was most welcome as I only had 1d [penny] left'.[133] When Bob became badly over-tired William wrote to reassure him and looked forward to a pleasant summer break in Leeds and at Bolton Abbey.[134] At the end of the school year, in July 1909, there were four days of celebrations in the new Great Hall, and 'Next morning early the Corps went off to camp'.[135] In the new year William told Rogers, 'I meet a lot of big engineering people here... I have not been over any works yet, but I want to go & take Rob, who quite intends to be an engineer... He has quite taken to public school life. He is in the Officers Training Corps and spent a week at Aldershot this summer'.[136]

There are two school reports extant, and the School has gleaned from its archives some information regarding Bob's wider activities.[137] The school reports are from his last year at Oundle, when he was in form ScVIA^2 on the Science and Engineering Side. In the report for the Lent term 1910, Bob was at or near the top of his class in mathematics, mechanics, physics, chemistry, and drawing, and first overall; while in the Summer term his results were not as good but still commendable.[138] He was the bow oarsman in the school crew

[125] Walker, n. 120, p. 533.
[126] Letter R C Bragg to G Bragg, 2 May 1909, Bragg (Adrian) papers.
[127] Ibid., 4 May 1909.
[128] Ibid., 9 May 1909.
[129] Letter R C Bragg to W H Bragg, 4 July 1909, RI MS WLB 37A/2.
[130] Ibid., 10 May 1909, Bragg (Adrian) papers.
[131] Letter L Jones to Mrs Jessie Todd, 8 June 1909, Bragg (Adrian) papers.
[132] Letter R C Bragg to W H Bragg, 14 June 1909, Bragg (Adrian) papers.
[133] Ibid., 18 June 1909.
[134] Letter W H Bragg to R C Bragg, 21 June 1909, RI MS RCB/36.
[135] Walker, n. 120, p. 524.
[136] Letter W H Bragg to A L Rogers, 5 January 1910, Bragg (Adrian) papers.
[137] I am grateful to the headmaster and his secretary for the school information, personal communications, February–March 1984.
[138] Oundle School, term reports for R C Bragg, April and September 1910, Bragg (Adrian) papers.

of 1910, and in 1911 was Captain of Boating and stoke of the first crew (a four). On 1 April 1911, at the School Sports, he tied for the long-jump title. In the autumn term Dryden won the house football (rugby) shield, and Bob was also a member of the School team, playing at three-quarter-back (centre). The School magazine, *The Laxtonian*, reported, 'R C Bragg: Is very useful in attack and has improved greatly in defence. He is still rather clumsy and bad at fielding the ball'. In 1911 Bob was successively appointed a House Prefect, Head of Dryden House, and then a School Prefect.

Bob left Oundle in December 1911, after a brief but successful career. That month there was also a letter from Rev. Barnes pointing out that he had been unsuccessful in the Trinity scholarship exams but would be excused the college entrance examination if he chose to come next year. Although some of Bob's work was 'somewhat favourably commented on', he had averaged only 35% on the physics, chemistry, and mathematics papers, whereas the best candidate had averaged 79%; 'our standard is very high and the competition for emoluments very severe', Barnes wrote.[139] He would have nine months at home before joining his brother for the Michaelmas term, reading engineering.

[139] Letter E W Barnes to R C Bragg, 16 December 1911, Bragg (Adrian) papers.

16
X-rays and crystals

We have reached the major point in our story, the summer of 1912, when Willie was home for the summer holidays and discussed with his father an extraordinary letter William had just received from Europe. William and Willie themselves, the wider circle of people involved, biographers, scientists, and historians of science have all told and retold the story that emerged from 1912, although none was aware of this pivotal letter.[1] There are more than thirty separate accounts by Willie himself, and at the end of his life he gathered together his thoughts on the matter; they were published posthumously.[2]

Willie remembered the years of his Cambridge undergraduate studies incorrectly; he graduated in mid-1912, not 1911. He did spend time in the Cavendish Laboratory from 1911 through early 1912, however, and it was during this period that he began some research ahead of his formal graduation; his signature is on a list of those attending the Cavendish research students'

[1] For William see W H Bragg and W L Bragg, *X Rays and Crystal Structure* (London: Bell & Sons, 1918); Sir William Bragg, *Concerning the Nature of Things* (London: Bell & Sons, 1925), Dover editions, 1925–; id., *The Crystalline State: The Romanes Lecture, 1925* (Oxford: OUP, 1925); id., *An Introduction to Crystal Analysis* (London: Bell & Sons, 1928); id., *The Universe of Light* (London: Bell & Sons, 1933); W L Bragg, *A General Survey*, being vol. I of Sir W H Bragg and W L Bragg (eds), *The Crystalline State* (London: Bell & Sons, 1933); W H Bragg and W. L. Bragg, 'The discovery of X-ray diffraction', *Current Science*, 1937, 7 (special number on Laue diagrams):9–10. For W L Bragg see above and below. For the German physicists see M von Laue, 'Physics for 1914', in *Nobel lectures...: Physics, 1901–1921* (Amsterdam: Nobel Foundation, 1967); id., 'Historical introduction', in *International Tables for X-ray Crystallography* (Birmingham: International Union of crystallography, 1965). For the wider group see P P Ewald (ed.), *Fifty Years of X-Ray Diffraction* (Utrecht: International Union of Crystallography, 1962) and the many articles therein, especially Parts I and II by Ewald; P P Ewald, 'Some personal experiences in the international coordination of crystal diffractometry', *Physics Today*, 1953, 6(12):12–17; id., 'William Henry Bragg and the new crystallography', *Nature*, 1962, 195:320–5. *For biographers* see those listed in the Preface (ns 12, 13). For historians of science see P Forman, 'The discovery of the diffraction of X-rays by crystals: a critique of the myths', *Archive for History of Exact Sciences*, 1969, 6:38–71; replies thereto by P P Ewald ibid., 1969, 6:72–81, and L D Gasman in *British Journal for the Philosophy of Science*, 1975, 26:51–60; J Teichmann, M Eckert, and S Wolff, 'Physicists and physics in Munich', *Physics in Perspective*, 2002, 4:333–9; L Hoddeson et al. (eds), *Out of the Crystal Maze* (New York: OUP, 1992), ch. 1; all these carry references to other relevant works.

[2] D C Phillips and H Lipson (eds), *The Development of X-ray Analysis by Sir Lawrence Bragg, C H, F R S* (London: Bell & Sons, 1975); W L Bragg, Autobiographical notes, provides a more personal account.

annual dinner on 8 December 1911.[3] Willie recalled:[4]

> It was a sad place at that time. There were too many young researchers... too few ideas for them to work on, too little money, and too little apparatus. We had to make practically everything for ourselves... we had to do our own glass-blowing and there was only one foot-pump for the blow-pipe. One poor lady student had managed to get hold of it after waiting for weeks; I passed her room shortly afterwards, saw... that she was not there, and pinched the pump. Later my conscience was smitten to see her bowed over her table in tears, but not sufficiently smitten to make me return the pump. [Willie regretted the episode; as a young graduate he was as socially inexperienced and uneasy as his father had been.] J J set me on some problem on the variation of ionic mobility with the saturation of water vapour... but with my self-made, crude set-up [the results] were meaningless. There were a few senior people who had built little kingdoms for themselves [C T R Wilson, for example]... but most of us were breaking our hearts trying to make bricks without straw... After a year of this, however, my golden opportunity came. Von Laue published his paper on the diffraction of X rays by zincblende and other crystals, and my father discussed it with me when we were on holiday at Cloughton on the Yorkshire coast, staying with Leeds friends.

This must have been the summer holidays of 1912, for the letter that alerted William to the German experiments, sent to him by Lars Vegard, his earlier Leeds visitor, is dated 26 June 1912. It reads:[5]

> Dear Professor Bragg
>
> During my stay in Germany this last year I have occasionally had the opportunity of discussing the Röntgen-ray problem. The current idea here is that they are ether pulses, and by several occasions I have attempted to put forward the difficulties involved in the wave theory, and which I think at present no one has been able to overcome by means of mechanically intelligible conceptions.
>
> Recently, however, certain new, curious properties of X-rays have been discovered by Dr Laue in Munich. As I thought the matter would interest you, I asked Dr Laue, who gave an account of his discoveries here at Würzburg, to give me a copy of one of his photographs to send to you [not now with the letter].
>
> Without entering into any special conception as to the nature of this phenomenon, it may be described in the following way: A narrow pencil of primary X-rays is made to pass through a crystal (C) and then to fall upon a photographic plate (P). Without crystal he gets a single black spot; when the crystal is introduced, however, a most curious 'scattering' of the primary beam takes place. He gets a number of very sharp, regularly arranged ray-bundles surrounding the primary beam. Absorption by

[3] A Pais, *Niels Bohr's Times, In Physics, Philosophy, and Polity* (Oxford, OUP, 1991), Plate 3.
[4] W L Bragg, Autobiographical notes, p. 28.
[5] Letter L Vegard to W H Bragg, 26 June 1912, RI MS WHB 7A/3; the diagram that accompanied this letter has not been found.

Aluminium has shown that the rays producing the surrounding spots have a penetrating power very much greater than that of the main bulk of primary X-radiation and very much greater than that of the radiation characteristic of the anticathode. The effect is obtained by crystals of zincblende and copper sulphate. The crystals are cut as plane-parallel plates, with the planes perpendicular to the principal symmetry axis of the crystal, and the rays must pass as exactly as possible along this axis. The plates may be many mm thick.

Regarding the explanation, Laue thinks that the effect is due to diffraction of the Röntgen rays by the regular structure of the crystal, which should form a kind of grating, with a grating constant of the order 10^{-8} cm, corresponding to the supposed wavelength of Röntgen rays. He is, however, at present unable to explain the phenomenon in its details, and there are several difficulties from the diffraction point of view. Let me call attention to the following:

1) According to Laue, the diffraction in a grating with regularities in three dimensions is most complicated and there is in such a grating a very little chance that a maximum may occur.
2) The deviated spots seem to be much more distinct than should be expected when the points were due to diffraction. It is also very difficult to understand how the scattered points can be smaller than the middle point due to the primary rays.
3) It is not easily understood how by diffraction a heterogeneous beam can give such sharp maxima—and *sharp* maxima only. If the scattered rays are at all due to diffraction, it must be from some homogeneous group of rays which are mixed up with the primary ones.

On the other hand, as the scattered ray bundles—according to Laue—are made up of very penetrating rays, it is not easy to see how the corpuscular theory of Röntgen-rays can explain the scattering into such sharp bundles of parallel rays. If the rays were very soft, it might be possible that the scattered bundles might be due to structures in the crystal, but such an explanation seems hardly possible for so very penetrating rays.

As you will see, the matter is not yet clear, and it is necessary to wait for further investigations. (The first publication by Laue will appear in Berichte der Münchener Akademie.) But whatever the explanation may be, it seems to be an effect of a most fundamental nature.

Since we met at Portsmouth [at the 1911 BAAS meeting], I have been for some weeks in Paris, and from the beginning of October I have been working by Prof. W Wien at Würzburg...

I am staying here for about five weeks more; then I am going back to Norway... Give my kind regards to Mrs Bragg and... to Mr Campbell...

This splendid letter, full of information and with threat for William's particle model of X-rays, is what he discussed with his sons during that seaside holiday. It contains several ingredients that plagued Laue's attempt to explain the experimental results in detail and that then became the focus for Willie's solution to the problem. Willie took the details back to Cambridge when he returned to continue research in the Cavendish Laboratory,[6] where he joined a growing list of colonial scholars, many with 1851 Exhibition scholarships, who were helping to modernize Cavendish physics and professionalize research.[7]

There have been disputed accounts of the German experiments. Paul Ewald was an important player in the events and Max Laue, around whom the story evolved, wrote about it, but their interpretations have been challenged by Paul Forman.[8] There is merit on both sides and a simple outline will suffice here.[9] On 21 April 1912, in Sommerfeld's Institute for Theoretical Physics at the University of Munich and despite his discouragement, Walter Friedrich and Paul Knipping acted on a proposal by Max Laue and observed the diffraction of X-rays by a crystal, as described by Vegard. To preserve the priority of the three authors, Sommerfeld deposited a sealed note with the Bavarian Academy of Sciences in early May. It noted that the crystal was copper sulfate (of no particular orientation), it provided two photographs of the results (now lost) and a diagram of the experimental arrangement (with dimensions), and it stated that the observed interferences were a consequence of the lattice structure of the crystal, because the lattice constants were of the same order of magnitude as the conjectured wavelengths of the X-rays.[10] A copper sulfate crystal was chosen because it contained elements that were thought to give the highest proportion of characteristic radiation and was readily available. When the early results were encouraging, the experimenters obtained a thin, accurately oriented crystal plate of zinc sulphide (zincblende) and achieved a startling, symmetrical pattern of spots on a photographic plate, as described by Vegard.

It now became necessary to have a theory of the phenomenon, which Laue was well placed to provide, since he had written an encyclopaedia article on wave theory and diffraction by an optical grating, in which he had applied the simple theory twice to obtain the result for a cross-grating. Writing out the equation three times—for the three dimensions of the crystal—now provided an interpretation of the new discovery. In particular, the observed rings of spots could be related to the cones of rays demanded separately by each of the three conditions of constructive interference. The details are complex and it is sufficient to note Laue's conclusion: assuming that zincblende was based on a simple cubic lattice, with one molecule at every corner, and that the

[6] Rev. E Barnes, ledger, Sir John Barnes, personal communication.

[7] K Dean, 'Inscribing settler science: Ernest Rutherford, Thomas Laby and the making of careers in physics', *History of Science*, 2003, 41:217–40; R Macleod and R Moseley, 'The "Naturals" and Victorian Cambridge: reflections on the anatomy of an elite, 1851–1914', *Oxford Review of Education*, 1980, 6:177–95, particularly pp. 186–90; A Warwick, 'Cambridge mathematics and Cavendish physics: Cunningham, Campbell and Einstein's relativity, 1905–1911', *Studies in the History and Philosophy of Science*, 1992, Part I, 23:625–56, and Part II, 24:1–25; D-W. Kim, *Leadership and Creativity: A History of the Cavendish Laboratory, 1871–1919* (Dordrecht: Kluwer, 2002).

[8] Ewald, von Laue, Forman et al., and references therein, n. 1.

[9] A detailed account of the German experiment and of the Braggs' reactions to it has been given by G K Hunter, *Light Is a Messenger: The Life and Science of William Lawrence Bragg* (Oxford: OUP, 2004), ch. 2; W L Bragg gave a brief summary of the German experiment and theory in Phillips and Lipson, n. 2, pp. 14–20.

[10] Forman, n. 1, pp. 38–9.

X-ray direction was parallel to an edge of the cubes, the diffraction spots were due to five discrete X-ray wavelengths, characteristic of the crystal and excited by the incident beam. A number of spots that this theory predicted were not present, however, partly because 'Laue at that time was entirely stuck in a rut with his "characteristic radiation of the crystal" '.[11] The experimental and theoretical results were communicated to the Bavarian Academy of Science by Sommerfeld on 8 June and 6 July 1912, and two papers were subsequently published in their *Proceedings*.[12]

Many scientists now accepted that X-rays were wave-like, but William was unwilling to abandon his particle model completely. He had a large wooden sphere made in the workshop and marked on its surface the position of the spots, 'because it made the geometry of their relationship clearer'.[13] Willie suggested that his father's model might be saved if it could be shown that the directions of the rays leaving the crystal corresponded to clear avenues between the atoms in the crystal, along which the X-ray particles ran. When the family returned to Leeds, therefore, Willie went to William's laboratory and attempted to confirm this suggestion. He placed a crystal slab and a photographic plate on opposite sides of a lead-lined box and tilted it in various directions so that X-rays fell in numerous directions within a significant solid angle. If they travelled along preferential avenues in the crystal they would appear on the plate, but no such effect was observed and no account was published.[14] William hoped the idea still had merit, however, and wrote to the journal *Nature* pointing out that it did explain at least some of the spots: 'a consequence of an attempt to combine Dr Laue's theory with a fact which my son pointed out to me, viz. that all the directions of the secondary pencils in this position of the crystal are "avenues" between the crystal atoms'.[15] He communicated these thoughts to Laue, who could not agree but was otherwise preoccupied with the second edition of his book on the principles of relativity.[16]

At the end of summer Willie returned to Cambridge and William began the new academic year in Leeds, both severely distracted by the new phenomenon. Willie, who now had a room on I staircase of the Great Court at Trinity,[17] abandoned his earlier project to concentrate on a better understanding of the German results. Although he later talked exclusively of the early publication by Laue as his initial inspiration, surely Vegard's letter was equally important, although

[11] Letter P P Ewald to W L Bragg, 8 February 1962, RI MS WLB 44C/42.

[12] W Friedrich, P Knipping, and M Laue, 'Interferenz-Erscheinungen bei Röntgenstrahlen', *Sitzungsberichte der Königlich Bayerische Akademie der Wissenschaften zu Munchen*, 1912, pp. 303–22; M. Laue, 'Eine quantitative Prüfung der Theorie für die Interferenzerscheinungen bei Röntgenstrahlen', ibid., pp. 363–73.

[13] Sir Lawrence Bragg and Mrs G M Caroe, 'Sir William Bragg, F.R.S., 1862–1942', *Notes and Records of the Royal Society of London*, 1962, 17:169–82, 176.

[14] Phillips and Lipson, n. 2, p. 20; P P Ewald, 'William Henry Bragg and the new crystallography', *Nature*, 1962, 195:320–5, 322.

[15] W H Bragg, 'X-rays and crystals', *Nature*, 1912, 90:219.

[16] Letters M Laue to W H Bragg, 15 October 1912, 10 November 1912, and 13 November 1912, RI MS WHB 4A/8, 9, and 10 respectively.

[17] Residence books, 1912–14, Trinity College Library, Cambridge.

he forgot it subsequently.[18] The letter's major emphasis was the distinct, sharp appearance of the diffraction spots, much smaller in size than the incident X-ray beam, suggesting a focusing effect; while the papers of Laue et al. showed the elliptical and changing shape of the spots as the photographic plate was placed at different distances from the crystal (although the incident beam was circular), and described changes in the intensity of the spots when the crystal was tilted from its fourfold symmetry position. Some of Willie's Tripos studies were directly relevant. For his Part II Natural Sciences examination there were four three-hour written papers—on mechanics, heat and thermodynamics, light and optics, and electromagnetism—and a six-hour examination in practical physics,[19] indicating that the study of light and optics received substantial emphasis. This was the subject taught by Wilson, which Willie remembered so fondly: 'I owed a tremendous amount to C.T.R.'s lectures. I remember them vividly, but very little of other lectures I attended...and I used my notes shamelessly for teaching optics during all my time as a professor...I also remember well his talk on white light as a series of formless pulses and white light as composed of a range of wave-lengths. His lectures, and the talks I had with him when my first ideas about X-ray analysis were brewing, meant just everything to me'.[20]

Willie continued to mull over Laue's results and soon convinced himself that they must, indeed, be due to diffraction but that von Laue's analysis was incorrect:

> My next advance was an interesting example of the way in which apparently unrelated bits of knowledge click together to suggest something new. J J had lectured us on the pulse theory of X-rays, which explained them as electromagnetic pulses created by the sudden stopping of the electrons [in the X-ray tube]. C T R Wilson, in his brilliant way, had talked about the equivalence of a formless pulse and a continuous range of "white" radiation. Pope and Barlow had a theory of crystal structure, and our little group had an evening meeting when Gossling read a paper on this theory. It was the first time that the idea of a crystal as a regular pattern was brought to my notice. I can remember the exact spot on the Backs where the idea suddenly leapt into my mind that Laue's spots were due to the reflection of X-ray pulses by sheets of atoms in the crystal.[21]

Further clarification was provided by Willie's later address to a conference on the use of X-ray analysis in industry, subsequently published in the 'Science in Britain' series of pamphlets (see Figure 16.1):[22]

> It is small clues that often lead to a solution, and perhaps I may be forgiven for repeating a figure [Figure 16.1] from my paper in the *Proceedings of the Cambridge Philosophical Society* (November 1912) which shows the

[18] M Laue, n. 12; the Vegard letter is not mentioned in any of Willie's later accounts.
[19] 'Natural Sciences Tripos Examination Papers 1912', Cambridge University Library.
[20] Letter W L Bragg to P M S Blackett, 20 June 1960, RI MS WLB 52A/85; also see ch. 15.
[21] W L Bragg, Autobiographical notes, p. 29.
[22] Sir Lawrence Bragg, *The History of X-ray Analysis* (London: British Council, 1943); also a report on 'X-ray analysis in industry', *Nature*, 1942, 149:503–4.

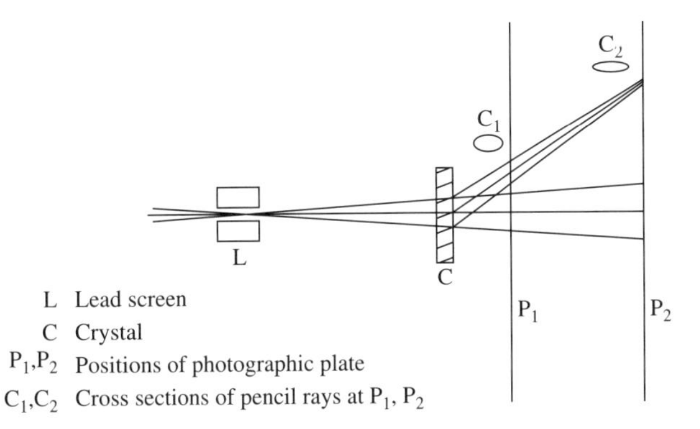

Fig. 16.1 Diagram used by Lawrence Bragg to illustrate his 1912 reflection explanation of X-ray diffraction by crystals. (From Sir Lawrence Bragg, *The History of X-Ray Analysis*, 1943, p. 7. Courtesy: The British Council.)

> clue I followed. When the plate was placed at P1 near the crystal the spots were almost circular like C1, but when placed further back at P2 they became elliptical (C2). Now Laue had ascribed his pattern to the diffraction of certain specific [characteristic] wave-lengths... [but] optical theory tells us that the diffraction must take place at a definite angle, and this means that the diffracted rays drawn in the picture should all have been parallel. I had heard J J Thomson lecture about Stokes' theory of the X-rays as very short *pulses*... [and] I worked out that such pulses of no definite wave-length should not be diffracted only in certain directions, but should be *reflected* at any angle of incidence by the sheets of atoms as if these sheets were mirrors. A glance at the geometry of [Figure 16.1], in which the rays are drawn as if reflected, shows that they close together again vertically while continuing to spread horizontally, thus explaining why the spots get more elliptical as the plate is placed farther away. It remained to explain why certain of these atomic mirrors in the zincblende crystal reflected more powerfully than others... Pope and Barlow had a theory that the atoms in simple cubic compounds like ZnS were packed together, not like balls at the corners of a stack of cubes, but in what is called cubic close-packing, where the balls are also at the centre of the cube faces. I tried whether this would explain the anomaly—and it did!

In other words, reflection gave combined focusing and non-focusing effects, leading to elliptical spots, and it also explained their change in intensity when the crystal was tilted.[23] This was not a correction to the German analysis, simply an alternative approach, but it did simplify the problem and provided the key to unlock it. Second, Willie realised that it was the continuum of X-ray wavelengths, otherwise described as the bremsstrahlung component of the X-ray beam, that gave the reflected spots, and not the characteristic radiation

[23] Letter W L Bragg to J D Bernal, 30 October 1932, RI MS WLB 49B/42.

of the crystal. Finally, using a face-centred cubic structure for ZnS rather than a simple cubic one, he obtained a good match to the spots on the experimental photograph, including their intensities, given by the different planes causing the reflected spectrum of X-rays.

Willie excitedly wrote up his findings and presented them to the Cambridge Philosophical Society at its meeting on 11 November 1912; after which they were communicated by J J Thomson and published in the Society's *Proceedings*.[24] The effective creators of the Cambridge Philosophical Society in 1819 were Adam Sedgwick and John Henslow and, though its interests were always broadly interpreted, the great majority of papers were devoted to mathematical, scientific, or technical topics. The Society was always keen that its work should be preserved, either in the regular issues of its *Transactions*, which were published from 1821 to 1928, or in its *Proceedings*, published from 1844. Membership in Willie's time was about 340.[25]

In his presentation to the Society Willie summarized the German work and noted that, while Laue's analysis appeared to explain the observed spots, it also predicted many more that were not present on the photographs. He began his own explanation by reference to Arthur Schuster's well-known text, *An Introduction to the Theory of Optics*, and its explanation of the action of an ordinary line diffraction grating, the incident light being regarded as a number of independent pulses that reflected from the grating to form an interference pattern.[26] Applied to a three-dimensional crystal with planes of atoms, it followed that waves (of wavelength λ) reflected from successive planes (separated by a distance d) reinforced each other and produced a spot provided that $n\lambda = 2d \sin \theta$, θ being the glancing angle at which the X-ray beam struck the crystal planes and n a (small) whole number and the 'order' of the diffraction.[27] The intensity of a given spot depended on the intensity of the X-rays at wavelength λ and on the number of atoms in the reflecting planes. Using a ZnS structure in which the basic element was a cube with an atom in the centre of each face as well as at each corner, as suggested by Pope and Barlow, and using the simplest and most likely sets of reflecting planes, Willie was able to conclude that, 'Every spot in the photograph is accounted for'.[28] In addition, when the crystal was tilted through 3°, the movement of the spots corresponded to a 6° deviation in the reflected beam, as Willie's theory required. His knowledge of crystal description and nomenclature was very limited at this time,[29] and his analysis was therefore awkward, but the result was to have huge ramifications.

[24] W L Bragg, 'The diffraction of short electromagnetic waves by a crystal', *Proceedings of the Cambridge Philosophical Society*, 1912, 17:42–57; the presentation was noted in *Nature*, 1912, 90:402; Willie deliberately omitted the term 'X-rays' from the title in deference to his father.

[25] A R Hall, *The Cambridge Philosophical Society: A History, 1819–1969* (Cambridge: Cambridge Philosophical Society, 1969).

[26] A Schuster, *An Introduction to the Theory of Optics* (London: Arnold, 1909).

[27] Most undergraduate physics textbooks show this simple derivation.

[28] W L Bragg, n. 24, p. 50.

[29] The history of crystallography before the 20th century is available in, for example, C S Smith, 'The prehistory of solid-state physics', *Physics Today*, 1965, 18(12):18–30; J G Burke,

William Lawrence Bragg had unknowingly founded a new field of physics—X-ray crystallography—that was destined to change the discipline and soon also transform inorganic, organic, and biochemistry, mineralogy, metallurgy, and later biology. The detailed internal atomic structure of materials of greater and greater complexity would successively yield to the new technique. Only twenty-two years old and in his first year of postgraduate study, Willie had largely done it alone. The central result—$n\lambda = 2d \sin \theta$—became known as 'Bragg's law', a part of every physics student's repertoire; but the fact that the law was due to Willie/Lawrence Bragg was soon only poorly known and later often attributed incorrectly either to his father or to them both jointly. As Willie lived to 81 years of age and remained a leading figure in the field until the end, he was increasingly invited to recall its early history; and since it also offered an opportunity to set the record straight he frequently obliged. Indeed, on one such occasion he wrote, 'I have gone into these early experiments in some detail because it is a story which I alone can tell, and which I wish to put on record'.[30]

In his annual Christmas letter to Arthur Rogers in Adelaide William wrote, 'My family are quite all right. Billy is coaching and demonstrating at Cambridge, and has just brought off rather a fine bit of work in explaining the new X-ray and crystal experiment... Bob is at Cambridge now, working at mathematics: also rowing hard and riding... The little girl is very bright and cheerful... she helps to fill the house now that both boys are away so much'. There was also mention of an Adelaide graduate who is to play an important role later in our story: 'Jauncey is using one of these new bits of apparatus and getting good results with it'. Eric Jauncey had been a student during William's last two years in Adelaide and had now won an 1851 Exhibition scholarship to work with him in Leeds on the development of a new instrument to study X-rays.[31]

Origins of the Science of Crystals (Berkeley: University of California Press, 1966); C J Schneer (ed.), *Crystal Form and Structure* (Stroudsburg: Dowden, Hutchinson and Ross, 1977); M E Lines, *On the Shoulders of Giants* (Bristol: Institute of Physics Publishing, 1994); P P Ewald, *Fifty Years*, n. 1, ch. 3.

[30] The quotation is from Sir Lawrence Bragg, *The History of X-ray Analysis* (London: British Council, 1943), p. 11; the first of such papers was W H and W L Bragg, 'The discovery of X-ray diffraction', n. 1, while some of the more extensive papers by W L Bragg were: 'Forty years of crystal physics', in J Needham and W Pagel, *Background to Modern Science* (Cambridge: CUP, 1938), pp. 77–89; 'X-ray analysis: past, present and future', *Proceedings of the Royal Institution*, 1945, 33:393–400; 'Acceptance of the Roebling Medal...', *American Mineralogist*, 1949, 34:234–41; 'The discovery of X-ray diffraction by crystals', *Proceedings of the Royal Institution*, 1953, 35:552–9; 'The diffraction of X-rays', *British Journal of Radiology*, 1956, 29:121–6; 'The diffraction of Röntgen rays by crystals', in O R Frisch et al. (eds), *Beiträge zur Physik und Chemie des 20 Jahrhunderts* (Braunschweig: Vieweg and Son, 1959), pp. 147–51; 'The Rutherford Memorial Lecture, 1960: The development of X-ray analysis', *Proceedings of the Royal Society of London*, 1961, 262A:145–58; 'A history of X-ray analysis', *Contemporary Physics*, 1965, 6:161–71; *The Start of X-ray Analysis* (Harmondsworth: Nuffield Foundation, 1967); 'Reminiscences of fifty years of research', *Journal of The Franklin Institute*, 1967, 284:211–28; 'The start of X-ray analysis', *Chemistry*, 1967, 40(11):8–13; and 'Half a century of X-ray analysis: Nobel guest lecture I [1966]', *Arkiv för Fysik*, 1969–74, 40:585–603.

[31] Letter, W H Bragg to A L Rogers, 21 November 1912, Bragg (Adrian) papers.

For the present Willie needed to do more work to confirm the new understanding. Wilson suggested that the reflection interpretation would be strengthened if a crystal with very distinct cleavage planes, such as mica, showed strong specular reflection of X-rays. As a result Willie allowed a narrow pencil of X-rays to fall at glancing angle upon a thin slip of mica, and this produced on a photographic plate both a marked reflected spot and another formed by the incident rays passing through the slip. 'Variation in the angle of incidence... left no doubt that the laws of reflection were obeyed',[32] Willie reported; 'I took the photograph, still wet, to J J and he was really excited'.[33] His father wrote excitedly to Rutherford: 'My boy has been getting beautiful X-ray reflections from mica sheets, just as simple as the reflections of light in a mirror'.[34]

'Harry' Moseley and Charles Darwin at Manchester confirmed that the reflected beam was, indeed, X-rays; and, referring to the earlier work of Professor Bragg, they further noted that X-rays now had the contrary properties of 'energy concentrated as if they were corpuscular' and 'some kind of a pulse with an extended wave-front'.[35] For some the mystery only deepened; for others there was now no doubt that X-rays were waves. Willie could not abandon his father's notion completely, however, and, in an article published in April 1913 in *Science Progress*, he recounted the German experiment and his own response, and also concluded, 'It is possible that there may be in the rays from the X-ray bulb two components, waves and corpuscles... There is perhaps the possibility of both these components having been hitherto classed together as one... the more paradoxical the case seems, the more interesting it becomes'.[36]

Anxious to test his new understanding on other crystals, Willie sought apparatus and advice regarding the best way to proceed. Regarding equipment he recalled: 'A young student [today], getting a fundamentally new effect... would have a really accurate apparatus rushed through the workshop for him, and be provided with the best available source of X-rays and other aids. I had to manage with bits of cardboard and drawing pins, and a very poor tube worked by an induction coil. I got so excited with the reflections that I worked this coil too hard and burnt out the platinum contact. Lincoln, the head mechanic who doled out the stores, was very angry. The contact cost ten shillings, and to "larn" me he made me wait about a month for a replacement'.[37] Better advice, regarding the simplest samples to use, came from William Pope, the Professor of Chemistry at Cambridge, whom Willie had already acknowledged in his initial publication. Pope had been appointed in 1908 and occupied the chair until his death in 1939. He was in the midst of an illustrious career.

[32] W L Bragg, 'The specular reflection of X-rays', *Nature*, 1912, 90:410.

[33] W L Bragg, Autobiographical notes, p. 30; Willie also wrote a number of excited letters to his father on the mica experiment, RI MS WLB 28A.

[34] Letter W H Bragg to E Rutherford, 5 December 1912, CUL RC B392; also ibid., 9 and 18 January 1913, CUL RC B393 and B394 respectively.

[35] H Moseley and C G Darwin, 'The reflection of the X-rays', *Nature*, 1913, 90:594.

[36] W L Bragg, 'X-rays and crystals', *Science Progress in the Twentieth Century: A Quarterly Journal of Scientific Work and Thought*, 1913, 7:372–89.

[37] W L Bragg, Autobiographical notes, p. 30.

A biography notes that, 'Pope and Kipping also investigated the crystallization of sodium chloride from aqueous solution... The investigation of this subject... was greatly aided by Pope's crystallographic skills', and the extensive bibliography of his publications includes many on crystallography.[38] Pope had also collaborated recently with the crystallographer, William Barlow, 'a privately educated genius... perhaps one of the last great amateurs in science', who had recently studied—and guessed a structure for—the alkaline halides.[39]

Pope advised Willie to examine crystals of the alkaline halides, and 'they proved to be so simple that it was possible to analyse their complete atomic arrangement'.[40] Willie used the German arrangement for these experiments: an X-ray tube and a series of stops to define a narrow pencil of rays, that then fell on a crystal and were collected on a photographic plate behind the crystal. He wrote excitedly to his father about it on at least two occasions: 'Such an exciting photo today, with rock salt! I have worked it out'; and 'My last photograph, taken with KCl, has turned out toppingly. It is perfectly characteristic of the point system with points [atoms] at the cube corners alone'.[41] By the time Willie reported this work to the Royal Society at its meeting on 26 June 1913,[42] it also involved a number of other crystals and an experimental arrangement that had changed radically, thanks to Willie's collaboration with his father.

William was more focused on the impact of the German work on his own theoretical position rather than the wider ramifications that his elder son took to Cambridge. William too consulted Schuster, this time in person: 'Dear Prof. Schuster, I enclose a drawing of the curious X-ray effect obtained by Dr Laue in Munich. It is claimed, I understand, that it is a diffraction effect due to the regular arrangement of the molecules in space... zinc blende and copper sulphate are the only two crystals mentioned in the letter I received. I have got some zinc blende specimens and am going to try the experiment... If you have any suggestions on the general question I should be very grateful... I wonder whether the rays producing the side spots are really "rays" proceeding in straight lines from some point in the crystal (say where the X-ray impinges or emerges), or are they sections of some loci by the photographic plate. It all seems most mysterious'.[43] No reply has survived, but the letter indicated William's immediate commitment to the problem, his particular interest in the nature of the rays emerging from the crystal, and his intention to try the experiment.

After William's initial letter to *Nature*,[44] A E H Tutton wrote a long commentary for the journal, summarizing the German experiment, referring at

[38] F G Mann, 'Pope, William Jackson', in C C Gillispie (ed.), *Dictionary of Scientific Biography* (New York: Scribner's Sons, 1975), vol. XI, pp. 84–92.

[39] W T Holser, 'Barlow, William', ibid.,vol. I, pp. 460–3.

[40] Phillips and Lipson, n. 2, pp. 27–8.

[41] Letters W L Bragg to W H Bragg, undated, quoted in Hunter, n. 9, pp. 38, 39.

[42] W L Bragg, 'The structure of some crystals as indicated by their diffraction of X-rays', *Proceedings of the Royal Society of London*, 1914, 89A:248–77.

[43] Letter W H Bragg to A Schuster, 21 July [1912], RI MS WHB [no number].

[44] W H Bragg, n. 15.

length to existing crystallographic knowledge, and concluding, 'These are the crystallographical facts which must be taken into account in any discussion as to the nature of these photographs, which does not appear to have been the case in a letter from Prof. W H Bragg'.[45] William dismissed the criticism, stressing his interest in the rays, but he did take the opportunity to refer readers to 'a paper read recently [in which] my son has given a theory which makes it possible to calculate the positions of the spots for all dispositions of the crystal and photographic plate'. He also added, concerning the nature of radiation, 'The problem then becomes, it seems to me, not to decide between the two theories of X-rays, but to find... one theory which possesses the capacities of both'.[46]

Two months later the scene had changed. William wrote to *Nature* in mid-January 1913 to report that, 'It is not at all difficult to measure the ionization produced by the radiation reflected by crystals... I find it possible to follow with an ionization chamber the movement of the reflected spot while the mirror [a sheet of mica] is rotated'.[47] During the latter half of 1912 William had seized on Willie's reflection idea to build his X-ray spectrometer (see Figure 16.2). This first instrument, pictured and described in Willie's posthumous account and elsewhere and 'constructed in the Leeds workshop by its clever head mechanic, Jenkinson', consisted of a modified optical spectrometer, probably taken from the undergraduate physics laboratory. An X-ray source and collimator (a lead block pierced by a hole that could be stopped down to slits of various widths) replaced the light source and focusing tube, the revolving table in the centre carried a crystal instead of a prism or optical grating, and the ionization chamber took the place of the optical telescope and rotated about the centre of the instrument. Using ionization rather than photography 'was more elastic and adaptable' and gave 'quantitative information... easier to analyse'. William was superb at the 'tricky ionization measurements' involved.[48]

Willie, home for Christmas, was now collaborating with his father. The subsequent paper bore both names and had several purposes. [49] It acknowledged Willie's foundational paper before the Cambridge Philosophical Society, described the new spectrometer, and reported its use to examine several crystals over a wide range of angles. Its principal conclusion, however, was not the various crystal structures but the repeated appearance in the ionization intensity-versus-angle spectra of the same pattern of peaks. William worried about their interpretation in letters to Rutherford,[50] and they were finally classified as arising from the characteristic radiations of the unchanging anticathode in the X-ray tube (in fact, the three members of the platinum L series). Two footnotes completed the paper: reports by Moseley and Darwin and by Barkla,

[45] A E H Tutton, *Nature*, 1912, 90:306–9, 308–9.

[46] W H Bragg, *Nature*, 1912, 90:360–1.

[47] W H Bragg, *Nature*, 1913, 90:572.

[48] Phillips and Lipson, n. 2, ch. 3.

[49] W H Bragg and W L Bragg, 'The reflection of X-rays by crystals [I]', *Proceedings of the Royal Society of London*, 1913, 88A:428–38.

[50] Letters W H Bragg to E Rutherford, 30 March and 3 April 1913, CUL RC B396 and B397 respectively.

Fig. 16.2 One of William's X-ray spectrometers, used at the Royal Institution, London (and later mounted for museum display), early 1920s. X-rays emerged from behind a lead shield (represented here by the backboard), were selected by narrow slits, met the crystal on a movable table, and after reflection/diffraction were detected by the ionization chamber (slender tube right), whose ionization was measured by the tilted gold-leaf electrometer (at the base of the apparatus). (Courtesy: The Science Museum, London.)

both apparently confirming this suggestion. Having failed to recruit William, the University of British Columbia now approached Willie for a more junior post, but with the same negative result.[51]

The spectrometer now opened up two separate lines of investigation: exploration of the X-ray spectra emitted by different elements, and the analysis of crystal structures. Initially the first was taken up by William, the second by his son. William simplified and quantified the understanding of characteristic X-rays, which Barkla had laboriously uncovered by absorption measurements. Given a value for '*d*' in Willie's equation, William could calculate the wavelengths (λ) of the X-rays from platinum, nickel, tungsten, and iridium anticathodes.[52] Willie had completed an extensive study of crystal structures, and the two reports were received and read at the Royal Society on

[51] Letter W H Bragg to E Rutherford, 16 February 1913, CUL RC B395.
[52] W H Bragg, 'The reflection of X-rays by crystals (II)', *Proceedings of the Royal Society of London*, 1914, 89A:246–8; also id., 'The reflection of X-rays by crystals', *Nature*, 1913, 91:477.

the same dates in June 1913 and published alongside each other in the Society's *Proceedings*.[53]

Willie used the photographs he had obtained in the Cavendish Laboratory as well as ionization spectra from his father's new spectrometer to deliver a stunning announcement of the power of the new technique. Notable are his rapid mastery of the complex crystallographic literature and his ability to visualise potential structures in three dimensions with apparent ease. Structures were suggested for the alkaline halides (KCl, NaBr, and NaCl), zincblende (ZnS), fluorspar (CaF_2), calcite ($CaCO_3$), and iron pyrites (FeS_2), the dimensions of the crystal lattices were calculated, and the wavelengths of the homogeneous components of the X-ray beam were determined. As Pope suggested, the alkaline halides had the simplest structure: simple cubic, with the atoms along the cube edges alternately metal and halogen so that each atom was bonded to six nearest neighbours of the other kind. There were no NaCl molecules in the structure. Professor Pope had supplied 'sympathetic interest and generous assistance', Dr Hutchinson (Cambridge mineralogist) 'with the greatest kindness... the necessary crystal sections', J J Thomson 'kind interest in the experiments', and his father 'the X-ray spectroscope... in the laboratory of the University of Leeds' (see Figure 16.3).

In yet a third consecutive paper in the Royal Society's *Proceedings*, father and son next reported a structure for diamond.[54] They began by comparing the two available approaches: 'the photographic method', in which the chosen crystal constructively reflected particular wavelengths from the heterogeneous or bremsstrahlung radiation of the X-ray source to produce spots on the plate, and 'the reflection method', where the crystal preferentially reflected the homogeneous or characteristic radiation of the source to particular angles in the ionization spectrum. 'The two methods throw light upon the subject from very different points and are mutually helpful', the authors suggested, and then illustrated it by application to the diamond crystal. The paper is fascinating for the meticulous way in which the authors recount their journey to the structure; an extended account that is not possible in today's scientific literature for want of space but that, back then, provided an excellent tutorial in the new field for its readers.

Their conclusions were best summarised in the letter to *Nature*: 'The structure is extremely simple. Every carbon atom has four neighbours at equal distances from it, and in directions symmetrically related to each other... These facts supply enough information for the construction of a model, which is easier to understand than a written description [and the longer paper had photographs of a model]... If the structure is looked at along a cleavage plane it is seen that the atoms are arranged in parallel planes containing equal numbers of atoms, but separated by distances which alternate and are in the ratio 3:1... Zincblende

[53] W L Bragg, 'The structure of some crystals as indicated by their diffraction', *Proceedings of the Royal Society of London*, 1913, 89A:248–77.

[54] W H Bragg and W L Bragg, 'The structure of diamond', ibid., pp. 277–91; also id., 'The structure of diamond', *Nature*, 1913, 91:557.

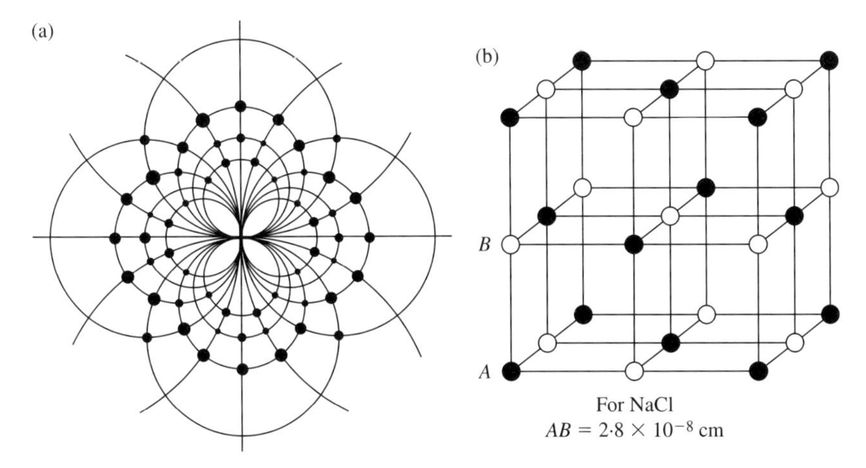

Fig. 16.3 (a) Lawrence's analysis of the X-ray diffraction spots from a common salt (NaCl) crystal, measured and analysed at Cambridge, 1912. (b) Resulting model of the atomic structure of NaCl deduced from (a), the sodium and chlorine atoms represented by white and black balls. (From Sir Lawrence Bragg, *The Development of X-ray Crystallography*, Bell & Sons, 1975, pp. 28, 29 respectively.)

appears to have the same structure, but the (111) planes contain alternately only zinc and only sulphur atoms'. Willie later remembered: 'The analysis of diamond aroused great interest and had a major effect in emphasizing the power of the new method of analysis, not only because of the direct way in which the structure explained the spectra but also because it was such an elegant illustration of the tetrahedrally arranged four bonds of a carbon atom'.[55]

This was a period of unprecedented success and closeness between father and son. Working in Cambridge during term time and at Leeds in vacations, Willie was now his father's scientific partner and collaborator; indeed, he spent the whole of Michaelmas term 1913 (October through December) in Leeds.[56] When he was at home they had breakfast together and then walked down Grosvenor Road, through an appropriate ginnel to Hyde Park Corner, then along Woodhouse Lane to the Physics Sheds, where William had his laboratories. In fine weather they surely crossed Woodhouse Moor and St George's Field rather than following Woodhouse Lane; or at least it seemed the natural thing to do when I visited in 1983. At night they reviewed the day's activities and the family was together in a way they had not enjoyed since Adelaide. The Leeds research notebooks clearly show when William was working alone and when Willie joined him, particularly during 1913; the former, immaculate

[55] Phillips and Lipson, n. 2, ch. 5; letter W Barlow to W L Bragg, 1 April 1914, RI MS WLB 86/7.

[56] Barnes ledger, n. 6 'W L Bragg, Michaelmas Term, 1913, working at Leeds this term'; confirmed by Trinity College Library, Cambridge

pages of data and graphs, unspoilt by correction and in William's hand, the latter, pages of mixed handwriting, often together on the same page.[57]

Willie remembered clearly: 'It was a wonderful time, like discovering a new goldfield where nuggets could be picked up on the ground, with thrilling new results almost every week, until the war stopped our work together'.[58] And, 'It was at this stage that we joined forces, which was extraordinarily fortunate; if I had struggled on alone at Cambridge I should have got nowhere... He pushed ahead with the investigation of X-ray spectra, while I was able to use the spectrometer results to solve crystal structures. In 1913 and 1914... we had a thrilling time, with new results tumbling out every week. It was only during vacations that I could work at Leeds, but I had a spectrometer of my father's design at Cambridge and investigated some structures there... Cecil Hopkinson's reaction to my new ideas about X-ray diffraction was typical. He was tremendously excited... He was the warmest-hearted and most loyal friend it was possible to imagine. We were at that time occupying the same set of rooms in Trinity. This went right against the college regulations, but... somehow Aunt Evelyn managed to wangle permission'.[59]

William was soon invited to speak publicly about the new work: in Leeds to the Workers' Educational Association, and to one of the Spring Holiday Courses at the university.[60] But now a problem arose. Willie again: 'It was not altogether an easy time, however. A young researcher is as jealous of his first scientific discovery as a kitten is with its first mouse, and I was exceedingly proud of having got out the first crystal structures. But inevitably the results with the spectrometer, especially the solution of the diamond structure, were far more striking and far easier to follow than the elaborate analysis of the Laue photographs, and it was my father who announced the new results at the British Association, the Solvay Conference, lectures up and down the country, and in America, while I remained at home. My father more than gave me full credit for my part, but I had some heart-aches'.[61] Willie, however, had been a joint author on many of these papers, including that for diamond, and he was mentioned in some reports; for example, in the *British Medical Journal* and in a newspaper item from the September meeting of the BAAS.[62]

His explanation for the biased recognition seems inadequate; it would appear that there were already suggestions in the air that William had been the primary instigator of the work, despite ample evidence to the contrary, and that his son had been carried along by family loyalty. Not only did William report

[57] W H and W L Bragg, twelve Leeds University quarto notebooks used to record research, February 1913—May 1915, Bragg papers, Archives of the Royal Institution, London.

[58] Bragg and Caroe, n. 13, p. 177.

[59] W L Bragg, Autobiographical notes, pp. 30–31.

[60] *Yorkshire Post, Leeds*, 20 April 1913, two items in 'Newspaper Cuttings, 1913–24', RI MS WHB Cuttings/1.

[61] Bragg and Caroe, n. 13, p. 177.

[62] 'Science notes', *British Medical Journal*, 11 January 1913, and unattributed newspaper report on BAAS meeting, 11 September 1913, both in 'Newspaper Cuttings, 1913–24', RI MS WHB Cuttings/1.

to the September 1913 Birmingham meeting of the BAAS on the diamond study,[63] but he was also invited to the second Solvay Conference in Brussels in October 1913.[64] Ernest Solvay had established the Solvay International Institute of Physics and the conferences bearing his name in 1912, in response to the profound problems of the new physics. The second conference took as its theme 'The structure of matter', of which the Laue et al. and Bragg developments formed the main focus.[65] Presentations were made by J J Thomson, Laue, W H Bragg, Barlow and Pope, and Brillouin, among others, and there were lively discussions after most papers. Following William's presentation, during which he referred to Willie's Cambridge Philosophical Society paper, there was an extended period of questions and comments, participants including Maurice de Broglie, Rutherford, Brillouin, Marie Curie, Sommerfeld, Lindemann, Einstein, and Nernst.[66] This was clearly a conference of major significance. Willie was not invited, but a postcard signed by Sommerfeld, Curie, Laue, Einstein, Lorentz, Rutherford, and others was sent, congratulating him for 'advancing the course of natural science'.[67]

At least the leaders of the field were aware, and William did try to give Willie due credit, as in a letter he wrote to Stark in response to another request for an article: 'I have so much writing on hand that I do not feel able to undertake a review article of the crystal work at present. But I think my son would do it for you, and I can assure you he is a far better crystallographer than I am'.[68] I believe the source of the difficulty lay elsewhere. In December 1927, for example, a leading British chemist, Henry Armstrong, wrote to the journal *Nature* regarding 'Poor common salt!':[69]

> 'Some books are lies frae end to end', says Burns. Scientific speculation would seem to be on the way to this state! Thus, on p. 405 of *Nature*... the statement is made... that a speculation by Prof. Lewis about the quantum 'is repugnant to common sense'. Again, on p. 414, Prof. W L Bragg asserts that, 'In sodium chloride there appear to be no molecules represented by NaCl. The equality in number of sodium and chlorine atoms is arrived at by a chess-board pattern of these atoms; it is a result of geometry and not of a pairing-off of the atoms'. This statement is more than 'repugnant to common sense'. It is absurd to the nth degree; not chemical cricket. Chemistry is neither chess nor geometry, whatever X-ray physics

[63] W H Bragg, 'Crystals and X-rays', in *Report of the Eighty-third meeting of the British Association for the Advancement of Science, Birmingham, 1913* (London: Murray, 1914), pp. 386–7.

[64] Letter H A Lorentz to W H Bragg, 22 November 1912, RI MS WHB 4A/37.

[65] P Marage and G Wallenborn (eds), *The Solvay Councils and the Birth of Modern Physics* (Basel: Birkhauser, 1999), second conference, pp. 112–17; J Mehra, *The Solvay Conferences on Physics: Aspects of the Development of Physics since 1911* (Dordrecht: Reidel, 1975), ch. 3.

[66] Institut International de Physique Solvay, *La Structure de la Matière* (Paris: Gauthier-Villars, 1921), pp. 113–40.

[67] G M Caroe, *William Henry Bragg, 1862–1942: Man and Scientist* (Cambridge: CUP, 1978), p. 76.

[68] Letter W H Bragg to J Stark, 15 January [1914?], Staatsbibliothek, Berlin.

[69] H E Armstrong, 'Poor common salt!', *Nature*, 1927, 120:478.

> may be... A little study of the Apostle Paul may be recommended to Prof. Bragg, as a necessary preliminary even to X-ray work... It were time that chemists took charge of chemistry once more and protected neophytes against the worship of false gods; at least taught them to ask for something more than chess-board evidence.

It was fourteen years since Willie had elucidated the non-molecular structure of crystalline salt, but animosity apparently festered. Some influential British chemists were to become positively hostile towards Willie in later years, and lack of recognition and undertones of jealousy dogged him from 1913 onwards.

William could have helped more, but his personality was not conducive to dealing with such things. His daughter examined the matter in some detail: 'It was difficult for the young WL; father and son never managed to discuss the situation, WHB being very reserved and WL inclined to bottle up his feelings. And WL had strong feelings—his mother had given him the dramatic and artistic nature which concentrated on a point, putting the rest of life somewhat out of focus... he felt things strongly but he could not hurt his father by telling him what he felt. However hard WHB tried to redeem the situation... the trouble lingered down the years'.[70] It has come to be seen as a problem solely between Willie and William,[71] but I see it differently. The unfolding story will reveal that there was a deep bond of affection between father and son and an understanding of each other's frailties. There were irritations from time to time, as between any parent and child trying to plough separate furrows in the same field, but there was no abiding animosity. What Willie lacked, what he craved, and what the conservative leaders of British science refused to give him, was full recognition of his many outstanding contributions, of which 'Bragg's law' was simply the first and most obvious example.

From now on, although father and son worked together when Willie was in Leeds, they published separately. Three papers again appeared close to each other in the Royal Society's *Proceedings*, two received and read in November 1913, the third in December/January 1914. William studied the X-ray spectra of osmium, iridium, platinum, palladium, rhodium, copper, nickel, silver, and tungsten, mostly reflected from a NaCl crystal, and precisely confirmed a number of general conclusions suggested earlier by Barkla and more recently by Moseley and Darwin. Namely, most emitted K and L radiations, their hardness (energy) increased with atomic weight, Planck's quantum law relating energy and wavelength was obeyed, and such characteristic radiations were emitted only when stimulated by radiations more energetic than themselves. Furthermore, it was noted that their general absorption increased rapidly with atomic weight, but that there was also a discontinuity in the absorption. As the energy of the radiation increased it was absorbed less and less by a given element, until it reached the energy of one of the element's characteristic radiations (an absorption

[70] Caroe, n. 67, pp. 77–8.

[71] Most biographical writings on father and son; for example, Hunter, n. 9, pp. 48–9; Sir David Phillips, 'William Lawrence Bragg,1890–1971', *Biographical Memoirs of Fellows of the Royal Society*, 1979, 25:75–143, 92.

edge), when the absorption rose abruptly, followed by a further steady decline. Multiple examples were given of these various effects.[72]

Particularly interesting is Willie's paper in this trio.[73] In order to highlight his own contribution and to be better identified, Willie now designated himself 'W Lawrence Bragg' for the first time; a method he continued for the rest of his life and that was reinforced when he later became Sir Lawrence Bragg. We, too, shall use this new designation, although his closest friends—such as George Thomson, J J's son, Willie's sailing companion during their early days, and close colleague thereafter—continued to use 'Willie'. This paper also contained the following acknowledgement: 'For many of the experimental results I am indebted to my father; the rest have been obtained in Leeds with one of the spectrometers which he has constructed'.[74] Yet the paper carries only Lawrence's name. Was his father more aware than previously understood of his son's disappointment, and did he insist that Lawrence publish this important paper alone, despite his own significant contribution? William's extraordinary skill in using the new spectrometer, with its unstable X-ray source and temperamental Wilson electroscope, was producing data of excellent quality and at an extraordinary rate.

The crystals discussed were not new and included some of the first Lawrence had studied, but they had now been examined with the spectrometer—'a much more powerful method of research into the structure of the crystal'—and there was a more sophisticated theoretical analysis. Using the preliminary structures he had obtained previously, Lawrence subjected the new data to a searching examination, in which 'we can obtain enough equations to solve the structure of any crystal, however complicated, although the solution is not always easy to find'.[75] This method of continual interchange between experiment and theory was to characterise the field for many years. Lawrence calculated the peak intensities in the ionization spectra as expected from the structures suggested earlier, and compared them with the new data. He assumed that the diffracting power of an atom was proportional to its atomic weight, that the spectra reflected from a simple series of identical planes had specified intensity ratios between the different orders, and that neighbouring atoms diffracted independently of each other; and he then presented tables of observed and calculated relative intensities. The results for the alkaline halides, zincblende, fluorspar, iron pyrites, and a number of calcite crystals showed impressive agreement. The simple structures had now been nailed down securely. William completed the trio of papers with a preliminary examination of the structures of sulfur and quartz, whose structures 'have not been completely solved, but which have nevertheless given interesting results'.[76]

[72] W H Bragg, 'The influence of the constituents of the crystal on the form of the spectrum in the X-ray spectrometer', *Proceedings of the Royal Society of London*, 1914, 89A:430–8.

[73] W L Bragg, 'The analysis of crystals by the X-ray spectrometer', *Proceedings of the Royal Society of London*, 1914, 89A:468–89.

[74] Ibid., p. 469.

[75] Ibid.

[76] W H Bragg, 'The X-ray spectra given by crystals of sulphur and quartz', *Proceedings of the Royal Society of London*, 1914, 89A:575–80.

Early in the new year (1914) the reduced family of William, Gwendoline, and their daughter Gwendy moved to another house in Cottage Road at the northern extremity of Headingley.[77] William returned to the study of X-rays and their various orders of reflection, for which the spectrometer was an ideal vehicle. It was anticipated that more accurate measurements would not only lead to greater usefulness in crystal structure determinations but also test existing theories of reflection and, in particular, determine the effect of temperature on the intensities. Peter Debye had recently predicted the last theoretically.[78] William began by defining the intensity of reflection. Because crystals were not perfect, an integration of intensity over a small range of angles around each proper angle of reflection was required, and this integration was soon automated by sweeping the crystal orientation mechanically, a method that 'has remained ever since the standard method of making quantitative measurements of X-ray diffraction'.[79] Lawrence gave credit to his father for the 'very great influence' of this contribution on subsequent developments, and William gave fulsome credit to Lawrence for suggesting the method's 'further development'.[80]

As in earlier experiments, a rhodium X-ray anticathode was used, since the stronger of its two prominent lines was easily selected and its bremsstrahlung background was especially low. Accurate relative intensity values were obtained for reflections from seven different sets of planes in a NaCl crystal, and it was found that the values all fell on a smooth curve, 'showing that all the sets of reflections conform to the same law'.[81] Regarding temperature variations, William enclosed the NaCl crystal in a small electric oven and measured intensity profiles at room and elevated temperatures. There was a diminution in the intensities and in the angles of reflection with temperature, the former due to the increased vibration of the atoms and a consequent loss in reflectivity of the crystal planes, particularly for the higher orders, and the latter due to the thermal expansion of the crystal. There was excellent and 'surprising' agreement with Debye's theory, although these were just initial examinations of the topic.[82]

Such was the interest in the new field that when William gave two lectures in March and June 1914 they were extensively reported in the pages of *Nature*. His address to the Manchester Literary and Philosophical Society was a general summary of progress so far: X-ray characteristics, cubic crystals, the alkaline halides, diamond, and more complex samples;[83] and following his

[77] The reason for the move is unknown; it is recorded in *Kelley's Directory of Leeds, 1915* as 'Cottage Rd., Far Headingley, no. 50, Bragg, Wm., Hy., M.A., D.Sc., F.R.S., pro-vice-chancellor and professor of physics, University of Leeds'.

[78] W H Bragg, 'The intensity of reflexion of X-rays by crystals', *Philosophical Magazine*, 1914, 27:881–99 and references therein; also W H Bragg, 'An X-ray absorption band', *Nature*, 1914, 93:31–2.

[79] Phillips and Lipson, n. 2, pp. 39–41, 41.

[80] Ibid., p. 39; W H Bragg, n. 78, p. 885.

[81] Phillips and Lipson, n. 2, pp. 40–1.

[82] W H Bragg, n. 78, p. 893; letters P Debye to W H Bragg, 30 January 1913 and 28 November 1913, RI MS WHB 2D/10 and 11 respectively.

[83] W H Bragg, 'Crystalline structures as revealed by X-rays', *Nature*, 1914, 93:124–6.

Friday Evening Discourse at the Royal Institution his text was reproduced in full.[84] In the second lecture William agreed that, 'X-rays consist of extremely short ether waves', but that 'X-ray energy travelled as a stream of separate entities or quanta'. He referred several times to the work of W Lawrence Bragg, in sympathy with his son's use.

Lawrence, too, was reviewing the output of the enormous exertion of the last eighteen months. He surveyed the field for the Röntgen Society and also prepared a very large aggregation of recent papers for German scientists. The first necessarily discussed X-rays as well as crystals and in considerable detail, for which he felt obliged to apologise.[85] He was new to lecturing. The second involved many figures and much editing, and must have provided a massive introduction to the Braggs' work for its German audience.[86] For a preliminary study of copper Lawrence had difficulty obtaining regular faces on his natural copper crystals, but he did achieve spectra sufficiently clear for him to suggest that 'copper crystals are arranged on a face-centred cubic lattice'. The paper also noted that Lawrence was the recipient of the Allen Scholarship for 1914–15 and that he was indebted to the Solvay International Institute of Physics for a grant to purchase the apparatus used in this experiment.[87] On 11 August 1914 Lawrence was elected a Fellow of Trinity College and appointed a College Lecturer in Natural Sciences. He was later granted leave of absence for the duration of the war that had just broken out, and the college subsequently made up his army pay to the full value of a lecturer's stipend.[88] These events concluded an extraordinary period of five years since the young Australian had arrived in England, and yet even more tension and even greater testing lay ahead.

Darwin and Moseley

The new Bragg methodology opened the fields of X-rays and crystals to others, and members of Rutherford's Manchester laboratory were some of the first to benefit. Charles Galton Darwin was a theoretical physicist, fourth Wrangler

[84] W H Bragg, 'X-rays and crystalline structure', *Proceedings of the Royal Institution of Great Britain*, 1914, 21:198–207; ibid., *Nature*, 1914, 93:494–8.

[85] W Lawrence Bragg, 'X-rays and crystals', *Journal of the Röntgen Society*, 1914, 10:70–8, 78.

[86] W L Bragg and W H Bragg, 'Die Reflexion von Röntgenstrahlen aus Kristallen', *Zeitschrift für Anorganische Chemie*, 1915, 90:153–296; see also letters W L Bragg to Max Ikle, 5 March and 7 June 1914, Staatsbibliothek, Berlin.

[87] W Lawrence Bragg, 'The crystalline structure of copper', *Philosophical Magazine*, 1914, 28:355–60, 359; the Allen scholarship was an annual award, given alternatively to arts and science men from the whole university, to aid research, yielding £250 for one year only (University of Cambridge Archives). Regarding the Solvay grant, see letters E Rutherford to H A Lorentz, 8 December 1913 and 12 January 1914, Lorentz papers, Rijksarchief in Noord, Haarlem, Holland, where Rutherford says 'I have just received an enquiry from Bragg junior, who wishes to make an application for a grant of about £50 from the Solvay Funds to obtain an X-ray spectrometer etc.... I hope there will be no difficulty in according him a subsidy', and 'I am in cordial agreement with all the proposals you make; viz. (1) to recommend 1250 francs to Bragg junior...'

[88] Trinity College Library, Cambridge.

of 1910, son of Cambridge professor and Astronomer Royal, Sir Charles Darwin, and grandson of the great naturalist. After graduation he had come to Manchester as the department's Mathematical Reader. Having participated in some research in radioactivity Darwin now had time for a new project.[89] Henry ('Harry') G J Moseley was the son of the Linacre Professor of Anatomy at Oxford and his clever wife, Amabel. Harry enjoyed a comfortable childhood, 'raised in upper-middle class luxury, both material and intellectual, attended by servants, exposed to professors, and surrounded by books, natural rarities, and foreign objets d'art'.[90] Despite the early death of his father, Harry's life continued smoothly, his natural ability shining at Summer Fields School and Eton, before his entry to Trinity College, Oxford, to study physics. He graduated in 1910 with a second-class degree after a disastrous examination period. He was fortunate to obtain a demonstrator position in Rutherford's department, which also offered the possibility of research training on interesting topics. Harry, too, served his academic apprenticeship on several radioactivity projects.[91]

By October 1912 Moseley had become aware of the German X-ray diffraction experiment and had decided to investigate it further. Anticipating mathematical difficulties he invited Darwin to join him. Together they arrived at an understanding of the German work, which Harry presented to a November 1912 fortnightly physics colloquium. William was in the Manchester audience and reported on his son's new, then unpublished, analysis in the discussion that followed. Moseley and Darwin 'thereupon decided to leave the matter [X-ray crystallography] to the Braggs'. They would 'give themselves entirely to the study of X-rays and to the attempt to find a theory that might reconcile the corpuscular properties... with the diffraction effects'.[92]

Rutherford needed to be persuaded, but he was encouraged when William 'generously invited Harry to Leeds to learn the tricks of experimental X-ray work', and Rutherford then let Harry have his way. He also consented to the necessary expenditure 'after a little "squeezing" from Harry', and Moseley and Darwin confirmed that the diffracted rays were X-rays, as reported in their letter to *Nature*.[93] The use of 'your son's device of reflection from a mica surface' was crucial to their success.[94] Moseley and Darwin then convincingly demonstrated the existence of both the continuous X-ray spectrum (bremsstrahlung) and homogeneous X-rays, the latter identified as Barkla's radiation characteristic of the platinum anticathode. Because of the narrowness of the slits in his apparatus, Moseley had initially missed the characteristic lines until alerted by

[89] J L Heilbron, *H G J Moseley: The Life and Letters of an English Physicist, 1887–1915* (Berkeley: University of California Press, 1974), pp. 70–1.

[90] Ibid., p. 9.

[91] Ibid., ch. 3; see also J G Jenkin, R C G Leckey, and J Liesegang, 'The development of X-ray photoelectron spectroscopy', *Journal of Electron Spectroscopy and Related Phenomena*, 1977, 12:1–35, 4–11.

[92] Heilbron, n. 89, pp. 71–2.

[93] Ibid., pp. 72–3, 75.

[94] Letter H Moseley to W H Bragg, 20 January 1913, ibid., letter 67, p. 198; William had told Rutherford about Willie's use of mica in a letter (W H Bragg to E Rutherford, 5 December 1912, CUL RC B392).

William;[95] but now he was able to resolve into their components the three main lines seen by the Braggs.[96]

It was at this time, Lawrence recalled, that Rutherford asked William 'to delay for a while the publication of his successful results, so that Rutherford's young men, Moseley and Darwin, could repeat their experiments and announce the spectra also. My father, I know, though he acceded to this request of Rutherford's, always felt it was not quite reasonable'.[97] Indeed, William not only accepted this bullying from Rutherford but also soon withdrew from the field entirely, leaving it to Moseley, and instead joined his son in exploring crystal structures. William had a 'love of quietude [and] was antipathetic to any kind of emotional strain'.[98] He was also extraordinarily generous to colleagues and competitors alike.[99]

Moseley now pushed ahead alone with the study of the X-ray spectra of a very wide range of elements, and he also moved to Oxford where he completed the work. He famously clarified the periodic table of the elements, showed that the spectra could be understood using Bohr's new model of the atom, and demonstrated that the frequency of characteristic X-rays was proportional to the square of atomic number (rather than atomic weight).[100] Seeking a more permanent position, Harry next applied for the chair of physics at Birmingham, and William wrote a generous testimonial: 'he possesses both insight and ingenuity to a remarkable degree; and he is sure, I think, to have many other achievements before him'.[101] But it was not to be. On the outbreak of The Great War Moseley rushed home from the 1914 British Association meeting in Australia and enlisted. His death at Gallipoli brought widespread sympathy for a promising career tragically cut short, and his story took on almost mythic status in the history of science.

What has been poorly acknowledged, however, is the central role William Bragg played in Moseley's success. William initially taught him 'the tricks' of X-ray work, including the use of mica, kept him informed of progress at Leeds and Cambridge, alerted him to the characteristic X-rays after he missed them initially, withheld publication until the Manchester pair could publish simultaneously, withdrew from the field altogether in Harry's favour, and finally wrote a generous testimonial for him. As Moseley and Darwin

[95] 'I remember vividly my father saying to me, "I should like you to remember that at first Moseley and Darwin missed the spectra which I discovered and only found them when I had shown they were there" ', letter W L Bragg to B S Page, 15 July 1963, RI MS WLB 52B/236.

[96] H G J Moseley and C G Darwin, 'The reflexion of the X-rays', *Philosophical Magazine*, 1913, 26:210–32; Heilbron, n. 89, pp. 76–7.

[97] Letter W L Bragg to L A Redman, 8 October 1964, RI MS WLB 53A/76; letter H Moseley to W H Bragg, 18 March 1913, in appendix to T J Trenn, 'Essay Review: Moseley and more Moseleyana', *Annals of Science*, 1976, 33:105–9, 108.

[98] Bragg and Caroe, n. 13, p. 181.

[99] Heilbron, n. 89, p. 77 and footnote 36.

[100] H G J Moseley, 'The high-frequency spectra of the elements [Part I]', *Philosophical Magazine*, 1913, 26:1024–34; id., 'The high-frequency spectra of the elements, Part II', ibid., 1914, 27:703–13.

[101] W H Bragg, testimonial for H Moseley, 9 June 1914, in Ludlow–Hewitt papers, Museum of the History of Science, Oxford.

generously acknowledged in their first substantial paper, 'We wish to express our warmest thanks to Prof. Bragg for the information which he has given us from time to time about his work on this subject. This information has been of great value to us'.[102] It is difficult to think of another young scientist who has enjoyed such extraordinary support from an active competitor.

Darwin, too, left an important legacy in the field. In 1914, after he and Harry had agreed to go their separate ways, Darwin published two long papers on 'The theory of X-ray reflexion',[103] which Lawrence later praised as 'landmarks in the history of X-ray analysis of crystals... X-ray crystallographers have always regarded this imaginative and original work of Darwin, produced at such an early stage of the subject, as one of his finest contributions to science'.[104]

Lawrence was finding that success could also bring frustration and increased pressure: to extend his analysis to more complex crystals and to write articles, reviews, and a book that he and his father planned. 'I find it impossible to do my experiments and write the book at the same time', he wrote, and 'I nearly faint when I think of the article'.[105] In his first letter to Max Ikle about the large article that appeared in the German *Journal of Inorganic Chemistry*, Lawrence pointed out, 'I find that I cannot finish it in time... I do not know whether my Father has told you that we are writing jointly a small book on "X-rays and Crystals"... I will be free in a week to commence to write the article'.[106] The book was *X-rays and Crystal Structure*, published in 1915, and in the Preface William wrote the following; a message that some people missed, others refused to heed, and still others never saw:[107]

> In this little book my son and I have first made an attempt to set out the chief facts and principles relating to X-rays and to crystals, so far as they are important to the main subject. We have devoted the remainder and larger portion of the book to a brief history of the progress of the work, and an account of the most important of the results which have been obtained.
>
> ... The publication of the book has been delayed by the difficulties of the times... The same circumstances have left me to write this preface alone. Probably, however, I should have demanded the privilege in any case. I am anxious to make one point clear, viz., that my son is responsible for the 'reflection' idea which has made it possible to advance, as well as for much the greater portion of the work of unravelling crystal structure to which the advance has led.

[102] Moseley and Darwin, n. 96, p. 232.

[103] C G Darwin, 'The theory of X-ray reflexion', *Philosophical Magazine*, 1914, 27:315–33 and 675–91.

[104] W L Bragg, untitled typed note about Darwin's early work, accompanying letter W L Bragg to G P Thomson, 10 July 1963, RI MS WLB 52B/233.

[105] Letter quoted in Phillips, n. 71, p. 92.

[106] W L Bragg to Max Ikle, n. 86, 5 March 1914.

[107] W H Bragg and W L Bragg, *X-rays and Crystal Structure* (London: Bell, 1918), pp. vi–vii.

In June 1914 Heinrich Rubens wrote to William inviting him to be the honoured guest of the Deutsche Physikalische Gesellschaft at its annual meeting, and to 'kindly speak to us about your latest research work concerning the dispersion of X-rays'.[108] William accepted, and Rubens then wrote again confirming arrangements and adding, 'Why not bring your son over? We would be delighted to make his acquaintance. Do try and induce him to come'.[109] William did indeed ask his son, and in July Lawrence wrote to his father to ask, 'Could you tell me our programme in Germany? Where is the meeting, and when?'.[110] That this journey did not take place was just one of the multitude of tragedies that unfolded in the next few years, for surely it would have introduced Lawrence to the best of German physics and helped to assuage his anxiety about recognition.

[108] Letter H Rubens to W H Bragg, 13 June 1914, RI MS WHB 11A/3.
[109] Letter H Rubens to W H Bragg, 30 June 1914, RI MS WHB 11A/4.
[110] Letter W L Bragg to W H Bragg, 19 July 1914, RI MS WHB 28A/4.

17
The Great War

Robert Bragg signed the Admissions Book at Trinity College, Cambridge, on 25 June 1912. Ernest Barnes became his college tutor, as he was for his brother. Bob completed the first year (Part I) of the Mathematical Tripos with a third-class result,[1] Barnes recording that he also took part in rowing and tennis.[2] Bob then spent the summer in Leeds, where he undertook a month's fitting and turning at one of the city's leading engineering firms.[3] Barnes noted, 'no LV [long vacation]—works in Leeds; Mech[anical] Sci[ences] Tripos next year'.[4] Bob had early intended to study engineering, and in November 1912 he had written home from Trinity to say, 'Dear Dad,...I am having a simply tophole term; never enjoyed myself so much in my life...I am getting quite a knut at drawing...I must bring them home for you to see. They are quite thrilling, cranes & swing bridges etc.'.[5] Bob completed his second-year engineering studies during the 1913–14 academic year, achieving a second-class result and a Trinity Exhibition.[6] He did not return to Cambridge after the summer of 1914, however, for by July he was in camp.[7]

Late in 1912 Bob had joined his brother in King Edward's Horse Territorial Force, continuing the army association begun in the cadet corps at St Peter's College and Oundle. He was discharged and re-enlisted on 1 March 1913 in King Edward's Horse (Special Reservists).[8] On 28 June 1914 Archduke Franz Ferdinand, the heir-apparent to the Austro-Hungarian throne, was assassinated in Sarajevo. On 1 August Germany declared war on Russia and on 3 August on France. On 4 August, after the German army had occupied Belgium, Britain declared war on Germany.[9] The next day Britain decided to send an

[1] *Cambridge University Reporter*, 14 June 1913.

[2] Rev. Ernest Barnes, ledger, courtesy of Sir John Barnes.

[3] Stephen Bragg, personal communication (from the records of the Engineering Department, Cambridge).

[4] Barnes, n. 2.

[5] Letter R C Bragg to W H Bragg, Bragg (Adrian) papers.

[6] Barnes, n. 2; *Cambridge University Calendar, 1914–15*, p. 726.

[7] Note card, R C Bragg to G Bragg, 29 July 1914, Bragg (Adrian) papers.

[8] Letter Ministry of Defence, Middlesex, England, to author, 22 June 1982, including the military service records of Major William Lawrence Bragg and 2nd Lieutenant Robert Charles Bragg.

[9] See, for example, H Strachan, *The First World War: A New Illustrated History* (London: Simon & Schuster, 2003), ch. 1.

expeditionary force to France and Robert Bragg was 'embodied', while on 26 August 1914 Lawrence Bragg was granted a commission as '2nd Lieutenant, Leicestershire Royal Horse Artillery (embodied Territorial Force)'.[10] Lawrence recalled: 'We were mounted infantry [in King Edward's Horse], trained in the tactics found useful in the Boer war. The emphasis was on marksmanship, riding, and care of our horses...It must have been an expensive unit to run, since we...took our horses to camp each summer; the expense was borne by the Dominions. Bob went into camp with this unit at the outbreak of war, but I applied for a commission and trained in Cambridge while waiting to be posted...I was sent to a Territorial battery, the Leicestershire R.H.A. [Royal Horse Artillery]. I was very much a fish out of water in the battery. The officers were hunting men, who talked and thought horse to the exclusion of most other interests'.[11] France and Britain faced the might of Germany, and other countries would soon become embroiled. Bob and Lawrence faced suspended study and war, just when they were on the threshold of academic fulfilment, and their father would soon become involved too. Gwendoline, wife and mother, was already worried.

On 1 August 1914 an item appeared in *The Times* reporting that a number of university professors and others had signed a document stating, 'We regard Germany as a nation leading the way in Arts and Sciences, and...War upon her in the interest of Serbia and Russia will be a sin against civilization'.[12] In Germany, however, most intellectuals joined the rest of the country in believing that it was forced into war, and that science was at one with militarism and other aspects of German life as part of German 'Kultur'. It quickly became true of both sides that 'Intellectuals, who distinguished opinions and value-judgements from demonstrable facts in their professional work, participated in the national euphoria and entered into the "Krieg der Geister" [war of the mind]'.[13] The best-known resulting document was the notorious Manifesto of 4 October, 'To the Cultural World', in which 93 German intellectuals asserted: 'It is not true that 1. Germany was guilty of this war, 2. we criminally violated Belgian neutrality, 3. the life and property of a single Belgian subject were interfered with..., 4. our troops behaved brutally in regard to [the university and library at] Louvain, 5. we disregarded the precepts of international law....' The signatories included the physicists Röntgen, Lenard, Wien, and Planck, and the text was very widely circulated.[14]

[10] War records, n. 8.

[11] W L Bragg, Autobiographical notes, p. 32.

[12] 'Scholars' protest against war with Germany', *The Times*, 1 August 1914, p. 6.

[13] S L Wolff, 'Physicists in the "Krieg der Geister": Wilhelm Wien's "Proclamation"', *Historical Studies in the Physical and Biological Sciences*, 2003, 33:337–68, 337–8; see also L Badash, 'British and American views of the German menace in World War I', *Notes and Records of the Royal Society*, 1979, 34:91–121 and references therein, and H Konig, 'General relativity in the English speaking world: the contribution of Henry L Brose', *Historical Records of Australian Science*, 2006, 17:169–95.

[14] G F Nicolai, *The Biology of War*, transl. by C A Grande and J Grande (New York: Century, 1918), pp. xi–xiii; W Zuelzer, *The Nicolai Case: A Biography* (Detroit: Wayne State University Press, 1982).

There were several other manifestos (proclamations, declarations, appeals), but two stand out. First the British response, in which 117 scholars published a 'Reply to German Professors', rejecting German accusations that England started the war, quoting Nietzsche and others regarding the attitudes of the German military and public, and, despite cooperation and friendships, supporting the war because it was a war for the defence of freedom and peace.[15] W H Bragg signed, as did Lamb, Lodge, Rayleigh, Sadler, Schuster, and J J Thomson. Again copies and a translation were widely distributed, and Wien wrote to William, regretting that the lack of understanding had destroyed any hope that personal ties could be healed in the foreseeable future. William replied kindly but accepted the breach: 'Yet you must believe me that the memories of our German friendships are still warm . . . even though it proves impossible to renew them in happier times'.[16] Second, very few German scientists had the courage to publicly oppose the appeal of the ninety-three signatories to the Manifesto; only four, including Einstein, signed Nicolai's significant 'Appeal to the Europeans' later in October 1914.[17]

Hereafter William tried to lead as normal a life as possible. The university's annual report for 1913–15 recorded that 'Professor Goodman and the staff of the Engineering Department, with the help of Professor Bragg and the members of the Department of Physics, have been engaged in special work in connection with the War',[18] but otherwise William pushed on with his teaching and research. With Sydney Peirce, an 1851 Exhibition scholar from the University of Sydney, he returned to the measurement of X-ray absorption, for the new spectrometer now enabled a much more precise study. Results obtained for 'the strictly homogeneous beams of X-rays which are now available'—those emitted by silver, palladium, and rhodium—were presented as an indication of the new method, each wavelength being 'isolated in turn by reflexion from a crystal of rocksalt'.[19] Atomic absorption coefficients for ten metal absorbers from aluminium to gold were tabulated, and they supported Barkla's suggestions that the ratio of coefficients for any two absorbers was independent of wavelength, and that for any absorber the coefficient was proportional to the fourth power of its atomic number. In a subsequent paper William had difficulty in understanding, except in general terms, the considerable absorption data now available, largely because the electronic structure of atoms was only poorly understood and quantum mechanics was in the future.[20] In the same month, October 1914, *Nature* published a detailed account of the construction and use of William's X-ray spectrometer,

[15] 'Reply to German Professors', *The Times*, 21 October 1914, p. 10.

[16] Wolff, n. 13, p. 345, including the Bragg quotation.

[17] Ibid., pp. 341–2; see also O Nathan and H Norden (eds), *Einstein on Peace* (New York: Schocken, 1960), ch. 1.

[18] University of Leeds Archives, *Annual Report, 1913–14–15*, p. 49.

[19] W H Bragg and S E Peirce, 'The absorption coefficients of X-rays', *Philosophical Magazine*, 1914, 28:626–30, 626.

[20] W H Bragg, 'The relation between certain X-ray wavelengths and their absorption coefficients', *Philosophical Magazine*, 1915, 29:407–12.

emphasizing its value in both determining crystal structures and analysing X-radiation.[21]

In the university's annual report for 1912–13 William had boldly asserted, 'A new crystallography is rapidly being created'.[22] He reiterated this suggestion in a paper in the journal *Scientia Bologna* entitled 'The new crystallography', in which he unknowingly glimpsed the future: 'The title of this paper seems ambitious', he wrote. 'Nevertheless, the analysis of crystalline structure by the X-ray spectrometer opens up an entirely new method of describing the characteristic features of crystals. The new stands to the old in much the same relation as biological to systematic botany'. Furthermore, 'The forces which bind atoms together are of course those which govern the formation of compounds, so that the new work is of prime importance to both chemistry and physics'.[23]

To introduce his readers to crystal structure and its analysis William employed analogies; he was now a master of their creation and use. 'If we stand in a vineyard we observe that the vines may be thought of as arranged in rows in many different ways...east and west [or] north-east and south-west', for example; and 'Let us imagine ourselves standing in front of a wall paper of some complicated [but repeated] pattern of flowers and leaves'.[24] In an article for the University of Leeds student magazine, *The Gryphon*, he wrote similarly.[25] William was also applying the method to a new group of crystals, the spinels, which he found were cubic in structure and possessed the highest possible number of symmetries. Magnetite—Fe_3O_4 = Fe(divalent) Fe_2 (trivalent) O_4—was William's first example, and he concluded that the structure was fundamentally the same as that of diamond, with a tetrahedron of Fe(divalent) O_4 replacing each carbon atom and the trivalent iron atoms placed half way between shared tetrahedra.[26] The complexity of the structures and their analysis was increasing rapidly.

The most notable crystallographic presentation in 1915, however, was William's Bakerian Lecture to the Royal Society in March. Instituted in 1775 by a bequest of Henry Baker, this annual lecture had become one of the Society's most prestigious events. William began by restating the basic principles of X-ray crystallography, and he immediately introduced Bragg's law and the case of rock salt, which he attributed to his son's Cambridge Philosophical Society publication. After discussing the structures of various cubic crystals as examples, William turned to the interpretation of the peak intensities in the reflection spectra. Here he used Lawrence's calcite work as the example, although he now had more precise data. He was seeking to explain the rapid

[21] W H Bragg, 'The X-ray spectrometer', *Nature*, 1914, 94:199--200.
[22] University of Leeds Archives, *Annual Report, 1912–13*, p. 61.
[23] W H Bragg, 'The new crystallography', *Scientia Bologna*, 1915, 18:378–85, 378.
[24] Ibid.
[25] W H Bragg, 'X-rays and crystals', *The Gryphon*, 1915, 18(3):34–6.
[26] W H Bragg, 'The structure of the spinel group of crystals', *Philosophical Magazine*, 1915, 30:305–15; ibid., 1916, 31:88; id., 'The structure of magnetite and the spinels', *Nature*, 1915, 95:561.

decease in the intensities of the different orders towards higher values. 'I think that an ample explanation... is to be found', he said, 'in the highly probable hypothesis that the scattering power of the atom is not localised at one central point in each, but is distributed through the volume of the atom'.[27] 'If we know the nature of the periodic variation in the density of the reflecting medium', he continued, 'we can analyse it by Fourier's method into a series of harmonic terms... We may even conceive the possibility of discovering from their relative intensities the actual distribution of the scattering centres... in the atom'.[28] William then applied this new insight to the simple case of a crystal of one type of atom. With a judicious choice of the distribution of scattering centres within each atom, he showed that it was possible to explain two things: first that, for a given spacing, the intensities fell off as the inverse square of the order, and second that, for the same order, the intensities of two spectra belonging to different spacings were proportional to the square of the spacings. There was much to be done to justify the method, but it was a fundamentally important contribution to the new field.

A young American physics student, Arthur Compton, wrote to *Nature* from Princeton University, where he was engaged in a PhD study of X-ray diffraction and scattering and was developing an improved Bragg spectrometer. He suggested that a distribution that fitted William's intensity data for rock salt was 'electrons in equally-spaced, concentric rings, each ring having the same number of electrons';[29] but William replied that the above two laws followed from an hypothesis 'which supposes the reflecting electrons to be distributed in space through the volume of the atom, and which imagines much overlapping to take place—atoms of one plane being thrust far into the interstices of the next'.[30] The insight and expertise needed to create this understanding, and similar evidence at other times, contradict Mrs Caroe's claim that 'WHB maintained his anti-mathematical bias for the rest of his life'.[31] Crompton will appear again in our story; but for now the war overshadowed everything else.

Late in 1914 the acceptance of two invitations had brought substantial change to William's life: the first to go to the USA and Canada on a lecture tour, the second to leave Leeds for University College London. On 24 October he wrote to his wife from the *SS Minnetonka* off Dover, noting the luxury of the ship and the presence of warships.[32] In mid-November he wrote from Boston and Providence, trying to reassure her about the boys: 'I cannot help thinking that Bob will not go just yet... We could not ask Bill [Lawrence] to chose any other course: he is making himself, and we have just to help all we can; dearest of boys we have had from the beginning. We may well be proud

[27] W H Bragg, 'X-rays and crystal structure', *Philosophical Transactions of the Royal Society*, 1915, 215A:253–74, 268.

[28] Ibid., p. 270.

[29] A H Compton, 'The distribution of electrons in atoms', *Nature*, 1915, 95:343–4.

[30] W H Bragg, ibid., p. 344.

[31] G M Caroe, *William Henry Bragg: Man and Scientist, 1862–1942* (Cambridge: CUP, 1978), p. 70.

[32] Letter W H Bragg to G Bragg, 24 October 1914, RI MS WLB 95F/1.

of them'.[33] He wrote to Lawrence to report on his trip and to reassure him that, 'Everybody asks after you: the Bumsteads sent you their regards'.[34] However, events in Europe were calling William home:[35]

> If Bob is liable to 24 hours notice I should like to come home as soon as ever I can. I have been asked to go to a great number of places... But I think I will have done my duty if I visit two or three of the principal places: I have accepted an invitation to go to Chicago because they ask me to address the Physical Society. As Chicago is some way to the west, that will give me a chance of meeting a number of men from the western universities... I can take Toronto and Cornell on the way back and be in New York again on the 5th [of December]... I see the situation in France is easier... Perhaps this may mean that Bob is not going just yet... Would you just send me a cable if Bob is called on?... I could cut out Baltimore & Washington... we'll be together again soon.
>
> [And a few days later:[36]]
>
> I must tell you a little more of my doings... I had the most extraordinarily successful time at the American Academy... They all looked so interested and pleased all the time and appreciated the models so. Of course I always tell them where Bill's work came in: this time I was particularly keen that they should understand Bill's achievements... One old boy who had come in from Worcester got up & said that it did not at all diminish their debt to the lecturer because...'What a man does through his son he does himself'... I assure you it was a triumph for the crystal work.

William then visited Harvard University and the General Electric Co. in Schenectady, where the latter promised to make him 'some special [X-ray] bulbs for the crystal work'. By 1 December the call home was irresistible: 'I got a wire from Sadler yesterday saying I had better not prolong my stay. A great many places have asked me to lecture, but I think it quite right I should go home. So I am going to get a berth... I expect the Lusitania will be the best... I am afraid Bob has gone: I see that so many territorials are on their way to the front. The news still seems pretty good'.[37] William did travel on the *Lusitania*, which was sunk five months later by a U-boat off Ireland. It was the second loss of a ship on which William had been a recent passenger.

On his return to Leeds William wrote a long letter to Vice-Chancellor Sadler about research, a cry from the heart very reminiscent of the letter he had written to the University of Adelaide on his return from leave in England in 1898. It read:[38]

[33] Ibid., 15 November 1914, Bragg (Adrian) papers.

[34] Letter W H Bragg to W L Bragg, 22 November 1914, RI MS WHB 28A/2; Henry Bumstead was an American physicist working on radioactivity at Yale University, a friend of Rutherford from the turn of the century.

[35] Letter W H Bragg to G Bragg, 22 November 1914, RI MS WLB 95F/2.

[36] Ibid., 27 November 1914, RI MS WLB 95F/3.

[37] Ibid., 1 December 1914, RI MS WLB 95F/4.

[38] Letter W H Bragg to Vice-Chancellor, 16 December 1914, Leeds University Archives, Registry: H Physics, H12.

> Since we discussed...the possibility of increasing the output of research from the Physics Department...I have come to the conclusion that I had better set down in writing some account of our present position...[A summary of recent X-ray and crystal work, including reference to 'the theoretical investigations of my son, W L Bragg']
>
> We have built five of the new X-ray spectrometers required for the work, one of which was taken to Cambridge...No one else has done any work on the new lines in England, with the exception of Mr Moseley in Manchester...Abroad, the amount of effective work has been small...The instruments are now being made for sale by W G Pye & Co. of Cambridge: two have already been sent to America. A book on the subject is to appear very shortly...I mention all these points in order to make clear the special nature of the position which we occupy...At present we have the field almost to ourselves, but presently of course there will be a large addition to the number of workers.
>
> [A review of the current staff position, with the best workers going into war service.] We can, of course, increase our staff, but this would undoubtedly be expensive if we were to engage assistants...We might do a good deal immediately by rendering the services of the present staff more effective...we are torn between the desire to do the clerical work of the laboratory efficiently and the desire to go on with the research work. I do think we might have a junior clerk to ourselves...
>
> Lastly, we must not, if possible, let ourselves suffer from lack of materials...we want about £100 worth now...but we could spend many times that sum with great advantage were it available.

William also wrote to Rutherford: he had thoroughly enjoyed his American visit but his staff were depleted by war service and his 'own boys are both gunners now...We had them home for a day or two recently, Bob at Christmas & the elder at New Year. They looked splendid'. He had a new motor, a ten-horsepower Austin.[39] The Pro-Chancellor and the chairman of the university's finance committee wrote to the Vice-Chancellor, responding cautiously to William's plea, and the university secretary was very enthusiastic, suggesting the appointment of additional staff, a liberal grant for apparatus, and sources of funds that could be used.[40] These recommendations were largely adopted, but they were soon overtaken by other events. There had been letters during 1913 asking whether William would consider appointment at King's College, London, or at Edinburgh, but William had declined to allow his name to go forward.[41] The invitation to University College he now considered more carefully, but on 10 March 1915 he wrote declining this offer too, pointing out

[39] Letter W H Bragg to E Rutherford, 15 January 1915, CUL RC B401.

[40] Letters, respectively, A G Lupton to Sadler, 18 December 1914, H I Bowring to Sadler, 20 December 1914, A E Wheeler to Sadler, 22 December 1914, all in n. 38.

[41] Letters A Schuster to W H Bragg, 22 July 1913, RI MS WHB 10A/9, and R Burrows to W H Bragg, 25 July 1913, RI MS WHB 10A/10, regarding King's College; letters J Walker to A Smithells, 13 June and 9 July 1913, Smithells' papers, University of Leeds Brotherton Library, MS 391, regarding Edinburgh.

that in Leeds his teaching commitments were modest, his research was well funded, he had an excellent workshop and instrument-maker, and minimum administrative work.[42]

Later that month, however, University College matched and surpassed what Leeds could offer, and William reconsidered. He consulted his close friend and colleague, Arthur Smithells, and on 26 March wrote him a long letter explaining his change of mind. He was reluctant to move, as he had been from Adelaide, and for the same reasons, being well established professionally and domestically; but one consideration was persuasive:[43]

> After the war, if it ends right, we know, as you say implicitly, that there is another struggle coming on, in which the organisation and efficiency and well being of England must be considered from the point of view of what science can do to improve them... it is up to all of us to help, each in the area he can do most... the physical questions that are involved in the textile trades to begin with... I could make a shot at it myself; but I am not so well equipped as many younger men, and I should have to give up all my own research work... On the other hand, another part of the *same* work can be done in London: and that is where I might hope to come in. The Government will want help... moreover, there is the Royal [Society]. Could we not make it a real advisory body, with more worthy work than that of reading papers?... I might help: I have already had personal hints to that effect... I think I could help all the more because I have been here, and lived in the centre of these industries.

Leeds later obtained the brilliant William Astbury to work on textile physics. William was offering his services to the country in response to the growing call for scientists and their science to enhance the productivity of industry and support the war effort.[44] At the University of Leeds Senate meeting on 2 June 1915 the Vice-Chancellor announced with deep regret that Professor Bragg 'has been pressed, especially by some of those who are chiefly responsible for the business of the Royal Society, to accept the Quain Chair of Physics at University College, London... The Senate will feel that no heavier loss could have fallen on the University'.[45] The University Council subsequently passed a formal resolution in which it noted, 'During his tenure of the Cavendish Chair, Professor Bragg has won international distinction through his researches into X-rays and crystals, which he has conducted in companionship with his son,

[42] Letter W H Bragg to Principal, University of London, 10 March 1915, Watson Library, University College, London, 'Professorship file—Physics, 1915', containing William's initial refusal and subsequent advertisement and applications but not his later acceptance (most of UCL's records dealing with the Bragg period were destroyed during WWII).

[43] Letter W H Bragg to A Smithells (draft), 26 March 1915, RI MS WHB 10A/11 (William's emphasis).

[44] J D Bernal, 'William Thomas Astbury, 1898–1961', *Biographical Memoirs of Fellows of the Royal Society*, 1963, 9:1–35; on the call for science see, for example, the many articles and editorials in *Nature*, October 1914– and references therein.

[45] M Sadler, untitled document, being announcement (typed) to Senate, 2 June 1915, University of Leeds Archives, W H Bragg personal file, CF03/0780.

and the Council appreciate in a high degree the value of the stimulus which his scientific achievements have given to his colleagues and to the intellectual life of the University'.[46] Professor Smithells wrote a valediction, which appeared in *The Gryphon*: 'Let us be quite clear about this; Professor Bragg has not left us for the honour of going to London... [he] has left us because he thinks that in London he can do better work. He has a perfect right to think so... what we are called upon to do is to acknowledge our debt for what we have received and to express our good wishes to a parting benefactor and a beloved friend... When he goes you can only say thank you and God speed... This notice would be incomplete if it did not mention Mrs Bragg and... here, happily again, everybody knows, everybody is grateful, and everybody is sorry'.[47]

William would complete the academic year in Leeds and begin at University College on 1 September 1915. He and Gwendoline went down to London in the summer to look for a house, and William gave a few lectures at the College before becoming embroiled in the war.[48] Ship's Lieutenant [Maurice] de Broglie wrote from the French Telegraph Station at Bordeaux, acknowledging William's work on X-rays and crystals and concluding, 'I have also heard that your son has volunteered for the ranks of the English army. May he be blessed with good luck and glory'.[49] The war was already a year old and there was no sign of victory.

Bob's war

During August 1914 the German army had advanced through Belgium, Luxembourg, and northern France with great speed, but as it threatened Paris it was halted by an Allied counter-attack. After Russian and French offensives and German and Allied outflanking attempts had all failed, with great loss of dead and wounded, the two sides dug in, literally, in opposing trenches that stretched from the English Channel to the Swiss border. The war had become a stalemate on the Western Front. Turkey's entry into the war on the German side offered the possibility of a major campaign in the Mediterranean: to relieve the pressure in the west, to open a route to the Mediterranean for the beleaguered Russians, to draw the Balkan States to the Allied cause, and to eliminate Turkey from the conflict. Winston Churchill, First Lord of the Admiralty, proposed a naval attack. A forced passage through the Dardanelles and capture of the slender Gallipoli peninsula, it was thought, would open the Sea of Marmara to the Allies and force the disintegrating Ottoman Empire to

[46] University of Leeds, Council resolution, 21 July 1915, RI MS WHB 10A/13.

[47] A Smithells, 'Professor Bragg', *The Gryphon*, 1915, 19(1):3–4; handwritten original at RI MS WHB 10A/14.

[48] Caroe, n. 31, p. 80; O Wells, 'About the Physical Department, University College, London, and those that worked therein, 1826–1950', typescript in Manuscripts and Rare Books Room, Watson Library, University College, London, p. 54.

[49] Letter [Maurice] de Broglie to W H Bragg, 18 July 1915, RI MS WHB 2B/44.

surrender its capital, Istanbul, without a fight. The Bosporus and the Black Sea would then complete a supply line to Russia.[50]

Although many had tried, however, few had succeeded in forcing this narrow strait between Eastern Europe and Asia Minor, and British intelligence was pessimistic. In the first of many blunders, the British Navy tried unsuccessfully to force the Dardanelles, losing several old battleships to mines and shore batteries. The baton was passed to General Sir Ian Hamilton and his Mediterranean Expeditionary Force, composed of British, French, and ANZAC (Australian and New Zealand Army Corps) troops. They would land at the toe (Cape Helles) and on the western shore of the Gallipoli peninsula and quickly subdue it, opening the sea-lane. Aged, inexperienced in mounting such an operation, and inadequately supported, Hamilton took many weeks to organize his plan. The Turks and their German advisors prepared for the inevitable invasion, which took place on 25 April 1915.

The Australians were landed before dawn at what became 'Anzac Cove', and by early afternoon were fighting for their survival, facing steep ridges and gullies, the most difficult terrain on the peninsula. The Anzac troops faced determined defenders led by the brilliant Mustafa Kemal (later Kemal Ataturk, modern Turkey's founding President), and barely held their precarious bridgehead. At Cape Helles, the main objective, there was fierce fighting and heavy casualties at the two main beaches. Across the peninsula there was divided command, inadequate artillery support, desiccating heat, poor food, inadequate water, high casualties, flies and lice, dysentery and diarrhoea. It was another stalemate. At Anzac there was a truce on 24 May to bury the dead littering no-man's-land. Churchill was replaced.

Robert Bragg's experience can be followed through the letters he wrote home. They begin before the declaration of war with a card sent from Canterbury to his mother at Deerstones: 'Camp is going well & my horse is ripping, quite one of the best'.[51] His high spirits continued, although he was missing his elder brother: 'Camp is going strong . . . though I wish old Bill was here'.[52] But uncertainty soon set in: 'we had hundreds of Generals down today to inspect us. There is a different war rumour every few hours . . . I jolly well hope we won't be called out'.[53] A day later reality was dawning: 'This war business is serious, isn't it! I don't expect they will call us out to fight actually,

[50] I have used the following sources for most of what follows: R R James, *Gallipoli* (London: Batsford, 1965); Br.-General C F Aspinall-Oglander, *Official History of The Great War, Military Operations, Gallipoli* (London: Heinemann, 1932), vols I & II; L Carlyon, *Gallipoli* (Sydney: Macmillan, 2001); H Broadbent, *Gallipoli: The Fatal Shore* (Melbourne: Penguin, 2005); K Fewster, V Basarm and H H Basarm, *Gallipoli: The Turkish Story* (Sydney: Allen & Unwin, 2003); P Taylor and P Cupper, *Gallipoli: A Battlefield Guide* (Sydney: Simon & Schuster, 2000); R Prior, 'The Suvla Bay tea party: a reassessment', *Journal of the Australian War Memorial*, 1985, 7:25–34.

[51] Letter R C Bragg to G Bragg, postmarked 29 July 1914, Bragg (Adrian) papers. *Note*: all the letters ns 51–79 are from the Bragg (Adrian) papers; I have added some punctuation to these letters to clarify meaning.

[52] Ibid., 29 July 1914.

[53] R C Bragg to W H Bragg, postmarked 1 August 1914.

Fig. 17.1 Bob Bragg in uniform, *circa* 1915. (Courtesy: Lady Adrian.)

but we now are all packed up & ready to move at a moment's notice'.[54] Later in August 1915 there were problems with the horses: 'We are still here you see... because last night the horses stampeded... About 10 had to be shot with broken legs etc.'.[55] There was still hope of a return to Cambridge: 'wouldn't it be best if you let Uncle Charlie have enough money to clear [my Cambridge bills]... That is, if I don't get back before the Long [vacation] is over'.[56]

Bob next became resigned to the situation: 'Things are looking pretty serious aren't they?'.[57] He was homesick: 'It was great luck I got leave for Saturday night and all today, Sunday. I had wild thoughts of dashing up to Bolton Abbey but gave it up in despair. I next thought of going to Cambridge but found that I couldn't get back; no trains'.[58] His cousin Vaughan Squires was in the same unit.[59]

[54] Ibid., postmarked 2 August 1914.
[55] R C Bragg to G Bragg, postmarked 14 August 1914.
[56] R C Bragg to W H Bragg, postmarked 29 August 1914.
[57] R C Bragg to G Bragg, n.d. [late August–early September 1914].
[58] Ibid., postmarked 7 September 1914.
[59] Ibid., postmarked 8 September 1914.

A long letter to his father talked of guard duty, sword allocation, field glasses, and pay, and then continued, 'Will this war affect your Research much? I mean what about your German correspondents?...How I should love to talk with you about all this'.[60] They were good questions that would increasingly weigh on William's mind. Bob's letters then lengthened: 'We were inspected by the King yesterday...We had a very fine lecture last night from a Lieutenant...just back from the front...All his regiment got cut up & he & another man spent 3 weeks in the German lines, hiding by day and travelling by night'.[61] 'At the end of the lecture our Colonel got up...& added that the regiment would be ready for service abroad by the middle of next month'.[62] He became resigned to action, but then their departure was postponed.[63] He also sent a photograph of his horse, 'Betty', to his young sister.[64]

By November action seemed inevitable: 'the division to which we are attached has been warned for active service abroad...It is frightfully thrilling for us', he told his mother, 'but I fear very worrying for you'.[65] Next day, 'I may get leave this weekend...How is Bill getting on?'.[66] 'I did love that weekend at Cambridge together. By Jove, won't we have a Beno when we come back finally from the war'.[67] He then wrote from London: 'I have just come up today to apply for a commission...I asked all about the R[oyal] F[ield] A[rtillery]. He said that he could almost certainly get me in...So couldn't you come & stay here too & help collect my kit?...I am feeling frightfully bucked tonight'.[68] His next dated letter was from Chapeltown Barracks, Leeds; he was now attached to the Royal Field Artillery there.[69] In late-March 1915 they were sent to Wales for firing practice, but the train broke down and the whole trip was a shambles: 'The battery fired one series and then chucked it'.[70] It was a sign of things to come. Later in the month Bob was on the move more seriously—from Leeds to Pembrey to Swindon to Newbury to Basingstoke and finally to Milford Camp in Surrey.[71] He was given two new horses but they lacked training and became 'crocked up' if overworked.[72] There were brief times of respite: 'Bill [Ellison] & I had such a cheery weekend in town'. They had a 'bath, bed & breaker' and an expensive dinner at a London hotel. They went to the theatre, had lunch with Sylvia and Gwendy, and chartered boats on a lake. The next week there were long marches, arduous gun manoeuvres, and firing practice in the rain.[73]

[60] R C Bragg to W H Bragg, [probably 11 September 1914].
[61] R C Bragg to G Bragg, 19 September 1914.
[62] R C Bragg to W H Bragg, 23 September 1914.
[63] R C Bragg to G Bragg, 4 October 1914.
[64] R C Bragg to G M Bragg, postmarked 11 October 1914.
[65] R C Bragg to G Bragg, 3 November 1914.
[66] Ibid., [4 November 1914].
[67] Ibid., n.d.
[68] Ibid., 25 November 1914.
[69] Ibid., 23 March 1915
[70] Ibid., 2 April 1915.
[71] Ibid, [9 April 1915]; R C Bragg to W H Bragg, 11 April 1915.
[72] R C Bragg to Mrs Gott, 29 April 1915.
[73] R C Bragg to G Bragg, n.d.

June brought decisions and Bob communicated the news to his father: 'A few days ago we got orders to provide ourselves with drill tunics & pith helmets! You know what that means... we are for the Dardanelles... we [are] getting new 5-inch Howitzers... Don't say too much about these things will you'.[74] He went shopping for necessities, for horse and rider, and went to Aldershot to draw ammunition. He told his mother, 'the horses are very fit... & do try & not worry'.[75] He wrote to his father, 'We have just got orders that we are to go to-morrow'; and to his mother, 'We are just off on the *Knight Templar*... My address will be: A Battery, 58th Brigade RFA, 11th Division, British Mediterranean Exped. Force'.[76] A senior officer's wife wrote to Gwendoline to say that the boat had sailed and to report that, 'My husband... said also "Ellison and Bragg are invaluable" '.[77] She wrote again later: 'Col. Drake has asked me to look after the collecting and sending out of socks & necessary comforts... I have got to write & worry all the parents as we are so bereft of wives'.[78] Britain was calling upon a new force of young men, many unmarried, the so-called 'New Army'.

Bob wrote to his father from 'On board', describing the steamy weather and the harbour of a 'sunbaked, shadowless city... By Jove, what a lot I shall have to tell you when I get back'.[79] It was probably Malta.[80] They spent two weeks in Alexandria, where 'We bathe every day... just like the bathing out in Australia'.[81] They sailed again late in July and arrived at the island of Lemnos on 1 August,[82] where Bob posted a letter to his brother: 'Well Cheeroh Bill, I expect the fun will begin in a day or two'.[83] He was part of three divisions of the New Army that were to make a major attempt to break the deadlock at Gallipoli in the 'August Offensive'. Bob was in charge of a battery of four guns of IX Corps, under Lieut.-General Sir Frederick Stopford. Stopford was 61 years old, in poor health, and had never commanded in battle; he was selected through an extraordinary, tragicomic process. The 11th Division was under the command of Major-General Frederick Hammersley, who needed to be watched for signs of psychological breakdown, Kitchener warned.

Hamilton had a complex plan to capture the high ground from Ejelmer Bay in the north to Cape Hellas in the south, thus subduing the peninsula and opening the Dardanelles. One of the new divisions would assist the existing Anzac troops to capture Chunuk Bair on the ridge behind the Anzac toehold, while the other two divisions would land at Suvla Bay further north, negotiate the dry salt lake and capture the hills surrounding the Suvla Plain, including the

[74] R C Bragg to W H Bragg, 19 June 1915.
[75] R C Bragg to G Bragg, [22 June 1915]; ibid., 26 June 1915.
[76] R C Bragg to W H Bragg, [29 June 1915]; R C Bragg to G Bragg, 1 July 1915.
[77] Mrs. E Crozier to G Bragg, 2 July 1915.
[78] Ibid., n.d.
[79] R C Bragg to W H Bragg, 10 July 1915.
[80] War Diary of the 58th brigade of the R.F.A., The National Archives (formerly Public Record Office), UK, WO 95/4298.
[81] R C Bragg to G Bragg, 15 July 1915, RI MS RCB/6.
[82] War Diary, n. 80.
[83] R C Bragg to W L Bragg, 30 July 1915, RI MS RCB/9.

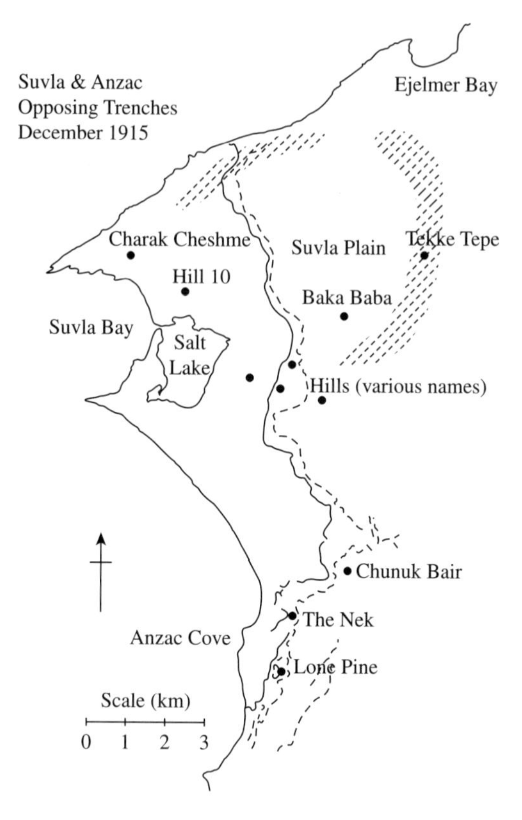

Fig. 17.2 Map of Gallipoli, showing places mentioned in the text (author).

Tekke Tepe Ridge to the east and the various hills to the south (see Figure 17.2). Unfortunately the existing troops were exhausted and ill, the new troops inexperienced, and the artillery wholly inadequate. At Suvla the will of the Turkish troops was again underestimated and the aim was reduced to securing a northern base. The fatal errors continued.

On 6 August 1915 there were assaults at Helles and at Lone Pine and Chunuk Bair in the Anzac sector, and in the evening the 11th Division went ashore at Suvla. Bob's brigade was not among them; the division was landed in stages, the artillery last. On 7 August there was an assault at The Nek, and the 10th (Irish) Division landed at Suvla. The fighting in the Anzac sector was fierce, with terrible casualties and much hand-to-hand combat. New Zealanders reached Chunuk Bair but were beaten back. The progress at Suvla was inexcusably confused and slow against scattered opposition; the opportunity to take the Tekke Tepe ridge and assure the security of Suvla Bay was lost. On 9 August there was a fierce counter-attack at Suvla by the newly arrived Turkish troops under Kemal and stalemates evolved everywhere on the peninsula: at Helles, Anzac, and Suvla. The Turks controlled all the heights and looked down on the invaders, now depleted and exhausted.

Nevertheless, Hamilton hoped the battle could be continued at Suvla if not elsewhere; but first the Suvla Plain had to be crossed. The 58th Brigade, including Robert Bragg, began landing just south of Suvla Bay on 9 August, but they were forced to camp there for several days while the rest of the brigade arrived and their guns were found at Anzac and brought along the coastal track by horses. Over the next few days the batteries moved north across the dry lake and set up near the Charak Cheshme well and on Hill 10, still close to the coast. The infantry brigades had attempted to gain the forward slopes of the Tekke Tepe range but had been repulsed with heavy casualties. The heat, the lack of water, and the prickly scrub, which caught fire and incinerated the wounded, compounded existing problems. Days were spent reorganizing, digging trenches, and planning the next assault; the men were tired and dispirited. Stopford (Corps commander), Mahon (10th Division), and Hammersley (11th) were all replaced. The attack had bogged down on the flat parched plain; the battle was already lost.

On 15 and 16 August the batteries fired most of the afternoon in support of an infantry attack that was unsuccessful. Bob's A Battery fired on the Turkish trenches and on the distant ranges; it had a wagon burnt and one gun failed. On 17 August the Turks returned heavy fire from howitzers. C Battery had three men wounded. 'Nothing much' happened 18 through 20 August but snipers were busy.[84] It was the first time Bob had a chance to write home: 'Well Mum dear, the last few days have been the most strenuous I have ever had in all my life'.[85] On 21 August the largest action of the campaign was attempted: capture of the hills to the south of the plain. 'Great battle today': the batteries separated, A into a lower valley of the Tekke Tepe range and to Baka Baba.[86] The guns were ineffective, however, being either inaccurate or too often defective. By 5 p.m. the 11th Division's infantry attack had failed, again with heavy casualties, as had those of other units, raked by Turkish shrapnel and machine-gun fire. The day was a failure. Since the opening of the offensive, total Allied losses to death, injury, and sickness were more than 140,000 men. Only defence and evacuation remained possible; fortunately the stock of Turkish shells was also low.

Bob wrote home on 22 August: 'We had an awful mix up when we landed; the arrangements here seem perfectly scandalous... someone ought to get the sack'.[87] Four days later he was sitting in the dugout doing nothing;[88] while on the 29th, 'Cheers! Today has been a day of days. We have just got our first parcel post. I scored four', with clothes, toiletries and 'many good things to eat; Ellison and I have been making pigs of ourselves all day & now we feel so ill'.[89] From 23 to 31 August the War Diary has little to say; ammunition was scarce and the guns were largely silent. Nevertheless, rifles barked and guns and howitzers boomed here and there; each side was holding its position. Then,

[84] War Diary, n. 80.
[85] R C Bragg to G Bragg, 19 August 12915, RI MS RCB/11.
[86] War Diary, n. 80.
[87] R C Bragg to W H Bragg, 22 August 1915, RI MS RCB/12.
[88] R C Bragg to G Bragg, 26 August 1915, RI MS RCB/13.
[89] Ibid., 29 August 1915, RI MS RCB/14.

on 1 September 1915, the War Diary recorded: '15.50. Received a message that two subalterns of A Battery were wounded (2/Lieut Ellison slight wound in elbow & 2/Lieut Bragg very serious in legs—one amputated)...Officers ordered to take cover in separate dugouts. Same shell that hit 2/Lieut Bragg also damaged 2/Lieut Ellison, direct hit. Shell did not burst, fuse came off'.[90]

Back home the news reached William by telegram in Leeds,[91] preceding a letter to Gwendoline from Gallipoli. Gwendy recorded that she and her mother were at Deerstones, and 'one morning as I was standing by the shallow sink in the kitchen, looking out into the garden, my father unexpectedly passed the window, came in, said to me quickly in a low voice "Bob's gone", and went upstairs to my mother. I heard her cry out. All the rest of the day she walked up and down the flowery meadow by the cottage, a dark veil hiding her bent head'.[92] On another occasion Gwendy wrote of her mother: 'She was always happy painting; nothing in life gave her so much direct pleasure...But the First World War and the loss of her younger son dealt her a grievous blow from which she never really recovered. Her gaiety went; her cheerfulness from then on was a fire lit for others, at which she warmed her own hands. She did so much for others, rushing in to help with Todd self-confidence'.[93] The handwritten letter from Gallipoli gave more details:[94]

> Dear Mrs Bragg
>
> It is with the deepest of regret that I sit down to write this letter. I feel that mere words are utterly too feeble to express my feelings of sympathy at your great loss. It was a most unfortunate accident, as your son & Ellison were sitting in the dug-out censoring letters when a shell came through the sand-bags, severing your son's left leg & injuring the other to such an extent that it would ever have been useless. Ellison escaped with a slight wound on his elbow. Your son was immediately taken down to hospital [on the edge of Suvla Bay] and expired on board a hospital ship early next morning. As his constant companion for nearly a year I am proud to think that we had made a lasting friendship, & I cannot tell you how he will be missed as the best of good companions.
>
> Yours, deeply sympathising R Ll. Peel

Bob was buried at sea from the hospital ship *Nevasa*.[95] Two letters to him, one from each of his parents, were returned with the inscription 'deceased'.[96]

[90] War Diary, n. 80.

[91] Letter E Rutherford to B B Boltwood, 14 September 1915, Yale University Library.

[92] Caroe, n. 31, p. 81.

[93] G Caroe, 'Bragg, Gwendoline', in R Biven, *Some forgotten...some remembered: Women Artists of South Australia* (Adelaide: author, 1976), no pagination.

[94] Letter R Ll. Peel to G Bragg, 6 September 1915, RI MS RCB/16; Lt Robert Lloyd Peel, MC, subesequently died of wounds sustained at Passchendaele (personal communication, Dr R Walding).

[95] War records, n. 8.

[96] Letters G Bragg to R C Bragg, 18 August 1915, and W H Bragg to R C Bragg, 22 August 1915, both Bragg (Adrian) papers; William's reaction to his son's death will be discussed below.

It was Gwendoline who told Lawrence of his brother's death: 'We have lost Bob... You will know what this means to your father and me... If you can just send a p[ost] c[ard] for your father's sake'.[97] William tried to be strong: 'Mum was very broken... She has... tried to pull herself together'; but he quickly weakened, 'But there; I can't talk about it all just now'.[98] In fact, he and his wife were never able to tell Lawrence precisely what had happened, despite repeated requests.[99]

Bob's death was recorded in Australian and English newspapers and letters of condolence flooded in: from J J Thomson, Jones and Sanderson at Oundle, and others in England; and from Charles Todd, University of Adelaide staff, Girdlestone at St Peter's College, and many acquaintances in Adelaide.[100] His Aunt Lorna Todd wrote to her sister: 'I think I have never shed more tears than over the death of anyone than Rob. Something inside one seems to melt... Rob, [his grandmother] said, was the first person who taught her the meaning of joie de vivre... & I think it's true, there was something heavy in Alice's nature which Rob quite dispelled'.[101] Gwendoline sent the battery a Christmas parcel in December.[102] Later, Robert's name appeared on the Helles Memorial in Turkey and in the crypt beneath the War Memorial on North Terrace in Adelaide, in the short list of 'South Australians who enlisted with or were transferred to British forces'.

Just three weeks before Bob's death Harry Moseley had also been killed at Gallipoli. Rutherford wrote to Boltwood: 'He was the best of the young people I ever had, and his death is a severe loss for science... You will also be very sorry to hear that Bragg's second boy, who had not yet graduated in Cambridge, died of wounds received in the Dardanelles'.[103] Like his colleagues Moseley had hurried to enlist, but now there was a groundswell to conserve such talented young men for war-related work at home.[104] At Gallipoli winter arrived and men were frozen to death in the ice-cold trenches. Hamilton was relieved of his commission and his successor recommended evacuation. Kitchener came to check and agreed. The evacuations in December and January were brilliant successes, the only ones for the Allies on Gallipoli. The brave and tenacious Turks had retained their homeland.

I visited Gallipoli in 2004. Productive farms dot the Suvla Plain, green and peaceful; the uplands are rocky and covered with prickly scrub. Very few visitors come this far; they stop at Anzac Cove, tour the nearby ridges and gullies, and stop to read Kemal Ataturk's extraordinary words: 'Those heroes that shed their blood and lost their lives... rest in peace. There is no difference between

[97] Letter G Bragg to W L Bragg, [early September 1915], Bragg (Adrian) papers.
[98] Letter W H Bragg to W L Bragg, 14 September 1915, Bragg (Adrian) papers.
[99] Lawrence Bragg's family, personal communications.
[100] Bragg (Adrian) papers; RI MS RCB/18.
[101] Letter L Todd to G Bragg, 1 December 1915, Bragg (Adrian) papers.
[102] Letter A Whyte (Btty Sgt Major) to G Bragg, 9 January 1916, RI MS RCB/24.
[103] Letter Rutherford to Boltwood, n. 91.
[104] J L Heilbron, *H G J Moseley: The Life and Letters of an English Physicist, 1887–1915* (Berkeley: University of California Press, 1974), chs 7 and 8, p. 125.

the Johnnies and the Mehmets to us, where they lie side by side, here in this country of ours... You, the mothers, who sent their sons from far away countries, wipe away your tears; your sons are now lying in our bosom and are in peace. After having lost their lives on this land they have become our sons as well'.[105]

Lawrence's war

Although he was older and more experienced in military affairs, Lawrence Bragg's passage to a battlefront was slower than Bob's. His first research student, Edward Appleton, had been continuing the crystal experiments, but he too soon enlisted.[106] Having been appointed to the Leicestershire Royal Horse Artillery in August 1914, Lawrence ate his last meal at Trinity College and was 'looking forward tremendously to starting my new job tomorrow'. He also started a moustache.[107] He would spend a year training at Diss in Norfolk, where he was billeted with two of the local identities.[108] He wrote very regularly to his parents, and especially to his mother, outlining his training, reporting domestic matters, and constantly reassuring her of his wellbeing.

He was soon complaining about his lack of horsemanship: 'I don't know enough about horses, it all takes a lot of learning';[109] and 'I get very sick sometimes because I am slack with the men, and get moments of awful despondency, and then again I feel I am much better than anyone else'.[110] His groom, Staniforth, and his batman, Colbey, were excellent, and Colbey later remembered that, 'When I first had to look after you, I thought you were the worst officer in the British Army, but after a month I'd have done anything for you'.[111] Lawrence's last extant letter for 1914 welcomed his father home from America and urged him to come to Diss, since leave for Christmas was unlikely. He apologized for poor proofreading of their book: 'I did my best with the book but was very stupid in correcting it... I could not think about it at all... crystals seem a long way away at present.' He also reiterated his wish to talk to his father: 'I'll tell you all about my experiences when you come here, and I am dying to hear all yours'.[112]

Early in the new year Lawrence had an infected throat, and the horses were a regular concern. He sent his father one hundred copies of his copper paper and asked him to distribute them, and was pleased to see a copy of *X Rays and Crystal Structure*: 'What fun it is getting the first copy of one's book'. He

[105] Fewster et al., n. 50, p. 24.
[106] Letter W L Bragg to W H Bragg, n.d. [August 1914], RI MS WLB 37A/6/46.
[107] Letter W L Bragg to G Bragg, 13 September 1914, RI MS WLB 37A/5/4.
[108] G K Hunter, *Light is a Messenger: The Life and Science of William Lawrence Bragg* (Oxford: OUP, 2004), p. 51; W L Bragg, Autobiographical notes, pp. 32–3.
[109] Letter W L Bragg to G Bragg, n.d., RI MS WLB 37A/6/48.
[110] Ibid., n.d. [late 1914], RI MS WLB 95E/1.
[111] Quoted in Hunter, n. 108, p. 51.
[112] W L Bragg to W H Bragg, 15 December 1914, RI MS WHB 37A/6/39.

supported his father's move to University College London, away from 'that deadly Leeds university atmosphere'.[113] He was living in a strange environment: constant reminders of his exploding research career but the impossibility of pursuing it, the excitement of the war but the inability to take part, and now extraordinary news from America. He and his father had jointly won the Barnard Gold Medal of Columbia University in New York, 'for meritorious service to science...awarded quinquennially to such person...as shall within the five years next preceding have made such a discovery in physical or astronomical science...as in the judgement of the National Academy of Sciences...shall be esteemed most worthy of such honor'. Earlier awards had been made to Rayleigh and Ramsay, Röntgen, Bequerel, and Rutherford. William and Lawrence were joining illustrious company.[114] Lawrence was informed by telegram and responded to his mother, 'I feel very proud...I can hardly realize that we have got that medal in such distinguished company...[a senior officer] has made me stand the mess a bottle of best port tonight, the old blighter'.[115]

By mid-year Lawrence was expecting to hear from his regiment about an active appointment but nothing came; he was at a loose end and frustrated. He had a day in London buying equipment and a week at Aldershot learning 'telephone work'. One of his horses was lame, he had 'a beastly field day', and he urged his parents to motor to Diss to see him. Trinity College was full of wounded and he recognized that 'The Germans will take some stopping'. His mother was increasingly anxious.[116] His friend Cecil Hopkinson 'was going out very soon', but there was a lot of tennis and some weekend leave.[117]

His parents had stayed on in London because, as Gwendoline told Bob in the letter that was returned, 'Poor old Bill has no luck. He got his orders to go off at 6 last night & succumbed to a cold an hour afterwards & is now in bed with a temperature; it's simply been this horrid life he has been leading for the last month, hanging around expecting to be off any minute & such a lot to worry over it all'.[118] William had written similarly, adding, 'I don't suppose it matters *very* much because the French people were really not ready for him'.[119] On 19 July 1915 Lawrence had been seconded out of the Leicestershire Royal Horse Artillery for other unspecified duties, although formally he remained a member,[120] attached to the Field Survey Company of the Royal Engineers.[121] It had taken him ten days to recover and catch a boat. His next letter home was

[113] Letters W L Bragg to G Bragg, early 1915, RI MS WLB 37A/6/1–6.
[114] Letter President, Columbia University, to W H Bragg, 3 May 1915, RI MS WHB 11A/5; *The Times*, 20 May 1915, p. 5.
[115] Letter W L Bragg to G Bragg, 19 May 1915, RI MS WLB 37A/6/7.
[116] Letters W L Bragg to G Bragg, n.d. [first half of 1915], RI MS WLB 37A/6/10–18.
[117] W L Bragg to parents, June–July 1915, Ri MS WLB 37A/6/no numbers.
[118] Letter G Bragg to R C Bragg, 18 August 1915, Bragg (Adrian) papers.
[119] Letter W H Bragg to R C Bragg, 22 August 1915, Bragg (Adrian) papers, William's emphasis.
[120] *The London Gazette*, 19 October 1915, p. 10289.
[121] *Army Book 439* for W L Bragg, RI MS WLB 37A/4/9.

headed *RMSS Vera* and said, 'We have so very nearly reached [blacked out] that perhaps this will do as an arrival letter...I'll finish this off when I get to Havre...We meet Lefroy at 1.15 this afternoon and probably go off to Paris at once'.[122]

It was already apparent to some that the way of waging war had changed substantially from previous practice, principally by the introduction of long-range weapons. The addition of improved rifles and powerful machine guns 'increased both range and accuracy and made frontal attacks out of the question, if simultaneous developments in artillery had not provided the firepower to support them...Armies would thus come under fire long before they could even see the enemy, let alone attack his position'.[123] A method of reducing the effectiveness of the unseen German guns became increasingly urgent. Accordingly, Lawrence had received a letter from the War Office instructing him to report to Colonel Hedley, who put to him a proposition that Lawrence described thus:[124]

> The French had started experiments with a method of getting the positions of enemy guns by measurements made on sound waves. Colonel Winterbottom, R E...was convinced that this method might be useful, and had persuaded the Army authorities to set up an experimental section under the direction of an officer who had scientific training. Colonel Hedley...asked me whether I would take on the assignment. I was thrilled...To have a job where my science was of use, after feeling so inefficient in the battery, seemed too good to be true...The R A had clearly been very doubtful that the method would be of any use, but finally agreed...Another young officer, Harold Robinson, was detailed to do the experimental work with me. Robinson was on Rutherford's staff at Manchester...We were sent out to France together, to report to Colonel Jack at G[eneral] H[ead] Q[uarters]...Our section was attached to Jack's office at St Omar while still in the experimental stage. We were first sent to a French section in the Vosges.

The French had been experimenting with 'sound-ranging' for some time, as Lawrence later described (see Figure 17.3):[125]

> There was an almost impassable barrier between the military and the scientific minds. It was into this rather unfriendly world that British Sound-Ranging was born...The principle is simple. A series of listening posts or microphones are situated in known positions along a base behind the

[122] Letter W L Bragg to G Bragg, 31 August 1915, RI MS WLB 37A/6/no number.

[123] M Howard, *The First World War* (Oxford: OUP, 2002), pp. 20–21.

[124] W L Bragg, Autobiographical notes, pp.33–34. Harold Robinson later became Professor of Physics at Cardiff and then East London College, now Queen Mary College, University of London, and finally its Vice-Chancellor; see E N da C Andrade, 'Harold Roper Robinson, 1889–1955', *Biographical Memoirs of Fellows of the Royal Society*, 1957, 3:161–72; and for Robinson's researches, J G Jenkin et al., 'The development of X-ray photoelectron spectroscopy, 1900–1960', *Journal of Electron Spectroscopy and Related Phenomena*, 1977, 12:1–35.

[125] Sir Lawrence Bragg, 'Sound-ranging', in Sir Lawrence Bragg, Maj.-Gen. A H Dowson, and Lieut.-Col. H H Hemming, *Artillery Survey in the First World War* (London: Field Survey

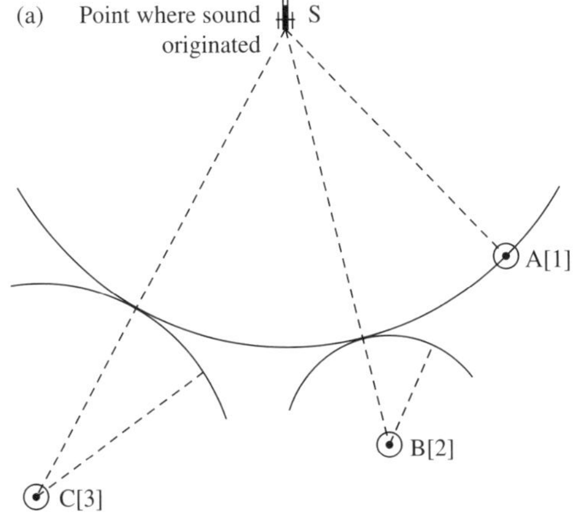

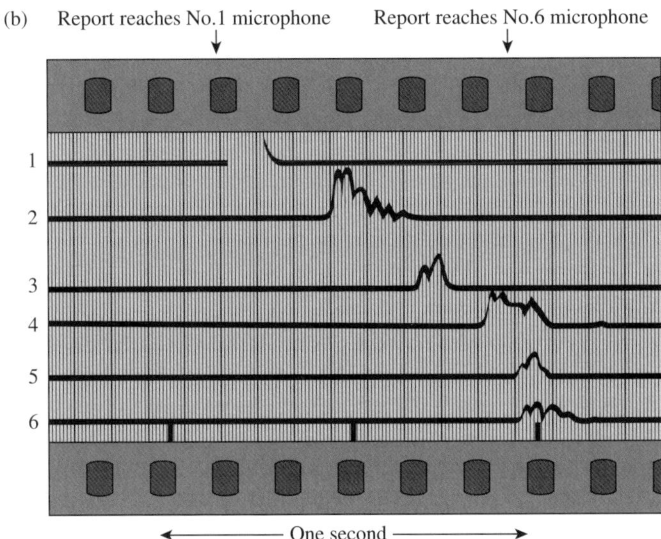

Fig. 17.3 (a) Schematic diagram of the sound-ranging method for locating enemy guns, as developed by Lawrence Bragg and the British Army during The Great War. (b) Film recording of a German howitzer firing, as detected by the sound-ranging apparatus. (From Sir William Bragg, *The World of Sound*, Bell & Sons, 1920, pp. 186, 188 respectively.)

Association, 1971), ch. 4. Other accounts are: J R Innes, *Flash Spotters and Sound Rangers* (London: Allen & Unwin, 1935), particularly ch. 6 by W L Bragg (text also at RI MS WLB 65A/104); G Hartcup, *The War of Invention: Scientific Developments, 1914–18* (London: Brassey's, 1988), ch. 5; M J G Cattermole and A F Wolfe, *Horace Darwin's Shop* (Bristol: Hilger, 1987), ch. 4; Lieut.-Col. E M Jack, 'Survey in France During the War', *Royal Engineers' Journal*, 1919, 30:1–27; P Chasseaud, 'Sound-ranging 1914–1918', *Stand To!*, 1990, 30:23-–7; W Van der Kloot, 'Lawrence Bragg's Role in the Development of Sound-ranging in World War I', *Notes and Records of the Royal Society*, 2005, 59:273–84; Hunter, n. 108; the official account is *Report on Survey on the Western Front 1914–1918* (London: HMSO, 1920).

> front line. The time difference between the arrival of the gun report at the posts is measured. Suppose the sound to reach post 1 first at time t_1, post 2 at time t_2 and so forth. Then, if one draws a circle on the map around post 2 with radius $v(t_2 - t_1)$, where v is the velocity [speed] of sound, and similar circles for the other posts; a great circle which passes through post 1 and touches the other circles represents the form of the report wave, with the gun at the centre.
>
> The French tried several systems. In the simplest, the arrival of the sound was registered by observers who pushed keys [but this method failed due to inaccurate timing]...In the TM [Télégraphique Militaire] method, which was finally adopted as standard by the French Army, currents from the microphones at the posts actuated pens, whose movements were recorded on smoked paper at the headquarters behind the base to which the microphones were connected. Yet a third system employed a recorder which had been designed by Lucien Bull, of the Institut Marey in Paris [a French physicist of Irish extraction, working on electrocardiography and heartbeat recording]. This was the most elegant and accurate of the recorders, but it was complex...Bull employed a six-string Einthoven galvanometer, in which the microphone currents were recorded [on a cine film] by the displacement of fine wires in a strong magnetic field...a toothed wheel governed by a tuning fork...[gave] time markings across the film 100 times per second. The apparatus was switched on and off by one or more forward observers...the operator at headquarters cut off the portion of the film which had run, developed and fixed it, and passed it to the reader, who measured the time intervals and deduced the position of the gun.

Once Lawrence arrived in France there was a nine-hour drive from Le Havre to Alsace, with contradictory impressions: 'It is most gorgeous country, all mountains and pines, one mass of forests. It seems ridiculous that there is a war going on so very close...One is told that that little village nestling in the valley is ours and that the next is German. You can just see the trenches in little clear patches...and every now and then a battery near us starts firing. The Germans reply very rarely...it rains the whole time...The French have been most awfully kind to us...we saw a lot of their infantry coming back from the trenches'.[126] The German guns did not fire, so Lawrence could not watch the Bull system in operation.

Then, just days after his arrival, he received a letter from his mother saying that his brother was dead. 'Dear Dad', he wrote from London, 'I got Mother's letter telling me about Bob just before I was leaving for England yesterday. I will come up to [Leeds] as soon as I have reported here and can get away...Give my very dearest love to Mother'.[127] Lawrence's emotions were on an unprecedented roller-coaster ride: daily boredom, death and destruction across the Channel, Barnard Gold Medal, wounded men clogging Trinity, imminent departure for France, illness, scientific work at the front line, and now Bob's

[126] Letter W L Bragg to G Bragg, 4 September 1915, RI MS WLB 37A/6/no number.
[127] Letter W L Bragg to G Bragg, n.d. [early September 1915], RI MS WLB 37A/6/47.

death. Bob was popular, with an outgoing personality and social ease. Gwendy thought her mother preferred her younger son, but I doubt Hunter's suggestion that Lawrence's grief was mixed with guilt.[128] It was an extraordinary time. He was back in Paris by 12 September, having driven 'down the valley... each village showing more signs of being knocked about than the last, until finally we ran through a big town where half the houses were in ruins'.[129]

Lawrence was soon in London again, to collect apparatus and visit his parents at their new London home.[130] He also wanted photos of Bob and Gwendy, 'to have with me'.[131] By 1 October he was back at Colonel Jack's St Omar headquarters, having determined to try the accurate but complex Bull system: 'All is going well; I'm very pleased with things'.[132] Next day he was in Paris to talk to Bull and collect one of his galvanometers: 'I walked back from the Invalides last night by moonlight, over the Seine. It did look lovely... You know I am always thinking of you and hoping you are feeling a little less sad about Bob'.[133] Still in Paris, he bought a necklace for Gwendy and called on Professor Cotton, who was developing yet another system:[134] 'I got an awful lot of useful information from him and some charts', and 'I want you to get me, if you can, some thin celluloid sheet about 60 cm wide and about 5 metres long' on which to trace maps. In addition, 'I am very fit. Our things are installed now... I'm feeling very bucked about that'. However, 'Tremendous complications unfold themselves as one penetrates further into our job; but I am sure it will work all right'.[135] They had been moved to La Clytte, eight kilometres southwest of Ypres, one-and-a-half kilometres from Kemmel Hill, near some of the bloodiest fighting. They were staying in the local Curé's house, 'a topping billet'. Robinson was proving an excellent colleague, capable, and a cheerful fellow to work with, and their French assistant, Bocquet, who had assisted in the development of the TM system, was helpful. Lawrence was grateful for letters and parcels.[136] Their initial work had been encouraging and they were now designated 'Sound Ranging Section, 18th Brigade, R.G.A. [Royal Garrison Artillery], B.E.F. France'.[137]

Later in October Lawrence sent a colleague back to England to get some sensitive microphones to use at the front and to get apparatus made, hoping his father would help in both cases. He also carried 'a letter to Mr Campbell at the N.P.L., who will tell him where to buy anything he wants'.[138] The Germans were shelling, and they were wiring up a set of microphones for a serious test to locate the guns at the front. 'Robinson and Albrecht went out to do some

[128] Hunter, n. 108, p. 53.
[129] Letter W L Bragg to G Bragg, 12 September 1915, RI MS WLB 37A/6/no number.
[130] Letter W L Bragg to W H Bragg, 16 September 1915, RI MS WLB 37A/6/19.
[131] Letter W L Bragg to G Bragg, 28 September 1915, RI MS WLB 37A/6/20.
[132] Letter W L Bragg to W H Bragg, 1 October 1915, RI MS WLB 37A/6/21.
[133] Letter W L Bragg to G Bragg, 2 October 1915, RI MS WLB 37A/6/22.
[134] Chasseaud, n. 125, p. 24.
[135] Letter W L Bragg to W H Bragg, 12 October 1915, RI MS WLB 37A/6/25.
[136] Ibid., 21 October 1915, RI MS WLB 37A/6/26.
[137] Letter W L Bragg to G Bragg, 5 November 1915, R MS WLB 37A/6/28.
[138] Letter W L Bragg to W H Bragg, 23 October 1915, RI MS WLB 37A/6/no number.

wiring this morning and came back at 5, being much too excited to have any lunch. Apparently they had been climbing trees in full view of Bosches, to put the wire up them'.[139] Conditions deteriorated, 'The weather has become atrocious, wet every day. The mud is unbelievable...The roads here consist of a narrow strip of pavé with a sea of mud on either side'.[140] Close to the front line they set up their microphones, with the recording apparatus in a lorry and with Bocquet, sometimes accompanied by Lawrence, as the forward observer in the thatch of a deserted cottage.[141] 'There are a couple of guns going off close to us; big fellows that make our poor old lorry jump into the air each time'.[142] In early November they succeeded in locating an enemy gun for the first time.

Now in the midst of artillery action and serious attempts to improve the technique, another bombshell arrived in Lawrence's mailbox: 'Just got Dad's letter and yours, with the cheery news in it. You can imagine how I felt; really I am the most lucky fellow in the world, I think. It is so awfully nice to be coupled with Dad in this way...I got many congratulations today; every one had seen it in the papers...Will Dad go over to Stockholm to get the booty?...I was sorry to see that a German had got the chemistry prize; that was tactless. Our section treats me with much more respect now than it did before'.[143] It was the 1915 Nobel Prize in Physics, awarded to William and Lawrence Bragg 'for their services in the analysis of crystal structures by means of X-rays'. The Chemistry Prize was awarded to Richard Willstätter of Berlin for pioneering research on plant pigments, especially chlorophyll. A few days later Lawrence reported, 'Today the Curé, who had seen my photo in the paper, came in and offered me a bottle of wine with his best bow, as a little present to felicitate the occasion. Generals humbly ask my opinion about things; it is great fun. The weather is much better now. There is the very dickens of a bombardment on today; a continual rumble of fat shells trundling along'.[144] Next day he wrote to his father: 'Just got your letter about the Nobel...I do hope they postpone the ceremonies. I would love to have some of them come off after the war...It's freezing hard now, and very misty. It is a tremendous relief to have all the mud frozen for a bit'.[145] Lawrence scratched his knee and it became infected and had to be lanced; there were no antibiotics and it took a long time to heal.[146] Knowledge of the circumstances surrounding the Nobel award, as well as its celebration, would have to wait.

As Christmas approached there was no sign of victory; attempts by both sides to break the stalemate had failed, with huge casualties. Lawrence's group was moved out of their billets into second-hand, prefabricated huts, but they made them cosy with furniture from deserted houses and parcels from home.

[139] Ibid., 26 October 1915, RI MS WLB 37A/6/no number.
[140] Letter, n. 135.
[141] Letter W L Bragg to [parents], 11 November 1915, RI MS WLB 37A/6/no number.
[142] Letter W L Bragg to G Bragg, 13 November 1915, RI MS WLB 37A/6/29.
[143] Ibid., 17 November 1915, RI MS WLB 37A/6/30.
[144] Ibid., 21 November 1915, RI MS WLB 37A/6/31.
[145] Letter W L Bragg to W H Bragg, 22 November 1915, RI MS WLB 37A/6/32.
[146] Letter W L Bragg to G Bragg, 29 November 1915, RI MS WLB 37A/6/34.

Their future was still unclear. Lawrence was also concerned about his close friend: 'Have you heard about Cecil [Hopkinson]; is he badly wounded? Do let me know as soon as possible'.[147] The huts were 'extraordinarily well ventilated; Robinson has been doing his best with newspaper and putty to stop up the worst ones but we still live in the open air... I am still very short-handed, with two men on leave out of a total of seven'.[148] A few days later there was good news: 'our show has been approved of', and 'We have at least got two extra servants'.[149] They had 'a very peaceful Xmas'. Staniforth cooked 'a wonderful effort', and 'our plumb pudding [from home] was great'. At the same time, 'Someone is getting an awful hammering down south; there is a continuous rumble of guns'.[150]

Although GHQ ordered seven more sets of the Bull apparatus to provide two sets to each Army in the field, there were a number of problems hampering the location of German guns on a regular basis. These included the effects of weather, particularly adverse wind, the lack of robustness in the wiring system, the inaccuracy of surveyed microphone positions, and the lack of manpower.[151] The most serious, however, was inadequate microphones. A field gun produced sound waves of very low frequency, and the diaphragm of a conventional microphone hardly moved. On the other hand, the ejected shell produced a high-frequency shock-wave 'crack' as it passed overhead, useless for gun location and masking the later barely audible 'boom' of the gun. The solution to this problem came with the arrival of men to staff the expanded Army Topographical Sections, including Corporal William Tucker from the Physics Department of Imperial College, London. He joined Lawrence's own section, which had been locally designated 'W Section', presumably W for Willie, by which name Lawrence was still known in the army. Officially he was 'O.C. Experimental Sound-Ranging Section and School', which was now not only responsible for developing the system and supporting the growing number of active sections, but also for the instruction and training of new staff. Tucker had been working on the cooling of fine hot platinum and tungsten wires by air currents, and Lawrence asked him to participate in seeking a solution to the microphone problem. Chasseaud explained what happened (see Figure 17.4):[152]

> Tucker... was the right man in the right place [at the right time]. Bragg had recognised, by being lifted off the privy at La Clytte when a six-inch gun fired, that the gun-report involved a large pressure change, and Tucker had noticed the jets of cold air which played on his face through holes in the hut wall as he lay in his camp bed, whenever the gun-wave arrived. In the classic 'eureka' act of creation, he related the pressure wave to the

[147] Ibid., 3 December 1915, RI MS WLB 37A/6/35.
[148] Letter W L Bragg to W H Bragg, 14 December 1915, RI MS WLB 37A/6/38.
[149] Letter W L Bragg to G Bragg, 16 December 1915, RI MS WLB 37A/6/40.
[150] Ibid., 24 and 28 December 1915, RI MS WLB 37A/6/41 and 42.
[151] W L Bragg, 'How sound ranging was done in theory and practice', in Innes, n. 125, ch. 6.
[152] Chasseaud, n. 125, p. 25.

Fig. 17.4 Pioneers and colleagues of British sound-ranging during The Great War, William Tucker standing left, Lucien Bull and Lawrence Bragg seated centre, no date. (Courtesy: Lady Heath.)

> cooling of a hot wire, and the idea of the Tucker microphone was born.[153] This was in June 1916. An electrically heated wire was stretched across an aperture in a container (rum jars and ammunition boxes were both used), and fitted in circuit to the galvanometer... the gun-wave cooled the wire and thus reduced its resistance... [which was] picked up by the galvanometer [in a Wheatstone Bridge circuit]... Bragg remembered the excitement of the first trial: 'I will never forget the thrill of seeing the first record, in which the shell-wave hardly made the galvanometer string quiver, while the gun-wave gave an enormous kick. The real success of our sound-ranging dated from that day'... New microphones on this principle were ordered at once, and by September 1916 all sections were equipped with them.

Lawrence always gave Tucker full credit for this breakthrough, but the carriage of the enterprise and the solution of other outstanding problems rested with Lawrence. In mid-1916 he was promoted to Lieutenant and remained seconded,[154] in September he was appointed 'Acting Captain whilst holding a special appointment',[155] and in April and November 1916 he was mentioned in dispatches.[156] While not in the trenches and able to retire to the rear in the evenings, Lawrence and his staff were very close to the front line. It could

[153] W S Tucker, 'Improvements in and relating to microphones', British Patent no. 138,368, application 15 September 1916, accepted but withheld from publication 22 May 1917, published 4 March 1920.

[154] *Supplement to the London Gazette*, 19 June 1916.

[155] The National Archives, London, WO 374/8505.

[156] *Second Supplement to The London Gazette*, 13 June 1916 and 2 January 1917; certificates dated 30 April and 13 November 1916, RI MS WLB 37A/4/3 and 4.

be very dangerous. His frequent letters home during 1916 gave few details, but he must have witnessed horrific scenes and experienced the agony. Harold Robinson, his colleague and deputy, confided to Rutherford as early as October 1915: 'A week or two ago I had the first...opportunity of finding out whether I was really brave or not, and I'm sorry to say that I wasn't at all happy; I managed to keep an anaemic sort of grin on my face, and to talk fairly lucidly, during the half hour or so that the little show lasted, but my nerves were a bit jumpy for a day or two afterwards. Bragg kept very cheerful through it all—by the way, I don't think he has said anything to his people about it, they would be worried if they knew: anyhow I don't think it will happen again for some time. P P S I regret to state that, as a result of the recent award, Bragg has acquired the unpleasing nickname of "The Nobbler", and that our Mess President has threatened to cut off his beer, as being too coarse a drink for one so exalted!'.[157] Lawrence and his close associates messed together in their cramped and poorly equipped hut, which suggests a commendable level of camaraderie between the OC and his staff.

Only rarely did his letters hint at the danger and front-line conditions: 'The battles are not near me so don't be worried, though we see and hear a good deal of them, and get very excited. The planes have been very lively, and we have seen lots of battles'.[158] And: 'I am writing this squatting in our O[bservation] P[ost]. There is a dickens of a row going on because our batteries are all blazing away at the Bosches and all the while there is a stream of fat shells trundling overhead. At the same time the Germans are firing at two of our planes that are observing for the batteries and the cracks up in the air add to the general excitement. Two of the German shells failed to explode just now and came sailing over and went off in the landscape just behind. They are chucking bombs like billyoh in the trenches. I had a long trip along the line the other day, seeing all sorts of things, and I am about to start for another'.[159]

The major Allied offensive during 1916 was the battle of the Somme.[160] Artillery was to open the way for British infantry, but it was vulnerable to enemy barbed-wire, machine-guns, and artillery, and the British artillery was inexperienced and poorly resourced. From the beginning casualties were huge and the gains in territory were lost to fierce counter-attacks. Some of the planning 'belonged in the realm of strategic Cloud-Cuckoo-Land'; and 'the grim process continued on the Somme of piecemeal advances along that blood-drenched crest of ground'.[161] Late in September rain and mud reduced the carnage and restored the stalemate.

A letter from Lawrence to his father in March 1916 commented on William's latest research, reported a visit to Paris, and discussed sound-ranging

[157] Letter H Robinson to E Rutherford, 29 November 1915, CUL RC R48.

[158] Letter W L Bragg to G Bragg, 22 February 1915 [1916], RI MS WLB 37A/6/4.

[159] Letters W L Bragg to W H Bragg, 1 April 1916, RI MS WLB 37B/1/24, and W L Bragg to G Bragg, 7 January 1917, RI MS WLB 37B/2/2.

[160] T Wilson, *The Myriad Faces of War: Britain and the Great War, 1914–1918* (Cambridge: Polity, 1986), Part 7.

[161] Ibid., pp. 322, 338.

apparatus. In Paris Lawrence reported to Bull and his colleagues on his progress, and he visited Maurice de Broglie: 'I saw de Broglie in Paris and he was very interesting. I saw a lot of his photos. The absorption experiments... are very good; the absorption bands are very sharp and... his method is a jolly good one for getting their positions... I enjoyed my lunch there; his wife and mother-in-law were very polite and the wife great fun'. Lawrence needed help to increase his pool of equipment, certainly not available commercially: 'Colonel Jack is madly keen to get it ready; we want all the sections badly, at once, and we were rather counting on having it at the end of the month... It is the prisms and recorders [for the six-string galvanometers] that are wanted... I don't want to worry Jenkinson [who had moved from Leeds to London with William], I know he is doing his damndest, but perhaps you could arrange to relieve him of a little of the work by having it done out'.[162] Later that month he reported that 'we have nearly finished our officers' class—they have been such a nice lot of fellows and so very keen'; and he noted his age, 'I can't imagine myself as 26... it seems a fearful age... we are all looking forward to the birthday cake'.[163]

Lawrence sympathized with his father's troubles with the navy (see below) and was frustrated with the extent of his own paperwork. Protection had been increased: 'I have a dugout now, with steel girders, two feet of concrete, four feet of earth, and then bags full of flints... it would stop several 5.9 inch shells easily. The men have two jolly good ones too'.[164] In November Tucker and other experienced men were sent back to England to staff an Experimental Section on Salisbury Plain, to expand the activities carried out by Lawrence's section.[165] W Section itself was now 'such a tremendous big place... 56 men and nine officers'.[166] In March 1917 its nucleus was transferred to GHQ to further expand its experimental, developmental, maintenance, and training programmes. Officially Lawrence was now 'Technical Adviser on Sound-Ranging'.[167] The technique was approaching a state of usefulness and acceptance, although problems remained and Lawrence was regularly driving between the sections along the front.[168]

When major hostilities resumed in 1917 the names Arras, Messines, 3rd Ypres, Passchendaele, and Cambrai were written into history. Historian Trevor Wilson called 1917 'The Killing Time': 'If one thing had changed markedly since the opening of the Somme battle, it was in respect to artillery... the quantity [and] effectiveness of artillery fire was improving, not least in suppressing the artillery of the enemy. This was consequent on the development by scientists, mathematicians, and gunners of such devices as flash-spotting and sound-ranging, the latter one of the most potent factors in the development

[162] Letter W L Bragg to W H Bragg, 17 March 1916, RI MS WLB 95G/1.
[163] Ibid., 27 March 1916, RI MS WLB 37B/1/20.
[164] Letters W L Bragg to G Bragg, 1 May 1916 and 25 July 1916, RI MS WLB 37B/1/29 and 43.
[165] *Report on Survey*, n. 125, ch. 3.
[166] Letter W L Bragg to G Bragg, 5 November 1916, RI MS WLB 37B/1/69.
[167] *Report on Survey*, n. 125, ch. 3.
[168] Letter W L Bragg to G Bragg, 7 January 1917, RI MS WLB 37B/2/2.

of counter-battery work towards the excellence it eventually attained...The sound-rangers, in addition, could follow the path of single British shells to their destination and correct the range accordingly. As a supplement to (or a substitute for) aerial observation, these innovations would prove of mounting importance'.[169] Overall, however, the results of the offensive were disastrous: 'Thanks to the weather, the...offensive involved a quantity of misery that almost beggars description'. For the infantry 'the results were calamitous...stopped dead by uncut wire and massacred by undamaged machine-guns...in the Ypres quagmire'. Finally a deadly depression settled on British officers and men.[170] The stalemate had returned yet again.

The improvements in sound-ranging were reflected in Lawrence's letters home: 'Our show is going famously, my only fear is lest the war should end before it has reached its full stage of perfection!',[171] and 'A lot of my little plans have been approved of and I am very pleased'.[172] The same month, however, Lawrence wrote to his mother in distress at the death of Cecil Hopkinson, who had been wounded and suffered a long illness: 'I have been trying ever so hard to write something for Mrs Hopkinson about Cecil [but] it is almost impossible. Do tell me what sort of thing I should write; I am in such despair over it'.[173] Lawrence wrote a long manuscript recollection of the many happy times of study and adventure he had enjoyed with Cecil; it is now amongst his papers but it is unclear if a copy was sent to Mrs Hopkinson or was otherwise published.[174]

At the front it was found advantageous to place the microphones evenly spaced on a straight baseline, and then later on a circle of about 8,000 yards (7.3 km) radius. An experienced film reader could then readily pick out individual guns on a noisy, complex record. The OP was about 2,000 yards (1.8 km) behind the front line, the microphones another 1,000 yards back, and the recoding galvanometer and associated equipment a further two miles (3 km) to the rear. For some time the wires connecting the microphones to the recoding equipment were strung on posts, but these were very vulnerable to damage by both friend and foe; well-insulated cable laid on the ground by the sections themselves eventually solved the problem. The location of a gun from a microphone record was not a trivial exercise. In practice a 'plotting board', with a detailed map marked with accurate microphone positions, was equipped with strings of gut that ran from the midpoints between the microphone positions to scales of time difference. The intersection of these strings then gave the position of the gun. The details can be found in the literature.[175] Detection of a gun discharge, its shell burst, and the time between, enabled the calibre of a gun to be determined. Corrections for temperature and wind were the most persistent

[169] Wilson, n. 160, Part 10, pp. 449–50
[170] Ibid., pp. 478, 479, 483.
[171] Letter W L Bragg to W H Bragg, 20 January 1917, RI MS WLB 37B/2/5.
[172] Letter W L Bragg to G Bragg, 18 March 1917, RI MS WLB 37B/2/11.
[173] Ibid., 31 March 1917?, RI MS WLB 37B/2/14.
[174] W L Bragg, untitled handwritten manuscript, 9 pp., RI MS WLB 37B/2/36.
[175] Sources in n. 125; a drawing explaining the method using strings is in Cattermole and Wolfe, n. 125, p. 90.

problem, solved empirically by building up a data set for most conditions.[176] Strong wind from the east, common in Flanders, lifted the sound above the microphones and rendered sound-ranging impossible. On the other hand it was excellent in mist and fog, when the breeze was light and other techniques became inoperable.[177]

Lawrence's letters confirmed often that 'I am very fit', and that the frequent parcels from home were most welcome. When the rain set in the mud became incredible: 'Imagine walking for three miles along a road with mud over your knees, up to the waist in parts, in the dark, and with tree trunks floating about in it... my poor pants are permanently na-pooed'.[178] Then the cold: 'The temperature in our room is below freezing in spite of the fire, it is the limit... 29° of frost the other night!'.[179] Ellison, Bob's friend at Gallipoli, was now quite close: 'I am going over to see him'.[180] Darwin, Tucker, Bull, Colonel Jack, Bob Chapman (his fellow mathematics student at St Peter's College), Cambridge fellow students, and numerous others visited Lawrence at GHQ. All were welcome and provided a 'cheery' interlude. At the end of the year (1917) the Americans came to watch the British sound-ranging sections in action. One section leader was not a welcome visitor, however, and was forced to leave: 'Andrade got jolly well kicked out of the show here as he became absolutely the limit. He is a hopeless chap. I am sorry for him too sometimes, but he has got a bad kink in him somewhere. There was an awful to-do about it all; the officers in his section refused to work with him any longer and told the colonel so'.[181] Lawrence and Andrade would cross swords again later (see chapter 19).

With the emerging success of sound-ranging, his developing management skills, growing recognition by his superiors, and increasing maturity, Lawrence was growing in confidence and assertiveness. In May 1917 he wrote to his father in a new vein: 'I was awfully pleased to hear about the Italian Medal [the gold medal of the physical section of the Societa Italiana delle Scienze]. Mother gave the show away, as you say, but it was very nice to get your copy of the Italian letter... Dad, if I were you I would not stand for a moment to let them [Navy personnel] criticize or watch your work at all. It is just rot... I do get so mad when I think they worry you. I do wish you could get hold of someone like Threlfall to help you and tell people off for you. Another thing that makes me mad is all those ruffians who work with you. They are damn lucky not to be in the trenches and they ought to hop around like anything... They just take advantage of it if you are nice to them... There is nothing like bullying people a bit for bucking one up... I should just chuck my weight about a bit if I were you'.[182] Later he wrote again to his father, 'I want for my workshop here

[176] *Report on Survey*, n. 125, ch. 3.
[177] W L Bragg in Innes, n. 125.
[178] Letter W L Bragg to G Bragg, 15 January 1917, RI MS WLB 37B/2/3.
[179] Ibid., 4 February 1917, RI MS WLB 37B/2/7.
[180] Ibid., 12 February 1917, RI MS WLB 37B/2/9.
[181] Letter W L Bragg to W H Bragg, 4 June 1917, RI MS WLB 37B/2/22.
[182] Ibid., 4 April 1917, RI MS WLB 37B/2/15; the medal was reported in *Nature*, 1917, 99:150.

some first-class instrument-makers, really good men who can be given a sketch and work out the finished article... What is Jenkinson doing now? Could I get him as a workman at G.H.Q?'.[183] This seems an unreasonable request, and perhaps his father told him so.

By 1918 the German U-boat threat had been averted and the entry of the USA seemed likely, but otherwise the war on the Continent was not encouraging. When German troops returned from the Eastern Front their numerical, tactical, and artillery superiority was decisive. They advanced on a broad front and won large territorial gains. It was not a total breakthrough, however, and the advance was halted. Further gains were bought at fearful cost and the German front became overstretched. Allied infrastructure remained intact and, when they counter-attacked with recuperated, re-equipped troops, and American soldiers, the battle turned for the last time. Coordination of all the necessary elements, including now excellent sound-ranging, proved decisive. Offence broke German defence when previously it had been self-destructive. Thus, for example, 'On 8 August 1918 two-thirds of British shells were directed at the German artillery... With devastating effects, the British gunners unleashed an unregistered bombardment on enemy batteries... Hence the Amiens battle opened, despite poor visibility, with British shells falling with deadly precision upon the enemy's carefully concealed artillery pieces'.[184]

In a more recent account and analysis of the role of British artillery during the war, Jackson Hughes offered a new perspective on the conflict. Previous accounts, he suggested, have fallen into two groups: one hostile to Sir Douglas Haig and his staff and attributing victory to British tanks, the other defending Haig and GHQ and claiming the war was won by attrition, the constant wearing down of German morale and resources. Both accounts were flawed, Hughes claimed, because they omitted the guns. In a detailed discussion of the war, culminating with the final attack that began on 8 August 1918, Hughes gave attention and credit to British artillery and its extensive support network, not least to the importance of sound-ranging: 'The sound-rangers found a mechanism for accurately locating the most troublesome of targets, German gunpits, [and] also introduced rapid and accurate calibration of guns'. In summary he said, 'The development of mapping, meteorological services, surveying techniques, sound-ranging, rapid calibration, and air-to-ground communication had meant that, in the autumn of 1918, the British artillery had finally arrived at the position they had desired since the Boulogne Conference of 1915'; with the result that, 'the Royal Artillery... proved to be a decisive force in dislodging the German armies from the Hindenburg Line'.[185]

The year 1918 also found new tones and perspectives in Lawrence's correspondence. He had about forty sound-ranging sections all along the British

[183] Ibid., 17 September 1917, RI MS WLB 37B/2/28.

[184] Wilson, n. 160, Part 12, pp. 585–6.

[185] J Hughes, *The Monstrous Anger of the Guns: The Development of British Artillery Tactics 1914–1918*, unpublished PhD thesis, History Department, University of Adelaide, 1992, pp. 337, 318, and i respectively.

battlefront, each with about fifty staff and equipment. It was quite an advance on the initial group of eight staff, Lawrence having recruited nearly all the subsequent officers by his famous dictum: 'When I was seeking recruits for sound-ranging, I had only to . . . say, "Bachelors of Science, one step forward", to get a generous response of eager aspirants to some job in which their knowledge could be used'.[186] The timing precision of Bull's galvanometer was crucial, and at the end of the war no other method of sound-ranging matched Lawrence's English system. The German system, for example, was inaccurate and slow by comparison.[187] Lawrence's contribution was acknowledged on a number of occasions. Having been mentioned in dispatches in 1916 and 1917, he was now awarded the Military Cross (on 1 January 1918, for which no individual citation was recorded)[188], and appointed an Officer of the Most Excellent Order of the British Empire (on 15 March 1918, 'for his services with the Field Survey Company')[189]. He was also mentioned in dispatches for a third time.[190]

These awards could not hide the hideousness of the war, however. Some of his sound-ranging officers were killed,[191] and in a letter asking his mother to sew OBE and MC ribbons on his home jacket, Lawrence reported that 'in the last push . . . a lot of our officers were wounded'.[192] He was having 'the laziest time just now, there is so little one can help . . . it feels rather rotten in the middle of this big show'.[193] 'Our people are getting good results', he reported, 'so I am very happy'; and he had met Harold Hemming for the first time. Hemming was a Canadian, who was 'very good fun; he does the same thing for flash-spotting as I do for S.R.' He was destined to become a life-long friend.[194]

Lawrence's letters make it clear that he was now travelling along the front line constantly, advising and encouraging his sound-ranging sections. Having been an Acting Captain for eighteen months, in June 1918 he was promoted to 'Temporary Major . . . whilst specially employed'.[195] In July he spent two days in Paris with Bull, and then had Bull back at his own headquarters. 'I think the Bosche will start squealing soon', he told his mother.[196] Occasionally he wistfully remembered the past: 'It does not seem like four years since I came up to Deerstones to see you . . . I remember so well going to see Bob and Vaughan at Alexandra Palace and taking them to tea on top of Selfridges just soon after that'.[197] And 'Weren't the Australians splendid in the last show [at Amiens in

[186] Quoted ibid., p. 217.
[187] Sir Lawrence Bragg, n. 125.
[188] *Supplement to The London Gazette*, 28 December 1917; W L Bragg, war service record, n. 8.
[189] *Fifth Supplement to The London Gazette*, 12 March 1918.
[190] *Fourth Supplement to The London Gazette*, 4 July 1919; certificate dated 16 March 1919, RI MS WLB 37A/4/5.
[191] Letter W L Bragg to G Bragg, 23 March 1918, RI MS WLB 37B/3/5.
[192] Ibid., 1 April 1918, RI MS WLB 37B/3/7.
[193] Ibid., 8 April 1918, RI MS WLB 37B/3/9.
[194] Ibid., 5 May 1918, RI MS WLB 37B/3/12.
[195] *Supplement to The London Gazette*, 30 November 1918
[196] Letters W L Bragg to G Bragg, 11 and 25 July 1918, RI MS WLB 37B/3/23 and 26 respectively.
[197] Ibid., 8 August 1918, RI MS WLB 37B/3/29.

early August]. I was so awfully bucked and felt very Australian again for a bit. I'll write a line to Dad tomorrow'.[198] This he did, asking about submarine sound-ranging, saying he was enjoying French lessons once a week and golf games on a crude course they had laid out, and then, for the first time in years, wondering about physics: 'Has any new work on our job come out lately? What do you mean to go on with? I don't know quite what line to take up'.[199]

When the battle moved: 'Hemming and I...crow frightfully in the mess just now over the show down south, which was a great feather in our respective colonial caps'.[200] With the Allies at last gaining the upper hand and pushing east relentlessly, German soldiers were captured in large numbers: 'James and I have been examining Bosche prisoners. They were very willing to oblige us with any little bit of information...James lived in Boschland for a few months but was terribly rusty...he ploughed through somehow [Lawrence had no German]...Isn't the news absolutely gorgeous'.[201] A few days later he wrote again, 'I got back two days ago from our last trip...we must have given the Bosche beans by the way his battery positions had been knocked about'. 'James' was Captain Reginald James, Lawrence's Cambridge fellow student, just back from Shackleton's perilous Antarctic Expedition, who joined the first experimental sound-ranging section in Flanders, helped develop the method, and became OC of the Sound-Ranging School.[202]

Sound-ranging was doing its work and Lawrence's thoughts turned more and more to life after the conflict. He wrote to his father: 'I simply can't think of what experiments to do after the war. I expect a year or two will be taken up with 1. Learning a little physics again...2. Trying to learn my job as Lecturer...It's awfully decent of you to say that I ought to chose what I want to do in the way of research...I'll tell you what I'm longing to do. Get on to a job again when I can boot everybody out of the room and lock the door, and depend on my own efforts alone for success or failure. I'm not built for a job which means, as part of it, getting hundreds of other people to do their jobs too'.[203] These were revealing words about his preference for solitary research work, like his father, and his forthcoming arrangement with William about their division of labour. It was also a portent of Lawrence's difficult early years at Manchester.

He was enjoying the success of the battle and of the sound-ranging: 'You would love to see the joy of the civilians who have got amongst friends again...an old lady riding back in front of one of our lorries...had a smile that nearly met round the back of her head'.[204] His next letter condemned the

[198] Ibid., 16 August 1918, RI MS WLB 37B/3/30.

[199] W L Bragg to W H Bragg, 17 August 1918, RI MS WLB 37B/3/31.

[200] W L Bragg to G Bragg, 22 August 1918, RI MS WLB 37A/3/32.

[201] Ibid., 4 September 1918, RI MS WLB 37B/3/34.

[202] Ibid., 12 September 1918, RI MS WLB 37B/3/35; for James see W L Bragg, 'Reginald William James 1891–1964', *Biographical Memoirs of Fellows of the Royal Society*, 1965, 11:115–25, 116.

[203] W L Bragg to W H Bragg, 13 September 1918, RI MS WLB 37B/3/36.

[204] W L Bragg to G Bragg, 13 October 1918, RI MS WLB 37B/3/42.

behaviour of the German occupying forces: 'They are absolute swine, you know, there's no doubt about it. It's the officers they all loath'. And Lawrence was in no mood to be generous to the enemy: 'It's awful rot, all this peace talk in England... It's just the time now to be more than ever determined to go on... It's all very fine to talk about the Bosche running, but I don't think they realize what it costs to make him do it. To slack off when you've nearly got a job finished is the stupidest policy out'.[205]

Late in October he was again thinking about the future: 'I got Dad's letter with the two offers of jobs. I don't want to go to America'.[206] He regretted that sound-ranging was 'going awfully well just as the war is over!... the old Bosche certainly has been beaten handsomely in the end. I hope I get a leave soon as I am just dying to see you'. He was looking forward to 'reorganizing and tidying up', and he had an offer from Campbell on which he wanted his father's advice. 'I feel bound to Trinity', he continued, but 'I wish I had the offer of a professor job somewhere in England; I think I would like that best'.[207] At last, on 11 November 1918 at 11 a.m., the eleventh hour of the eleventh day of the eleventh month, the guns on the Western Front fell silent, leaving both sides to mourn their dead, 3,410,000 for the Central Powers, 4,585,000 for the Allies.[208] In December Lawrence was 'very busy writing books on Sound Ranging'.[209] As Christmas approached he noted that, while there were celebrations in the UK regarding the armistice, it was quiet at the front and 'hard to realize that it's all over'.[210] There was a formal dinner, then a dance in the chateau, 'with the room lit up with candles', and 'a final beano for Colonel Jack'.[211]

Lawrence was still in Europe when the new year dawned. He was 'just off on a trip to Germany', on which he reported at length when he returned ten days later.[212] With Hemming and another Canadian he travelled by car—'swathed with coats and blankets as it was jolly cold'—to Brussels, Namur, Liège, Aix-la-Chapelle (Aachen), and finally Cologne and Bonn. They were fortunate to find replacements for a broken tie-rod and a fractured steering arm along the way. The Germans were 'frightfully keen to be polite'. All the parts of Germany they saw were 'quite untouched by the war and looked very prosperous... but they are an awful looking crowd'. One day the leader of one of the squadrons took him up in his own private aeroplane: 'it was the greatest treat I have ever had'. He hoped to be demobilized soon.[213]

A few days later Lawrence wrote to his father again about the future. His ambition had cooled a little. If he returned to Cambridge he would have a light load, have friends there, repay Trinity's generosity, and be able to find his feet

[205] Ibid., 19 October 1918, RI MS WLB 37B/3/43.
[206] Ibid., 21 October 1918, RI MS WLB 37B/3/44.
[207] Ibid., 8 November 1918, RI MS WLB 37B/3/46.
[208] Howard, n. 123, pp. 135, 146.
[209] W L Bragg to W H Bragg, 14 December 1918, RI MS WLB 37B/3/48.
[210] W L Bragg to G Bragg, 16 December 1918, Bragg (Adrian) papers.
[211] W L Bragg to G Bragg, 20 December 1918, RI MS WLB 37B/3/49.
[212] Ibid., n.d. [ca 12 January 1919], RI MS WLB 37B/3/53.
[213] Ibid., 1 January 1919, RI MS WLB 37B/3/50.

'without having too great a strain right at the start'. The vacant professorship at Birmingham was attractive, but he had no lecturing experience and it was too soon. He sought his father's advice on preparing an academic application and listed his qualifications: Adelaide BA, Cambridge scholarships, BA, MA, and Trinity lectureship and fellowship; the Barnard Medal, Nobel Prize, and Italian Medal; research on crystalline structures, articles, and a book in collaboration with his father; and war service: military commission, to the front in charge of the first British SR section, OC Experimental SR Section and School, Technical Advisor on SR at GHQ, Major, MC, OBE, and mentioned in dispatches.[214] Quite a record!

What effects did the war years have on Lawrence? There were a number of positive outcomes. The war was a great adventure, with high excitement, constant danger, and times of fun. His health was good. He made lifelong friends and enjoyed the camaraderie of a wide group of officers and men. He was charged with a very difficult scientific task. Nearly all his superiors predicted failure, but he succeeded, and to such an extent that sound-ranging played a central role in the last year of the war. He was widely honoured. His self-confidence was boosted. The raw young graduate was now a mature man.

But there were negative influences too. Lawrence was very close to the front-line trenches for about 18 months—September 1915 to March 1917—in the notorious Ypres salient in Flanders, the site of hellish and constant trench warfare. He drove along the trenches many times thereafter. He saw the gun flashes, the shattered bodies, the rotting corpses, the rats; he must have smelt the stench; he heard the deafening noise; he experienced the filth, the bitter winter cold, the water and mud that turned soldiers' feet to mush (the so-called 'trench foot'); and he must have been bothered by the flies grown fat on the putrefying flesh of humans and horses all around. He saw frostbite, dysentery, and VD everywhere.[215] He lost his beloved brother, his dearest friend, and many colleagues. No one came away unscathed, whatever they may have said later. 'Shell shock' it was called, but neither doctors nor psychologists, neither armies nor governments, neither soldiers nor civilians knew what it was or how to handle it. The veterans suffered—a little or hugely—until the day they died.[216]

One of Lawrence's daughters told me she sometimes woke in the night to hear her father, only half awake, banging his head on the head of his bed. Was he trying to exorcise the scenes of war that came to him in the night? He could

[214] W L Bragg to W H Bragg, 15 January 1919, RI MS WLB 37B/3/51.

[215] For accounts of the horrors of WWI see, for example, D Winter, *Death's Men: Soldiers of the Great War* (London: Lane, 1978); J Ellis, *Eye-Deep in Hell* (London: Croom Helm, 1976); B Gammage, *The Broken Years: Australian Soldiers in the Great War* (Melbourne: Penguin, 1974); A W Wheen, *All Quiet on the Western Front* (Boston: Little Brown, 1929).

[216] There is a growing literature on PTSD and associated neuroses; for example, P Leese, *Shell Shock: Traumatic Neurosis and the British Soldiers of the First World War* (Basingstoke: Palgrave Macmillan, 2002); E Showalter, *The Female Malady: Women, Madness, and English Culture, 1830–1980* (London: Virago, 1985); R Leys, *Trauma: A Genealogy* (Chicago: University of Chicago Press, 2000); I am grateful to Professor Mark Creamer of the Australian Centre for Posttraumatic Mental Health for valuable advice.

be like a volcano, she said, threatening to erupt at any moment; finally exploding in a rage, and then falling into a period of deep remorse.[217] He was awarded a Military Cross and an OBE but no one understood. Lawrence did not suffer from unending post-traumatic stress disorder (PTSD), but it seems he did experience war-related stress from time to time, especially when other factors increased the tension in his life. Lawrence was not yet thirty years old. It had already been an extraordinary life. Could the future hope to match the past?

William's war

William remained in England, working on projects related to the war, and he experienced upheaval, annoyance, frustration, and profound sadness: upheaval because he moved his work and his family from Leeds to London, with all its attendant disruptions; annoyance because his work for the Admiralty was handicapped by a deep-seated suspicion and consequent lack of cooperation between Navy personnel and civilian scientists; frustration because his crystal research was substantially interrupted and because the lack of cooperation and the difficulty of the navy problems caused his research there to be less productive; and profoundly sad because of Bob's death. For William and his wife the war encompassed four deeply unhappy years.

Britain was heavily dependent on imports—and therefore shipping—for foodstuffs and raw materials, a dependence that became critical in wartime and made its navy a crucial element of defence. At the outbreak of the war Germany's expanding navy therefore posed a threat to Britain's survival, but thanks largely to the foresight of its pugnacious First Sea Lord, Admiral Sir John Fisher, the British Navy was generally superior in size, technology, and seamanship. There were blind spots, but most were overcome as a result of early skirmishes at the Dogger Bank and elsewhere. From then on the British Navy blockaded Germany by closing the English Channel and by patrolling the wide gap between northern Scotland and Scandinavia with its Grand Fleet based at Scapa Flow in the Orkney Islands.[218] The one ingredient in the war at sea that had not been addressed was the menace of the German U-boats.

During 1915 the submarine assault on Britain's merchant shipping became serious. Given sufficient U-boats it seemed that Germany could blockade Britain, but their methods of attack—whether on the surface by gunfire or underwater using torpedoes—would violate accepted modes of waging war, would alienate neutral countries, and threatened to bring America into the war. Despite these disincentives, however, when its initial naval attacks failed Germany declared the waters around Britain a war zone and merchant shipping

[217] Patience Thomson (née Bragg), personal communication.

[218] See, for example, Wilson, n. 160, Parts 2, 6, and ch. 39; A J Marder, *From the Dreadnought to Scapa Flow: The Royal Navy in the Fisher Era, 1904–1919* (London: OUP, 1961–1970), vols I–V; G Jordan (ed.), *Naval Warfare in the Twentieth Centry 1900–1945: Essays in Honour of Arthur Marder* (London: Croom Helm, 1977).

became liable to attack. By August 1915 the losses exceeded the replacement rate of British shipyards. Yet the U-boat toll declined thereafter, due to British counter-measures and to American responses, particularly after the sinking of the liners *Lusitania* and *Arabic*. In the longer term, however, 'the main British strategy to beat the U-boat blockade was misconceived and doomed to failure... the essential point, that the way to counter submarines was not by going in search of them but by standing between them and their quarry, escaped British navy strategists throughout 1915, and for a good while after that'.[219]

At the same time the cry for more scientific and technological input into the war effort reached a crescendo. The government established the Advisory Council for Scientific and Industrial Research (later the Department of Scientific and Industrial Research),[220] and also the Board of Invention and Research (the BIR) specifically related to the Royal Navy.[221] The function of the BIR was to provide expert scientific advice to the Navy 'on definite problems, by encouraging research, and by considering schemes put forward by inventors or the general public'.[222] Lord Fisher, after taking premature retirement following a violent disagreement with Churchill's Dardanelles campaign, was appointed its first Chairman. Fisher's had been a lone, pre-war voice warning of the submarine threat,[223] but his outspokenness could be counter-productive and worked against the BIR's acceptance by the Navy.[224] Furthermore, the BIR was a small and informal advisory body with limited facilities.[225]

Initially the Board consisted of a central committee of three eminent scientists (J J Thomson, notable engineer Charles Parsons, and industrial chemist George Beilby), and a consulting panel of twelve, including Pope, Lodge, Rutherford, W H Bragg, and Strutt (later Lord Rayleigh). Its business was divided into six areas, of which the largest and most important was 'Section

[219] Wilson, n. 160, ch. 9, p. 92.

[220] I Vercoe, 'Comment: practical proposals by scientists for reforming the machinery of scientific advice, 1914–17', *British Journal for the History of Science*, 2000, 33:109–14 and references therein; F M Turner, 'Critiques & contentions: public science in Britain, 1880–1919', *Isis*, 1980, 71:589–608 and references therein.

[221] R M MacLeod and E K Andrews, 'Scientific advice in the War at Sea, 1915–1917: the Board of Invention and Research', *Journal of Contemporary History*, 1971, 6(2):3–40; R M MacLeod, 'Secrets among friends:..., 1916–1918', *Minerva*, 1999, 37:201–33.

[222] Index to Admiralty Board of Invention and Research Minutes and Papers, ADM 293, TNA (PRO), London.

[223] Extracts from Memorandum Lord Fisher to Prime Minister, *circa* February 1914, CAB 21/7, TNA (PRO), London.

[224] The anti-submarine story has been told in W Hackmann, *Seek & Strike: Sonar, Anti-submarine Warfare and the Royal Navy 1914–54* (London: HMSO, 1984), for The Great War see particularly the Introduction and chs I through III; see also id., 'Underwater acoustics and the Royal Navy, 1893–1930', *Annals of Science*, 1979, 36:255–78; id., 'Sonar research and naval warfare 1914–1954', *Historical Studies in the Physical and Biological Sciences*, 1986, 16(1):83–110; Hartcup, n. 125, chs 2 and 7; J Terraine, *Business in Great Waters: The U-boat Wars, 1916–1945* (London: Cooper, 1989); the items in n. 218, particularly J K Gusewelle, 'Science and the Admiralty during World War I: the case of the BIR', in Jordan, pp. 105–17. There was an Army equivalent to the BIR called the Munitions Inventions Department (see M Pattison, 'Scientists, inventors and the military in Britain, 1915–19', *Social Studies of Science*, 1983, 13:521–68).

[225] MacLeod and Andrews, n. 221, p. 13.

II: Submarines and wireless telegraphy'. Of the 14,000 suggestions received from the general public concerning submarines and wireless, only a handful were worthy of study; more important were the Board's own research projects. The subcommittee for Section II consisted of the Duke of Buccleuch as chairman and W H Bragg, Rutherford, Glazebrook, Threlfall, and two others. In the small world of British physics William was working with colleagues of long standing. They considered all feasible anti-submarine measures. Bizarre projects included the training of seagulls, hawks, and sea lions, but their most extensive work concerned underwater acoustics, which received by far the largest government grant.[226]

William had become captivated during his undergraduate years at Cambridge by Glazebrook's course on wave motions, including sound. He included such material in his Adelaide curricula and no doubt communicated his love of these subjects to Lawrence. Among the potentially overwhelming dangers to Allied forces during The Great War were German artillery on the battlefields and U-boats in the oceans. Lawrence was using sound to overcome the artillery threat; his father would attempt to locate German submarines by listening for the sound of their engines and their movement through the sea. These were clear examples of William's central focus in the science-versus-applied-science debate at that time: 'It is not realised that the fruit comes at the end of a long process, and that even a little application of science may be the result of many years of unseen growth and labour', and that, 'in the relation between science and applied science... no one can foretell what scientific research will enable us to do'.[227] There were strong echoes of his Adelaide days in each of the essays from which these extracts are taken.

The anti-submarine work was carried out at a number of experimental stations around Britain, some under naval control and others under the BIR. Within the Navy underwater acoustic devices were being developed on a trial-and-error basis by Commander Ryan at the Admiralty's Experimental Station at Hawkcraig, near Aberdour, on the Firth of Forth opposite Edinburgh. The passage of sound through water was not well understood. At sea there were substantial problems because of its significant inhomogeneities in density, temperature, salinity, and foreign bodies, as well as the presence of its surface and seabed, leading to complex reflection, refraction, scattering, and absorption effects. In addition the sea is a very noisy place, due to seismic activity, storms, waves, marine life, ship traffic, and reverberations. An underwater microphone, christened the 'hydrophone', had been created in the nineteenth century. It consisted of a carbon button microphone attached to a thick metal diaphragm

[226] Hackmann, n. 224, Introduction and ch. I; D A H Wilson, 'Sea lions, greasepaint and the U-boat threat: Admiralty scientists turn to the music hall in 1916', *Notes and Records of the Royal Society of London*, 2001, 55(3):425–55.

[227] Respectively, W H Bragg, 'Physical research and the way of its application', in A C Seward, *Science and the Nation: Essays by Cambridge Graduates* (Cambridge: CUP, 1917), pp. 24–48, 25; W H Bragg, 'Physical science and its applications to industry', *Journal of the Textile Institute*, 1916, 7(3):185–93, 188–9 (a lecture to the Autumnal Congress of the Institute at Leeds, 1915) .

inside an open metal case, so that variations in water pressure caused by underwater sound waves vibrated the diaphragm, varied the electrical resistance of the button, and thereby induced audible sound in a telephone receiver on board ship. But much research and development was required to make it useful; when hostilities began the Admiralty had not tried hydrophones nor found any other way of detecting submarines.[228]

Rutherford flung himself into the anti-submarine work with his usual energy, and within three months he had outlined the principle feature of this unknown terrain and correctly identified where the main attack had to be.[229] He produced three secret reports in which he reviewed all the options that seemed to be available. He suggested that the only practical way of detecting submarines using existing technology was by sound, but a 'scheme of acoustic research' was needed, and this formed the basis of the BIR's work for the rest of the war. He noted that the amplitude of vibration of a sound wave in water was only one-sixtieth of that in air, but that for the same energy and frequency the pressure variation was therefore sixty times, so that a metal diaphragm would indeed vibrate perceptively in water. He also reported on numerous experiments that he and his Manchester team had already conducted in line with these findings.[230] However, Rutherford was constantly anxious to continue his own physics research and, as the war dragged on, he left the subsequent anti-submarine work to William and his colleagues, with one exception. Rutherford played a leading role in the invention of asdic/sonar with his research on the piezoelectricity of quartz crystals and their use in producing high-frequency sound.[231]

Commander Ryan had served in the Royal Navy until his retirement in 1911, but had re-enlisted in 1914 and begun work on the detection of U-boats using simple hydrophones. He developed both moored hydrophone stations at important coastal locations and portable hydrophones for use at sea. He would be promoted, take out several patents, and remain at Hawkcraig until it closed early in 1919. In September 1915 it was suggested that Ryan's practical skills should be augmented by the theoretical and research expertise of BIR university scientists. The first to arrive were Albert Wood and Harold Gerrard, who had been working with Rutherford at Manchester. Wood was amazed to discover that Ryan had never heard of Rutherford, emphasizing Ryan's purely empirical approach and the general state of science in the Navy. Wood would later have a distinguished career in its post-war scientific service.[232]

Tension between the Navy and the scientific community had existed in earlier times,[233] and now it developed again, between the practical Navy staff

[228] Hackmann, n. 224, Introduction and ch. I.

[229] D Wilson, *Rutherford: Simple Genius* (London: Hodder and Stoughton, 1983), ch. 12 ('Rutherford at War'), p. 347.

[230] Ibid.

[231] Ibid.; for asdic/sonar see n. 242 below.

[232] Hackmann, n. 224, ch. II; for Wood see 'Albert Beaumont Wood, O.B.E., D.Sc.: Memorial Number', *Journal of the Royal Navy Scientific Service*, 1965, 20(4):188–283, and A B Wood, *A Textbook of Sound* (London: Bell, 1930).

[233] F A J L James, 'Davy in the dockyard: Humphry Davy, the Royal Society and the electrochemical protection... of His Majesty's ships in the mid-1820s', *Physis*, 1992, 29:205–25.

and the academic scientists. Wood was subjected to regular petty constraints and frequently complained to Rutherford. The Duke of Buccleuch sent a formal complaint to the Admiralty. Rutherford also attacked certain Admiralty views and Sir Richard Paget, Secretary of Section II, took the complaints to the First Lord of the Admiralty, Arthur Balfour. After a meeting at the Admiralty on 30 March 1916 it was agreed that the Navy and the BIR would cooperate more closely, that the BIR's budget would be augmented, BIR staff at Hawkcraig increased, and that Professor W H Bragg would be appointed Resident Director of the civilian scientists there. William had been a frequent visitor to Hawkcraig and in April took leave from University College to take up the new position.[234] Gwendoline and Gwendy followed. William was already corresponding with Rutherford on BIR matters,[235] and at Hawkcraig he, too, was soon complaining about the lack of facilities and support staff.[236]

Section II of the BIR pushed ahead with hydrophone research, and cooperation with the French was finally approved, against the Navy's wishes, with Maurice de Broglie attached to the BIR. William, however, was unable to resolve the major difficulties between BIR staff and Ryan and his colleagues. A definitive example of the impasse was the confinement to barracks of the skipper of the experimental ship, *Hiedra*, after he had obeyed orders from the Resident Director that were unknowingly against Ryan's wishes, even though William personally apologized to Ryan. Indeed, the mild-mannered Bragg was moved to write to Rutherford in exasperation, saying he felt as if he was in a Gilbert and Sullivan opera instead of a war and that his position was becoming untenable. He also wrote to Fisher about inadequate facilities and spoke to Balfour, with the result that the Navy and BIR groups were separated.[237]

The battle of Jutland in mid-1916 again focused public attention on the Navy, particularly when the British Grand Fleet under Jellicoe lost more ships (fourteen) than the German High Seas Fleet (eleven). Nevertheless, Britain retained her naval superiority, reinforced by post-Jutland improvements.[238] Unable to defeat the Grand Fleet, when Admiral Scheer reported to the German Emperor he saw a new U-boat offensive as the only remaining strategy available to his navy: 'A victorious end to the war within reasonable time can only be achieved through the defeat of British economic life—that is, by using the U-boats against British trade'.[239] Thus, through the last months of 1916 there was an alarming increase in the destruction of British merchant shipping: about forty ships per month. In December Lloyd George became Prime Minister and separate units of the British Navy were combined to form the Anti-Submarine Division (ASD). This was to be responsible for all aspects of U-boat warfare but was inevitably focused on short-term goals. Ryan's group became part of

[234] Hackmann, n. 224, ch. II.
[235] Rutherford–Bragg correspondence, late-1915 to late-1916, ADM 212/157, TNA(PRO), London.
[236] MacLeod and Andrews, n. 221, p. 21.
[237] Hackmann, n. 224, ch. II.
[238] Marder, n. 218, vol. III, ch. 6, p. 205.
[239] Ibid., p. 206.

the new Division and Hawkcraig remained the centre of Navy hydrophone research throughout the war.[240]

At Christmas 1916 William, his family, and his BIR staff were relocated to a new laboratory complex at Parkeston Quay, Harwich, where destroyer and submarine flotillas were based, where greater facilities were provided, and where staff numbers grew. William was clearly distressed by his lack of success at Hawkcraig: 'The work here has been perplexing and very disappointing in lots of ways...I feel I have fallen short of lots of people's expectations'; but he was upbeat to Rutherford: 'direct contact with the officers on active service...is the thing that I have always felt myself was the greatest essential'.[241] Their work embraced many aspects of submarine warfare: hydrophones, acoustic and other mines, sound-ranging, indicator loops on the sea-floor, and 'asdics', named 'sonar' by the Americans.[242] Yet despite these changes, the strain between the BIR and the Admiralty continued. In February 1917 the BIR presented a formal memorandum to Balfour, cataloguing instances of non-cooperation between BIR and Navy personnel. A noteworthy instance was Ryan's application for a personal patent for a directional hydrophone that was under joint development and for which an underlying principle had been enunciated by William. Fisher complained to the Prime Minister and Cabinet. When, in May and without comparative trials, the Navy cancelled an order for 200 of the BIR's portable directional hydrophones Mark I in favour of 700 of Ryan's untested Mark II, the civilian scientists were especially irate.[243]

This was precisely the wrong time for such deep-seated antagonisms, for 1917 was the 'Year of Crisis', when the submarine menace reached its height.[244] Desperate to change the course of the war and hoping to starve Britain into submission before America could intervene, Germany launched an unrestricted U-boat campaign, sinking without distinction and without warning. During the period January through April 1917 the loss of (British/total Allied and neutral) merchant shipping to German submarines peaked at (155/354) ships per month, a truly frightening figure for which Britain had few answers. There were a number of offensive anti-submarine measures, including depth charges, bomb howitzers, and mine nets, but none was adequate to stop the carnage.

'What appeared at the time to be the most important of the A/S devices... was the hydrophone',[245] but the early hydrophones had serious limitations. At sea they could only be used on a stationary vessel, because otherwise the noise of the ship's engine and of the sea washing past the hull drowned the noise

[240] Hackmann, n. 224, ch. II

[241] Letter W H Bragg to G Bragg, 16 December 1916, RI MSA WHB 95H/1; letter W H Bragg to E Rutherford, 1 November 1916, n. 235.

[242] A list of staff and a discussion of their activities is given in the Wood Memorial Number, n. 232; Hackmann, n. 224, p. xxv discusses the origins of these acronyms, leading to 'Anti-Submarine Division-ics' for asdic or asdics, while sonar is now interpreted as 'sound navigation and ranging'.

[243] Memorandum Fisher to Hankey, 30 March 1917, with attachments from BIR complaining of lack of cooperation and other problems, CAB 21/7, TNA (PRO), London.

[244] Marder, n. 218, vol. IV (1917: Year of Crisis).

[245] Ibid., chs III–V, p. 75.

made by the U-boat; but a stopped vessel was a perfect target for a U-boat's torpedo![246] And if detected, they gave no indication of a submarine's location. Two forms of a Portable Directional Hydrophone (PDH) emerged, and most of Ryan's energies were directed towards this end. His first bi-directional model was based on the pioneering work of engineering professor John Morris and his student Adrian Sykes at East London College.[247] Its later developed form was named PDH Mark II. It was essentially silent to sound arriving edge-on but was equally sensitive to sound from the front or the back. It was only about one-quarter as sensitive as the earlier models but its directional properties were essential (see Figure 17.5).[248]

The first BIR scientists had begun some essential basic and theoretical research; for example, on the sound spectrum produced by submarines, the distance these sounds could travel, and the influence of sea conditions. The problems were complex. Other research was carried out at various university laboratories, notably at Manchester.[249] The BIR scientists also began work on a directional hydrophone. Rutherford developed his own design, and his staff discovered that the directional effect was largely due to the symmetry of the arrangement. William watched these developments and suggested that an ellipsoid body would have bi-directional properties, but it was eventually determined that this form offered little advantage.[250] In addition, 'velocity' hydrophones, responding to the movement of water over them, were found to be less sensitive than 'pressure' or diaphragm hydrophones.[251]

William then noticed another report by Morris and Sykes concerning their attempt to obtain a sharper minimum response by destroying the symmetry of a pressure hydrophone.[252] He foresaw that such an asymmetry, introduced by the addition of a sound screen or baffle in front of one face of the diaphragm, might destroy the sensitivity of that side and therefore make the hydrophone unidirectional from the open side. Experiments discovered that it was the thin film of air trapped inside any baffle that made it effective. The first design produced in numbers consisted of a wooden disc covered with lead sheet which, when combined with a single diaphragm, formed a hydrophone designated PDH Mark I. However, William's unidirectional Mark I proved to be less accurate in determining the bearing of another ship than Ryan's bi-directional Mark II, and during the last year of the war a combination of the two, mounted together, gave the best results.[253]

[246] Ibid., pp. 75–6.

[247] A F Sykes and J T Morris, 'Improved means for detecting and locating subaqueous sounds', British patent no. 15,320, 1915, Patent Office, London.

[248] Hackmann, n. 224, ch. III.

[249] Ibid.

[250] Ibid.; A B Wood and F B Young, 'On "light body" hydrophones and the directional properties of microphones', *Proceedings of the Royal Society of London*, 1921, 100:252–60.

[251] A B Wood and F B Young, 'On the acoustic disturbances produced by small bodies in plane waves transmitted through water', *Proceedings of the Royal Society of London*, 1921, 100:261–88.

[252] See correspondence between W H Bragg, Paget, Morris, and Sykes at ADM 212/1, TNA (PRO), London.

[253] Hackmann, n. 224, ch. III.

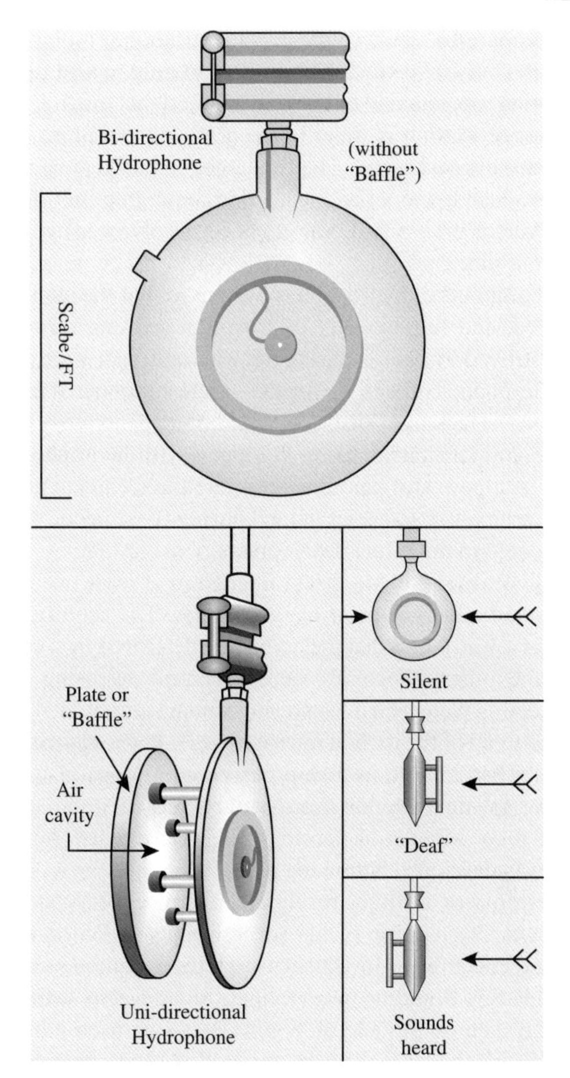

Fig. 17.5 Two forms of the portable directional hydrophone developed by the British during The Great War, (top) Mark II developed by Commander Ryan and the Navy, (below) Mark I due to Professor Bragg and the BIR. (From Sir William Bragg, *The World of Sound*, Bell & Sons, 1920, p. 175.)

William generously invited Ryan to join him in applying for a patent for the Mark I PDH,[254] and he also honoured the earlier work of the BIR by joining Rutherford in a patent application for the simple hydrophone.[255] In May

[254] W H Bragg and C P Ryan, 'Improvements in apparatus for detecting the direction of sounds in water', British patent no. 304,067, 1916, Patent Office, London.

[255] W H Bragg and E Rutherford, 'Improvements in apparatus for detecting the direction of sound in water', British patent no. 125,446, 1916, Patent Office, London.

1917 William prepared a report on the general question of 'submarine hunting' to summarize the existing situation and in an attempt to win further funding. Acoustic detection was found to be the only method that made it 'worthwhile to hunt'.[256] A major problem was water-borne noise, and attempts to overcome it were only moderately successful. The fundamental problem was the weakness of the signal produced, and an adequate electronic amplifier did not become available until late in 1917. A wide variety of forms of towed hydrophones were also tried without success.[257]

Substantial numbers of hydrophones were used, but the number of U-boats detected, located, and subsequently destroyed at sea was very small. 'Of all the U-boats destroyed by patrols during the war, only four were definitely sunk because of hydrophone contact; by far the greatest number were hunted after visual sightings'.[258] Hackmann also concluded more generously that, 'the presence of a huge anti-submarine force, equipped with hydrophones and depth charges, acted as a powerful harassment; while the technical and operational knowledge in underwater acoustics gained during these hectic years was to be of enormous benefit to the future development of asdics'.[259]

During 1917 William took a direct interest in at least two other projects, acoustic mines and a loop detector of submarines. The acoustic mines used a hydrophone, but it was found that they could be destroyed by a counter-mining charge. This and other difficulties were overcome by using a thicker diaphragm, a protecting plate, and a 'chattering contact amplitudemeter' designed by William[260] instead of the carbon microphone.[261] It was also discovered early in the war that when a large iron ship passed over loops of wire laid in the water a sensitive galvanometer on shore could detect the small electric current induced in the loop. Months of development were needed, however, and no useful system was developed before the end of hostilities.[262]

Other research was a direct result of Lawrence Bragg's sound-ranging work in Flanders. During one of his several visits to William at Parkeston Quay, Lawrence encouraged his father to try the technique underwater, and William asked R S H Boulding to investigate and develop it, first at the Quay during 1917 and then on the Isle of Wight. Boulding used a base line of six hydrophones, an Einthoven galvanometer, and photographic recording, mirroring Lawrence's method. He showed that underwater explosions could be located very accurately, and four stations were established to keep watch over a large section of the North Sea.[263] As William foresaw, the development of this technique had important implications for hydrography after the war.[264]

[256] W H Bragg, 'Memorandum on submarine hunting', 15 May 1917, ADM 212/159, TNA (PRO), London, p. 5.

[257] Hackmann, n. 224, ch. III.

[258] Ibid., p. 69

[259] Ibid., p. 71.

[260] Wood textbook, n. 232, p. 455.

[261] Wood Memorial Number, n. 232, pp. 31–2; Bragg–Wood correspondence, RI MS WHB 37A.

[262] Ibid., p. 34; ibid.

[263] Hackmann, n. 224, p. 76; A B Wood and Captain H E Brown, 'A radio-acoustic method of locating positions at sea', *Proceedings of the Physical Society*, 1923, 35:183–93.

[264] MacLeod and Andrews, n. 221, p. 26.

Late in the war the Royal Navy was advised that a very different technology, based on echo-ranging and later named 'asdic', might offer a superior method of detecting submarines and distant surface vessels. It was clear that generating a sound and listening for its echo over a narrow frequency range would eliminate the majority of water and ship-borne noises that had plagued the hydrophones and related methods. Three sound sources were investigated, of which the supersonic used in the British system proved the most fruitful. Initial efforts had several deficiencies, but trials late in 1918 with improved apparatus were more encouraging, and post-war work with electronic amplifiers, swept narrow beams, and better detectors ensured that the technique would have a long and useful life.[265]

As is well known, the solution to the submarine crisis was reversion to a method from the sailing era of forming ships into groups protected by armed escorts (convoys).[266] Earlier the Navy had been firmly against its deployment for merchant shipping, arguing that the variety of ships and destinations was too large, that the congestion at ports was unacceptable, and that faster vessels would be delayed. Whatever factors caused the change of heart—and a number have been suggested—in April 1917 the Admiralty decided to trial ocean convoys. Their success showed the objections to be either erroneous or overstated and (British/total) losses due to U-boat attacks dropped steadily from (155/354) ships in April, to (88/210) in July, to (56/113) in November 1917. Although not completely eliminated, 'it is evident that the convoy system was mainly responsible for the spectacular reduction in shipping losses'.[267] The threat that the unrestricted U-boat campaign would force Britain to sue for peace was over.

At the BIR, Fisher's suggestion that the Board should be reorganized was entertained by the First Lord of the Admiralty, who commissioned a report on the matter.[268] This reiterated all the well-known complaints and was critical of both the BIR and the Navy. It recommended that the BIR be abolished and that the Navy establish a central department for all naval experiment and research. After some delay a new Department of Experiments and Research (DER) was created at the Admiralty early in 1918, with an electrical engineer as Director (Charles Merz). William Bragg became a scientific adviser to the Director of the Anti-Submarine Division of the DER.[269]

In a final 'Report on the position of experiment and research for the Navy', Merz included three sections written by William: I—On the usefulness under peace conditions of certain devices developed during the war, II.4—Indicating loops, and II.7—Sound-ranging under water.[270] The first contained sections on sound-ranging, acoustic devices, and echo methods, including asdic, although

[265] Hackmann, n. 224, ch. 4; Wood Memorial Number, n. 232, pp. 39–40.

[266] Marder, n. 218, vol. IV, p. 114; also Wilson, n. 160, ch. 39.

[267] Ibid. (Marder), pp. 276–92, 285.

[268] R S Holland, H R Skinner, and A Egerton, 'Report on the present organisation of the Board of Invention and Research', 21 September 1917, ADM 212/158, TNA (PRO), London.

[269] See ns 221 and 224.

[270] C H Merz, 'Report on the position of experiment and research for the Navy', 31 December 1918, ADM 218/3, TNA (PRO), London; the Bragg sections are at pp. 9–12, 32–3, and 42–4 respectively.

few details were given. The second and third entries were longer and more informative. 'Indicating loops' outlined the history of their development, including the use of balancing methods to overcome disturbances, man-made and natural. 'Sound-ranging under water' reported the successful adoption of Lawrence's method at sea (without mentioning his name), with explosion pulses travelling 'immense distances', giving 'extraordinarily sharp and definite records', and leading to far greater accuracy in fixing positions across the ocean.

There are now few indications of how much William contributed directly to the BIR research under his direction. Some correspondence between Wood and William has survived, and it makes clear that William was heavily involved in Wood's projects, offering regular advice and suggestions on everyday technical matters. Surely this occurred with his other project leaders too. There are also reports that show that William carried out investigations himself.[271] In addition, the Wood correspondence makes it clear that when Wood came to publish his findings he relied heavily on William's guidance, although in typical fashion William simply 'communicated' the papers to the publishers.[272]

William gave several public lectures after the war on the detection of submarines by sound, and he included a lecture on 'Sound in war' in his famous 1919 Christmas lectures at the Royal Institution; but he was careful to restrict himself to topics that had a post-war security clearance.[273] Lawrence spoke to the Manchester Literary and Philosophical Society on sound-ranging in October the same year.[274] The most noteworthy address was that given by the President of the British Association, Hon. Sir Charles Parsons, at its delayed Bournemouth meeting in September 1919. In surveying 'Science in the war' Sir Charles spoke of sound-ranging and methods of locating submarines, in the course of which he said that the sound-ranging apparatus was developed by 'Professor Bragg and his son', and that 'The successful development of sound-ranging... led to the suggestion by Professor Bragg that a modified form could be used to locate under-water explosions'.[275] Six days later William wrote to the editor of *The Times* to point out that 'Sir Charles has been too kind to me in giving me credit for a share in the development of the sound-ranging methods... used by the Army during the war. I should have been proud indeed if that had been the case, but the credit belongs to others'.[276] This is an early example of the confusion, that continues to the present day, surrounding the identities of William and Lawrence Bragg, and of the tendency to attribute too much to William and too little to Lawrence. Lawrence developed sound-ranging and, in fact, first

[271] For example, in ADM 293, TNA (PRO), London.

[272] RI MS WHB 37A: the papers are shown in ns 250, 251, and 263.

[273] For example, 'Listening under water', *Engineering*, 1919, 107:776–9; 'Science in industry', *Nature*, 1919, 103:393; 'Sound under water', *The Times Engineering Supplement*, 18 July 1919, p. 220; 'Prince of Wales, F.R.S.: Detection of submarines by sound', *The Times*, 23 January 1920, p. 15; the RI Christmas Lectures were published as Sir William Bragg, *The World of Sound* (London: Bell & Sons, 1920) and (New York: Dover, 1968).

[274] Report in *Nature*, 1919, 104:187.

[275] 'British Association, Bournemouth Meeting', *The Times*, 10 September 1919, p. 15.

[276] Letter W H Bragg to the Editor, *The Times*, 16 September 1919, p. 6.

suggested its use underwater. Why William did not mention him by name is unclear. There is no evidence that he ever tried, here or elsewhere, to enhance his own position at his son's expense. Was it, perhaps, another example of William's excessive modesty, that he now extended to his son?

Between November 1918 and early 1919 the BIR and the stations at Hawkcraig and Parkeston Quay were closed, the ASD was demobilized, and the DER further established as the Navy's scientific centre. William returned to University College London. One later commentator has written: 'Because of the failure to perceive the changing relationship between science and technology, and the mutating nature of each unto itself, the scientists of the BIR were grossly misused... For the above reasons alone the BIR was doomed to failure... in spite of the considerable number of august persons associated with it, the BIR accomplished virtually nothing that had a direct and immediate bearing on the war effort'.[277] But there were some longer-term gains. In his report entitled 'On the usefulness under peace conditions of certain devices developed during the war', William noted, 'principles have been established and methods as well as apparatus... which will be of great service under peace conditions'; and other 'less tangible but very real benefits... may well prove to be some of the most lasting and important consequences... of the war'.[278] But in noting this quotation Hackmann then added, 'Alas, this particular lesson in cooperation had to be almost totally relearned during the Second World War'. William, ever modest and reserved, had been unable to moderate the mutual antagonisms between the Navy and civilian groups. If he had adopted his son's suggestion, being more assertive and demanding, would things have been very different? Probably not; Fisher fulfilled such a role without success. The mistakes and agonies of The Great War only began the process of changing centuries of tradition.

On 24 August 1917 *The London Gazette* announced that King George V had instituted a new order of knighthood, largely in recognition of services rendered during the war: 'The Most Excellent Order of the British Empire'.[279] In the initial lists William received a CBE as a Member of the Panel of the BIR,[280] and on 30 March 1920 he was elevated to KBE 'for services in connection with the war' and particularly as 'Superintendent of Admiralty Experimental Station at Parkeston'.[281] The war was over, but after more than four years and Bob's death it would not be easy for the Bragg family to resume the academic lives that had been so rudely interrupted. The award to William (alone) of the Royal Society's Rumford Medal for 1916—for 'the most important discovery... in any part of Europe during the preceding two years on Heat or on Light'—seemed long, long ago.[282]

[277] Gusewelle, n. 224, pp. 111–12.

[278] W H Bragg in Merz, n. 270, pp. 9, 12 (quoted in Hackmann, n. 224, pp. 38–9).

[279] *Second Supplement to The London Gazette*, 24 August 1917, p. 8796; *The Times*, 25 August 1917, p. 7.

[280] Ibid. (*Gazette*), p. 8796.

[281] *Third Supplement to The London Gazette*, 26 March 1920, pp. 3757–8.

[282] *The Record of the Royal Society of London, Fourth Edition* (London: Royal Society, 1940), pp. 348–9.

18
Post-war separation: Manchester and London

William, Gwendoline, and Gwendy returned to their home at 32 Ladbroke Square, London, and William renewed his appointment at University College. For the third time he faced the task of building a teaching and research department, this time in a city and a country that was tired and war-torn: 'A research school of physics in the sense in which it would be understood today, does not go back earlier than the twenties and owes its origin to Sir William Bragg and his two successors'.[1] Lawrence returned to Trinity College, Cambridge, which had kept his position open. Tidying up the loose ends of the Nobel Prize award was one of the first tasks he and his father undertook.

As political tensions had grown, the Prize increasingly acquired a nationalistic perspective. The Nobel Foundation first petitioned the Swedish government to defer the 1914 decisions until 1915, but when the war dragged on it was decided to award both the 1914 and 1915 prizes. The rules envisaged the reward of important discoveries of recent date and of benefit to mankind, and in physics there was an emphasis on experiment, or theory closely involved with experiment.[2] Nominations for the 1914 Physics Prize included: Laue and W H Bragg jointly, Laue separately, Planck (by nine nominators), and a number of others. For 1915 the most significant nominations were: Moseley or W H Bragg; Laue or W H and W L Bragg jointly or Planck; Laue or W H Bragg; W H and W L Bragg jointly; Laue; and Planck again.[3] Moseley's work was very recent and not fully evaluated, so finally the Physics Committee reduced the list to four nominees: Laue, the two Braggs, and Planck. By 1915 Planck had accumulated fifty-four nominations, but the committee found his quantum hypothesis so difficult that it felt unable to honour him. Allvar Gullstrand, a

[1] N Harte and J North, *The World of University College London, 1828–1978* (London: University College, 1978), p. 149.

[2] E Crawford, *The Beginnings of the Nobel Institution: The Science Prizes, 1901–1915* (Cambridge: CUP, 1984); E Crawford, *Nationalism and Internationalism in Science, 1880–1939* (Cambridge: CUP, 1992); R M Friedman, *The Politics of Excellence: Behind the Nobel Prize in Science* (New York: Time Books, 2001), parts I and II.

[3] E Crawford, J L Heilbron, and R Ullrich, *The Nobel Population, 1901–1937* (Berkeley: Office of History of Science and Technology, 1987).

member of the committee, wrote a detailed report on the work of Laue and W. H. Bragg for 1914, and an even longer survey of the joint Bragg work for 1915, highlighting Lawrence's contribution.[4] These events left the Physics Committee with a very limited choice, and fortunately there could be political balance: the German Laue for 1914 and the Braggs of Britain for 1915.[5]

Congratulations flowed in, chiefly to William as the senior partner and because Lawrence was known to be actively engaged in the war. In addition to their English colleagues and friends there were letters from Adelaide and from other places in Australia that remembered them with affection.[6] Charles Barkla, William's vigorous opponent in the battle over the nature of radiation, wrote generously: 'I most heartily congratulate you and your son…this is a tremendous honour and a substantial reward. I sincerely hope that you will live long to continue your work and to enjoy its pleasant fruits'.[7] Max Laue's response was both generous and nationalistic. He wrote to his friend Wilhelm Wien to acknowledge his own congratulations and added, 'This news [of the Bragg award] also pleases me, and I think the Academy has shown the British, in a very subtle way, that the undoubted great advances by their physicists have been based on the findings of German science on quite a few occasions'.[8] A month later Laue wrote again to Wien to clarify rumours that the young Bragg was dead: 'It is indeed a fact that a son of Professor Bragg was killed in action, but it wasn't the one who participated in the research on crystal structures'.[9]

The chemistry award for 1915 went to Richard Willstätter of Berlin.[10] There were no awards for some of the prizes (they were 'permanently reserved'), but in 1917 Charles Barkla won for Physics, in 1918 Max Planck for Physics and Fritz Haber for Chemistry, and in 1919 Johannes Stark for Physics.[11] Having previously deferred the award ceremony to June 1916 and then indefinitely,[12] the Royal Swedish Academy of Sciences determined to present all these prizes at a ceremony in Stockholm in June 1920.[13] The five Germans attended. Of the British, Barkla accepted but William declined, outwardly because of 'all the Cambridge Tripos examinations and…several other engagements', but in truth,

[4] Copies of documents held in the archives of the Nobel Committee for Physics and the Centre for History of Science of the Royal Swedish Academy of Sciences, Stockholm, kindly made available to the author in 1984.

[5] J L Heilbron, 'H Moseley and the Nobel Prize', *Nature*, 1987, 330:694.

[6] See RI MS WHB 10B for the letters of congratulation; UAA S200, docket 173/1916.

[7] Letter C G Barkla to W H Bragg, 14 November 1915, RI MS WHB 10B/5.

[8] Letter M Laue to W Wien, 14 November 1915, Wien papers, Deutsches Museum, Munich.

[9] Ibid, 15 December 1915.

[10] R Willstätter, *From My Life: The Memoirs of Richard Willstätter*, Lilli Hornig, transl. (New York: Benjamin, 1965).

[11] Crawford et al., n. 3; see also R M Friedman, 'Text, context, and quicksand: method and understanding in studying the Nobel science prizes', *Historical Studies in the Physical and Biological Sciences*, 1989, 20(1):63–77; S Widmalm, 'Science and neutrality: the Nobel Prizes of 1919 and scientific internationalism in Sweden', *Minerva*, 1995, 33:339–60; E Crawford, 'Nobel population 1901–50: anatomy of a scientific elite', *Physics World*, 2001, 14:31–5.

[12] Letters C Aurivillius to W H Bragg, 11 December 1915 and 13 March 1916, RI MS WHB 11A/8 and 9 respectively.

[13] Letter C Aurivillius to W H Bragg, 1 May 1920, RI MS WHB 11A/23.

as he told Rutherford, because 'I believe that several Germans are going'.[14] Nor did Lawrence attend: 'a series of unfortunate circumstances made it impossible for me to accept your invitation'.[15] Presumably his father had persuaded him to refuse. Certainly William had assumed that Barkla would refuse: 'I wondered if the Swedes were sufficiently Machiavellian to have asked us when they knew we could not come'.[16] William never did give a Nobel Lecture; the pain of the war had destroyed the joy of the award, and his previous internationalism was also a casualty.[17] His son was mystified: 'My father never went to Stockholm... Why this was so I have never been able to understand'.[18]

Lawrence gave his Nobel Lecture in the Hall of the Technical University of Stockholm in September 1922, accompanied by his new wife, Alice. In reference to his earlier refusal he said: 'I have always profoundly regretted this, and it was therefore with the very greatest satisfaction that I received the invitation from Prof. Arrhenius a few months ago, and arranged for this visit. I am at last able to tell you how deeply grateful I am to you, and to give you my thanks in person'.[19] The Swedish journalists 'all reported on our youth, which seemed to make a great impression', Alice recalled, and the couple enjoyed themselves 'immensely'.[20] However, Lawrence's acceptance as a Nobel laureate was questioned by a number of English scientists. Had he simply ridden along on his father's coat tails? Rutherford reflected the thinking of many when he wrote to William in regard to an earlier award: 'I was delighted to see in this morning's paper that you and your boy have been granted the Barnard Medal... It is very early for your boy to be getting these distinctions, but it is a great advantage that it is all in the family and is not shared with a German colleague'.[21] Such thoughts were still in the air in 1965 when, uniquely honouring the fiftieth anniversary of his Nobel award, Lawrence returned to Sweden to deliver the first Nobel Guest Lecture, entitled 'Half a century of X-ray analysis'. He again took the opportunity to stress that initially he alone, and not in company with his father, had started X-ray analysis with the reflection interpretation, Bragg's law, and the alkali halide structures. He concluded his survey of the subsequent years with an arresting photograph showing models of the myoglobin protein molecule and of rock salt on the same scale.[22]

In Lawrence's wartime correspondence there are several references to Elaine Barran of Leeds, apparently his first girlfriend and the young woman

[14] Letter W H Bragg to E Rutherford, 12 May 1920, RI MS WHB 11A/24; see also letter W H Bragg to President, Swedish Academy, 17 May 1920, RI MS WHB 11A/25.

[15] W L Bragg, 'The diffraction of X-rays by crystals', in Nobel Foundation, *Nobel Lectures: Physics, 1901–1921* (Amsterdam: Elsevier, 1967), pp. 370–82, 370.

[16] Letter W H Bragg to C Barkla, 18 May 1920, RI MS WHB 11A/26.

[17] E Crawford, *Nationalism and Internationalism in Science, 1880–1939* (Cambridge: CUP, 1992), particularly ch. 3.

[18] W L Bragg, Autobiographical notes, p. 50.

[19] W L Bragg, n. 15, p. 370.

[20] A G J Bragg, Autobiographical notes, pp. 160–2.

[21] Letter E Rutherford to W H Bragg, 20 May 1915, RI MS WHB 26A/27.

[22] W L Bragg, 'Half a century of X-ray analysis: Nobel Guest Lecture I', *Arkiv för Fysik*, 1974, 40:585–603, photo p. 602.

the family expected him to marry.[23] Undated wartime letters to his mother record that he had two letters from Elaine, one 'amusing' but 'written on a typewriter with fearful mistakes in spelling'.[24] Lawrence apparently concluded the relationship hurtfully early in 1918, for he was overcome with shame and wrote three long letters to his mother, pouring out his heart and pleading for her understanding and guidance: 'I always take my troubles to you'.[25] He went into a period of depression: 'At the present mo I'm very much in disgrace here because I was so ashamed I couldn't think of any work and I neglected all my jobs hopelessly'.[26] Perhaps his behaviour was not as bad as he first imagined, however. He soon had 'an awfully jolly letter from Elaine and we are just not worrying about anything for the moment...though I get so fed up with myself, Mum, for being such a stupid unbalanced sort of person'.[27] The two families became close friends and Lawrence and Elaine continued to correspond, but only as friends.[28] Periods of depression would return in the future.

On 30 June 1919, Lieut. W L Bragg, OBE, MC, of Trinity College, Cambridge, was informed that, 'in consequence of the demobilisation of the Army you have been disembodied as from the 24th January 1919',[29] although he was not decommissioned until 30 September 1921, retaining the rank of Major.[30] Cambridge was both tense and extraordinarily alive: 'Cambridge had reason to remember the First World War. To fight in it as an officer meant probable death, mutilation, or injury. Of those who graduated in the five years before the outbreak of the war, more than a quarter were killed, more than a half wounded. The slaughter was the massacre of a class that expected to rule. At one time in the war, the survival time of a second lieutenant in the front line was estimated at six weeks'.[31] On the other hand, those who had survived were determined to make up for lost time, educationally, socially, and in every other way. They flooded the universities and the dance halls, and student and civilian life generally. During this period Lawrence 'had some grand holidays...with George Thomson in his yacht "Fortuna", 4$^1/_2$ tons. We sailed her by degrees from Harwich around to Dartmouth...We had endless adventures as one always does when sailing'.[32]

[23] G K Hunter, *Light is a Messenger: The Life and Science of William Lawrence Bragg* (Oxford: OUP, 2004), p. 60.

[24] Letters W L Bragg to G Bragg, undated, RI MS WLB 37B/4/2 and 6.

[25] Letters W L Bragg to G Bragg, 6, 8, and 20 February 1918, RI MS WLB 37B/3/1–3, quotation from no. 1.

[26] Ibid., 8 February 1918.

[27] Ibid., 20 February 1918.

[28] Letter W L Bragg to G Bragg, 16 August 1918, RI MS WLB 37B/3/30.

[29] Letter War Office to W L Bragg, 30 June 1919, RI MS WLB 37A/4/7.

[30] Memorandum OC 239th (Leicester) Battery RFA to W L Bragg, 23 November 1921, RI MS WLB 37A/4/8.

[31] A Sinclair, *The Red and The Blue: Intelligence, Treason and the Universities* (London: Weidenfeld and Nicolson, 1986), p. 9.

[32] W L Bragg, Autobiographical notes, p. 43.

Lawrence: Manchester and marriage

Lawrence also remembered demonstrating in Searle's Part I laboratory classes in the Cavendish Laboratory: 'Searle was a really extraordinary character...he despised research [but] took infinite pains over the practical class...He was a terrific tyrant...another time, holding up the large sheet on which the student's experiments were registered, [he shouted] "Bragg, what do you think these marks are? Human tears", pointing scornfully to a wretched girl sobbing her heart out over the desk'.[33] There was one event of lasting importance during this brief interlude in Cambridge, 'I met for the first time my wife-to-be, Alice Hopkinson':[34]

> I had heard a good deal about her. She was a first cousin of my great friend, Cecil Hopkinson. Her father...had been a doctor in Manchester. Alice was always referred to by the relations as an extremely pretty girl, a statement generally accompanied by a slight shaking of the head. She was then up at Newnham reading History and having a whale of a time. A large number of naval officers had been sent to Cambridge after the war, most attractive young men who gave endless parties with their war gratuities, and the undergraduates were many of them demobilized officers, so that the girls of Newnham and Girton were in great demand. We met first at a thé dansant...I think it was on the fourth occasion we met that I proposed; but Alice was having much too good a time—she was only nineteen—and she had an ambition to complete her University studies. Her family moved from Manchester to Adams Road, Cambridge...Her father had a serious illness at the end of the war...and decided to give up his practice...he was appointed a demonstrator at the Anatomy laboratory.

Alice Grace Jenny Hopkinson's family background was very different from that of Lawrence Bragg. Her father, Albert, was the tenth child in the family of Alice and John Hopkinson, Engineer, Mayor of Manchester, and a Governor of Owen's College. Amongst Albert's brothers were a Senior Wrangler, a lawyer, MP, and Principal of Owen's College, and several prominent engineers. The Hopkinsons had 'a tremendous sense of duty...a passion for hard work and self-discipline...an emphasis on service to the public...They had to achieve whatever they set their minds to...in recreations they showed a glorious sense of adventure and recklessness'.[35] Her mother, Olga, was a daughter of Jenny and Philip Cunliffe-Owen, a Director of the South Kensington (now the Victoria and Albert) Museum and a member of a family Alice described as: 'happily confident...quite uninhibited...they said what they thought...cosmopolitan in outlook, all spoke French and German fluently...charming and accomplished...and full of character...they all had a great love of life'.[36] Much of this described Alice herself.

[33] Ibid., pp. 41–42.
[34] Ibid., p. 41.
[35] A G J Bragg, Autobiographical notes, pp. 8–10.
[36] Ibid., pp. 36–37.

Educated at Ladybarn House School in Manchester and St Leonards at St Andrews, Alice 'hated it...I was bad at games...and most of all disliked being a fag'.[37] There were redeeming features, however: acting, the beauty of St Andrews, and being in the A stream academically. When the war came 'we made bandages and knitted, we were cold and soon badly fed...[her brother] Eric was reported seriously wounded and missing, mentioned in dispatches and awarded the Military Cross. He had only been in France six weeks'.[38] He died. In 1918 Alice completed her exams and wept at leaving St Leonards. She asked the headmistress for advice. 'Go to Cambridge and read history, get a degree, then marry a man older than yourself', she said; to which Alice added, 'These instructions I carried out to the letter in time. For the moment I was intoxicated by the sense of freedom'.[39]

Alice spent a year at home before going up to Newnham College, Cambridge. 'Young men began to come around. An alarming number of my brothers' Rugby and Cambridge friends had been killed, but those who were left came on leave and took me out...One or two of the young men wanted to marry me...but I had to refuse'.[40] At Cambridge her friends were 'all serious-minded, and later, when Cambridge life became very gay and went to my head, they were a salutary and an accepted influence on me'.[41] 'The other colleges were almost empty...Suddenly in November it was Armistice Day... Everyone was violently emotional, laughing and kissing...the colleges filled with people...there was a craze for dancing...The thé dansant was introduced...I became very excited and restless, wandering in and out of people's rooms in college in the fear of missing something'.[42] Many years later the unfolding events were still vivid to Alice:[43]

> One night I went to a dance in the Master's Lodge at Downing College, where I met a young medical, Vaughan Squires, who asked me to join a party for a thé dansant at the Guildhall. When I arrived there he said he would like me to meet his cousin, Major Bragg...I thought him good looking. We danced together and then sat out. Of course we knew something about each other, as he had been the great friend of my cousin, Cecil Hopkinson,...The Hopkinsons had told me with bated breath that he was brilliant scientifically. He turned this aside and told me what he had heard about me...that he was a Fellow of Trinity and...[had been] doing Sound Ranging with the Army...The Victory Ball took place at the end of term...Mr Bragg was there and, in spite of the protest at going 'out of party', I danced with him...
>
> Now I met Mr Bragg a third time...He told me that he was shortly leaving Cambridge to succeed Rutherford in the Chair of Physics in

[37] Ibid., p. 86.
[38] Ibid., pp. 88–89.
[39] Ibid., p. 93.
[40] Ibid., pp. 100–101.
[41] Ibid., p. 108.
[42] Ibid., pp. 110–111.
[43] Ibid., pp. 111–117.

> Manchester...I told him rather fiercely that he would find it wet and ugly there after Cambridge and asked him why he wanted to leave Trinity...I politely hoped that he would be happy there. I did not see him again till May week...
>
> Meanwhile...life had been too exciting, I had not done enough work, and in our May examinations or 'Prelims' I was the only one of my history year to get a third class...[but] I had promised my father that whatever happened I would stay the course at the university...
>
> With relief I embarked on my first May week. There came a note from Mr Bragg asking me to have lunch in his rooms in Nevile's Court...and meet the family...His Australian mother seemed a very exuberant woman... She seemed to me to be amazingly young to be a professor's mother...She talked a great deal...W L B's twelve-year-old sister, Gwendy, had whooping cough and could not come...W L B was devoted to his little sister and now he was really disappointed...Sir William Bragg was a large man, beaming genially but rather silent...At the Masonic Ball I had a dance with W L B and as we walked in the moonlight in the sitting-out time he asked me to marry him. It was agreed that we did not know each other as yet...[and] proved to be a rather stormy time for us. I was nineteen years old and immature...W L B was in a hurry; when he really wanted something he always was...I was overwhelmed and confused, so that at the end of the summer we finally parted...My own family were deeply disappointed, as they all loved W L B His parents, however, were assured by my Aunt Evelyn in Cambridge that their son had had a merciful escape. I was 'too young, unstable, a sad flirt, and totally unsuitable for a professor's wife'.

Lawrence's appointment to Manchester followed J J Thomson's decision to resign as Cavendish Professor and spend more time as Master of Trinity College while retaining a presence in the Laboratory. His wish to see Rutherford in the Cavendish chair was granted and the Langworthy Professorship of Physics at Manchester therefore became vacant. Lawrence's desire to have a place of his own became possible. No doubt he consulted his father, but he also sought other advice. The Vice-Chancellor was Sir Henry Miers, who had been Professor of Mineralogy at Oxford and who held a special Chair of Crystallography at Manchester. He had pleaded publicly for the creation of a department of pure crystallographic research and Lawrence's appointment enabled him to realize much of his dream.[44] In addition Lawrence sought Rutherford's advice: 'I was hoping that I might have a chance of seeing you when I was in London, and thanking you personally for your letters and for advising the Senate to make me this offer...I would be tremendously influenced in deciding by what you thought was the right thing to do'. He then listed the several advantages of staying in Cambridge and wondered about the existing staff at Manchester, his lack of teaching experience, and the chance of employing James to assist him.

[44] Sir David Phillips, 'William Lawrence Bragg, 1890–1971', *Biographical Memoirs of Fellows of the Royal Society*, 1979, 25:75–143, 95–6.

He was anxious not to be overwhelmed by administration and teaching.[45] He talked to Rutherford in person,[46] and then wrote to him to say: 'I would just like to tell you how proud I feel to think that I am succeeding you at Manchester. It is a tremendous honour for anyone who has still his spurs to win'.[47] Lawrence also wrote to his father:[48]

> Thanks very much for your letter and all the good advice in it... This is a fine laboratory... It is fitted up in the most wonderfully convenient way... [but] it is a very different thing when one comes to look at the apparatus used for teaching. The elementary class work with the most jerry-built home-made stuff I have ever seen. I think it is because there has never been a real mechanic like Jenkinson in this place. So much research has been done here that all the money has been spent on it and none on the teaching stuff... My own apparatus is nearly set up and James and I are eager to get going... I have been wondering what you were intending to go on with. I do hope you will never keep from doing any bit of work, Dad, because you think that may be the line I am going on... I had a letter from Newton & Wright accusing me of having a coil and switchboard of theirs. I replied pointing out that you were the culprit... I am always signing my name Lawrence Bragg now to avoid confusion.

It rapidly became clear, however, that the new professor would have the worst possible introduction to his new job, a welcome that might have destroyed any young man of lesser ability and worldly experience. Lawrence's time at the front stood him in good stead. He recalled:[49]

> I was only 29 when I took up my duties as head of the Physics Laboratory in Manchester, and was handicapped by not having served my apprenticeship in a junior position in a department. Further, we had forgotten most of our physics during the war... Finally, the students were largely older men who had been demobilized and were a tough crowd. The staff consisted of [those who remained from earlier times and sound-rangers Dickson, Nuttall, Robinson, and James, who became his loyal deputy]. We had a strenuous and difficult time... It was not made easier by a vile series of anonymous letters which started soon after I came... directed mainly at myself and James, and abused us bitterly as incompetent and useless. They were the worse because it was clear that the writer had access to laboratory gossip and knew of every mistake we made; there was just a small element of truth in every criticism... in the end it drove me into what was really a nervous breakdown. Curiously enough, I recovered when the letters began to attack my father and Rutherford as well... I shall never cease to be grateful to colleagues in Manchester who... helped me through the worst of it. My recovery was hastened also by James and myself getting some first-rate results in our research.

[45] Letter W L Bragg to E Rutherford, 17 April 1919, CUL RC B413.
[46] Ibid., 24 April 1919, CUL RC B414.
[47] Ibid., 12 May 1919, CUL RC B415.
[48] Letter W L Bragg to W H Bragg, undated, RI MS WHB 28A/5,
[49] W L Bragg, Autobiographical notes, pp. 42–3.

> As a bachelor in Manchester I kept house with Drew, who had been a sound-ranger and who was a lecturer in the classics department. We were looked after by old Charlotte, an arrangement made by my mother.

The returned servicemen had no mercy on the novice lecturers: 'There were rowdy, boisterous goings on in the lecture room... One could hear this... and there was visible evidence in the fact that panels of the benches were kicked into matchwood during the lecture periods taken by Bragg, James, and Dickson. In one dramatic episode a student set off a firework under the reading desk and Bragg boxed his ears'.[50] His wife concluded later that some of the staff 'deeply resented a young man of 29, with no experience of lecturing or running a department, being made their professor'.[51]

In 1921 things changed for the better. In May Lawrence was elected a Fellow of the Royal Society at his first nomination, and amongst the letters he received 'was one in handwriting which made my heart turn over. It was Alice writing to congratulate me. We had never met or written to each other since I left Cambridge. My troubles in Manchester had been such that I think I felt my chances in that direction were nil... I replied asking when I could come to see her and she arranged a tea date in Newnham... We became engaged the next day'.[52] Alice went up to London to see Lawrence's parents: 'We were all a little nervous... Lady B was reduced to tears of emotion, Sir William grunted and chuckled and... Gwendy positively squeaked with excitement... Next day Lady B saw me off at the station... and said that life would not always be easy, "You must make the running, my dear, and hold his hand as I have always had to do with Dad"'.[53]

Alice continued: 'just before my Tripos, W L B saw a little house in Didsbury and had to buy it... My Newnham days were over, my Tripos done... there was no planning a career, my future was settled'.[54] They were married at Christmas—20 December 1921—in Great St Mary's Church, Cambridge, with Gwendy amongst the bridesmaids and Vaughan Squires as best man. They honeymooned in Somerset and in France. Alice was 'enchanted with everything, [but] early in January we had to go to Manchester for the beginning of term[55]... We arrived there on a dark, wet winter afternoon, to be greeted by a curious old woman in black, muttering in German and clearly none too pleased to see us. This was Charlotte, who... had been persuaded by my mother-in-law to settle us in. She disliked me on sight and, during the four or five days she remained in the house, she hardly spoke to me but went about muttering "Ah, poor Mr Villy, God help him"'.[56]

[50] Phillips, n. 44, p. 96.

[51] Letter Lady Bragg to Sir David Phillips, 8 June 1979, D C Phillips Collection, Bodleian Library, Oxford.

[52] W L Bragg, Autobiographical notes, p. 43.

[53] A G J Bragg, Autobiographical notes, pp. 125–6.

[54] Ibid., pp. 126–7.

[55] Ibid., pp. 131–2.

[56] Ibid., p. 149.

Fig. 18.1 Lawrence and Alice Bragg on their wedding day, 20 December 1921. (Courtesy: Mrs R. Staughton and Lady Heath.)

Initially isolated and without much to do during the day—they had servants—Alice's life was dull after Cambridge. I was fortunate to meet and speak with Lady Bragg on two occasions: in September 1983 and October 1987. She told me she was particularly attracted by her husband's sense of humour, his love of puns and spoonerisms, his genuine modesty, his extreme difference from other young men, and their shared love of nature. She said he was quite tall (1.7 m) and handsome, and she made him laugh. He could be preoccupied with his science and oblivious to his surroundings, and he came to her many dinner parties as a guest rather than as the host, although he loved parties, dancing, and entertainments. He was not good at meetings, and he did not think quickly on his feet.[57] She also made her own fun. Sometimes she put

[57] Conversations with Lady Bragg, 7 September 1983 and 16 October 1987, Cambridge, England; recordings were not made but I wrote extensive notes immediately after each conversation.

a gold fish in each finger bowl at a dinner party; once she told the university treasurer that she thought one of his clever business deals was dishonest.[58]

Lawrence began research soon after arriving in Manchester. There are relevant letters to his father from mid-1920 onwards,[59] and he spoke on 'Crystal structure' at both the Royal Institution and to the Geology Section of the British Association meeting in Cardiff the same year.[60] In reviewing the structures determined pre-war, he illustrated his Royal Institution lecture with several ball-and-stick models and noticed, as he later recalled, 'that the distances between atoms in the alkali halides were additive... and this suggested that all atoms had definite sizes... In 1920 I published a list of radii but unfortunately got my datum line wrong, assigning too large a contribution to all the cations and too small a one to the anions... The distinction between ionic and homopolar bonds was not clear at the time, at any rate to me... so I missed the chance of establishing an authoritative set of ionic radii at a very early date. They were of the greatest importance in the later analysis of more complex crystals'.[61] He also revised the honours physics course and ordered £250 worth of apparatus to equip the advanced undergraduate laboratory.[62]

From 1920, too, Lawrence embarked upon a closer investigation of the reflection of X-rays at crystal surfaces, the core of his initial inspiration that had led to the creation of X-ray crystallography. He discussed the apparatus and his initial results in correspondence with his father and wrote, 'Do spare a few days to run up here and see everything';[63] for 'Whenever I have a good talk about the work with you it bucks me up like anything and gives me lots of new ideas for the work'.[64] He began with the faces of NaCl crystals and planned to do a number of others, but later realized that 'KCl and NaCl are going to keep us fully occupied for a long time yet', because 'it's so impossible to get any time off during the term'.[65] Late in 1920 he was annoyed by a visitor who suggested that the explanation Lawrence and James had proposed for some recent results was his idea. Lawrence was angry and confided to his father, 'we have always discussed everything together and I owe so many ideas to you... Any outsider butting in, however, makes me see red! The trouble is that I always lose my temper... and say more than I mean to'.[66] William replied at once, and Lawrence then wrote: 'Ever so many thanks for your letter... one gets rather

[58] A G J. Bragg, Autobiographical notes, pp. 152, 153.

[59] Letters W L Bragg to W H Bragg, RI MS WHB 28A.

[60] W L Bragg, 'Crystal structure', *Proceedings of the Royal Institution of Great Britain*, 1920, 23:190–205; id., 'Crystal structure', in *Report of the Eighty-Eighth Meeting of the British Association for the Advancement of Science, Cardiff, 1920* (London: Murray, 1920), pp. 357–8.

[61] W L Bragg, 'Manchester days', *Acta Crystallographica*, 1970, A26:173–7, 173; the papers were: W L Bragg, 'The arrangement of atoms in crystals', *Philosophical Magazine*, 1920, 40:169–89; id., 'The arrangement of atoms in crystals', *Nature*, 1921, 106:725; W L Bragg and H Bell, 'The dimensions of atoms and molecules', *Nature*, 1921, 107:107; W L Bragg, 'The dimensions of atoms and molecules', *Science Progress*, 1921–2, 16:45–55.

[62] Letter W L Bragg to W H Bragg, 18 June 1920, RI MS WHB 28A/8.

[63] Ibid.

[64] Letter W L Bragg to W H Bragg, no date, RI MS WHB 28A/9.

[65] Ibid., 17 October [1921?], RI MS WHB 28A/10.

[66] Ibid., 3 November 1920, RI MS WHB 28A/14.

worked up about any original ideas, and a little while later the whole thing appears unimportant'.[67]

In a long letter to his father Lawrence complained about the lack of research funds and the shortage of first-rate men.[68] Late in the year he sent a rough sketch of a scheme for a new edition of their earlier book, *X Rays and Crystal Structure*,[69] and in January 1921 reported that he had four chapters done.[70] William replied at length with wise advice and numerous suggestions for improving the text.[71] The fourth revised and enlarged edition appeared in 1924, not a comprehensive treatise but an introduction to the usefulness of the methods.[72]

In February Lawrence reported that he had rediscovered the relevance of Darwin's earlier papers for the intensity question,[73] and by October he had published the first two of the series of 'BJB' papers. He wrote later, 'James, Bosanquet and I made a series of experimental investigations into the optics of X-ray diffraction on a sounder quantitative basis. This in its turn made it possible to attack much more complicated crystal structures because quantitative measurements gave far more information. So we started the "Manchester School" of X-ray analysis'.[74] He described this research more fully during a symposium to mark his eightieth birthday:[75]

> James and I, with Bosanquet from Oxford, who worked with us in the vacations, made an attack from 1921 [*sic*, 1920] onwards on the quantitative aspects of X-ray diffraction... We used the technique my father had developed of sweeping the crystal through the reflexion angle so that all elements of the crystal mosaic made their contribution to the integrated reflexion. I had an X-ray spectrometer made in my father's workshop in Leeds, and we added certain refinements such as using a string electrometer (a delightful instrument) and a potential divider to bring the string to zero. The crystal was turned by a capstan... in time with a metronome. It was a perpetual thrill to see the string barely moving, then much displaced as one went through the reflecting position, then almost coming to rest again; it made X-ray diffraction seem very real. The X-ray tube was activated by an induction coil and my break, and keeping the gas in the tube at constant pressure was an art. It was a great boon when Coolidge tubes became available. The results, however, were surprisingly accurate, and we really sorted out primary and secondary extinctions, using Darwin's formulae, and established standards for absolute intensity.

[67] Ibid., 4 November 1920, RI MS WHB 28A/15.
[68] Ibid., no date, RI MS WHB 28A/16.
[69] Ibid., 6 December [1920?], RI MS WHB 28A/17.
[70] Ibid., 15 January 1921, RI MS WHB 28A/19.
[71] Letter W H Bragg to W L Bragg, no date, RI MS WHB 28A/23.
[72] Sir William Bragg and W L Bragg, *X Rays and Crystal Structure* (London: Bell, 1924).
[73] Letter W L Bragg to W H Bragg, 25 February 1921, RI MS WHB 28A/20.
[74] W L Bragg, Autobiographical notes, p. 51.
[75] W L Bragg, 'Manchester days', n. 61, p. 174

> The result of these investigations was that, in examining a crystal, one could assign to any reflection (hkl) an absolute value of *F(hkl)* [the structure factor] in terms of the scattering by a single electron. Simultaneously, the *f* factors of atomic scattering at different angles were being established, first empirically and later by calculation by Hartree.[76] Armed with this precise information one could tackle complex crystal structures... The extension of crystal analysis on a more ambitious scale was, I think, a major contribution of the 'Manchester School'.

Darwin's early work provided the theoretical basis for the analysis and gave the effect in terms of the scattering of single electrons. The total effect was less than the sum of these individual scatterings because, in a perfect crystal there were multiple reflections and interference between them ('primary extinction'), while in imperfect crystals (such as their NaCl) there were similar but smaller reductions ('secondary extinction').[77]

Although the thought arose naturally from his crystallographic research, Lawrence did not subsequently pursue the possibility of determining the arrangement of electrons within the scattering atoms, and I do not agree with Hunter that Lawrence made a concerted attempt at this time to determine them.[78] Lawrence was slower than Linus Pauling to appreciate the importance of the chemical bond, but to be second to Pauling in chemistry was not a sign of ignorance. To suggest that Lawrence was 'always weak on chemistry' and 'quite ignorant of chemistry and biology and had little interest in either discipline', is a severe misjudgement and a curious contradiction in view of Hunter's text and his further assessments that, for example, Lawrence's work 'had profound implications for both inorganic and organic chemistry', and that 'Biology was, if anything, even more profoundly affected than chemistry'.[79] Lawrence's chemistry teacher at St Peter's College strongly enhanced the latent interest in science planted by his father, and Lawrence had an excellent school and undergraduate record in chemistry. No one could change the history of a discipline, as Lawrence and his father did, without having some interest in it. He knew enough chemistry to recognize the importance and impact of X-ray crystallography upon it, and when he later needed advice he sought and obtained the best.

Early in 1921 Lawrence attended the Third Solvay Council on Physics in Brussels on the subject 'Atoms and electrons', but William did not attend.[80]

[76] The relevant papers were W Lawrence Bragg, R W James, and C H Bosanquet, 'The intensity of reflexion of X-rays by rock-salt', *Philosophical Magazine*, 1921, 41:309–37, and 42:1–17; id., 'The distribution of electrons around the nucleus in the sodium and chlorine atoms', *Philosophical Magazine*, 1922, 44:433–49; there were also related publications: W L Bragg, 'Uber die Streuung der Röntgenstrahlen durch die Atome eines Kristalles', *Zeitschrift für Physik*, 1921, 8:77–84; W L Bragg and R W James, 'The intensity of X-ray reflection', *Nature*, 1922, 110:148.

[77] K Lonsdale, *Crystals and X-rays* (London: Bell, 1948), chs 5 and 6.

[78] Hunter, n. 23, p. 66.

[79] Ibid., pp. 103, 249–50.

[80] P Marage and G Wallenborn (eds), *The Solvay Councils and the Birth of Modern Physics* (Basel: Birkhäuser, 1999), pp. 112–115, 129.

In his October letter to his father Lawrence also reported, 'I have been reading Sommerfeld's book. It is very good, and not a bit abstruse except in certain chapters at the end. He is convinced of the truth of Bohr's atomic model, and pictures the atoms as planet-systems. The numerical checks he gets, the explanation of the "fine-structure" etc., is very convincing, but do you really think these orbits have any existence? I can't believe it'.[81] Lawrence was reading the new theoretical physics literature as soon as could be expected, but later he was savagely criticized: 'The term "classical physicist", often applied to Bragg by his contemporaries, was not, of course, meant as a compliment. As a classical physicist in the age of quantum theory, he was a scientific dinosaur in a world now dominated by mammals'.[82]

This statement is nonsense. Parts of modern physics, and of applied physics in particular, are still classical, and there is no shame in that. Lawrence had no need of quantum mechanics for much of his research, and as Hunter himself acknowledged, 'Bragg had several weapons... first was his mastery of classical optics... second was his ability to visualize three-dimensional objects in space... [and] a third factor... was his ability to see the essential point of a problem and to strip away the inessentials'. Hunter also supported Pippard's words: 'For [Lawrence] beauty and economy were the touchstones of a physical argument or an experiment, and unless one sympathized with his quest for these ideals one missed his intellectual power and subtlety'.[83] These are 'classical' values that served Lawrence extraordinarily well.

Lawrence was not, in fact, ignorant of quantum mechanics, and indeed he fostered its teaching in the institutions he led. Thus, in his Foreword to Tolansky's 1942 book, *Introduction to Atomic Physics*, which arose from an important undergraduate course at Manchester, Lawrence wrote: 'An undergraduate must spend most of his time in mastering the classical treatment of the main divisions of physics... At some point, however, the student has to be introduced to the fundamental change in outlook which has taken place since the beginning of this century... He should begin to appreciate the power of the new quantum mechanics, and something of the symmetry and beauty of its conceptions... The course at Manchester was the result of a belief that the student should get this introduction in his second rather than his third year... This course became a tradition in the Manchester Physics Laboratory... It has been attended by quite a number of PhD candidates from the chemical and other departments'.[84]

Nevill Mott recalled his time in Manchester similarly: 'In 1929 quantum or wave mechanics was three years old, and few people in the UK understood it. [Lawrence] Bragg wanted me to give lectures to his staff, including himself,

[81] Letter W L Bragg to W H Bragg, 17 October [1921?], RI MS WHB 28A/10; A Sommerfeld, *Atombau und Spektrallinien* (Braunschweig: Vieweg, 1921); the English translation had not yet appeared, so Lawrence either struggled through the German edition alone or sought assistance from a colleague (James?).

[82] Hunter, n. 23, p. 249.

[83] Ibid., p. 251.

[84] W L Bragg, 'Foreword', in S Tolanski, *Introduction to Atomic Physics* (London: Longman Green, 1942).

and this I did...I remember Bragg vividly—how interested and approachable he was'.[85] In 1931 Lawrence deliberately chose to spend three months at Sommerfeld's Institute of Theoretical Physics in Munich, and while there he accompanied Sommerfeld on his regular weekend skiing trips; while in 1934, as a visitor to the Baker Chemistry Laboratory at Cornell University in New York, he presented an 'elegant' lecture on the Heisenberg Uncertainty Principle and indeterminism in physical processes. During his tenure of the Cavendish chair in Cambridge (see chapter 19), Lawrence regularly gave an impressive introductory lecture to the quantum mechanics course.[86] These are hardly the words or actions of 'a scientific dinosaur'!

Lawrence and Alice's first child, Stephen Lawrence, was born in November 1923,[87] and his arrival was the centrepiece of my first and longest conversation with Lady Bragg. Expecting me to ask especially about her father-in-law, she had chosen one particular story to tell me. She and her husband had arrived at William and Gwendoline's London home with the exciting news that they were to become grandparents for the first time. Entering the drawing room and noticing William standing in front of the fire, she had opened her arms and hurried forward to embrace him and tell him the news. But as she drew close he recoiled, extended one hand and clasped his breast pocket, preventing her embrace and saying, 'careful, you'll break my slides'.[88] He was frightened of human contact, she said, and could not bear people saying unkind things about others. He was often withdrawn and never let himself go; his wife made all the social running. Gwendoline was strong and very sociable; William was courteous but very reserved. He liked his daughter-in-law, and she, in turn, found him a very useful sounding board later, but the Braggs were quite unlike the Hopkinsons.[89]

Lawrence's life was now much more settled, and although he would always stay in touch with his parents on both family and professional matters, his research was diverging from that of his father. He also had, in addition to loyal academic colleagues, Rutherford's brilliant laboratory steward, William Kay, and an excellent secretary, Mair Jones.[90] The painstaking work of measuring individual reflections and their dependence on crystal perfection and temperature was not his forte, however, and he left this work to James and concentrated himself on the analysis of crystal structures. He returned to aragonite, begun before the war and the most complex crystal yet attempted, and for the first time discussed the symmetry in terms of formal space-group theory. This work was followed by a long and intense research programme on the structures

[85] Sir Nevill Mott, 'Manchester and Cambridge', in J M Thomas and Sir David Phillips (eds), *Selections and Reflections: The Legacy of Sir Lawrence Bragg* (Northwood: Science Reviews, 1990), p. 96; Hans Bethe was also at Manchester and remembered Lawrence's interest in the new theory with similar admiration (S S Schweber, 'The Happy Thirties', *Physics Today*, 2005, 58:38–43, 40).

[86] Hunter, n. 23, pp. 110, 116, 141.

[87] A G J Bragg, Autobiographical notes, p. 170.

[88] Glass slides were used in a projector to illustrate lectures.

[89] Conversations with Lady Bragg, n. 57.

[90] Hunter, n. 23, p. 66.

of silicate minerals, and, while the Manchester group could not define the field alone, Phillips has concluded rightly that, 'no subsequent analysis can diminish [Lawrence's] achievement in guiding the studies of these complex mineral structures to a fruitful conclusion. He was clearly the guiding hand, and his colleagues of those days have written of the excitement of working with him as more and more complex arrangements of atoms yielded to their attack. Silicate chemistry was shown to be inherently a chemistry of the solid state, intelligible only in terms of the three-dimensional structures, and [he] never tired of using the story... to illustrate his conviction that the analysis of increasingly complicated structures can lead to the discovery of unimagined new principles'.[91] His book, *Atomic Structure of Minerals*, contains Lawrence's retrospective view of the field and resulted from his tenure of the Baker Lectureship in Chemistry at Cornell University during 1934.[92]

The relationship between Lawrence and his father, personally and professionally, has been a topic of confused and unresolved discussion in accounts of their lives and achievements. Most authors have suggested that there was a tension and coolness between them, based upon their competitiveness in research and Lawrence's belief that his father had not done enough to ensure that he received the credit he deserved. I have made a closer study of both men than most authors, and I disagree; the evidence is otherwise. William spoke warmly and often about his son's pivotal contributions, and the private letters that Lawrence wrote to his father throughout his life were very warm. Only one conclusion seems possible: that Lawrence had great respect and a deep affection for his father. Of course disagreements arose from time to time, as between any father and son, especially when they work together in the same business, or within the same organization, or on the family farm. In this case it could arise, for example, over the origin of an idea that emerged during one of their many discussions; and they were quite different in temperament.[93] As Lady Bragg also recorded: 'I am myself inclined to think that difficulties, which were inevitable and natural, were exaggerated... They were both in the same line of research, and their minds worked in a similar way'.[94] It was a tragedy that William and Lawrence could not talk to each other, face-to-face, in the same spirit they showed in writing to each other, but this does not invalidate my conclusion. The origin and seat of the basic problem of inadequate acknowledgement and recognition lay elsewhere.

William: London and wave–particle duality

At University College there was a great deal to be done. The war ended during the first term of the 1918–19 academic year and special arrangements were

[91] Phillips, n. 44, p. 102.
[92] W L Bragg, *Atomic Structure of Minerals* (Ithaca: Cornell University Press, 1937).
[93] Conversations with Lady Bragg, n. 57.
[94] A G J Bragg, Autobiographical notes, p. 171.

made to allow returned servicemen to begin or resume their studies. In January 1919 an appreciable number did so, and yet again William bore the burden. The existing Carey Foster laboratory was appropriated for research, the Senior Laboratory was expanded, and the deconsecrated All Saints Church, purchased by the College in 1912, was fitted out for another class. New support-staff were employed and Jenkinson made all the new equipment with the help of just one or two assistants. When the church was converted to a Great Hall to commemorate the College students who had lost their lives during the war, the anatomy dissecting room, the Birkbeck Chemistry Laboratory, and several adjoining rooms were all reallocated and renovated for the Physics Department.[95]

Dissatisfaction with the existing constitution of the University of London had led, in 1909, to the appointment of the Haldane Commission, but its recommendations were suspended. Post-war conditions were still unsatisfactory and William angrily asserted, 'It is time that something was done'.[96] He and his fellow correspondent continued: 'The question at issue is—Can the University of London be so reconstituted as to give the desired freedom [in teaching and research] to its constituent elements?...Academic freedom is essential to progress. There will be no real University in London until the colleges are released from the paralyzing influence of the present so-called University of London...by complete separation of the colleges...and their establishment either as separate universities or as a federal university'. In the face of major post-war changes the Haldane Report was not revived, and it was 1924 before the British government appointed a departmental committee to recommend what changes were required. The University of London Act was not passed until 1926, and the Privy Council did not approve the new statutes until 1929. William had left the university long before this.[97]

William soon attracted a brilliant young research team to University College, including William Astbury, Ivor Backhurst, Reginald Gibbs, Alexander Müller, Sydney Peirce, George Shearer, Walter Stiles, and Miss Yardley (later Dame Kathleen Lonsdale and the first female FRS), appointed as research assistants or as demonstrators on the teaching staff. At this time William was still actively experimenting himself as well as directing research.[98] Initially he still used an ionization chamber as detector for the diffracted X-rays, but he progressively replaced it with a photographic plate, particularly when it was found advantageous to use powdered samples rather than single crystals.[99] Indeed, William's first paper in the new area demonstrated that such minute crystals (powder)

[95] O Wood, 'About the Physical Department, University College London, and those who worked therein, 1826–1950', typescript in Manuscripts and Rare Books Room, Watson Library, University College London, pp. 55–8.

[96] Letter W H Bragg and E H Starling to *The Times*, 22 December 1919, p. 8.

[97] N Harte, *The University of London, 1826–1986: An Illustrated History* (London: Athlone Press, 1986), pp. 182–213.

[98] E N da C Andrade, 'William Henry Bragg, 1862–1942', *Obituary Notices of Fellows of the Royal Society*, 1942–44, 4:276–300, 284.

[99] W H Bragg, 'Application of the ionisation spectrometer to the determination of the structure of minute crystals', *Proceedings of the Physical Society of London*, 1921, 33:222–4.

could be examined and their structure determined, and he reproduced three spectra of reflected intensity against angle—for aluminium, silicon, and lithium fluoride—to demonstrate the method's viability.[100] There were generous grants from the Department of Scientific and Industrial Research but it was not easy; new equipment had to be developed and X-ray tubes were a continual source of difficulty. Following an agreement with his son, who was examining the structure of inorganic materials, William turned his attention to organic crystals, upon which he made the first concerted attack. It was to prove very productive, both at University College and especially when he and his team moved to the Royal Institution a few years later.[101]

In 1920 William was elected President of the Physical Society of London, the learned society for physics in Britain. Yet again it was a challenging appointment for, during the war and the upsurge in the importance of science, there was agitation for an improvement in the professional status of physicists, and this now came to a head. There was an Institute of Chemistry but not one of physics, and physicists could gain official status only by being termed 'research chemists'. The Physical Society decided to sponsor the formation of an independent Institute of Physics to safeguard the interests and professional status of physicists, and it was incorporated on 1 November 1920. William would have to smooth the inevitable early tensions between the two organizations.[102] As President of the Society William also thought it his duty to publish in its journal, and he did so generously. His presidential address, delivered on 11 November, described his new research for the first time and produced a large paper entitled 'The structure of organic crystals'.[103]

His published summary read: 'the structure of crystals of organic substances invites examination by the methods of X-ray analysis; but their molecular complexity would seem to throw great difficulties in the way. It is possible, however, that the difficulties in the case of aromatic compounds may be surmounted by adopting a certain hypothesis; viz., that the benzene and naphthalene ring is an actual structure... built as a whole into the organic substances in which it occurs... A more complex molecule such as either of the naphthols is not to be regarded as an addition of one oxygen atom to these 18 [the 10 carbon and 8 hydrogen atoms of naphthalene], an idea on which nothing can be built, but as a naphthalene double ring of the same size and form as before, except that one particular hydrogen has been replaced by a hydroxyl group'.[104] Starting with the diamond and graphite structures, William then built carefully crafted arguments for the structures of naphthalene, anthracene,

[100] Ibid.

[101] Andrade, n. 98, p. 285.

[102] J L Lewis (ed.), *125 Years: The Physical Society and The Institute of Physics* (Bristol: Institute of Physics, 1999); the two organizations continued side by side until 1960, when they amalgamated; on the award of a Royal Charter in 1970 they were renamed simply 'The Institute of Physics'.

[103] Prof. Sir W H Bragg, 'The structure of organic crystals', *Proceedings of the Physical Society of London*, 1920, 34:33–50.

[104] Ibid., pp. 33–4.

acenaphthene, and other organic crystals. Models assisted the analysis, and William acknowledged 'Mr Shearer for his assistance in examining crystals by the powder method, which method he has greatly improved; and Mr J Reid, for the labour and skill which he has devoted to making models'.[105]

William now took his findings to the chemistry community, with a presentation and paper on the structure of ice,[106] and an important lecture to the Chemical Society.[107] In the first he added his own deepening understanding of crystal science to existing X-ray analyses and other knowledge of ice crystals to clarify a clouded situation. In the discussion that followed, Professor Armstrong asked about the structure of liquid water, which he thought might contain oxygen ring systems, and also if William's structure of ice should not be 're-arranged' to maintain the oxygen chains instead of having the oxygens separated by hydrogen atoms. William responded with a message that he and Lawrence would have to repeat time and time again in the years ahead: 'The method tells us nothing about the liquid... but it is quite clear that the structure is *not* close-packed, and that the unit in the crystal is the atom, not the molecule. It cannot be said that a given hydrogen atom belongs to one oxygen atom more than any other. The oxygens are entirely separated by hydrogens'.[108] Using a number of organic crystals as illustrations, William repeated and stressed this message to the Chemical Society: 'It has sometimes been assumed that the unit [the unit cell of the crystal and its contents] is the chemical molecule, and much good experimental work has been described in terms of that conception. But such work can never bear its full fruit as long as it is rooted in an unsound idea. There is really no reason why the crystal unit should be identical with the chemical molecule; there is, in fact, every reason to expect the contrary'.[109] A year later he had to repeat the point yet again.[110]

In the Sixth Trueman Wood Lecture to the Royal Society of Arts in January 1923 William rehearsed the basis of the new method, and he illustrated it with numerous challenging examples from organic chemistry, with the spiral left- and right-handed forms of quartz, and with brief reference to the structures of iron and steel, aluminium and kaolinite.[111] If the printed version corresponds

[105] Ibid., p. 50.

[106] Prof. Sir W H Bragg, 'The crystal structure of ice', *Proceedings of the Physical Society of London*, 1922, 34:98–103.

[107] Sir William H Bragg, 'The significance of crystal structure', *Journal of the Chemical Society of London: Transactions*, 1922, 121:2766–87; William had, in fact, first introduced the Chemical Society and the Institute of Metals to the rudiments of X-ray crystallography during the war (W H Bragg, 'The recent work on X-rays and crystals and its bearing on chemistry', *Journal of the Chemical Society of London: Transactions*, 1916, 109:252–69; id., 'X-rays and crystal structure, with special reference to certain metals', *Journal of the Institute of Metals*, 1916, 16:2–13).

[108] W H Bragg, n. 106, pp. 102–3; regarding the liquid state and its transition on freezing, see letter C V Raman, 'The nature of the liquid state', *Nature*, 1923, 111:428, and William's reply, ibid.

[109] W H Bragg, n. 107, p. 2766.

[110] W H Bragg, 'X-rays and crystal symmetry', *Nature*, 1923, 112:618.

[111] Sir William Bragg, 'New methods of crystal analysis and their bearing on pure and applied science', *Supplement to Nature*, 1923, 111:iii–x.

to that delivered orally, one wonders what the audience made of it. Surely his address was more focused and gentler, as when he reported recent work by his colleagues, Müller and Shearer, on the linking of atoms in organic compounds, to a Faraday Society discussion on 'The electronic theory of valency'.[112]

These publications were the first results of a new period of research, but there were other immediate post-war events that gave William an even higher profile, both within the scientific community and with the wider public: namely, his many invited lectures on the contemporary state of physics. They signalled his impending arrival as 'not only one of the great figures of English science but also something of a national figure'.[113] The first made William a household name and arose from a Royal Institution invitation to deliver their 1919 Christmas Lectures. The series was entitled 'The World of Sound', and the six lectures—dealing in turn with the nature of sound, with music, and with the sounds of the town, the country, the sea, and war—showed his power of simple exposition based upon deep understanding and his affection for young people. He found their unaffected warmth and openness irresistible; he was a splendid grandfather too.[114] Later Lawrence would experience the same feelings and be an equally superb lecturer to children.

Much had happened in physics during the war and, indeed, since 1895 there had been a scientific revolution of a severity not seen since the seventeenth century. More recently Rutherford had 'spit the atom' in his Manchester laboratory for the first time.[115] Most physicists were disoriented and bewildered, and some were rumoured to have committed suicide, so severely had their stable and comfortable classical world been rent apart.[116] There was a desperate need for explanation, clarification, and reassurance, delivered with understanding. In Britain William Henry Bragg met this need. It is not clear when he found time to assimilate the new knowledge and convert it into the superb presentations he made during the years immediately after the war. His apprenticeship in Australia and the skills he learnt there now bore a splendid harvest.

William first accepted a joint invitation from the Faraday and Röntgen Societies to introduce an April 1919 discussion on 'The examination of materials by X-rays', particularly in engineering and metallurgy. He surveyed the basic features of X-ray photography and the use of X-rays to examine crystalline materials, the structures of which were extremely important in determining the quality of substances, especially iron and steel. He added: 'Hardly anything has so far been done, but it seems to be on the direct road for the would-be investigator of the properties of metals and alloys'.[117] In August he gave a

[112] W H Bragg, *Transactions of the Faraday Society*, 1923, 19:478–9.

[113] Andrade, n. 98, p. 289.

[114] G M Caroe, *William Henry Bragg, 1862–1942: Man and Scientist* (Cambridge: CUP, 1978), photograph facing p. 133.

[115] H A Boorse and L Motz, *The World of the Atom* (New York: Basic Books, 1966), vol. I.

[116] R McCormmach, *Night Thoughts of a Classical Physicist* (Cambridge, Mass.: Harvard University Press, 1982).

[117] W H Bragg, 'The examination of materials by X-rays', *Transactions of the Faraday Society*, 1919–20, 15:25–31.

lecture on 'Atomic Theories' as part of a summer school series devoted to recent developments in European thought. He noted that science, too, had a history of its own and that its progress made a connected story. The core of his lecture was the proposition that in science there were three main subjects of study—matter, electricity, energy—and that each could be measured quantitatively. Furthermore, although science had its own units for these quantities, nature herself has already chosen units for them: the atom for matter, the electron for electricity, and the quantum for energy. 'I will only add', he concluded, 'that the whole position of physics is indeed at this time of extraordinary interest'.[118]

In March the following year (1920) William delivered the Third Silvanus Thompson Memorial Lecture of the Röntgen Society on the topic, 'Analysis by X-rays'. He focused on an explication of the two new methods that X-rays provided for the detailed study of matter, X-ray spectroscopy and X-ray crystallography, and particularly the potential use of the powder method to discover the structures of 'ordinary materials of construction', of 'all the solid materials that live, grow and decay', of 'the substances of chemistry, inorganic and organic', and to understand 'the treatment of steel to various ends'.[119] In May he opened a discussion of 'X-ray spectra' for the Physical Society, during which he discussed the precision of the X-ray spectrometer, the accuracy of its measurements, and a few notable results, and concluded with references to the work of Moseley and of Sommerfeld.[120] In October the same year he gave a public lecture at University College to introduce a new course on the history of science. He emphasized that there was no finality in science and that its history was interesting because it traced the evolution of the great scientific conceptions, it told the story of its fruitful application, and it had great human interest.[121]

Early in 1921 William delivered the Twelfth Kelvin Lecture of the Institution of Electrical Engineers in an address entitled simply 'Electrons', and he returned to a concept that he had first conceived and developed in Adelaide; that there was 'the most remarkable connection between moving electrons and electromagnetic waves. One, it seems, can always call up the other, and the action obeys certain precise numerical laws'. The particle electron and the X-ray wave were intimately related. He began in characteristic style: 'As knowledge grows, the importance of the part played by the electron in the mechanics of the world becomes even clearer... we find ourselves able to express, quantitatively and with confidence, laws and relations which have been matters of vague surmise... While knowledge grows by experiment, theory is also busy. The attempts to coordinate the new discoveries are of singular interest because of their daring, their width, and their strength: because they are so often fruitful in prediction: and, not least perhaps, because they seem so

[118] W H Bragg, 'Atomic theories', in F S Marvin, *Recent Developments in European Thought* (Oxford: OUP, 1920), pp. 216–28.

[119] Sir W H Bragg, 'Analysis by X-rays', *Journal of the Röntgen Society*, 1920, 16:127–33.

[120] Sir W H Bragg, Opening remarks to discussion on 'X-ray spectra', *Proceedings of the Physical Society of London*, 1921, 33:1–9.

[121] Report of the lecture in *Nature*, 1920, 106:250.

often to be irreconcilable with each other'.[122] He noted that progress had been possible only because high vacuum and large voltages had become available. He focused on the energy relationship between X-rays and their initiating electrons, and he instanced wartime American work that had confirmed that the ratio of electron energy to maximum X-ray frequency was always the same, namely Planck's quantum constant. Furthermore, 'Just as the swiftly moving electrons excite X-rays, so X-rays when they strike any substance lose their energy, which now appears as the energy of moving electrons...Not only in the case of X-rays are these effects observed, but also in the case of light'.[123] He wondered what electron states gave rise to the X-ray lines emitted by hydrogen. 'Why not, as Bohr suggests, so many different orbits in which electrons can move round the central positive nucleus of the atom?', he answered, for 'During the last few years Bohr and Sommerfeld have led an inquiry into the possibilities of this theory which has produced very remarkable results'.[124] He had made similar points more briefly two years earlier.[125]

A few months later William delivered the 1921 Robert Boyle Lecture in Oxford and, in discussing 'Aether waves and electrons', he used one of his most delightful and arresting analogies to illustrate the dilemma he had been wrestling with for so long. He recalled Newton's corpuscular theory of light and its apparent overthrow by the work of Huygens, Young, Fresnel, and others, and then turned to the intimate connection between the particle electron and the X-ray wave. He suggested that:[126]

> the ultimate explanation of all optical problems must involve the recognition of corpuscular radiations, at times replacing and being replaced by the waves. Thus once more the corpuscular theory appears as a working hypothesis.
>
> But in its relation to the wave theory there is one extraordinary and, at present, insoluble problem. It is not known how the energy of the electron in the X-ray bulb is transferred by a wave motion to an electron in...any other substance on which the X-rays fall. It is as if one dropped a plank into the sea from a height of 100 ft and found that the spreading ripple was able, after travelling 1,000 miles and becoming infinitesimal in comparison with its original amount, to act upon a wooden ship in such a way that a plank of that ship flew out of its place to a height of 100 ft. How does the energy get from one place to the other?

This was, indeed, a good question. Arthur Compton had been studying the scattering of X-rays for more than five years and had reached a similar impasse.[127] He had received his AM and PhD degrees at Princeton University,

[122] Sir William Bragg, 'Electrons', *Nature*, 1921, 107:79–82, 109–11.
[123] Ibid., pp. 81 and 82 respectively.
[124] Ibid., p. 109.
[125] W H Bragg, 'Recent work on the spectra of X-rays', *Science Progress*, 1919, 13:569–85.
[126] Sir William Bragg, 'Aether waves and electrons', *Nature*, 1921, 107:374.
[127] R H Stuewer, *The Compton Effect: Turning Point in Physics* (New York: Science History Publications, 1975); J G Jenkin, 'G E M Jauncey and the Compton Effect', *Physics in Perspective*, 2002, 4:320–32.

where his interest in X-ray diffraction and scattering had begun. Three short-term appointments followed, including at the Cavendish Laboratory, where he developed 'an attitude of complete confidence in the universal validity of classical electrodynamics', built upon a 'negative predisposition towards the quantum theory'.[128] He proposed that the scattering electrons were, in turn, large spheres, flexible spherical shells, and finally rings. The two forms of secondary rays excited by energetic γ-rays falling upon a scattering material Compton defined as 'truly scattered' (for those of unchanged wavelength) and 'fluorescent' (for those of longer wavelength). Seeking a place to 'do thinking' and pursue his own research, he accepted an invitation to become Wayman Crow Professor of Physics at Washington University, St Louis, USA.[129] 'To help him in his work, Compton soon brought in G E M Jauncey', Stuewer records.[130] It was a fateful decision.

Eric Jauncey had an unusual background. Born in 1888 into humble circumstances in Adelaide, South Australia, he had won a scholarship to Prince Alfred College for his final years of secondary education. He completed the Senior (matriculation) public examination of the University of Adelaide in 1904, being first in the state in physics and fourth in chemistry. Aged just sixteen and rather too young for university life, Jauncey returned to school, taking the Higher public examination and being placed first in applied mathematics and second in both physics and chemistry. Facing significant tuition fees at the university, Jauncey was granted a Physics Department cadetship, which exempted him from fees and paid a small living allowance in return for part-time work in the department's workshop and laboratories. Here there was daily contact with academic and technical staff, and for Jauncey this was particularly fortunate, for his first two years coincided with the last two years of William Bragg's tenure in Adelaide, when William was developing his neutral-pair hypothesis for the material nature of radiation. Jauncey completed a first-class honours degree in physics in 1910.[131]

Two periods of research under an 1851 bursary and scholarship followed: the first in Adelaide, the second for one year (1912–13) under William Bragg at Leeds.[132] This was again an extraordinary important time, when William was first developing his X-ray spectrometer. Eric Jauncey was assigned two projects: to measure 'the absorption coefficients of homogeneous X-rays in various gases', and 'the properties of X-rays regularly reflected from mica'. These were also the titles of the two papers Eric presented to the 1851 Scholarships Committee in an application for an extension to his scholarship, both projects being part of William's program to develop the instrument and begin a new research programme. The Committee declined, principally because William judged that 'Jauncey is an enthusiastic, persevering student [but] I cannot say that he

[128] Stuewer, ibid., pp. 97, 98.
[129] Ibid., p. 158.
[130] Ibid., p. 160.
[131] Jenkin, n. 127, pp. 325–7.
[132] *Record of the Science Research Scholars of the Royal Commission for the Exhibition of 1851* (London: the Commissioners, 1961).

has originality or exceptional merit', as was required for the extension.[133] The judgement now seems harsh, and from 1913 until 1920 Jauncey held only a series of minor academic appointments in North America, where he published very little but did complete an MS degree at Lehigh University in 1916.[134] Now his luck changed.

Jauncey was an ideal appointment for Compton's new department: he was similarly youthful, with formative English experience, had taught widely, had done research with X-rays, and was in America and available. He arrived at St Louis in mid-1920 and his research productivity soon surged alongside Compton's, especially concerning X-rays and their interaction with matter (see Figure 18.2). Many of his papers, like Compton's, were single-author with no acknowledgements, but on notable occasions they thanked each other.[135] Compton's early experiments at St Louis on X-ray scattering confirmed, he thought, that 'the quantum relat[ion] does not enter', that the 'truly scattered' radiation was unchanged in energy, and that both X- and γ-rays produced his new type of 'fluorescent' radiation, dependent on the exciting radiation but not the scattering material. Further experiments followed, many prompted by a large literature review of 'secondary radiations produced by X-rays' requested by the National Research Council. Compton was thus led to consider again the relative merits of the classical and quantum theories of radiation, and he came to recognize that Thomson's classical theory and his own large-electron models could not encompass all the existing date. He recognized a small but distinct change in wavelength between primary and secondary spectra but still refused to embrace a quantum interpretation.[136]

Then suddenly Compton changed his mind. Carl Eckart, a graduate student in the department, explained what happened.[137] He was assigned a desk in Jauncey's office, which communicated with Compton's office next door, and each afternoon he joined Compton and Jauncey while they discussed the research work of the department, which consisted almost entirely of the study of scattered X-rays. A detailed examination of Schrödinger's theoretical work was undertaken, and during one afternoon in November 1922 Compton and Jauncey recognized that the essential element in Schrödinger's papers was the assignment of a momentum to the light quantum. In the morning of the very next day Compton took over Jauncey's class on electrostatics and, with Jauncey and Eckart in the audience, presented his derivation of the quantum scattering of radiation for the first time. An X-ray quantum of radiation underwent a discrete change in wavelength when it experienced a billiard-ball-like

[133] Jauncey file (no. 355), Archives of the Royal Commission for the Exhibition of 1851, Imperial College, London.

[134] Jenkin, n. 127, pp. 325–7.

[135] See ibid., p. 327, for details.

[136] Ibid., pp. 321–3.

[137] J L Heilbron interview with C H Eckart, 31 May 1962, Archive for the History of Quantum Physics, Niels Bohr Library, American Institute of Physics, Maryland, USA; for Eckart's life and work see W Munk and R Preisendorfer, 'Carl Henry Eckart, 1902–1973', *Biographical Memoirs of the National Academy of Sciences*, 1976, 48:195–219.

Fig. 18.2 George Eric Macdonnell Jauncey (1888–1947), no date. (Courtesy: Mrs H. Burns.)

collision with a single atomic electron, a phenomenon that became known as 'the Compton effect' and for which Compton shared the 1927 Nobel Prize in Physics.[138]

Eckart was convinced that 'in retrospect the discussions between Compton and Jauncey seem to have been necessary preliminaries to the derivation of these equations'.[139] When Jauncey later developed a serious illness and died in May 1947, a memorial service was held in Wilson Hall at Washington University, at which four of the university's most senior scholars spoke. Especially notable were the words of Arthur Compton, now Chancellor of Washington University. He spoke of Jauncey's 'simple, straightforward and direct' approach to achieving 'results that were understandable and convincing', and of his 'extraordinary

[138] Jenkin, n. 127, pp. 328–30.
[139] Heilbron interview, n. 137.

courage'. He continued:[140]

> Then there was his live concern with the corpuscular interpretation of X-rays. In this discussion he took a central part from the very beginning. He had studied with Sir William Bragg in England when Bragg had declared that, while on Mondays, Wednesdays and Fridays we talk of X-rays as waves, on Tuesdays, Thursdays and Saturdays we use them as particles. He brought these ideas into the Washington University school of physics of the 1920s, when elsewhere in our country the problem seemed of little importance. Out of the discussion and experiments to which he thus contributed a vital share came the first clear evidence that the X-rays did indeed have, at the same time, the properties of waves and particles. If X-rays, then also light... Thus was laid the basis for... perhaps the most dramatic change in scientific thought... since the time of Newton.

Compton was here making two important acknowledgements: first to his considerable personal debt to Jauncey for his contribution to the X-ray experiments and their interpretation, and second to a clear line of intellectual descent from William Bragg through Eric Jauncey to the quantum interpretation of the Compton effect. Neither Eric nor William ever claimed a major role in the discovery, although Eric later remarked wistfully to Eckart that the preceding evening he had promised to take his wife to the movies, and he 'often wondered whether, if Compton had gone to the movies and he [Jauncey] had stayed at home, he would have derived these equations'.[141] For William, his insistence on the dual nature of radiation—'wave–particle duality' as it came to be called—had finally been vindicated.

[140] 'Addresses made at the Memorial Service honoring Mr George Eric Macdonnell Jauncey, in Wilson Hall, Washington University, Saint Louis, May 23, 1947', Washington University Archives, St Louis, Missouri, USA.

[141] Heilbron interview, n. 137.

19
Epilogue

Nearly twenty years of William's productive life remained and there was more than half of Lawrence's equally fruitful journey to be travelled, but their close collaboration was at an end. Lawrence, mature and approaching the zenith of his powers, was following his own star, although conversations with his father would still form an important part of his family and professional life. My account of this most extraordinary collaboration is at an end; only an outline of the rest of this remarkable story, together with a few concluding remarks, remains.

William: Royal Institution and Royal Society

William was not happy at University College: 'WHB, used... to having the vast resources of the Admiralty and the BIR behind him, saw the obstruction of University College finances as petty and ridiculous. He campaigned against the restrictions and this caused jealousy from those who were competing for funds. College politics depressed him'.[1] The Australian days of 'sunshine and fresh, invigorating air' were gone; he had not been really happy or contented since returning to England. He would move once more, seeking the family atmosphere he remembered. Sir James Dewar, the Director of the Royal Institution (RI), died early in 1923. The vitality of the Institution had ebbed away, and when Rutherford declined the position he recommended his long-standing colleague and friend. William sensed an opportunity: to consolidate his research in the Davy–Faraday Laboratory, to continue the honoured tradition of lectures that worked for understanding between the sciences and for enlightenment of the general public, to rebuild the Institution according to his own vision, and to regain control of his own destiny. There was a rent-free Director's flat within the Institution, although the salary was low and William could not have accepted had Uncle William not left him money 'to found a family'. Nor would the Institution's affairs be free of the politics he so wished to avoid.[2]

[1] G M Caroe, *William Henry Bragg, 1862–1942: Man and Scientist* (Cambridge: CUP, 1978), p. 92.
[2] Ibid., ch. 7, p. 95.

William did succeed in fulfilling his ambitions, however (see Figure 19.1). The Friday Evening Discourses were reinvigorated, and in the flat Gwendoline introduced dinner parties that were a great success. The Christmas Lectures for children were further enhanced to considerable acclaim and William gave three more himself, each of which became an influential book: *Concerning the Nature of Things*, *Old Trades and New Knowledge*, and *The Universe of Light*.[3] Indeed, it is astonishing how often the first two inspired later generations to a life in science. Copies of them were purchased for the teenage Dorothy Crowfoot (later Hodgkin)—biochemist, crystallographer, and Nobel laureate in Chemistry (1964)—and its 'elegant introduction... excited the impressionable Dorothy beyond measure... "I was fascinated by the way this knowledge was acquired... I began to see X-ray diffraction as a means to exploring many of the questions raised but left unanswered by school chemistry"'.[4] Similarly, Primo Levy, professional chemist, great Italian writer, and one of the most profound and haunting commentators on the Holocaust, recalled, regarding the first book:[5]

> I owe a great deal to this book. I read it by chance at the age of sixteen. I was captivated by the clear and simple things that it said, and I decided I would become a chemist. Between the lines I divined a great hope: the models on a human scale, the concepts of structures and measurement, reach very far, towards the minute world of atoms and towards the immense world of the stars; perhaps infinitely far? If so, we live in a comprehensible universe, one accessible to our imagination, and the anguish of the dark recedes before the rapid spread of research.
>
> I would become a chemist: I would share Bragg's faith (which today seems very ingenuous). I would be bound up with him, and with the legendary atomists of antiquity, against the discouraging and lazy herd of those who see matter as infinitely, fruitlessly, tediously divisible.

Staff and research workers at the RI were entertained by William, Gwendoline, and Gwendy at the family's Watlands country cottage and did, indeed, become part of an extended family. William pioneered radio broadcasts on science for the British Broadcasting Corporation,[6] and he organized a major rebuilding programme at the RI. However, at Easter 1929 Gwendoline fell seriously ill, and she died in September. It was another body blow, and William could not bear to talk about it or even show his grief, he concealed his feelings so completely. He was fortunate, however, to have thereafter the companionship and social skills of his daughter to soften the impact.[7] William commissioned a

[3] Sir William Bragg, *Concerning the Nature of Things* (London: Bell & Sons, 1925); id., *Old Trades and New Knowledge* (London: Bell & Sons, 1926); id., *The Universe of Light* (London: Bell & Sons, 1933).

[4] G Ferry, *Dorothy Hodgkin: A Life* (London: Granta, 1998), pp. 29–30.

[5] P Levi, *The Search for Roots: A Personal Anthology*, P Forbes transl. (London: Penguin, 2001), ch. 4 ('To see atoms'), pp. 31–7, 31.

[6] J Faraday, 'The Braggs and broadcasting', *Proceedings of the Royal Institution of Great Britain*, 1974, 47:59–85.

[7] Caroe, n. 1, ch. 7.

Fig. 19.1 William Bragg, a Bassano portrait, 1926. (Courtesy: State Library of South Australia, SLSA: B 3991.)

large sculpted marble plaque for the entrance hall of the RI, whose text speaks of Davy and Faraday and whose symbols are a child surrounded by ears of corn and birds in flight. Perhaps it represents the fertile imagination of the child, leading to discoveries of practical and intellectual scientific fruits; I see it as a tribute to Gwendoline, who gave him their children, was the centre of their home, and lifted his spirits to the sky.[8]

Hunter has discussed the relationship between Gwendoline and her surviving son. Lawrence appreciated his mother's social skills and the important role she played within the family and in her husband's institutions, and he admired her art and her desire to maintain family harmony. But he noted her lack of education and her limited belief in it; and his sister recalled that Lawrence was

[8] The plaque was made by Ernest Gillick, but I have not been able to find his or William's rationale for its features.

'inept with her, too earnest, would take her exaggerations too seriously when he ought to have laughed and challenged them. They gave me a lot of trouble'.[9] Hunter concluded that, 'With the exception of painting, she had little in common with her intellectual and introspective son'.[10] I believe this is a narrow and distorted view. Lawrence was particularly annoyed that his mother, deliberately or unwittingly, denied his sister a university education, and that his father refused to intervene to correct the situation;[11] and his independent personality did not fit easily with his mother's. On the other hand, he did share her artistic ability and temperament, and as a young man he wrote to her often to seek her advice and to reassure her of his wellbeing. He was especially conscious of her anxiety during The Great War, writing to her frequently, and he hurried home on the death of his brother to comfort her. When things got especially difficult Lawrence wrote to his mother at length saying, 'I always take my troubles to you'.[12] It was a complex relationship, not a simple one, and there were difficulties; but there was, I believe, a deep, loving, and life-long bond.

William continued to study, publish, and speak at the Royal Institution; Kathleen Lonsdale lists 140 items from this period. Andrade mentions specifically the provision of new experimental facilities, William's interest in liquid crystals and the extra reflections due to irregularities in crystal structure, and his invitation to Schrödinger to deliver a series of lectures on the new wave mechanics.[13] In 1931 William became a member of the Order of Merit, the most prestigious British honour still in regular use and the personal gift of the Sovereign, restricted to twenty-four members. The Royal Society honoured him with its highest award, the Copley Medal, in 1930 and its Presidency for 1935–40. He accepted extensive leadership roles during the early years of the Second World War and fulfilled them with 'grace, dignity and efficiency' in the eighth decade of his life.[14] On 10 March 1942 he went to bed feeling unwell; he died two days later, in his eightieth year. Andrade gave a warm summary of William's life and personality:[15]

> Bragg's nature was simple, straightforward and tenacious—incidentally, of course, he was a man of genius... [his] long period of apprenticeship probably had a profound influence on his work... He was a very great experimenter, who never wasted time on the trivial or hid difficulties under the graceful veils of mathematical obscurity... His work, like his personality, was simple yet profound, sincere and compelling... Bragg

[9] Letter G M Caroe to Sir David Phillips, 17 May 1979, D C Phillips Collection, Bodleian Library, Oxford.

[10] G Hunter, *Light is a Messenger: The Life and Science of William Lawrence Bragg* (Oxford: OUP, 2004), p. 105.

[11] Caroe, n. 1, pp. 100–1.

[12] Letter W L Bragg to G Bragg, 6 February 1918, RI MS WLB 37B/3/1.

[13] E N da C Andrade, 'William Henry Bragg, 1842–1942', *Obituary Notices of Fellows of the Royal Society*, 1942–44, 4:276–300, with a bibliography compiled by K Lonsdale.

[14] Ibid., p. 290.

[15] Ibid., pp. 291–2; there were also a number of heartfelt obituary notices in *Nature*, 1942, 149:346–52.

> preserved all through his fame many of the more admirable characteristics of youth... [He] was a man of very strong family feeling, who was never happier than with his children, and later his grandchildren. He took a particular pride in the achievements and career of his brilliant son. It was always a delight to his hearers to note the affection that came into his voice when he found occasion to deal with work which had been carried out by 'my boy'... There was nothing narrow about Bragg's interests... He was an ornament, not only of English science, but of English learning, a great teacher and a good man... With him went an outstanding representative of a great period of English physics.

William could never understand, his daughter said, how he had reached such an exalted position. 'He got very tired', she continued, 'but carried on determinedly. He would walk to Burlington House from Albemarle Street with his slow old countryman's walk, the slight lurch copied perhaps from the men on his father's farm long ago'.[16] Lawrence wrote: 'Gwendy and I were invited by the Royal Society to write our personal memories of him in 1962, the hundredth anniversary of his birth, and I put into that all my dearest and most vivid impressions of him'.[17] After rehearsing his achievements, the tribute speaks of 'his simplicity and a cool wisdom and gentleness... his diffidence about his own achievements... his love of quietude... he always felt a bit apart; he had no intimate friends and seemed not to need them. A pleasant family atmosphere was enough'. It continued, 'As for the things he enjoyed, after family life, work and his country garden, a sea voyage gave him most pleasure... He enjoyed simple music... he read widely... he painted also... He was a craftsman always in big things and small... Always on Christmas Eve he went out to buy the family the largest pineapple he could find'. Are these the words of a son permanently at odds with his father? I think not. The article concluded with William's own words: 'in all our lives, in all we work and strive for, it is of first importance to know as much as we can about what we are doing, to learn from the experience of others and, not stopping at that, to find out more for ourselves so that our work may be the best of which we are capable. That is what Science stands for'.[18]

Lawrence: Cambridge and London

Lawrence's career still had rocky roads to travel. In 1930 the pressures of overwork, in teaching, research, and administration, coupled with conflicting loyalties to Manchester University and his wife, who wanted to escape the city's industrial grime, all conspired to precipitate 'some kind of nervous

[16] Caroe, n. 1, p. 109.

[17] Sir Lawrence Bragg and Mrs G M Caroe, 'Sir William Bragg, F.R.S. (1862–1942)', *Notes and Records of the Royal Society of London*, 1962, 17:169–82; Lawrence also wrote a fine tribute to his father a little earlier: Sir Lawrence Bragg, 'William Henry Bragg', *New Scientist*, 1960, 7:718–20.

[18] Ibid., pp. 180–2.

breakdown'.[19] He was planning a new building, writing a new book—published in 1933 as *The Crystalline State: A General Survey*[20]—and preparing new lectures for industrial physicists. There was also the onset of The Great Depression and its effects on industrial England. He went to see his father, 'but it was one of those times when it was quite impossible to get him to talk'.[21] Characteristically, however, William, with Alice, devised a plan for Lawrence to leave Manchester for a term, and he spent the first three months of 1931 in Munich, which 'did a great deal to put me right again'.[22]

It was also at this time that the place of crystallography in the scientific world came into focus and under question: was it physics or chemistry or an entirely new discipline? Lawrence felt the pressure acutely, writing to Harold Robinson, his close Great War colleague, 'I always wish I were in a field which had more right to be called Physics',[23] and to Rutherford, 'I do not want to label myself a crystallographer as against a physicist'.[24] Other writers have seen this in terms of Lawrence's insecurity, but I see it differently. In the second half of the twentieth century physics spawned new fields of research in the boundary areas between it and kindred sciences: the transistor and the electronic revolution, computing, crystallography, biophysics, materials science, and medical imaging, to name only some. Established physicists saw such emerging fields as 'not really physics', and all these areas have now left the discipline to build separate homes for themselves. Lawrence did, indeed, wish to be regarded as a physicist, but in a number of ways to be outlined below it rejected him. Immediately after World War Two crystallographers found it necessary to form the International Union of Crystallography, followed by the inauguration of the international journal, *Acta Crystallographica*, the first of an ongoing series of international congresses, and the first volume of the *International Tables for X-ray Crystallography*.[25] The British Crystallographic Association was inaugurated in April 1982.[26]

Lawrence now reconsidered his academic priorities. He encouraged his colleagues to continue existing lines of research, began new investigations himself, and became more interested in scientific affairs outside the university. He served on the Council of the Royal Society from 1931 to 1933 and was awarded its Hughes Medal for a major discovery in the physical sciences, Raman, Geiger, and Maurice de Broglie being his immediate predecessors.[27] In his

[19] W L Bragg, Autobiographical notes, p. 68.

[20] W L Bragg, *A General Survey* (London: Bell & Sons, 1933), being vol. 1 of Sir W H Bragg and W L Bragg (eds), *The Crystalline State*.

[21] W L Bragg, n. 19.

[22] Ibid., p. 69.

[23] Letter W L Bragg to H R Robinson, 31 January 1930, RI MS WLB 77D/13.

[24] Quoted in Sir David Phillips, 'William Lawrence Bragg, 1890–1971', *Biographical Memoirs of Fellows of the Royal Society*, 1979, 25:75–143, 104.

[25] H Kamminga, 'The International Union of Crystallography: its formation and early development', *Acta Crystallographica*, 1989, A45:581–601.

[26] D Blow and S Wallwork, 'Prehistory of the British Crystallographic Association', *Notes and Records of the Royal Society of London*, 2004, 58:177–86.

[27] *The Record of the Royal Society of London* (London: Royal Society, 1940), 4th edn, pp. 353–4.

work on diopside, $CaMg(SiO_3)_2$, Lawrence had obtained a complete series of reflection measurements for all the crystal planes parallel to each axis. He had already used the orders of reflection from a plane to form a one-dimensional Fourier series, which represented the density parallel to that plane and, with the help of his father's 1915 Bakerian lecture and new personal discussions, he now extended the analysis. As a result, 'behold, there were the calcium, magnesium, silicon and oxygen atoms clearly shown by peaks in the Fourier summation'.[28] Lawrence was later sorry that he did not insist on including his father's name on the resulting paper; the method 'has been an essential feature of X-ray crystallography ever since'.[29] The major new direction of Manchester research was now the study of the structures of metals and alloys and of alloy phase diagrams, and for Lawrence himself the order–disorder transformation, all of which he later assessed as: 'It is no exaggeration to say that the principles of metal chemistry for the first time began to emerge'.[30]

Lawrence also became increasingly involved in the presentation of science to the public, including his first radio broadcasts—six lectures on 'Light'—and the Royal Institution's Christmas Lectures for 1934 on 'Electricity'.[31] In his lecture to the Royal Society of Edinburgh in 1935 he mentioned the attraction of 'X-ray investigation of structures produced by living matter', and that it would be 'perhaps the most interesting field of all'.[32] It was a sign of things to come.

In 1937 Lawrence was invited to accept appointment as Director of the National Physical Laboratory (NPL), and he agreed; it was time for a change and Alice would have the pleasure of a splendid house in the south of England. He left Manchester with some regret, however. He remembered with gratitude the kindness of his colleagues and the inspiring atmosphere of the university.[33] It, in turn, acknowledged his research achievements, his kindness and modesty, his recognition of the work of others, and the approachability that made him a trusted friend of his staff, research colleagues, and students.[34] But Lawrence's time at the NPL was brief and disappointing; he stayed less than a year. Rutherford died unexpectedly in October 1937, as Lawrence was leaving Manchester, and there was much speculation regarding who would succeed him as Cavendish Professor at Cambridge, 'the supreme position in British physics'.[35]

The electors made a bold decision and offered the position to Lawrence. William opposed the move; he was not comfortable at Cambridge—'clever talk alarmed him and smutty wit worried him . . . Trinity High Table conversation,

[28] W L Bragg, Autobiographical notes, p. 64.

[29] Phillips, n. 24, p. 104.

[30] W L Bragg, 'The development of X-ray analysis', *Proceedings of the Royal Society of London*, 1961, A262:145–58, 153.

[31] W L Bragg, *Electricity* (London: Bell & Sons, 1936).

[32] Quoted in Hunter, n. 10, p. 121.

[33] W L Bragg's final Manchester Physics Department Annual Report, quoted in Phillips, n. 24, p. 109.

[34] University of Manchester Senate, minutes of meeting held on 4 November 1937, John Rylands University Library.

[35] Quoted in Hunter, n. 10, p. 127.

with its brilliance and College politics, silenced him'.[36] Lawrence, however, accepted the position, although his reception was again mixed. Succeeding Rutherford for a second time was a daunting task. Phillips said, 'There was undoubtedly some consternation, especially among nuclear physicists';[37] and Hunter was particularly blunt, 'Under Rutherford, the Cavendish Laboratory had become a leading centre for nuclear physics...The selection of a non-nuclear physicist—indeed, someone who was felt not to be a physicist at all—could be seen as jettisoning Rutherford's legacy'.[38] Brian Pippard, who would succeed to the chair later, recalled: 'W L Bragg's election...was taken by many as a threat to the great tradition of fundamental physics research established by J J Thomson and especially Rutherford...The choice of a crystallographer, however distinguished, was a blow to many hopes'.[39]

On the other hand, some welcomed the appointment, including Douglas Hartree, Patrick Blackett, Edward Appleton, and J D Bernal. An editorial in the journal *Nature* was very positive: 'The Cavendish Laboratory is now so large that no one man can control it all closely, and Bragg's tact and gift of leadership form the best possible assurance of the happy cooperation of its many groups of research workers, while his brilliant lectures and personal charm ensure his success as a teacher of undergraduates'.[40] Most commentators had forgotten that J J Thomson's appointment to the chair had been equally controversial, although the electors were subsequently seen to have been extraordinarily perceptive.[41] Time would tell whether they had judged the situation correctly this time.

Lawrence's first task was to deal with major staff changes. Philip Dee and Norman Feather continued the nuclear physics research, but Appleton and Bernal and his team of biological crystallographers left the Cavendish. John Cockroft took the Jacksonian chair of natural philosophy, and Ratcliffe, who would later become Lawrence's indispensable right-hand man, took over atmospheric research. Lawrence brought in Alfred Bradley, Henry Lipson, Egon Orowan, and Paul Ewald to continue the work in crystallography, joining Max Perutz, who was the sole remaining member of Bernal's group. Perutz, a refugee student from Vienna, was working on the structure of haemoglobin protein crystals and Lawrence obtained a grant from the Rockefeller Foundation to enable him to continue: 'Some fortunate intuition made me feel that this line of research must be pursued, although it seemed absolutely hopeless to think of getting out the structure of so vast a molecule'.[42] A little later

[36] Caroe, n. 1, p. 107.
[37] Phillips, n. 24, p. 110.
[38] Hunter, n. 10, p. 128.
[39] Sir Brian Pippard, 'Bragg—Cavendish Professor', in J M Thomas and Sir David Phillips (eds), *Selections and Reflections: The Legacy of Sir Lawrence Bragg* (Northwood: Science Reviews, 1990), pp. 97–100, 97–8.
[40] Discussed and quoted in Hunter, n. 10, p. 128.
[41] G P Thomson, *J J Thomson and the Cavendish Laboratory in his Day* (London, Nelson, 1964), p. 73.
[42] W L Bragg, Autobiographical notes, p. 92.

Lawrence began to develop his bubble-raft model of the structures of metals and their deformation and dislocation, a model that illustrated dramatically Lawrence's 'uncanny gift for visualising atomic arrangements and expressing their essence in simple models'.[43] Richard Feynman highlighted the model in his famous *Lectures on Physics*.[44]

For a second time a major world war now interrupted Lawrence's research career and the work of the Cavendish Laboratory, although teaching continued. During 1941 Lawrence spent eight months in Canada as the British Scientific Liaison Officer, he was an adviser to the British Navy on asdic, and as early as 1937 he had been asked to report on the state of sound-ranging in the Army. He was 'appalled by the "whiskers" which had grown on the Sound Ranging gear' and 'made the mistake, of which I have often repented, of throwing my weight about far too much. A little gentle pressure would have been much more effective... Once the apparatus got to the front', however, 'it reverted very much to the simple affair we had used in World War I'.[45] There, despite the anticipation that it would not be of much use, sound-ranging proved to be a winner, and 'both in the Sicilian and D-Day landings the units were sent in'.[46] In the New Year Honours for 1941 Lawrence was knighted, to the delight of his father, who wrote to Lorna Todd in Adelaide: 'Isn't that fine?... He will have to be Sir Lawrence... I am so very glad for his sake. In spite of all care, people mix us up and are apt to give me a first credit on occasions when he should have it'.[47] But William's wish to end the confusion was in vain; scientists, historians, and the wider public still fail to distinguish them. Perhaps the most notable example is the magisterial *Sources for History of Quantum Physics*, where the main 'Author Catalog' lists W H Bragg but not W L Bragg, the 'Index of Names' includes W L Bragg but not W H Bragg, and in each there is, in fact, a mixture of references to both father and son![48]

Lawrence's wife, Alice, was heavily involved in the Women's Voluntary Services Association during the war, and later took a very active part in public affairs. She became a town councillor and then Mayor of Cambridge, a magistrate and a member of the Lord Chancellor's Commission on Legal Aid, and Chairman of the Cambridge Marriage Guidance Council, a member of Lord Denning's Royal Commission on Marriage and Divorce, and then Chairman of

[43] W L Bragg, 'A model illustrating intercrystalline boundaries and plastic flow in metals', *Journal of Scientific Instruments*, 1942, 19:148–50; Sir Lawrence Bragg and J F Nye, 'A dynamical model of a crystal structure', *Proceedings of the Royal Society of London*, 1947, A190:474–81; the quotation is from Phillips, n. 24, p. 113.

[44] R P Feynman, R B Leighton, and M Sands, *The Lectures on Physics* (Reading, Mass.: Addison-Wesley, 1963–5), vol. 2, section 30–9, which reproduced the 1947 paper in full and 18 of its 21 plates.

[45] W L Bragg, Autobiographical notes, pp. 91–92.

[46] Ibid., p. 97; during WWII the equipment and methods of sound-ranging were apparently not greatly changed from WWI forms (G Lennon, Adelaide, personal communication, a man who was there).

[47] Quoted in Caroe, n. 1, p. 177.

[48] T S Kuhn, J L Heilbron, P L Forman, and L Allen (eds), *Sources for History of Quantum Physics: An Inventory and Report* (Philadelphia: American Philosophical Society, 1967).

National Marriage Guidance. She sat on numerous committees and boards and was awarded the CBE in 1973.[49] She built a happy home for the family in West Road, Cambridge, 'a place of which we became very fond, and to the children it has always figured as the family home... They brought their friends to West Road, which became quite a centre for young people in Cambridge'.[50]

'I had a very difficult time indeed at the Cavendish when the war ended', Lawrence remembered, when Cockroft and Fowler, for different reasons, were absent but had not resigned, so that the two key posts in the laboratory were essentially unfilled and likely candidates accepted positions elsewhere. Fortunately 'Ratcliffe... was magnificent, and did the main work of organising the classes', and he was excellent in negotiations with boards and committees. Furthermore, 'it was he who foresaw the immense importance of the radio waves coming from space, which had just been discovered', and 'he infected me with his enthusiasm'. When Martin Ryle came to consult him about his future, therefore, Lawrence was able to persuade him to stay with Ratcliffe, 'so he went on to explore the possibilities of radio-astronomy and became a world pioneer in this fascinating field [Nobel Prize for Physics, 1974]'.[51]

Lawrence then recalled, 'It was a strange time at the Cavendish because Rutherford's influence had been removed... It was as if some mighty forest tree had fallen, and saplings hitherto starved of light and nourishment were beginning a more normal development', with radio-astronomy, electron microscopy, and low-temperature physics blossoming. And then, in what is now seen as a remarkable understatement, he wrote, 'But probably the work which in future years will be regarded as the outstanding contribution of the Cavendish Laboratory in these after-war years was the start of the investigation of biological molecules by X-rays'. The Medical Research Council agreed to support the protein research, and for a long time the results were 'very meagre indeed... Perutz at times became quite discouraged... Why I continued to be optimistic I shall never understand'.[52] In 1946 John Kendrew joined the group but, having realized that a helical structure was likely, they failed to solve the protein structure because of an ignorance of one piece of chemistry and an erroneous assumption regarding the helix. Lawrence never forgave himself and determined in future to seek the best chemistry advice before publishing. By the early 1950s a major part of the haemoglobin puzzle had been solved, and Lawrence had played a major part; it was the first substantial result of the Medical Research Council's Unit for the Study of Molecular Structure of Biological Systems at the Cavendish Laboratory.[53]

In 1949 Francis Crick came to the Cavendish to complete the PhD he had nearly finished at University College London before the war and a German

[49] Biographical notes prepared by her elder son, Stephen Bragg, and text of funeral oration delivered by John Keast-Butler, April 1989, in the possession of the author.

[50] W L Bragg, Autobiographical notes, p. 90.

[51] Ibid., pp. 111–12.

[52] Ibid., pp. 117–18.

[53] Phillips, n. 24, pp. 120–2.

bomb destroyed his laboratory, and in October 1951 James Watson knocked on Crick's front door. Much has been written about the subsequent events, leading to the discovery of the structure of deoxyribonucleic acid, DNA. Phillips has described it succinctly:[54]

> Bragg played no direct part in the study of DNA; indeed, at one stage he actively discouraged Crick and Watson from working on it in an attempt to avoid competition with the M.R.C. Unit at King's College London, but Watson has given a colourful and irreverent account of his growing appreciation of its importance, his encouragement at a critical stage, and his quick comprehension of the result.[55] Watson also noted Bragg's concern that the chemistry underlying the model should be checked by A R Todd... the leading expert on the chemistry of nucleic acids. When Todd approved, he was more than willing to promote rapid publication. As Watson saw it: 'The solution to the structure was bringing genuine happiness to Bragg. That the result came out of the Cavendish and not Pasadena [by Pauling] was obviously a factor. More important was the unexpectedly marvellous nature of the answer, and the fact that the X-ray method he had developed forty years before was at the heart of a profound insight into the nature of life itself'.

Lawrence retired at the end of 1953 and, as at Manchester, his tenure of the Cavendish chair ended in triumph. It was crowned by the Nobel awards in 1962, when Perutz and Kendrew shared the chemistry prize for the structures of the haemoglobin and myoglobin protein molecules, and Crick, Watson, and Maurice Wilkins (from King's College) shared the prize in medicine and physiology for the discovery of the structure of DNA. Lawrence was seriously ill at the time and had undergone a major operation; these awards greatly lifted his spirits and hastened his recovery, his family said.[56]

There has been a great deal of discussion of Watson's personal account of these events, *The Double Helix*, but far less of the role Lawrence played in its publication. Initially entitled *Lucky Jim*, the book was refused publication by Harvard University Press when Crick and Wilkins raised serious objections and numerous others, close to and distant from the story, were also severely critical. It was, they said, historically inaccurate and self-glorifying and, more particularly, gratuitously hurtful in its characterization of, and offhand remarks about, many people, not least Rosalind Franklin. Watson then modified some of the offending passages and added an epilogue, and the book was published by Atheneum.[57] Lawrence had, in fact, suggested in 1965 that Watson write

[54] Phillips, n. 24, p. 122; the classic accounts are: R Olby, *The Path to the Double Helix* (Seattle: University of Washington Press, 1974), and H F Judson, *The Eighth Day of Creation: Makers of the Revolution in Biology* (London: Jonathan Cape, 1979).

[55] J D Watson, *The Double Helix: A Personal Account of the Discovery of the Structure of DNA* (New York: Atheneum, 1968).

[56] Personal communications; also see comment by Lady Bragg in '50 years a winner', BBC TV programme, 1965.

[57] G S Stent, *James D Watson, The Double Helix, A New Critical Edition, including Text, Commentary, Reviews and Original Papers* (London: Weidenfeld and Nicolson, n.d.).

his version of the discovery, and Hunter has noted that he did not make the same suggestion to Crick or Wilkins.[58] Lawrence seems to have preferred the younger man, since Crick clearly irritated him from time to time, but his suggestion probably related more directly to his own experience. He was advising Watson to record his own contribution so that it would not be overlooked in future in favour of his older, more senior, and better known collaborator.

It was suggested that Lawrence write the foreword, but Watson feared his reaction to the manuscript. Alice Bragg was appalled by the book and was furious that her husband and others were held up to criticism and ridicule, although Lawrence himself was far less troubled. He saw in Watson—and in Crick—some of his own personality and academic history. Deep down he shared their dislike of the conservatism of the English scientific establishment, although he would never have expressed it so strongly. He was asked by Crick and Pauling to withdraw his support and his foreword but he declined. When Watson responded to his request to revise the worst sections Lawrence allowed his words to appear; indeed, Watson doubted that the book would have been published without the support they represented.[59] Lawrence wrote the following very generous and brave foreword; he was again putting himself at odds with many senior scientists in Britain, who abhorred the book:[60]

> This account of the events which led to the solution of the structure of DNA...is unique in several ways. I was much pleased when Watson asked me to write the foreword. There is in the first place its scientific interest. The discovery of the structure by Crick and Watson, with all its biological implications, has been one of the major scientific events of this century. The number of researchers which it has inspired is amazing...I have been amongst those who have pressed the author to write his recollections...The result has exceeded expectations. The latter chapters, in which the birth of the new idea is described so vividly, are drama of the highest order...I do not know of any other instance where one is able to share so intimately in the researcher's struggles and doubts and final triumph.
>
> Then again, the story is a poignant example of a dilemma which may confront an investigator...he has good reason to believe that a method of attack which he can envisage...will lead straight to the solution...Should he go ahead on his own?...When competition comes from more than one quarter there is no need to hold back. This dilemma comes out clearly in the DNA story.
>
> Finally, there is the human interest of the story...He writes with a Pepys-like frankness. Those who figure in the book must read it in a very forgiving spirit...The issues were often more complex, and the motives of those who he had to deal with were less tortuous, than he realized at the time. On the other hand, one must admit that his intuitive understanding of human frailty often strikes home.

[58] Hunter, n. 10, p. 242.
[59] Ibid., pp. 243–5.
[60] Watson, n. 55, pp. vii–ix.

Lawrence had made visionary changes at the Cavendish and set it on a path on which it continues. He apologized for his shyness, his inability to remember names, and his discomfort with administrative duties,[61] but others have applauded his leadership. More recent successors who enjoyed his legacy have written: 'Bragg performed a notably excellent job in decentralizing the work of the Cavendish and thus effectively breaking away from what would have ultimately become the dead hand of the Rutherford tradition'; and 'today Bragg must be given eternal credit for recognising where the new world of science was to be, and for having the courage to take his colleagues... into the second half of the twentieth century'.[62] Pippard also wrote: 'Yet when one looks back on Bragg's tenure of the chair one sees that the seeds sown then had already, by the time he left, shown great potential, and shortly afterwards came to fruition in advances and discoveries that rivalled, perhaps even eclipsed, any from Rutherford's Cavendish'.[63] After Lawrence left, the new Cavendish Professor, Nevill Mott, 'let it be known that he did not intend to keep the MRC unit in his department', which led to the establishment of a new and independent MRC Laboratory of Molecular Biology.[64] In its premier British home, physics had finally cast out X-ray crystallography and biophysics.

It is now appropriate to deal with the most outrageous accusation made against Lawrence by some of his jealous contemporaries and successors, to which Hunter has given expression: 'Despite Bragg's lifelong wish to be thought of as a physicist, and despite his tenure of numerous senior positions in British and international physics, it must be said that his work had no great influence on the physics of his time—or after'.[65] This statement is not only manifestly wrong but also poisonous. In the second half of the twentieth century, as physics came to terms with the revolution that had occurred in the first half, old fields were reinvigorated and new ones emerged. Solid-state (or condensed-matter) physics was perhaps the most notable; certainly it was the most productive. I was personally involved in the field for a quarter of a century from the late-1960s. The pace and breadth of discovery and innovation was unprecedented. The field broadened and expanded so that it quickly became complex and convoluted, and sub-fields emerged. There were numerous practical and industrial applications of the new knowledge.

By the end of the century the task of writing a history of the field had already become difficult. The American Institute of Physics published a *Guide to Sources for History of Solid State Physics* to assist.[66] Restricted works

[61] J G Crowther, *The Cavendish Laboratory 1874–1974* (New York: Science History Publications, 1974), p. 328.

[62] A B Pippard, as quoted in Phillips, n. 24, p. 117; Lord Porter, 'W. L. B. at the R.I.', in J M Thomas and Sir David Phillips (eds), *Selections and Reflections: The Legacy of Sir Lawrence Bragg* (Northwood: Science Reviews, 1990), pp. 126–9, 126.

[63] Pippard, n. 39, p. 98.

[64] S de Chadarevian, *Designs for Life: Molecular Biology After World War II* (Cambridge: CUP, 2002), p. 172.

[65] Hunter, n. 10, p. 248.

[66] J Warnow-Blewett and J Teichmann, *Guide to Sources for History of Solid State Physics* (New York: AIP, 1992).

appeared,[67] but when the British and American physics communities encouraged the writing of overview volumes it was necessary to invite a range of writers to cover the breadth of the subject. For Britain: 'The articles in this book sketch the history, present achievements and future potential of the science of the solid state'.[68] The larger American volume began: 'Future generations, looking back on the twentieth century, with its tremendous variety of intellectual and social developments, may find none of these more significant than the way in which solid-state physics has become of major importance... It is evident that the story of this great development must be an important component of the history of our times'.[69]

What do these works say regarding the origins of this explosion in knowledge and application? In the first, after a Preface and an Introduction highlighting the impact of quantum mechanics on the science of materials and a chapter on the prehistory of solid-state physics, the first extensive treatment of the modern subject begins with a discussion of the German X-ray diffraction experiment and the Braggs' explanation and exploitation of it.[70] The American volume also notes the importance of the combination of quantum mechanics and experimental information, and concludes: 'Solid-state physics stood on three pillars. The earliest to be erected was x-ray crystallography, as developed after 1912 in many laboratories. It was this technique and the study of crystal lattices in general that gave physicists the precise, geometric, atomic picture of solids they required before they could get anywhere at all. The second essential common element was, of course, quantum mechanics...The third common element, rather more subtle, was an appreciation for..."structure-sensitive" properties. In the mid-1930s, as W H Bragg remarked in 1934, interest began to focus on the question of which properties of a solid depended on its idealised crystal pattern, and which on "accidents" of the surface or interior arrangements'.[71] All these works agree that the beginning of the modern field of solid-state physics is to be found in the pioneering work of Lawrence and William Bragg. Lawrence's work had no great influence on the physics of his time—or after? Nonsense!

Hunter similarly asserts that 'Nor, despite his early training in mathematics, was Bragg a mathematical physicist. Crick said, "I don't think he was very powerful mathematically; I think some of the other physicists rather looked down on him"'.[72] I do not believe Lawrence ever claimed to be a mathematical (or theoretical) physicist, a very specific type of scientist. He had obtained a

[67] For example, M Eckert and H Schubert, *Crystals, Electrons, Transistors: From Scholar's Study to Industrial Research*, transl. T Hughes (New York: AIP, 1986).

[68] D L Weaire and C G Windsor, *Solid State Science: Past, Present and Predicted* (Bristol: Adam HIlger, 1987), Preface.

[69] L Hoddeson, E Braun, J Teichmann, and S Weart (eds), *Out of the Crystal Maze: Chapter from the History of Solid-State Physics* (New York: OUP, 1992), Foreword.

[70] Weaire and Windsor, n. 68.

[71] Hoddeson et al., n. 69; the quotation is from ch. 9, S Weart, 'The solid community', pp. 617–69, 623.

[72] Hunter, n. 10, pp. 248–9.

BA degree in Adelaide with first-class honours in mathematics, but he abandoned it for physics in his second year at Cambridge. Thereafter he was an experimental physicist with an excellent knowledge of the theory behind his research, a characteristic of the best experimentalists. He was the Cavendish Professor of Experimental Physics and had no ambition to be the Plummer Professor of Mathematical Physics. That he was not a mathematical physicist is an empty accusation and a further indication of the unfounded, underground criticisms ranged against him.

The final chapter in Lawrence's professional career continued the difficulties he had faced at Manchester and Cambridge. Indeed, the circumstances of his final appointment were so painful for the people concerned that they have been publicly discussed only recently.[73] Briefly stated, when Sir Eric Rideal resigned from the RI in 1950 he was succeeded by Edward Andrade, as the Director of the Davy–Faraday Research Laboratory, Fullerian and Resident Professor of Chemistry, and Superintendent of the House. Initially he seemed an ideal appointment, but Andrade was autocratic in an institution with a very complex management structure and conflict became inevitable. He wanted to be overall Director of the Royal Institution, but no such position existed. He dismissed long-serving staff, and soon the members, managers, visitors, and staff all became heavily involved. Lord Brabazon, the President of the RI, had an impossible task in trying to mediate between the parties. In March 1952 meetings were called to consider contradictory resolutions: first to secure the retirement of Professor Andrade and to declare the posts of Secretary and Treasurer vacant, and second to affirm full confidence in Andrade. Lawrence wrote to Brabazon early the same month, no longer able to remain aloof from the affairs of the institution in which he had often lectured and which had such a strong family connection. He had seen Andrade during the war and he suggested that the Institution would be 'killed' if Andrade continued.[74]

The motion to dismiss the Secretary and Treasurer was lost, as was that of confidence in Andrade. Andrade asked about compensation and the members agreed to pay any amount awarded by an arbitrator appointed by the Lord Chancellor. Andrade resigned in May and the Treasurer and Secretary followed. At Christmas the arbitrator's decision was a blow to the RI's finances (£7,000 and costs) and to its reputation; the arbitrator said that the roles of its officials lacked definition and its constitution was outdated.[75] The relationship between the RI and the Royal Society, which generally supported Andrade, degenerated into 'icy estrangement', with speculation that the RI would die.[76] It needed a new head and a new start. Brabazon asked Lawrence if he would

[73] F A J L James and V Quirke, 'L'affaire Andrade or how *not* to modernise a traditional institution', in F A J L James (ed.), *The Common Purposes of Life: Science and Society in the Royal Institution of Great Britain* (London: Ashgate, 2002), pp. 273–304; see also Lord Porter, n. 62, pp. 127–8.

[74] Ibid., pp. 273–91.

[75] Ibid., pp. 292–6.

[76] Porter, n. 62, p. 127.

accept an invitation. Although he and most of his colleagues realized the dangers, two of Lawrence's closest friends encouraged him to accept: Lord Adrian, Master of Trinity College and now President of the Royal Society, and George Thomson. It was, they said, the only way to save the Institution. His father had been very successful and much loved there, and Lawrence was equally committed and suited to its major roles of research and taking science to the people. He agonized but finally agreed, commencing duty on New Year's Day 1954.

Scars remained, however. Brabazon showed Lawrence's letter criticizing Andrade to a number of people, and his succession to the position therefore rankled: 'after Bragg had gone to the Royal Institution, he was never in the running to be President of the Royal Society, which, as a Nobel Prize winner, he might have reasonably expected, especially as, following Adrian, it should have been the turn of a physicist'.[77] Alice wrote: 'A group of chemists amongst its Fellows boycotted the Royal Institution, refusing all invitations to come there and showing their displeasure in many ways now best forgotten... I vowed that one day in the future I would say what I thought about it to Sir Robert Robinson, who might, I suppose, be called the doyen of the leading chemists; and I did'![78]

Gwendy recorded: 'But in the outcome he lived down the criticism to win admiration instead, and with strong support from his wife, Alice, gave new life to the RI after completing the sore job of nursing it back to normality'.[79] Lawrence slowly reorganized the structure so that he and his successors became 'Director of the Royal Institution', but it was a long and delicate process. 'One general strand that emerges from this [story] is the power of the administrative conservatism of British scientific institutions'.[80] Based upon the success of the Christmas lectures, Lawrence introduced lectures for London school children, featuring the experiments for which the RI was famous. They were an immediate success. Like his father Lawrence loved being with children and explaining things to them; he shared their transparent curiosity about the natural world. The Friday Evening Discourses continued, one being given by his wife—on marriage and divorce—and one by his elder son, Stephen—on jet engines.[81] Through his own lectures, some of which were televised for the first time, he became a recognized and admired public figure. Research blossomed under a new group of crystallographers and the advice and assistance of Perutz, Kendrew, and Hodgkin, principally studying proteins. The most notable discovery was the first high-resolution structure of an enzyme, lysozyme, by David Phillips and his colleagues, which Lawrence greeted with much excitement, as described delightfully by George Porter.[82]

In 1960 Lawrence went to New Zealand to deliver a Rutherford Memorial Lecture at the University of Canterbury, and this gave him the opportunity to visit Australia and show Alice the scenes of his childhood and youth. He gave

[77] James and Quirke, n. 73, p. 303.
[78] A G J Bragg, Autobiographical notes, p. 249.
[79] G M Bragg, *The Royal Institution: An Informal History* (London: Murray, 1985), p. 116.
[80] James and Quirke, n. 73, p. 301.
[81] Phillips, n. 24, p. 127.
[82] Porter, n. 62, pp. 128–9.

several lectures on the eastern seaboard but was disappointed by Adelaide. After so many years it was no longer the city of his memories. He retired for the last time in November 1966, at the age of seventy-six, and was appointed Emeritus Professor at the RI, where he continued to lecture. Just before he died he completed his last book, *The Development of X-ray Crystallography*, displaying again his ability to reduce complex problems to enlightening paragraphs of lucid prose.[83]

In 1962 the International Union of Crystallography held a large meeting in Munich to commemorate the fiftieth anniversary of von Laue's discovery and the two branches of physical science it spawned, X-ray crystallography and X-ray spectroscopy. The massive volume resulting from the meeting included innumerable references to Lawrence, William, and their scientific productivity and leadership.[84] The fiftieth anniversary of their Nobel Prize was marked in 1965 by several celebrations, most notably a large party at the Royal Institution, a BBC television programme entitled '50 years a winner', and Nobel celebrations in Sweden. At the RI some twenty British Nobel Prize winners joined other distinguished guests and members of the RI to acknowledge and applaud Lawrence's achievements and to present him with a message from the Queen and an illuminated address.[85] The TV programme was excellent and included a long interview with Lawrence and shorter appearances by members of his family, crystallographers Perutz, Kendrew, Crick, Watson, and Hodgkin, astronomer Ryle, long-term friend and fellow physicist George Thomson (see Figure 19.2), and Great War associates Lucien Bull and Harold Hemming. His colleagues spoke repeatedly of Lawrence's 'enthusiasm', that constantly encouraged their research, and of his modesty. His wife noted that he was a shy, private, family man, whose mind was partly child-like in its ability to see the simple behind the complex and to relate so well to children.[86] Gwendy watched the programme with Lawrence and wrote to him afterwards: 'You should take great joy from the way you have been able to light candles for other people... its lovely to be your sister'.[87]

In December, in Stockholm, Lawrence and Alice were fêted as part of the annual Nobel award ceremonies. Alice recalled that the TV programme had been shown there the night before and that 'When we went out shopping next day... men stopped us to shake hands and take off their fur hats, and women blew kisses... Our room was full of flowers'.[88] Lawrence delivered the first Nobel Guest Lecture. Now seventy-five years old and widely honoured, he still found it necessary to begin as he had done so many times before: 'It

[83] Sir Lawrence Bragg, *The Development of X-ray Crystallography*, edited by D C Phillips and H Lipson (London: Bell & Sons, 1975).

[84] P P Ewald (ed.), *Fifty Years of X-ray Diffraction* (Utrecht: International Union of Crystallography, 1962).

[85] 'Royal Institution notes: evening party at the Royal Institution', *Proceedings of the Royal Institution of Great Britain*, 1965, 40:482–97.

[86] '50 years a winner', BBC TV programme, 1965, written by A Jay and compared by B Westwood.

[87] Letter G M Caroe to W L Bragg, n.d., RI MS WLB 23B/143.

[88] A G J Bragg, Autobiographical notes, p. 265.

Fig. 19.2 Lawrence Bragg (L) and George Thomson (R), close, long-term friends and Nobel laureates, with laureate fathers, 1965. (Courtesy: The Royal Institution, London.)

is sometimes said that my father and I started X-ray analysis together, but actually this was not the case', and he went on to point out that *he alone* had first analysed the Laue photographs using a reflection model, that *he* had devised Bragg's law, and that *he* had determined the first structures using the new technique.[89] His father had said it so often, crystallographers knew it very well, and the Nobel laureates who had gathered at the RI also understood the magnitude of his contribution. From whom, then, was Lawrence still seeking recognition?

In 1970 *Acta Crystallographica* published a special issue to celebrate Lawrence's eightieth birthday, and tributes and recollections again flowed to the man now regarded as the father and enduring fulcrum of the discipline. Lawrence contributed a paper himself entitled 'Manchester Days'.[90] Other honours came. In 1966 he was awarded the Royal Society's Copley Medal, its highest honour, and Bernal wrote to him: 'This is only to congratulate the Royal Society for giving you at last the Copley that you have deserved many times over... Crystal structure may seem now an old story, and it is; but you, its only begetter, are still with us. Three new subjects, mineralogy, metallurgy, and now molecular biology, all first sprang from your head, firmly based on applied optics. You can afford to look back on it all with justified feelings of

[89] The lecture was not published until later and carries the strange note 'Read 1 June 1966': W L Bragg, 'Half a century of X-ray analysis', *Arkiv för Fysik*, 1969–1974, 40:585–603.

[90] W L Bragg et al., *Acta Crystallographica*, 1917, A26:171–96.

pride and achievement'.[91] Early in 1967 Lawrence was made a Companion of Honour, an order restricted to sixty-five members appointed for services of national importance.

Lawrence's personality was more complex than his father's, although they did have many similarities. William was modest, humble, and gentle; he was happiest with his family and in his work, which was simple yet profound. As a leader he worked through persuasion; he abhorred arguments and passionate exchanges. Lawrence shared his father's humility and love of private family life, and like him depended on his wife in both public and family affairs, but there were differences too. Hunter's suggestion that Lawrence, like his father, 'did not have a flamboyant personality' is incorrect.[92] Lawrence *was* emotional, with an artistic temperament, but he kept his emotions under stern self-control. Lady Phillips, née Diana Hutchinson and Lawrence's secretary at the Royal Institution before she married his favourite scientific son, David Phillips, confirmed to me what I had heard elsewhere. Lawrence regularly returned to his office upset by a situation or meeting in which he had been involved and, following a long-established custom, he would sit down and write a blistering letter to express his anger and frustration. He would then put it in his desk drawer, however, only to recover it days later and rewrite it.[93] No wonder his voluminous surviving correspondence is rather dull. Only his family and very closest friends ever saw below the armour-plated exterior.

Arriving in England as a young man, he had tried valiantly to blend in fully to Cambridge and English life, but he did not have an English public-school education, he was socially inexperienced and naive, and he was largely unaware of the subtleties of English social and academic practices and traditions. Unlike Harrie Massey, Howard Florey, and Ernest Rutherford, who retained characteristics of their Australian or New Zealand origins and visited their homelands often, Lawrence found few reasons to return. He also experienced episodes of depression that were sometimes severe, and he suffered the vivid horrors of The Great War. In some ways he remained Australian, as in his directness and his determination to go where angels fear to tread. He saw the need to rebuild the physics department at Manchester after the war, to restructure and diversify the Cavendish Laboratory in Cambridge after Rutherford, and personally to rescue the Royal Institution in London after the Andrade affair. He did them all, and history has judged him kindly in all three cases.

I also spoke to Sir David Phillips.[94] In preparing his Royal Society memoir he had spoken to a number of Lawrence's contemporaries, whose responses fell into two groups. Many remembered him fondly, even Francis Crick. They were grateful for his guidance and advice, were conscious of his brave leadership of their institutions, and were aware of his central role in the birth and development of the field.[95] Others had a permanent dislike because of rumoured

[91] Quoted in Phillips, n. 24, p. 131.
[92] Hunter, n. 10, p. xiv.
[93] Interview with Lady Phillips, Oxford, 30 September 1983.
[94] Interview with Sir David Phillips, Oxford, 29 September 1983.
[95] F Crick, 'W L Bragg: a few personal recollections', in Thomas and Phillips, n. 39.

weaknesses and the fact that he had not promoted their membership of the Royal Society or other honours. Lawrence argued that such honours were to be awarded on merit, and that the road to fellowship of the Royal Society, through its discipline-based process, was largely closed to X-ray crystallographers because they were not 'real' members of any one of them. Sir David also told me that Lawrence was diffident and insecure in interpersonal relations and lacked social subtlety and skills; he was, for example, unattractively obsequious to British nobility and royalty. Unlike his father, to many of the leaders of British science Lawrence remained an outsider, whom they thought was honoured too young, was not a 'real' physicist (or chemist or biologist), did not adequately respect English traditions and forms, and lacked his father's equanimity. He constantly sought recognition from the leaders of London's scientific elite, but it came from very few.

In his science Lawrence was both unusual and a man of his times. From poor initial photographs he was able to put his father's influential ideas to one side and see the correct interpretation, and thereafter he had a hand in almost every major step in the development of X-ray crystallography. Lawrence had a profound grasp of classical physics, Sir David told me, and in determining crystal structures he was an unparalleled problem-solver, with a wonderful sense of space. He liked to tackle each new structure individually, and he was somewhat saddened by the automation of structural analysis by computers. 'He was very quick, even when he was old', Phillips said. The cancer that had been excised in 1962 recurred in 1971. He survived another operation but suffered a relapse, and he died on 1 July 1971 in his eighty-second year.

Although he spoke regularly and fondly about Australia to his family and closest associates, Lawrence tried very hard to be an Englishman to others. But to his contemporaries he was not, and the conservative scientific establishment refused to embrace him as one of their own. (William) Lawrence Bragg was, nevertheless, one of the great scientists of the twentieth century. With his father he influenced a wider range of disciplines more profoundly than anyone else, and their achievements transformed our understanding of both the natural and the manmade worlds. He will be remembered and honoured around the globe long after his detractors and their words are forgotten.

Index

Note: page numbers in *italics* refer to Figures. As in the main text, W L Bragg is referred to as 'Willie' in the early years, and 'Lawrence' from 1914 onwards.